Practical Process Research and Development

To Hope
For all her love and support

Practical Process Research and Development

A guide for organic chemists

Second Edition

Neal G. Anderson

Anderson's Process Solutions
Jacksonville, Oregon

AMSTERDAM • BOSTON • HEIDELBERG • LONDON
NEW YORK • OXFORD • PARIS • SAN DIEGO
SAN FRANCISCO • SYDNEY • TOKYO

Academic Press is an imprint of Elsevier

Academic Press is an imprint of Elsevier
The Boulevard, Langford Lane, Kidlington, Oxford, OX5 1GB, UK
225 Wyman Street, Waltham, MA 02451, USA
Radarweg 29, PO Box 211, 1000 AE Amsterdam, The Netherlands

Notice
No responsibility is assumed by the publisher for any injury and/or damage to persons or property
as a matter of products liability, negligence or otherwise, or from any use or operation of any
methods, products, instructions or ideas contained in the material herein

British Library Cataloguing in Publication Data
A catalogue record for this book is available from the British Library

Library of Congress Cataloging-in-Publication Data
Anderson, Neal G.
  Practical process research and development : a guide for organic
chemists / Neal G. Anderson. – 2nd ed.
      p. cm.
  Rev. ed. of: Practical process research & development. c2000.
  ISBN 978-0-12-386537-3 (hardback)
  1. Chemical processes. I. Anderson, Neal G. Practical process
research & development. II. Title. III. Title: Practical process
research and development.
  TP155.7.A55 2000
  541'.39–dc23

                                                    2011051049

ISBN: 978-0-12-386537-3

For information on all Academic Press publications
visit our website at books.elsevier.com

Printed and bound by CPI Group (UK) Ltd, Croydon, CR0 4YY

Transferred to digital print 2012

Working together to grow
libraries in developing countries

www.elsevier.com | www.bookaid.org | www.sabre.org

ELSEVIER   BOOK AID
            International   Sabre Foundation

Contents

Neal G. Anderson, Ph.D., has worked for over 30 years in chemical process R&D in the pharmaceutical industry. He earned a bachelor's degree in biology from the University of Illinois, and a Ph.D. from the Medicinal Chemistry Program at the University of Michigan. After two-year industrial post-doctoral studies at McNeil Laboratories he began working in 1979 on chemical process R&D at E. R. Squibb & Sons, in New Brunswick, New Jersey. There he had extensive hands-on experience with chemical process development in the lab, pilot plant, and manufacturing, including 12 manufacturing start-ups and process development for four major drugs and many new drug candidates. His final position at Bristol-Myers Squibb was principal scientist.

In 1997 Neal left BMS and began consulting on the development of practical, cost-effective processes for the pharmaceutical, biotech, and fine chemicals industries. He has presented short courses internationally on process chemistry R&D for "small molecules" to over 1300 participants from more than 160 companies. At home he has brewed beer, and he practices process chemistry in the kitchen.

Preface to Second Edition

"We have a tendency in science to become experts about something very narrow. We're so focused on minutiae that we miss the big picture."

– Linda S. Birnbaum, Director, National Institute of Health Sciences
[Hogue, C. *Chem. Eng. News* **2009,** *87(5),* 26.]

"Discovery consists of seeing what everybody has seen and thinking what nobody has thought."

– Albert Szent-Gyorgyi, Hungarian biochemist, 1937 Nobel Laureate

In the last ten years or so there has been a flood of information on process chemistry, from case studies in articles, to reviews and books, to e-books, blogs and technical information on the internet. With so much information available the foundation of knowledge of process chemists and engineers has expanded and deepened. Accordingly I have presented material in this edition at a more advanced level than in the first edition.

The focus of this edition is the same as the first: to provide a comprehensive, step-by-step approach to organic process research and development for the preparation of "small molecules." The approach is reflected in the order of the chapters. The text discusses how and why operations are carried out, and then illustrates principles with examples – mostly successful ones. There are few equations, mainly those that have valuable implications and may be unfamiliar to readers. By understanding the "hows" and "whys" of processing and operations readers may be able to devise practical approaches to current challenges. The examples that I selected illustrate practical points, and many other interesting aspects of processing could be mined from the references cited. Most of the material in this edition is new is relative to the first edition: almost 85% of the references were published since 1999.

This edition includes many new or expanded topics. Chapter 2, SAFETY considerations, is new. Rather than focusing on SAFETY as restrictive or negative, I have tried to present SAFETY considerations as enabling, leading to the development of SAFE, reliable, and efficient processing. Chapter 3, route selection, is expanded with green chemistry as a framework, and highlights biocatalysis as a means to devise more cost-effective routes. Chapter 4, reagent selection, emphasizes the applications of biocatalysis and phase transfer catalysis. Solvent considerations are expanded in Chapter 5. Chapter 8 focuses on practical considerations for developing processes for scale-up in the pilot plant and manufacturing, and in the kilo lab. Because those working in early process research may find Chapter 8 to be too applied I have separated it from other material; however, understanding the considerations in this chapter should speed the scale-up and development of new chemical entities (NCEs) and active pharmaceutical ingredients (APIs). Chapter 10 is new, focusing on approaches to optimize organometallic reactions. Chapter 11, workup, is greatly expanded. More examples of crystallization are detailed in Chapter 12. Chapter 13, final product form, includes a section on genotoxic impurities and an expanded section on polymorphs. Chapter 14 is devoted to continuous processing. Chapter 15, refining the process, is new. Chapter 16 outlines process validation and implementation. As chiral syntheses are woven into the fabric of process chemistry, operations to incorporate or minimize the loss of chirality are found throughout this edition.

Green chemistry considerations have been highlighted in the recent literature. For decades process chemists and engineers have tried to minimize waste, decrease the cost of manufacturing, and devise inherently SAFE operations for scale-up and for routine manufacturing of APIs. Joining these considerations with others as "green chemistry" can only advance the development of responsible processes. Examples of green chemistry are sprinkled throughout the book. I look forward to the day when green chemistry considerations will be so much a part of the fabric of process R&D that we won't label those efforts as green chemistry.

SAFETY considerations are paramount in the laboratory, pilot plant, and manufacturing facilities. In this edition I have tried to draw attention to SAFETY and SAFE operating conditions by capitalizing those words in the text. In the figures I have often shown the quench reagents, because quenching excess reagents can be hazardous; ideally for each synthetic transformation the quench and workup conditions should be described below the arrow in a synthetic scheme. Workups and isolations are not described where they don't seem relevant to the discussions at hand. By reviewing the operations mentioned in this book and in cited references the reader may begin to develop SAFE operating conditions for the processes under consideration. Suitable SAFETY studies (chemical hazard assessment, toxicology studies, and potential environmental impact) should be performed before any reaction is carried out on scale. The reader must assume such responsibility for his or her own SAFETY, along with the SAFETY of those nearby.

Many people have helped me in preparing this book. Among others, thanks to Tim Ayers, Didier Benoit, Dave Burdick, Rich Carter, Ray Conrow, Mike Cruskie, Dave Ennis, Zhinong Gao, Marc Halpern, Peter Harrington, Gjalt Huisman, John Knight, Trevor Laird, William Leong, Jyanwei Liu, Hans Maag, Ann Newman, Bill Nugent, Wayne Slawson, David Snodin, Mike Standen, Andy Teasdale, Alan Tse, Elizabeth Vadas, Rob Waltermire, Steven Weissman, and Gregg Federowicz of De Dietrich for helpful discussions and information. Thanks to the people who participated in my short courses, for their feedback and for sharing their experiences. I thank PerkinElmer Informatics (previously CambridgeSoft Inc.) for providing a copy of ChemOffice with ChemDraw, which I used to represent the structures and reactions shown in this book. Finally, I thank Adrian Shell and Louisa Hutchins of Elsevier for their belief in this project and their continued support of this second edition.

The information in this book should be regarded as guidelines, not rigid rules, and I am always interested in examples contrary to these guidelines. I welcome the mention of the inevitable errors.

Neal Anderson

When I first became interested in the subject of process research, development and scale up of chemical processes in the 1970s there were no books or dedicated journals to guide me, although there were isolated publications, particularly from the process R&D group at Merck, USA. It wasn't until the 1990s that books such as Lee and Robinson [1] and Repic [2] were published and a single journal appeared. Neal Anderson's first edition of *Practical Process Research & Development* was published in 2000 and on first reading I was immediately attracted to his approach, namely a pragmatic and systematic method of carrying out process R&D, but always with an aim towards large scale manufacture.

Many other books have appeared since 2000 covering aspects of process R and D (see Introduction for a comprehensive list) but many focus on synthetic chemistry, without paying enough attention to optimisation, safety, work up and product isolation, physical form of solid product, factors which are vitally important in scale up and manufacture. After all process R&D is on the interface between organic chemistry, physical chemistry, and chemical engineering, so those organic chemists within the profession need to understand chemical engineering principles and how they affect processes on plant, just as chemical engineers must understand the basics of organic chemistry. Organic chemistry knowledge alone is not sufficient to develop processes for manufacture.

It is therefore gratifying to see that in this second edition of *Practical Process Research and Development*, the emphasis is again on practicality, with lots of tips for the inexperienced process chemist, though even old-timers like me can still learn a few tricks from the author. Although the author focuses on developing processes for the pharmaceutical industry, readers from other industries, such as fine chemicals, agro-chemicals, colour chemicals and flavours and fragrances will benefit so much from the lessons, the case studies and the discussions within.

But what of readers from the university world, both students and teachers? What can they learn from an experienced process chemist such as Neal Anderson? First of all they will discover an organic chemistry-based text with Chapter headings which are quite unfamiliar to them; the emphasis on safety in an early chapter is unusual, but clearly the author has chosen this to reflect its importance in process R&D. The recent examples of accidents at US universities indicate that perhaps more emphasis on this subject is needed in the academic world. But safety is so important for scale up, it has to come first. Students will learn that hazardous chemistry can be scaled up, but that it is HOW it is scaled up that is the crucial factor.

Secondly students will learn about understanding reactions and the importance of reaction mechanisms in development and scale up. Only by understanding the chemistry, and the sequence of reactions that comprise a single chemical step within a reaction vessel, can you control it, and manufacture is all about maintaining control to ensure reproducibility and high quality products.

Thirdly students will learn that analysis, particularly in-process analysis, and analytical data are critical to the understanding of a process. Knowing the impurities likely to be formed helps to devise conditions for their minimisation, or to design a work up or crystallisation strategy for their removal, ensuring a product within specification. Raw material and solvent quality can vary from laboratory to plant and these apparently trivial issues need to be checked using use-tests; the student is thus taught to think ahead and plan.

Fourthly, students will soon realise, after reading this text, that chemistry doesn't end when the reaction is finished, and one cannot on large scale simply evaporate the reaction mixture to dryness and chromatograph the product. Reaction mixtures often need to be quenched and the work up and product isolation are fundamental to the whole process and must be designed carefully, then optimised. Neal teaches readers how to do this to best effect.

The fifth lesson students will learn is that time is an extremely important factor in scale up; extended reaction times or work up times can lead to differences in yield and quality of product and the experienced process chemist is always aware of this as the process moves from laboratory to pilot plant to manufacture. Carrying out a laboratory process using projected pilot plant timings can therefore prevent surprises. Process chemists soon learn that "whatever can go wrong, will" if he/she is not well prepared. Neal, in every chapter, gives the reader practical tips to avoid problems so that the philosophy could be expressed as "forewarned is forearmed". Nevertheless problems do occur on plant and there can occasionally be surprises, so Neal devotes a whole chapter to "troubleshooting" to give the reader his experiences of problem solving. With the launching of the journal *Organic Process Research and Development* in 1997, there are now many more problems – and solutions – published in the literature but often these are buried in a single paragraph in a paper and need to be highlighted and referenced as in this and other chapters in the book.

Experienced practitioners of process R&D will enjoy reading the author's approach to the subject and the case studies and examples he has chosen to illustrate it, as well as the practical tips. All will enhance their knowledge and be stimulated by the discussions in the chapters.

Although I and my colleagues (and Neal Anderson too) make a living from teaching courses on process chemistry, it is my dream that one day there will be many university departments (instead of the one or two now) teaching the fundamentals of process chemistry at the organic chemistry–chemical engineering interface, and carrying out fundamental research on chemical reactions in a process-friendly way. But that ideal is a long way off, and meanwhile we have to rely on experienced process chemists to write books and pass on the knowledge. In 2000 I opined that Neal's book was the best book on process chemistry and having read the second edition 12 years later, I have not changed my opinion.

It is my impression that active pharmaceutical ingredients (and agrochemicals and other active molecules) are becoming more structurally complex and this poses increased challenges for the modern process chemist. Timelines for delivery of the first kilogram are becoming shorter, and human resources are often spread over many projects and so most process chemists face increased pressure early in a project. Pressure to retain the existing discovery route, not designed for scale up, is immense but the risks (to safety, quality and failure to synthesise on larger scale) can mean a new synthetic approach is warranted and the best option for the project in the long term. However "short-termism" often prevails.

For later stage projects there is much more opportunity to spend time on a full understanding of reaction chemistry, particularly if a quality by design (QBD) approach to understanding design space (the effect of changing reaction variables on yield and product quality) is adopted. The use of statistical methods of optimisation, rarely taught in university chemistry courses, rather than one variable at a time optimisation (usually carried out in universities) is essential for the characterisation of design space for each step in the synthesis.

But whatever area of process R&D is being undertaken, the need is to do it better, faster and at lower cost with fewer resources. The practical knowledge gained from reading this book and the understanding of the tricks of the trade that can be employed will surely help all readers to achieve these demanding targets, and to enjoy the fascinating subject of organic process research and development.

Trevor Laird
Scientific Update LLP, UK
Editor; Organic Process Research and Development

REFERENCES

1. Lee, S.; Robinson, G. *Process Development: Fine Chemicals from Grams to Kilograms*; Oxford University Press: Oxford, 1995.
2. Repic, O. *Principles of Process Research and Chemical Development in the Pharmaceutical Industry*; Wiley-Interscience: New York, 1998.

Introduction

"The world doesn't move because of idealism…. [i]t moves because of economic incentives."

– Fernando Canales Clariond, formerly Mexico's secretary of the economy [1]

"It is well-known that there are no technical optima in industry, only economic optima…."

– G. Guichon et al. [2]

"Today, green chemistry is simply a good business choice."

– Paul Anastas [3]

I. INTRODUCTION

The driving forces of the pharmaceutical industry are to develop medicines that maintain or improve health and the quality of life for people, and to provide a reasonable return for investors. The many unknowns of the business, especially our imperfect understanding of biology, make for a high-risk environment. Yet the rewards are high as well. In Table 1.1 are presented some statistics associated with developing drugs, and these statistics explain some of the pressures of the business.

Significant financial gains are possible by developing drugs, as indicated by the penalties and fines levied against firms and people recently. For instance, in 2010 AstraZeneca was fined $520,000,000 for promoting Seroquel for off-label uses [4] and GlaxoSmithKline was fined $150,000,000 for violations of current Good Manufacturing Practices (cGMPs) [5]. In 2009, Pfizer was fined $2,300,000,000 for illegally promoting

Practical Process Research and Development. DOI: 10.1016/B978-0-12-386537-3.00001-0

TABLE 1.1 Some Statistics Relevant to the Pharmaceutical Industry

Value	Factor
$1,300,000,000	Cost to bring a drug to market (1)
5–20% of drug product price	CoG of API
30% of the CoG for drug product	Cost of QC (2)
Millions of dollars	Cost of failed drug formulation (3)
As high as market will bear	Price of drug product to consumer (higher for US than for most countries) (4)
$75,000	The value of one additional year of life, set in 2005 by health economists (5)
$200,000–$300,000	Annual cost of US chemist or engineer for an employer
About 95%	Portion of drug candidates that fail in pre-clinical or clinical studies
About 30%	Portion of approved drugs that recoup development costs (6)
8 years	Average time of development (goal is 5 years)
20 years	Period for exclusive sales of a patented drug (US)
20 – Development time	Years to recoup investment costs
$1,000,000	Sales lost for every day that a filing is delayed, if drug sales are $400 MM/year

(1) Undoubtedly includes the cost of advancing drug candidates that failed. Jarvis, L. M. Chem. Eng. News **2010**, 88(23), 13. May be higher: Vertex developed Incivek (telaprevir) over 20 years at the cost of around $4,000,000,000: Jarvis, L. Chem. Eng. News **2011**, 89(22), 8.
(2) Mullin, R. Chem. Eng. News **2009**, 87(39), 38.
(3) May be higher for formulations involving spray drying. A huge cost may be incurred if the physical form of an API is not controlled and a dosage form fails specifications, thus interrupting clinical trials or discontinuing sales of a drug (Chapter 13). Thayer, A. M. Chem. Eng. News **2010**, 88(22), 13.
(4) http://www.economist.com/node/4054095 (June 16, 2005).
(5) The Washington Post, July 17, 2005: http://www.washingtonpost.com/wp-dyn/content/article/2005/07/16/AR2005071600941_pf.html.
(6) Mod. Drug Discov. **2001**, 4(10), 47.

Bextra and three other medications [6], and Bristol–Myers Squibb (BMS) was fined $2,100,000 for making agreements with Apotex to delay the launch of generic Plavix [7]. In 2009, the FDA stopped reviewing applications from Ranbaxy and prohibited Ranbaxy from importing 30 generic drugs, due to falsified QC data [8]. In 2007, a former top official of China's state organization approving drugs was executed for taking bribes [9]. In 2004, BMS was fined $150,000,000 for "channel stuffing," an accounting practice that artificially boosted the sale of drugs [10]. The reputation of the pharmaceutical industry has been sullied over the past few decades, but many people, including this author, enter this industry because they want to be able to help others.

The development and sales of drugs are influenced by an interplay of financial, political, governmental and personal considerations. For example, Bayer reduced the cost of a pill of ciprofloxacin from $1.77 to $0.75 after the horrific crashes of September 11, 2001 [11]. As of August 2009 the pharmaceutical lobby had 1544 lobbyists, or almost three lobbyists per congressman in Washington DC [12]; this is a sizeable increase from the 625 registered lobbyists in 2001 [11]. No major US pharmaceutical company would develop RU-486, an abortifacient, due to backlash anticipated from conservative groups. The sale of anti-AIDS drugs at reduced prices to the third world is a nice example of philanthropy; probably some income tax write-offs are also involved. Philanthropic efforts exist, such as Merck's gifts of ivermectin to prevent river blindness and the development of tenofovir by the Clinton Health Access Initiative to treat AIDS in the developing world [13]. Pharmaceutical industries may need support from government to continue developing drugs for the third world, such as compounds to treat malaria [14]. Efforts to minimize wastes and decrease impacts on the environment have increased, due to sensible and altruistic reasons, and due to penalties imposed. Although bacteria resistant to powerful antibiotics are continually emerging, the development of antibiotics has slowed due to the anticipated longer times to recoup development costs from drugs that are not taken on a chronic basis. The development of new chemical entities (NCEs) is becoming more difficult with increased scrutiny by regulatory authorities; for instance, the third or fourth entry into a therapeutic category may have to demonstrate superior benefits to win FDA approval [15], and control of potentially genotoxic impurities at the ppm level demands additional efforts. And everything is influenced by people striving for personal advancement.

As a result of increasing pressure to bring compounds to market, business trends within the pharma sector have been changing. Working smarter and faster is stressed, for instance, using high-throughput screening and statistically designed experiments. Pharmaceutical industries are being pressured to develop more efficient processes [16]. The FDA has advanced process analytical technology (PAT), and this may decrease manufacturing costs [17]. The importance of solid process development efforts has been recognized [18].

The complexity of compounds that has emerged as active pharmaceutical ingredients (APIs) may be increasing, as shown in Figure 1.1, and structural complexity increases the cost of development. Drug development is expensive: Vertex's first approved drug was developed in-house over 20 years at the cost of around $4,000,000,000. Incivek (telaprevir) will treat hepatitis C, at a projected price of $49,200, competing with Merck's Victrelis (boceprevir) at a projected price of $31,000–$44,000. Each company has set up assistance programs to help with the co-payments of insured patients [19].

The foundation of thorough processes is the work carried out by academicians. Some brilliant total syntheses have been described by academic chemists, such as the syntheses of codeine by the groups of Stork [20] and Magnus [21]. Total synthesis in an academic sense [22,23] is often not suitable for scale-up on an industrial scale; scaling up in an academic setting to millimoles or grams may not be enough to uncover and address processing difficulties. Prof. Hudlicky has described many practical considerations for syntheses [24] that are applicable to process R&D efforts in the pharmaceutical industry.

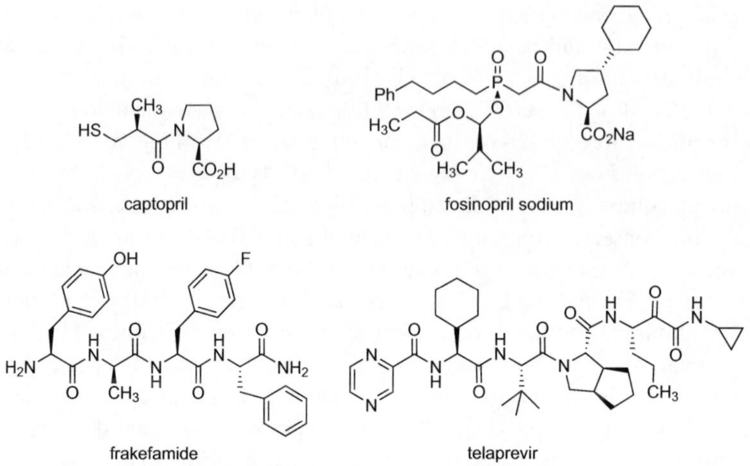

captopril

fosinopril sodium

frakefamide

telaprevir

FIGURE 1.1   Structures of APIs and NCE in Table 1.2.

TABLE 1.2 Details on Development of Four APIs and an NCE

	Captopril (1)	Fosinopril (2)	Frakefamide (3)	Enfuvirtide (4)	Telaprevir (5)
Market date	1981	~1988	Halted after Phase 2	2003	2011
Steps	5	14	7	109	NA
Isolations	5	13	3	7	NA
Preparative chromatographies	0	0	0	2	NA
CoG	NA	NA	$2500/kg	$36,000/kg	NA
Therapeutic area	High blood pressure	High blood pressure	Pain treatment	AIDS	Hepatitis C
Patient cost/year	NA	NA	NA	$25,000	$49,200

(1) Anderson, N. G.; Bennett, B. J.; Feldman, A. F.; Lust, D. A.; Polomski, R. E. US Patent 5,026,873, 1991 (to E. R. Squibb & Sons.)
(2) Grosso, J. A. US Patent 5,162,543, 1992 (to E. R. Squibb & Sons).
(3) Franzén, H. M.; Bessidskaia, G.; Abedi, V.; Nilsson, A.; Nilsson, M.; Olsson, L. Org. Process Res. Dev. **2002,** 6, 788.
(4) Enfuvirtide is a synthetic peptide made of 36 aminoacids: Bray, B. Nat. Rev. Drug Discov. **2003,** 2, July, 587.
(5) Jarvis, L. M. Chem. Eng. News **2011,** 89(22), 8.

Perhaps the first description of the principles of process research and development as applied to pharmaceutical products was written by Trevor Laird in 1988 [25]; this article covered a breadth of topics, including design of experiments (DoEs), automation and the importance of polymorphism. Caron recently detailed the responsibilities of the process chemist [26]. In the past 11 years there have been many insightful books and many detailed articles on process development [27–49].

Process chemists from AstraZeneca, GlaxoSmithKline, and Pfizer critically assessed the parameters for route selection, developing the SELECT acronym: for SAFETY, Environmental, Legal, Economics, Control, and Throughput [50]. SAFETY is of primary importance. The environmental impact of processing falls under the heading of green chemistry. Legal matters are important in the pharmaceutical business, especially regarding the issuing of patents and freedom to operate. Economics are expressed in the cost of goods (CoG), as influenced by the cost of raw materials, labor, quality control (QC), and waste disposal. A primary goal is to minimize the cost of an API to \$1000–\$3000/kg. To minimize the CoG chemistry must be efficient and robust, and stringent in-process controls (IPCs) are crucial to ensuring routine operations to produce high-quality products. Starting materials and reagents must be subject to competitive bidding [51]. Multi-purpose equipment is used in development efforts in order to minimize capital investment; when sales of a compound seem more certain purchasing dedicated and specialized equipment may be justified. Throughput is reflected in not only isolated yields, but also in the amount of product that can be made per unit of volume or per unit of time (space–time yield). Manufacturing complexity is minimized; for instance, eliminating the use of hazardous or toxic chemicals eases SAFETY concerns, and by minimizing processing bottlenecks and the proportion of rejected batches that need to be reworked productivity is improved. Throughput and productivity are expressions of economics, and these are some of the key drivers for the manufacturing of APIs. The SELECT criteria can be considered throughout this book.

A perspective on process research and development activities is shown in Figure 1.2, in which optimization of process development increases going to the right. Based on results of screening, a compound may be designated as an NCE and advanced on the pathway to becoming an API. As with any general guideline, the development of an API may not follow this sketch. For instance, quantities may be prepared in the kilo lab after research efforts have identified an optimal route and preferred reagents; in other instances supplies of an API suitable to meet the demand for several years may be manufactured in one 5-gm batch. The time to progress from an idea to an API may be 5–13 years, or longer.

Efficient process development begins with drug discovery, and continues through routine manufacturing of the drug product [52]. The roles in drug discovery, process

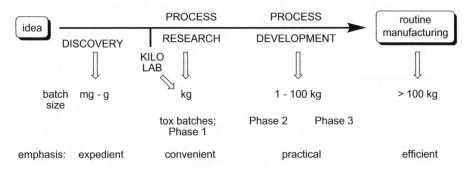

FIGURE 1.2 Perspective on the drug development continuum.

research and process development are different. Drug discovery chemists design expedient, diversity-oriented syntheses to prepare many compounds from a common intermediate. Process research chemists focus on convenient routes to prepare one or perhaps a few similar compounds; route selection and reagent selection are usually prime considerations. Process development chemists optimize the processes to prepare an NCE or API, focusing on minimizing impurities, streamlining work-ups, avoiding chromatography if possible, and isolating the desired final form of one product [53]. Of course these general descriptions may overlap. Almost inevitably process chemists change the route developed by the discovery chemist; for a new project one of the initial efforts of the process chemist is to consider possible benefits in reordering steps in a route. At some point process chemists consider changing the route, using different starting materials or taking advantage of biocatalysis. The reactions commonly encountered by researchers within pharmaceutical process R&D [54] and drug discovery [55] are relatively similar, not surprisingly. The process chemist speeds the development of an NCE by considering the "hows" and "whys" of processing. While carrying out the present duties one can consider the needs of the next step and thus smooth the development of an NCE. Time invested early to make material by processes that can be readily scaled up will shorten overall development time.

Of course the problem with investing time in downstream efforts is that most compounds fail to make the marketplace; in 2003, Federsel estimated that less than 0.03% of all active compounds prepared were approved by regulatory authorities and 70% of drugs failed to recoup the cost of investments [56]. Perhaps 95% of compounds passing through process R&D operations fail. Optimizing processes too early may waste resources, but if more material is quickly needed, scale-up may be difficult unless some optimization was in place. By Phase 3 the route should be established and process optimization should be well underway; if not, supplies for Phase 3 may be at risk. Furthermore, the pressure for filing an NDA may result in advancing processes needing substantial optimization. Suffice it to say that not all processes are optimized before scale-up, and processes may be rushed into production to provide material for clinical trials or manufacturing. The conundrum of when to invest in process optimization often pivots on the GO/NO GO decision points (Table 1.3).

"Fit for purpose" has been used and overused to describe efforts to make NCEs. All process R&D should be fit for purpose, but usually this phrase is applied to the early efforts to make sufficient supplies for toxicology and Phase 1 studies in an expeditious fashion. Once proof of safety in humans is established there is usually a need to redevelop processes to make them more efficient and lower the CoG such efforts are also fit for purpose [57]. Almost inevitably processes are changed to prepare material for Phase 3 if not for Phase 2 clinical studies.

Serendipity should always be exploited. (As Pasteur said, "Chance favors the prepared mind.") For example, the desired dibenzoylated ribonolactone epimer crystallized and was converted to gemcitabine (Figure 1.3). The benzoyl protecting group had been selected for ease of following the reactions by HPLC [58]. Another example is the fortuitous epimerization of the methine alpha to the ketone, shown in Figure 1.3. After condensation with the ester acid the *trans:cis* ratio was 35:64, but the ratio improved somewhat during acidification for workup. A study showed that the best selectivity for the *trans*-isomer occurred after the basic aqueous phase was held at

TABLE 1.3 Using GO/NO GO Decision Points in Process R&D (1)

GO/NO GO Studies	Amount of API Required	Suitable Route
Toxicology studies Bioavailability (Phase 1)	Low Low	Expedient: consider using discovery route and preparative chromatography. SAFETY concerns may be addressed using routine laboratory operations. Scale-up can be grueling with only minimal experience
Efficacy (Phase 3)	High	SAFETY concerns must be thoroughly addressed for scale-up Cost-effective synthesis desired Controls needed for reliability and filing Productive operations desired

(1) Grabowski, E. J. J., Reflections on Process Research II, in Fundamentals of Early Clinical Drug Development; Abdel-Magid, A. F.; Caron, S., Eds.; Wiley: Hoboken, NJ; 2006; Chapter 1.

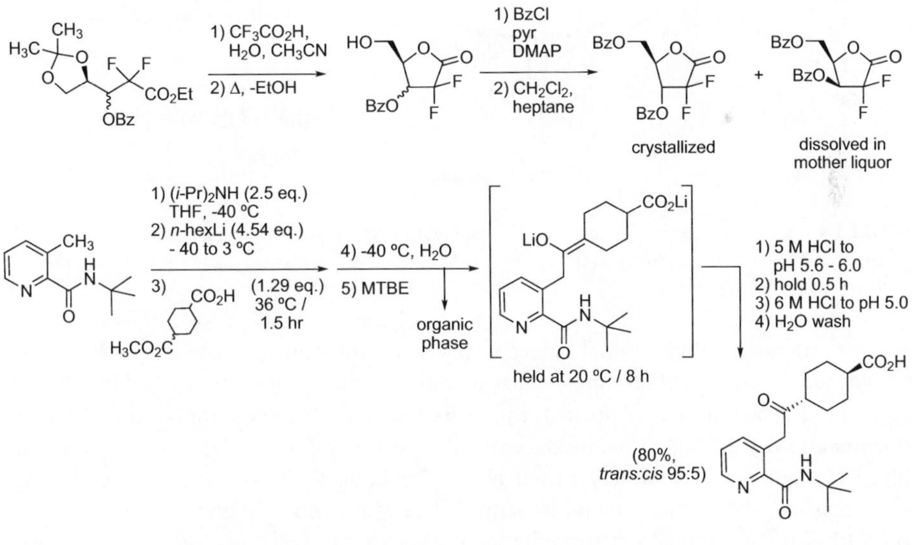

FIGURE 1.3 Two examples of serendipity that were exploited.

room temperature for 8 hours [59]. Researchers exploit observations such as these to optimize their processes.

II. EQUIPMENT CONSIDERATIONS ON SCALE

Basic operating differences for equipment used on scale are centered on the fact that the equipment is opaque and immobile, and there is limited access to the contents. Most multi-purpose reactors (Figure 1.4) are made of glass-lined steel, as that material is more

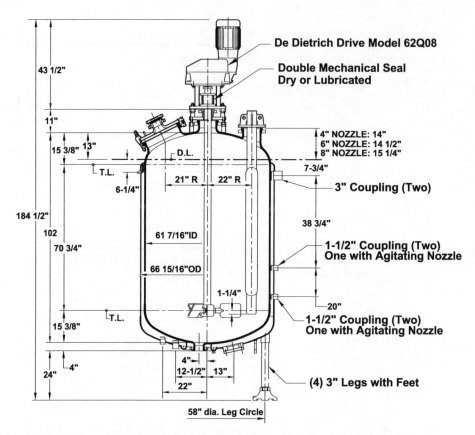

FIGURE 1.4 Section view of a cylindrical 1000 gallon multi-purpose glass-lined reactor. Heat transfer fluid can circulate inside the jacket for cooling/heating capability. Agitator blades are exchangeable. *Courtesy of De Dietrich Process Systems.*

inert than metal. Baffles, vertical blades near the reactor walls, are used to increase the turbulence of the agitated mixtures. Once vessels in a pilot plant or manufacturing have been charged they are rarely opened; this is to ensure SAFE operations and to prevent contamination of the batch. All mixtures must be stirrable for transfer through a bottom valve. The agitator is routinely raised above the bottom surface of the reactor, and volumes below the agitator cannot be stirred. The minimum agitation volume (V_{min}) is about 10% of the nominal reactor volume. The maximum volume a vessel can contain (V_{max}) may be 100–110% of the nominal reactor volume. In a large vessel heat transfer by circulating heat transfer fluids through an external jacket is slower, due to the high volume and relatively low surface area of the reactor. A concern of using glass-lined reactors is that they may fracture due to thermal shock; other reactors may be preferred for cryogenic or high-temperature processes. Childers has presented an excellent pictorial perspective of large-scale syntheses from the standpoint of a chemical engineer [60].

Most scale-up costs are associated with mass transfer and heat transfer. Heat transfer affects SAFETY and impurities. In semi-batch operations, heat transfer rate often determines the time required for operations. Continuous operations may be chosen for fast reactions due to improved heat transfer rates on scale. Mass transfer often limits the time for operations. Some operations cannot be readily speeded up, e.g., the time needed to separate immiscible phases during extractive work-ups, or the time required to transfer liquids through a 1-inch line. Efficient mixing is needed for heterogeneous reactions, such as liquid/liquid, gas/liquid, solid/liquid, and multiply heterogeneous processes. Efficient mixing may be needed for homogeneous reactions

> Most operations on scale are possible if time is not an issue, and money is not an issue. Rarely are there excesses of either. The best approach to develop processes that can be readily scaled up is to mentally scale down operations to the laboratory, and mimic them in the laboratory.

also; the rate and mode of addition can be crucial, and controlling micromixing and mesomixing may be necessary to minimize impurity formation in fast reactions.

The limitations of heat transfer through external cooling of spherical flasks in the laboratory and cylindrical multi-purpose reactors are noticed with the scale-up of batch operations. For cylinders and spheres as the radius increases the volume increases faster than does the surface area, hence the ability to remove heat by circulating external fluids progressively decreases as the volume increases. For a 10-fold scale-up about twice the amount of time will be required [61]. Hence rapid additions on scale can lead to uncontrollable exotherms, with degradation of the reaction mixture and possible runaway reactions. Dose-controlled additions are used to moderate exothermic response from additions. On scale additions typically require 20 minutes to hours; pumps are available for extended addition times. Longer operations are not necessarily bad for scale-up. Smaller reactors, frequently used for continuous operations, provide more rapid heat transfer.

The scale-up of a Swern oxidation (Figure 1.5) illustrates the extended time required for heat transfer and the impact on processing. The benchmarks for the oxidation were the laboratory conditions with additions at $-15\ °C$. Unfortunately when these temperatures were employed in the pilot plant the isolated yield was about 60% of the expected yield. In the pilot plant about 2 hours was required to add both oxalyl chloride and triethylamine, allowing time for the reactive intermediates to decompose. By making the additions at $-40\ °C$ the expected yield was achieved [62]. While the additions at the colder temperature may not have been any faster, the colder temperature slowed the decomposition of the reactive intermediates.

> For rapid scale-up, design homogeneous reaction conditions whenever possible. Check the stability of process streams under extended conditions in the laboratory before scale-up, because processing on scale will be extended.

The Swern oxidation in Figure 1.5 was successfully scaled up later through semi-continuous processing, affording essentially the same yield, but with improved process control [63] (Figure 1.6). The reaction was carried out at about 40 °C, in contrast to the colder temperatures used for batch oxidations. The higher reaction temperature was made possible by the extremely short residence time (τ) of 0.1 seconds for the reaction stream in the reactor. (This is an example of the high-temperature short-time (HTST) practice for continuous operations, as is used for pasteurizing milk [64].) The reactor

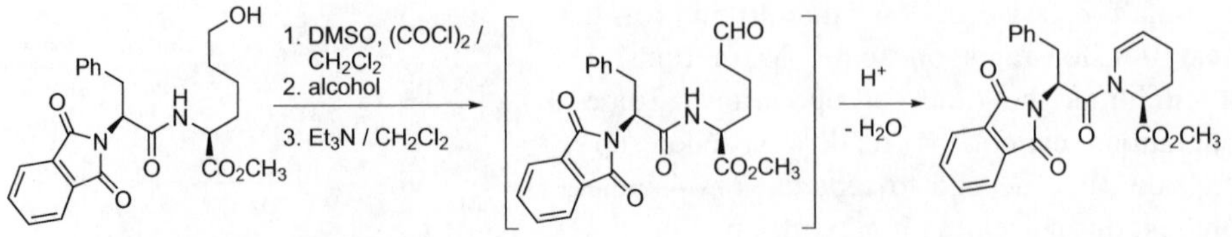

FIGURE 1.5   Scale-up of a Swern oxidation through batch operations.

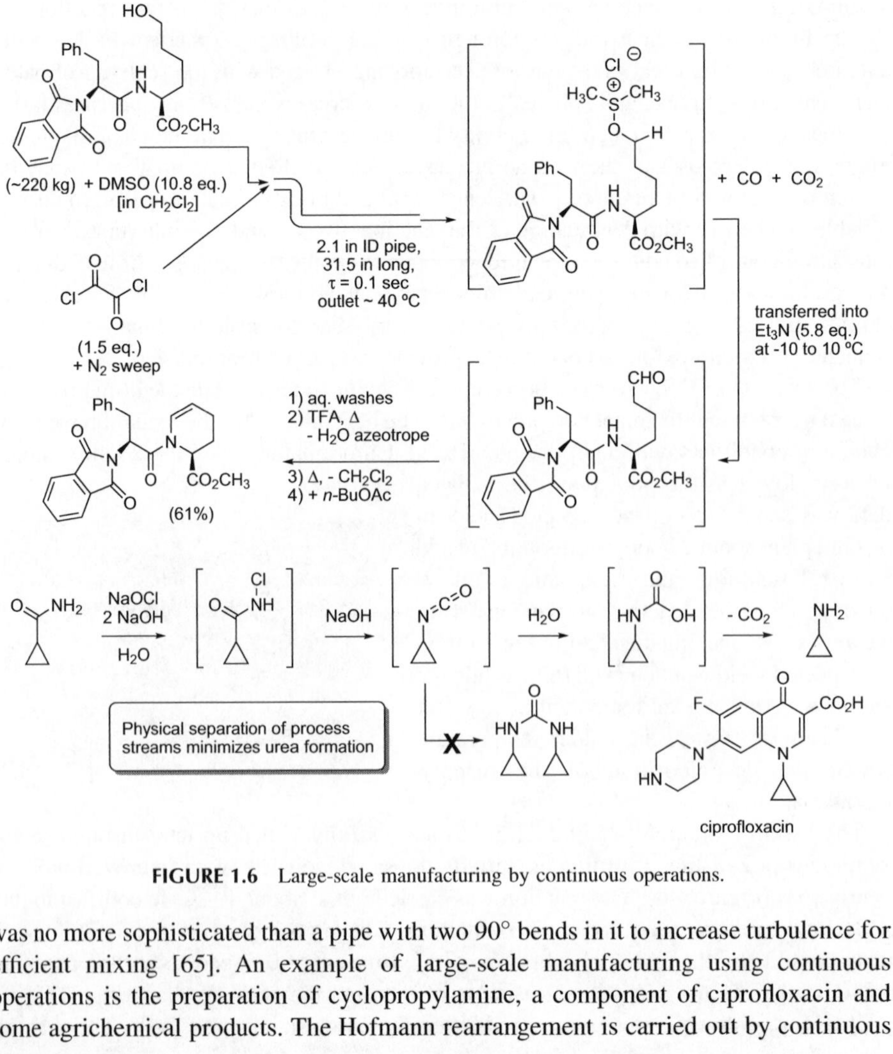

scale	addition temperatures	yield	comments
laboratory	-15 °C	55%	
pilot plant	-15 °C	31%	2 h each to add (COCl)₂ and Et₃N allowed intermediates to decompose
pilot plant	-40 °C	52%	faster additions gave higher yield

FIGURE 1.6   Large-scale manufacturing by continuous operations.

was no more sophisticated than a pipe with two 90° bends in it to increase turbulence for efficient mixing [65]. An example of large-scale manufacturing using continuous operations is the preparation of cyclopropylamine, a component of ciprofloxacin and some agrichemical products. The Hofmann rearrangement is carried out by continuous

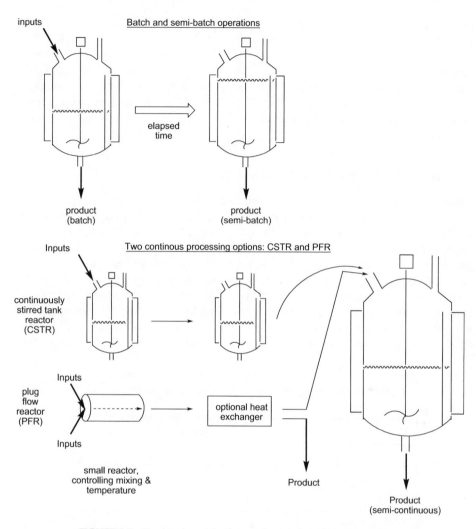

FIGURE 1.7 Batch and semi-batch operations, and continuous operations.

operations with either continuously stirred tank reactors [66] or static mixers [67] (Figure 1.7). Chemical engineers in the fine chemicals industry often have a great deal of experience with continuous operations.

The two primary options for operations, batch and semi-batch (or semi-continuous) operations, and continuous operations, are shown schematically in Figure 1.7 [68].

Continuous operations have been used on scale for decades and are slowly being incorporated into processes for APIs. Continuous operations are particularly useful for fast reactions, those that take place in seconds or minutes as opposed to hours. One attractive benefit of these operations is the SAFETY afforded by conducting a reaction with only a small amount of a reaction stream at any given moment (Chapter 14). Continuous operations are part of process intensification efforts, efforts to improve the space–time productivity of processing [69]. Federsel has estimated that continuous operations are applied in 10–20% of the processes in the fine chemicals and pharmaceutical industry [70]. Roberge and coworkers feel that 50% of the processes used in the pharmaceutical industry could benefit from continuous operations [71]. Chris Dowle, of the Center for Process Innovation, has stated that "Continuous processing can reduce operating expenses by at least 90% and capital expenses by at least 50%" [72]. Continuous operations are not appropriate for every process, but sometimes are the only practical approach to make more material.

III. OPERATIONS PREFERRED ON SCALE

Perhaps the first advice that a seasoned industrial process chemist will give to someone from academia is that it is not necessary to dry extracts over desiccants such as Na_2SO_4 or $MgSO_4$. This operation may be unnecessary if water is removed as an azeotrope during the concentration. As detailed in Table 1.4, a considerable amount of time may be expended in drying over desiccants. The cost of this operation can be calculated in several ways. The total time required of 16 operator-hours may be conservative; this assumes there are no problems, and no supervision is required. During the stir-out time of 2 hours other tasks can be undertaken, but the remaining steps require some attention. Considering that the cost of plant time may be as high as $500–750/h, a substantial amount of money may be spent unnecessarily. But

TABLE 1.4 Estimated Processing Time on Scale for Na_2SO_4 Drying Step

Operation	Estimated Operator-hours
Set up and test filter	2
Charge Na_2SO_4	0.5
Stir suspension	2
Filter off solids	2
Deliver rinse to site	0.5
Apply rinse to solids	0.5
Suction dry solids	0.5
Pack up solids for disposal	0.5
Rinse for "SAFE to clean"	0.5
Clean equipment	4
Test equipment for cleanliness	2
Store equipment	1
Total	16

foremost, as mentioned earlier, this operation may be redundant if a water-solvent azeotrope removes water while the extract is being concentrated. The cost of an unnecessary step is opportunity cost, for the opportunity lost to accomplish something else or learn something else. Hence on scale extracts are rarely dried over desiccants such as Na_2SO_4.

Table 1.5 compares operations run in the laboratory and on scale. These are guidelines suggested for convenient, productive, and cost-effective operations, but most of the operations can be run on scale if desired.

TABLE 1.5 Comparison of Process Operations in the Laboratory and on Scale

Process Operation	In the Laboratory		In Stationary Equipment, at Least 50 L Volume		Comments
	Commonly Carried Out	Easy to do	Commonly Carried Out	Easy to do	
Drying extracts over desiccants, e.g., Na_2SO_4, $MgSO_4$	X	X		X	Often redundant if concentrating removes H_2O by azeotroping
Concentrating to dryness (1)	X	X (2)		(3)	Thermal decomposition is a concern
Triturating, lixiviating (4)		X			Reslurrying solids is carried out on scale
Use of highly flammable solvents, e.g., Et_2O	X	X		X	More time and care are necessary for SAFE operations
Decanting	X	X			Siphoning may be an option
Column chromatography	X	X		X	Large-scale equipment may be purchased
Rapid transfers	X	X		X	Narrow-diameter openings limit transfer rate. Applying pressure of N_2 will speed transfer
Maintain cryogenic temperature	X	X		X	Large-capacity chillers may hold temperatures better
Extended additions	X	X	X	X	Wide selection of pumps is available for controlled additions
Maintain constant pH	X	X (5)	X	X	Enzymatic processing

(Continued)

TABLE 1.5 Comparison of Process Operations in the Laboratory and on Scale—cont'd

Process Operation	In the Laboratory		In Stationary Equipment, at Least 50 L Volume		Comments
	Commonly Carried Out	Easy to do	Commonly Carried Out	Easy to do	
Fine control of heating and cooling	X	X	X	X	Equipment with a high surface area to volume ratio can remove heat quickly and efficiently
Use of dangerous reagents, e.g., n-BuLi, CH$_3$I, LDA	X	X		X	More time and care are necessary for SAFE operations
Solvent displacement by distillation (chasing or stripping)		X	X	X	
Azeotropic drying		X	X	X	
Activated carbon treatment		X		X	Cleaning equipment becomes an issue; activated carbon is usually constrained to in-line filters
Suction filtration	X	X	X	X	Common in pilot plant
Centrifuge filtration		X	X	X	Common in manufacturing
Tray drying	X	X	X	X	Common in pilot plant
Tumble dryer or agitated dryer				X	Common in manufacturing
Spray drying				X	For manufacture of API
Lyophilization		X (6)		X	For manufacture of API

(1) Concentrating to dryness is rarely carried out on scale because lengthy times are required, and some decomposition may occur if poor mass transfer does not remove solids from the heated surface.

(2) Rotary evaporators are commercially available up to 100 L.

(3) Evaporation of solvents to solid residues did not afford processing difficulties on 50–250 kg scale, in the miniplant or in production [73].

(4) To triturate is to crush or grind solids, and to lixiviate is to wash soluble matter from other solids or oils. Trituration is rarely carried out on scale because reactors are rarely opened, due to SAFETY concerns, to preclude contamination, and the absence of suitably long spatulas. Other meanings applied to "triturate" are to mash an oil with extracting solvents, and to reslurry.

(5) A pH meter with an autoburette (pH-stat) may be used to maintain a narrow pH range by automatically adding acidic or basic solutions. This technology has been used for decades to stabilize enzymes used on large scale for fermentations and biocatalytic conversions.

(6) Lyophilization may be used to convert oils to solids in the laboratory. On scale it is used to prepare powders that will be dissolved in water for injection.

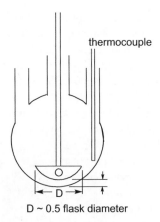

thermocouple

D ~ 0.5 flask diameter

FIGURE 1.8 Typical round-bottomed flask useful for process development in the laboratory.

A useful setup of laboratory glassware in the process R&D laboratory is shown in Figure 1.8. A thermocouple is preferred over a thermometer, as there is no liquid to clean up as with a broken thermometer; in addition, temperatures can be recorded electronically with a thermocouple. The thermocouple also provides a vertical element similar to the baffles in stationary equipment that increase the turbulence of reaction mixtures. (The indentations of a Morton flask also increase turbulence in the laboratory.) A mechanical stirrer is preferred over a magnetic stirbar for several reasons. First of all, a mechanical paddle stirrer can provide more power to resume stirring than a magnetic stirbar if a solid mass in the flask stops the agitation. Secondly, magnetic stirbars can also grind solids on the bottom of the flask, changing reaction rates by particle size reduction. Third, iron leached from magnetic stirbars has catalyzed reactions [74]. Furthermore a mechanical paddle stirrer more closely mimics the agitator in a multi-purpose reactor, by using a crescent-shaped Teflon paddle with a length about 50% of the diameter of the flask, and raising the blade above the bottom of the flask. A round-bottomed flask such as that depicted in Figure 1.8 would be fitted for a nitrogen sweep or static pressure of nitrogen for SAFETY reasons.

IV. PATENT CONSIDERATIONS

Patents for chemical inventions are based on new compositions of matter or new processes. New molecules may be easily identified for composition of matter patent applications, and can become part of the intellectual property (IP) portfolio of a company. Patent infringement can be easily determined if a competitor markets a compound protected by a composition of matter patent, but infringement of a process patent is more difficult to determine; detailed analyses of the process-related impurities or of the isotopic ratios of the API may be necessary (Chapter 3). Some companies feel that disclosing process details in a patent is tantamount to divulging for free the research of many person-years [70].

Researchers in industry generally sign away their rights to receive any financial gains from patents, and are obliged to propose any advances they have made for potential patent applications. Ideas and accomplishments should be registered through the appropriate channels in a timely manner, through witnessing and recording the date and time, as cases contesting patent issues may be decided based on a difference of days in recording an invention. Patenting innovations supports employers' positions for intellectual property, and is the *quid pro quo* for being able to conduct research.

Publications and patents acknowledge the creativity and accomplishments of researchers. To be recognized as an inventor one must have contributed in a creative way, as in designing a compound (composition of matter) or devising a non-obvious way to make a compound (processes). The first person to make a compound reduces a theory to a practice, and may become an inventor if he or she was not instructed on how to perform the tasks [75]. Universities and university employees, including graduate students and post-docs, may reap significant gains from patents [76].

> Propose compounds and processes for patents. Write up your work, publish, and keep your resume current. The best job security is the assurance that you can find another job if needed.

V. SUMMARY AND PERSPECTIVE

Process chemists are driven to make material that is fit for purpose. A route designed by medicinal chemists may be scaled up with no more optimization than what is needed to make the first 2 kg of material for toxicology or Phase 1 studies, as the timeline is important, not the route and technology. For instance, microwave carousels might be used to make more material if this technology offered significant advantages. The cyclization in Figure 1.9 afforded a better yield in less time when run under microwave conditions instead of thermal [77]. In this case the shorter reaction may be irrelevant, since a chemist could undertake other efforts while a reaction was taking place in the fume hood. The 19% improvement in yield may not be enough to persuade a process chemist to undertake a microwave-mediated scale-up, unless the purification was significantly easier. A curious person may experiment with a new technology and champion its use. Leadbeater has described the successful scale-up of chemistry using

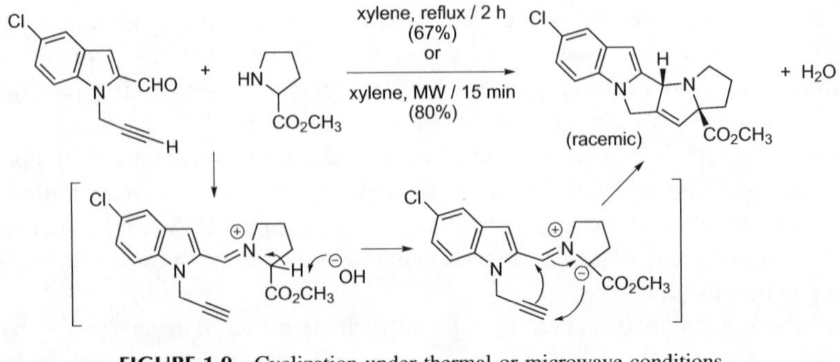

FIGURE 1.9 Cyclization under thermal or microwave conditions.

TABLE 1.6 Comparison of Time Considerations for Projects

Reference compound	Most expeditious route is used, with minimal optimization Chromatographic purification OK
Kilo lab batches	Invest time to decrease the risk of poor yields Meeting the deadline is critical
Toxicology batches	May prepare desired final form, but not necessarily the optimal polymorph High-quality material is preferred, perhaps 95–97% by HPLC, but not ultra-pure. The goal is to qualify the impurities
Pilot plant batches	Expensive to occupy equipment, therefore reasonably smooth processing desired Poor yields and quality affect campaigns Preferable to solve any processing problems in the laboratory before entering the pilot plant
Manufacturing	±2% in yield/quality can have a significant impact, so fine-tuning is readily justified

microwave technology [78]. Optimization requires time and money. It is important to know how much time is available. Some examples are shown in Table 1.6.

There are additional considerations for preparing material for toxicology studies (the tox batch) that affect processing for subsequent batches. Since the purpose of the tox batch is to test and qualify impurities likely to be present in NCEs and APIs administered to humans, the tox batch should be crystallized to high quality, perhaps 95–97% for the compound of interest, but not ultra-pure. While it is against the training of chemists not to prepare compounds as pure as possible, various approaches are undertaken for this purpose to prepare material that is less than pristine. For instance, only one aqueous wash instead of multiple washes may be used to remove impurities, or a key intermediate may not be recrystallized. A crystallization may be cooled rapidly in the hope of entrapping some process impurities in the tox batch. Some companies have crystallized a tox batch in the presence of impurities added from the mother liquor of a previous batch – the resulting batch could be considered "blended" and have uncharacteristically high impurity levels, but after all the tox batch is not intended for human use. The batch submitted for toxicology screening will hopefully have the same impurities as subsequent batches intended for human use, which must have impurities that are not different from those of the tox batch. The impurity levels in cGMP batches must be not higher than those of the tox batch; otherwise those lots must be rejected, or bridging tox studies must be carried out to qualify the impurities now routinely found in batches. The opportunities to change physical characteristics and impurity profiles may be limited to the first year of drug development.

Set purity specifications reasonably high for an API: the FDA will readily allow purity specifications to be *upgraded*. To minimize changes to impurity profile in commercial batches "freeze" the final steps early in development, and make changes only to the steps leading up to the penultimate [79].

As mentioned earlier, processes to prepare NCEs and APIs must be SAFE, and reliably produce high-quality APIs with the expected yield. Processes must be as simple as possible, but rugged, with effective ranges for operations well understood. At the end of a development cycle, the process scientist will have developed a streamlined process with no unnecessary steps. He or she will know the critical steps, and any action that is necessary, and know the steps which can afford some flexibility. Possible processing difficulties have been anticipated. The process chemist will have an exquisitely thorough knowledge of the process.

Efficient process development is carried out through teamwork. The observations of those involved in drug discovery, process research, and process development are crucial.Thus the discovery chemist can help by choosing reagents that are inexpensive, but still convenient, and by paying attention to any scale-up details. Process R&D departments have helped by regularly giving input to drug discovery chemists on selecting routes and inexpensive reagents, even suggesting or preparing preferred reagents. Any observation may be valuable; for example, exotherms point out the need for hazard investigations, and identifying an impurity may simplify process R&D work later. Noticing a different polymorph can guide investigations into crystallizing an API, and may be an important opportunity for IP. Identifying crystalline intermediates can preclude chromatography, making a route generating these intermediates very attractive. Generating and recording analytical data, even solvents for TLC systems, can help. All scientists can recognize processing limitations. For instance, steps where SAFETY is a big concern or where the stability of the product is low might call for continuous operations; while these steps may be run on a 22 L scale, perhaps they should be developed by experienced CROs/CMOs. Paying close attention to the details of processing and sharing those findings will speed development.

To be valuable to an employer a process chemist or engineer must adapt to the changes facing the pharmaceutical industry. For instance, petroleum feedstocks to prepare chemical intermediates may become more costly as the amount of easily extractable petroleum declines, driving further development of other technologies. The production of syngas from pyrolysis of organic matter may increase, but will probably provide only simple compounds [80]. Plant tissue culture may become a more important source of complex molecules [81]. Transition metal catalysts are used heavily today, but the availability of many transition metals (including precious metals) and rare earth elements is becoming restricted, and demand is growing [82]. Biocatalysis is increasingly attractive to the pharmaceutical and fine chemicals industries. Biodegradation or supercritical water may be used to treat waste streams [83]. Membranes will probably be used more often to purify reaction streams on scale [84]. Overall, being knowledgeable in more than one area increases the value of an employee.

The development and manufacture of drug substances and drug products are rather like a tapestry hung on a wall: a pull in one direction creates ripples across the remainder. We would like to think that drug development and manufacture are guided by principles of good science, but factors including quality (control of impurities), SAFETY (hazards and uncontrolled releases of chemicals and energy), personal politics (striving for promotions and the "NIH syndrome, for "Not Invented Here"), and economics can pull the fabric in different directions and at different times. In the final analysis, economic considerations drive process development.

The attitudes for chemists in process R&D may differ from those of drug discovery chemists.

1. Teamwork is key. Optimize interactions between discovery chemists, engineers, analysts, microbiologists, operators, Environmental Health & Safety officers, pharmaceutical chemists, QA/QC personnel, attorneys for patent applications, out-sourcing (CROs, CMOs), your manufacturing counterparts, and regulatory personnel. Don't trivialize the work of others, especially if you don't understand their job.
2. Recognize that the need to minimize impurities drives process development.
3. Know the edge of failure for your operations. For rapid development, stay away from the edge. For cost-effective processes, you may want to operate close to the edge.
4. Reproducible, accurate analytical data are necessary for effective process development.
5. If an operation is extremely difficult, reconsider. Perhaps additional information will simplify the operation.
6. Heed the advice of ecologists: "We're all in this together." Humor helps defuse the gravity of most difficult situations.
7. During process introduction and scale-up boring situations can be good — that means there are no problems to address.

REFERENCES

1. The Los Angeles Times, http://www.latimes.com/news/science/environment/la-fi-seafarm10-2008jul10,0,1092501,full.story; July 10, 2008.
2. Guichon, G.; Shirazi, S. G.; Katti, A. M. *Fundamentals of Preparative and Nonlinear Chromatography*. Academic Press: San Diego, 1994; p 596.
3. Anastas, P. *Chem. Eng. News* **2011,** *89*(26), 62.
4. Anon., *The Washington Post*, April 28, 2010.
5. Mullin, R. *Chem. Eng. News* **2010,** *88*(44), 9.
6. Johnson, C. *The Washington Post*, 9/2/09.
7. Hess, G. *Chem. Eng. News* **2009,** *87*(14), 22.
8. Erickson, B. E. *Chem. Eng. News* **2009,** *87*(9), 30.
9. Tremblay, J.-F. *Chem. Eng. News* **2007,** *85*(29), 9.
10. http://www.sec.gov/news/press/2004-105.htm
11. Bradsher, K. *The New York Times*, October 25, 2001.
12. http://www.opensecrets.org
13. Halford, B. *Chem. Eng. News* **2010,** *88*(39), 58.
14. Jack, A. *Financial Times*, Sept.30/Oct.1, **2006.**
15. Class, S. *Chem. Eng. News* **2005,** *83*(49), 30.
16. Abboud, L.; Hensley, S. *Wall St. J.* 9/03/03; Mullin, R.; Jarvis, L. M. *Chem. Eng. News* **2008,** *86*(42), 24.
17. Schmidt, C. W. *Mod. Drug Disc.* **2003,** *6*(6), 51.
18. 50% of in-licensed compounds fail due to scalability issues: D'Antuono, J., in-pharmatechnologist.com, February 2008.
19. Jarvis, L. *Chem. Eng. News* **2011,** *89*(22), 8.
20. Stork, G.; Yamashita, A.; Adams, J.; Schulte, G. R.; Chesworth, R.; Miyazaki, Y.; Farmer, J. J. *J. Am. Chem. Soc.* **2009,** *131*, 11402.
21. Magnus, P.; Sane, N.; Fauber, B. P.; Lynch, V. *J. Am. Chem. Soc.* **2009,** *131*, 16045.
22. Nicolaou, K. C.; Vourloumis, D.; Winssinger, N.; Baran, P. S. *Angew. Chem. Int. Ed.* **2000,** *39*, 44.
23. Gaich, T.; Baran, P. S. *J. Org. Chem.* **2010,** *75*, 4657.

24. Hudlicky, T. *Chem. Rev.* **1996,** *96,* 3.

25. Laird, T. *Comprehensive Medicinal Chemistry* 1990, Vol. 1, p 32.

26. Caron, S. In *Fundamentals of Early Clinical Drug Development*; Abdel-Magid, A. F.; Caron, S., Eds.; Wiley: Hoboken NJ, 2006; p 101.

27. Harrington, P. J. *Pharmaceutical Process Chemistry for Synthesis; Rethinking the Routes to Scale-Up.* Wiley-VCH, 2011.

28. Caron, S., Ed.; *Practical Synthetic Organic Chemistry: Reactions, Principles, and Techniques*; Wiley, 2011.

29. Reichardt, C. In *Solvents and Solvent Effects in Organic Chemistry,* 4th ed.; Wiley-VCH, 2011.

30. Tao, J.; Kazlaukas, R. J., Eds.; *Biocatalysis for Green Chemistry and Chemical Process Development*; Wiley-VCH, 2011.

31. Jiménez-González, C.; Constable, D. J. C. *Green Chemistry and Engineering: A Practical Design Approach;* Wiley, 2011.

32. Matlack, A. S. *Introduction to Green Chemistry,* 2nd ed.; CRC Press: Boca Raton, 2010.

33. Blaser, H. U.; Federsel, H. J., Eds.; *Asymmetric Catalysis on Industrial Scale: Challenges, Approaches and Solutions,* 2nd ed.; Wiley-VCH, 2010.

34. Li, J. J.; Johnson, D. S., Eds.; *Modern Drug Synthesis*; Wiley-VCH, 2010.

35. Yasuda, N., Ed.; *The Art of Process Chemistry*; Wiley-VCH, 2010.

36. Shiori, T.; Izawa, K.; Konoike, T., Eds.; *Pharmaceutical Process Chemistry*; Wiley-VCH, 2010.

37. Dunn, P.; Wells, A.; Williams, M. T., Eds.; *Green Chemistry in the Pharmaceutical Industry*; Wiley-VCH, 2010.

38. Leitner, W., Jessop, P. G., Li, C.-J., Wasserscheid, P., Stark, A., Eds.; *The Handbook of Green Chemistry – Green Solvents*; Wiley-VCH, 2010.

39. Tao, J., Lin., G.-D. L., Liese, A., Eds.; *Biocatalysis for the Pharmaceutical Industry*; Wiley-VCH, 2008.

40. Gadamasetti, K.; Braish, T., Eds.; *Process Chemistry in the Pharmaceutical Industry,* 2nd ed.; CRC Press: Boca Raton, Florida, 2008.

41. Rao, C. S. *The Chemistry of Process Development in Fine Chemical & Pharmaceutical Industry,* 2nd ed.; Wiley: New York, 2007.

42. McConville, F. *The Pilot Plant Real Book;* 2nd ed.; FXM Engineering & Design: Worcester MA; 2007; http://www.pprbook.com; 2007.

43. Abdel-Magid, A. F.; Caron, S., Eds.; *Fundamentals of Early Clinical Drug Development*; Wiley: Hoboken, NJ, 2006.

44. Ager, D., Ed.; *Handbook of Chiral Chemicals,* 2nd ed.; CRC Press, 2006.

45. Li, J.-J.; Johnson, D. S.; Sliskovic, D. R.; Roth, B. D. *Contemporary Drug Synthesis;* Wiley: New York, 2004.

46. Blaser, H. U.; Schmidt, E., Eds.; *Asymmetric Catalysis on Industrial Scale*; Wiley-VCH: Weinheim, 2004.

47. Cabri, W.; Di Fabio, R. *From Bench to Market: The Evolution of Chemical Synthesis;* Oxford University Press: New York, 2000.

48. am Ende, D. J., Ed.; *Chemical Engineering in the Pharmaceutical Industry: R&D to Manufacturing*; Wiley, 2010.

49. Turton, R.; Bailie, R. C.; Whiting, W. B.; Shaeiwitz, J. A. *Analysis, Synthesis and Design of Chemical Processes,* 3rd ed.; Prentice Hall, 2009.

50. Butters, M.; Catterick, D.; Craig, A.; Curzons, A.; Dale, D.; Gillmore, A.; Green, S. P.; Marziano, I.; Sherlock, J.-P.; White, W. *Chem. Rev.* **2006,** *106,* 3002.

51. Bray, B. *Nat. Rev. Drug Disc.* **2003,** *2,* 587, July.

52. Mullin, R. *Chem. Eng. News* **2007,** *85*(4), 11.

53. Davies, I. W.; Welch, C. J. *Science* **2009,** *325,* 701.

54. Carey, J. S.; Laffan, D.; Thomson, C.; Williams, M. T. *Org. Biomol. Chem.* **2006,** *4,* 2337.

55. Roughley, S. D.; Jordan, A. M. *J. Med. Chem.* **2011,** *54,* 3451.

56. Federsel, H-J. *Chirality* **2003,** *15,* S128.

57. Lakoff would be proud of using "fit for purpose" to justify current activities as a nice example of "framing the argument": Lakoff, G.; Dean, H.; Hazen, D. *Don't Think of an Elephant!: Know Your Values and Frame the Debate – The Essential Guide for Progressives;* Chelsea Green, 2004.

58. Chou, T. S.; Heath, P. C.; Patterson, L. E.; Poteet, L. M.; Lakin, R. E.; Hunt, A. H. *Synthesis* **1992,** 565.

59. Jiang, X.; Lee, G. T.; Villhauer, E. B.; Prasad, K.; Prashad, M. *Org. Process Res. Dev.* **2010,** *14,* 883.

60. Childers, J. H., Jr. In *Fundamentals of Early Clinical Drug Development*; Abdel-Magid, A. F.; Caron, S., Eds.; Wiley: Hoboken, NJ, 2006; p 163.

61. Anderson, N. G. *Org. Process Res. Dev.* **2004,** *8,* 260.

62. Horgan, S. W.; Burkhouse, D. W.; Cregge, R. J.; Freund, D. W.; LeTourneau, M.; Margolin, A.; Webster, M. E. *Org. Process Res. Dev.* **1999,** *3,* 241.

63. Rouhi, A. M. *Chem. Eng. News* **2003,** *81*(28), 37.

64. For instance, see, http://www.abprocess.com/index.cfm?pid=54&pageTitle=HTST-%28High-Temperature-Short-Time-Pasturizers%29; 1999.

65. McConnell, J. R.; Hitt, J. E.; Daugs, E. D.; Rey, T. A. *Org. Process Res. Dev.* **2008,** *12,* 940.

66. Diehl, H.; Blank, H. U.; Ritzer, E. US 5,032,687, 1991 (to Kernforschunganlage Julich Gesellschaft).

67. Kleemiss, W.; Kalz, T. US 5,728,873, 1998 (to Huels Aktiengesellschaft).

68. Roberts, G. W. *Chemical Reactions and Chemical Reactors;* Wiley, 2009.

69. Reay, D.; Ramshaw, C.; Harvey, A. *Process Intensification: Engineering for Efficiency, Sustainability and Flexibility.* Butterworth Heinemann, 2008.

70. Federsel, H.-J. *Acc. Chem. Res.* **2009,** *42,* 671.

71. Roberge, D. M.; Ducry, L.; Bieler, N.; Cretton, P.; Zimmermann, B. *Chem. Eng. Technol.* **2005,** *28,* 318.

72. Rouhi, A. M. *Chem. Eng. News* **2005,** *83*(3), 41.

73. Federspiel, M.; Fischer, R.; Hennig, M.; Mair, H.-J.; Oberhauser, T.; Rimmler, G.; Albiez, T.; Bruhin, J.; Estermann, H.; Gandert, C.; Göckel, V.; Götzö, S.; Hoffmann, U.; Huber, G.; Janatsch, G.; Lauper, S.; Röckel-Stäbler, O.; Trussardi, R.; Zwahlen, A. G. *Org. Process Res. Dev.* **1999,** *3,* 266.

74. Eisch, J. J.; Gitua, H. N. *Organometallics* **2007,** *26,* 778.

75. Richard, C. R. *Chem. Eng.* **2007,** *114*(8), 63.

76. Halford, B. *Chem. Eng. News* **2009,** *87*(21), 39.

77. Bashiardes, G.; Safir, I.; Barbot, F. *Synlett* **2007,** *11,* 1707.

78. Schmink, J. R.; Kormos, C. M.; Devine, W. G.; Leadbeater, N. E. *Org. Process Res. Dev.* **2010,** *14,* 205.

79. Shutts, B., *'Freezing' the Commercial Process – Issues and Challenges*, talk contributed at The Third International Conference on Process Development Chemistry, March 26, 1997.

80. Levenspiel, O. *Ind. Eng. Chem. Res.* **2005,** *44,* 5073.

81. For example, in 2004 Bristol-Myers Squibb received a US Presidential Green Chemistry Award for the manufacture of paclitaxel through plant cell tissue culture: http://www.epa.gov/greenchemistry/pubs/docs/award_recipients_1996_2011.pdf

82. Jenkins, S. *Chem. Eng.* **2010,** *117*(10), 17.

83. Ondrey, G. *Chem. Eng.* **2011,** *118*(6), 11.

84. Székely, G.; Bandarra, J.; Heggie, W.; Sellergren, B. Castelo Ferreira, F. J. *Membr. Sci.* **2011,** *381,* 21.

Process Safety

"...and the addition was only slightly exothermic...."

– Anonymous

I. INTRODUCTION

Safety considerations require some specific values: How much heat per reaction mass is evolved? What is the temperature at which a runaway reaction begins? Can the coolant in the jacket of a batch reactor contain the temperature at 100 °C below that critical temperature, or below the temperature at which side products form?

The goal of process safety is to control the release of hazardous material and energy from process operations. Of course all concerns can be traced back to the reactive nature of chemicals, which is required to construct the molecules we require. Helpful references on safety include an updated edition on laboratory safety from the National Research Council [1], a section in McConville's book [2], the safety section in volumes of *Organic Syntheses* [3], the on-going special section on safety in the sixth issue of *Organic Process Research & Development*, and others [4]. Safety letters are available in *Chemical & Engineering News*. Safety management of plants has also been discussed [5]. This chapter will focus on how safety considerations can facilitate process development.

With any operation there is always a risk, but with increased scale-up the greater is the liability or cost of accidents [6], through damage to operators, equipment, and communities; reduced time to meet deadlines; financial losses; and tarnished corporate

Practical Process Research and Development. DOI: 10.1016/B978-0-12-386537-3.00002-2

reputations. Classic examples are the 1984 release of methyl isocyanate in Bhopal that killed thousands of people [7] and the effects on plants and humans from long-term exposure to dioxins at Love Canal [8,9].

In general we are concerned with three types of hazards from the reactivity of compounds: health hazards that damage the organs of (primarily) humans; physicochemical hazards; and environmental hazards. Compounds designated as health hazards may be labeled as toxic, harmful, corrosive, or irritating; or as carcinogens, mutagens, or teratogens (reproductive toxins). (The safety aspects of drug candidates and APIs are determined by toxicological testing, clinical trials, and ultimately are controlled by controlling dosage and impurities.) Compounds designated as physicochemical hazards may be labeled as flammable, explosive, oxidizing, or corrosive; the hazardous or toxic nature of such compounds may lead to restrictions on transport and storage. Environmental hazards may present immediate or delayed danger to the environment [10]. Butters and coauthors have detailed the impact of safety considerations and how to develop safe processes [11].

As scientists who routinely work with chemicals we sometimes overlook the risk of compounds and processes that are familiar to us. Other times we try to anticipate difficulties, but unfortunately lack of knowledge in just one aspect of operations can lead to "incidents." For instance, the build-up of diazidomethane has led to explosions in the laboratory [12] (see Chapter 11 for a detailed discussion). Oxidations using or generating O_2 can lead to the combustion of the solvent vapors above the reaction mixture due to electrostatic discharge; for this reason N_2 sweeps are employed to keep the $[O_2]$ in the headspace of a reactor below 6–10% [13,14]. Electrostatic discharge between a syringe needle and a digital thermometer may have led to the explosion on a 2-L scale in an oxidative coupling of trimethylsilylacetylene (Glaser–Hay reaction) [15]. Virtually any release of H_2 has the potential for ignition, as H_2 is flammable in concentrations of 4–75% [16]. "Spent" streams may react during storage. We need to be aware and educate ourselves about safety problems that could arise.

Many references are available on the toxicity and chemical hazards of compounds.
- Material Safety Data Sheets (MSDSs) are routinely included in shipments of chemicals and are available on-line: http://www.sigmaaldrich.com/safety-center.html
- International Chemical Safety Cards: http://www.ilo.org/legacy/english/protection/safework/cis/products/icsc/dtasht/index.htm
- The Carcinogenic Potency Project Database: http://potency.berkeley.edu/cpdb.html
- TOXNET, Toxicology Data Network Hazardous Substances Data Bank (HSDB): http://toxnet.nlm.nih.gov
- ToxRefDB (US EPA): http://www.epa.gov/toxrefdb
- Bretherick's Handbook of Reactive Chemical Hazards
- CHETAH calculations for explosivity of compounds: Glaesemann, K. R.; Fried, L. E. https://e-reports-ext.llnl.gov/pdf/322927.pdf
- Internet searches for "(compound X) + toxicity" or "(compound X) + hazard"

Initiate safety analyses before problems could occur. With so many resources available we have little excuse to be naïve or blasé regarding the compounds we encounter.

Respect chemicals, but don't fear them — fearing a compound could lead to the wrong reaction in an emergency.

SAFETY — Deaths have been caused upon exposure just once to very small amounts of compounds, such as methyl mercury [*Chem. Eng. News* **1997,** *75*(19), 7; http://pubs.acs.org/cen/safety/19970512.html], trimethylsilyldiazomethane [Barnhart, R. W.; Dale, D. J.; Ironside, M. D.; Vogt, P. F. *Org. Process Res. Dev.* **2009,** *13*, 1388], and methyl fluorosulfonate ["Magic Methyl": http://www.ncbi.nlm.nih.gov/pubmed/484483]. Minimize contact with these and other toxic compounds by following SAFE laboratory procedures, including use of suitable gloves, and by making sure that the fume hood is functioning. Best of all, avoid highly toxic compounds, and design processes that are inherently SAFE.

II. REPERCUSSIONS OF USING COMPOUNDS REGARDED AS ENVIRONMENTAL HAZARDS

The long-term influence of environmental hazards is attracting greater attention. Matlack's book contains some valuable reflections and specific examples of the impact of chemicals on the environment [17]. Concerns of CO_2 and halocarbons on atmospheric chemistry have been discussed [18]. The toxicity of quaternary ammonium salts on aquatic ecosystems has probably discouraged the use of ionic liquids in preparative organic chemistry [19]. Greater focus has been placed recently on the cumulative impact of pharmaceuticals on the environment [20]. We owe it to ourselves and future generations to take care of our environment.

Many laws have been enacted to protect the environment. Some selected laws that are relevant to the United States are shown in Table 2.1. More information is available from the US Environmental Protection Agency (EPA, http://www.epa.gov).

TABLE 2.1 Selected Laws Enacted to Protect the Environment and People

Act	Responsibility
Comprehensive Environmental Response, Compensation, and Liability Act (CERCLA)	Federal "Superfund" to clean up uncontrolled or abandoned hazardous-waste sites
Emergency Planning and Community Right-to-Know Act (EPCRA)	Community safety, for firefighters and others Every state must do RTK (right to know) Reporting threshold varies per chemical
Occupational Safety and Health Act (OSHA)	Worker and workplace safety
Resource Conservation and Recovery Act (RCRA)	Control hazardous waste from the "cradle-to-grave"

TABLE 2.2 Some Entries From New Jersey's Extraordinarily Hazardous Substances List (1)

Extraordinarily Hazardous Substance	New Jersey Threshold Quantity (lb) (2)	EPA Threshold Quantity (lb) (2)
Acrolein	200	5000
Bis-chloromethyl ether	100	10,000
Furan	80	1000
Methyl isocyanate	200	5000
SOCl$_2$	250	–

(1) Permit Identification Form, New Jersey Department of Environmental Protection, July 2006: http://www.state.nj.us/dep/pcer/docs/pifupdate.pdf.
(2) The threshold quantity is that quantity in a process at any time or that quantity generated by a process within 1 h.

In accordance with right-to-know laws states may limit the amounts of hazardous chemicals that can be used in processing to quantities less than those mandated by the EPA. Some compounds restricted by New Jersey are shown in Table 2.2. The hazardous nature of these materials and the paperwork involved make such compounds less attractive as starting materials or reagents.

III. TOXICOLOGICAL HAZARDS OF COMPOUNDS

The toxicity of some compounds is familiar, such as burns to skin from NaOH or HCl, lachrymation from fumes of warm AcOH, irritation to the lungs and throat from breathing ammonia, and the metabolic poisoning from HCN, CO and H$_2$S. With all compounds the toxicity is related to the exposure, i.e., how much of the compound someone contacts, and how it is contacted. (After all, even water can be toxic.) Unfortunately serious health problems may become evident hours or even days after exposure [21]. Chemicals should always be treated with respect.

SAFETY — Note that H$_2$S may exert effects at concentrations below what humans can smell [Chem. Eng. News **2006,** 83(29), 5]. If one cannot detect a compound by its smell or irritating characteristics one may still be at risk. It is best to assume that toxic compounds could be released into your environment, and introduce precautions accordingly. Be aware for chance encounters with chemicals!

Chemicals with some reactive functional groups can be expected to cause toxic effects. Any compound that can form covalent bonds with reactive functional groups in tissues, such as the amino groups of lysine in proteins, is likely to be toxic, and any substance that can cross-link functional groups will be especially toxic (Figure 2.1). The

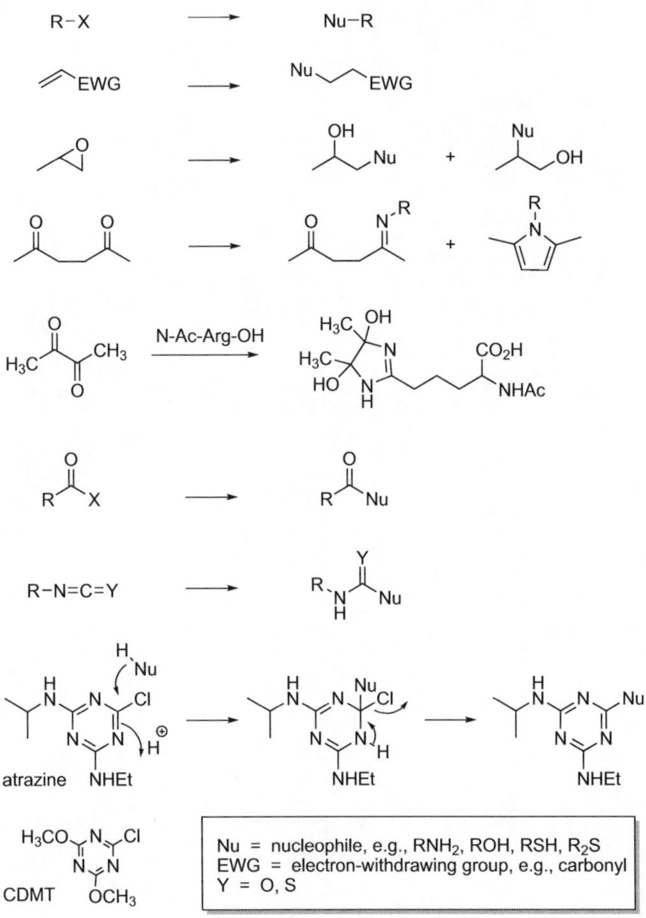

FIGURE 2.1   Potentially toxic electrophiles.

toxicity of butanedione may be associated with condensation of the terminal group of arginine residues [22]. If atrazine, a popular herbicide to control weeds, were present in an API it would be considered a potentially genotoxic impurity (PGI) as an arylating agent (see Chapter 13). (Recently the genotoxic potential of atrazine on fish was demonstrated [23], and with some studies connecting atrazine to birth defects and other anomalies the EPA is re-examining its approval of atrazine [24].) Aldehydes are also marked as PGIs, as discussed in Chapter 13.

Other chemicals may seem harmless enough, but metabolism (primarily oxidation and hydrolysis [25,26]) may convert them to reactive substances (Figure 2.2). Oxidative metabolism produces the toxic chloroacetaldehyde from chloroethanol [27] and toxic methoxyacetic acid from methoxyethanol [28]. Ethylene glycol affects first the central nervous system, then the cardiovascular system, and finally the kidneys, which may be consistent with successive metabolism to glycolic acid and oxalic acid [29]. Oxalic acid can lead to kidney failure through the formation of insoluble calcium oxalate [30].

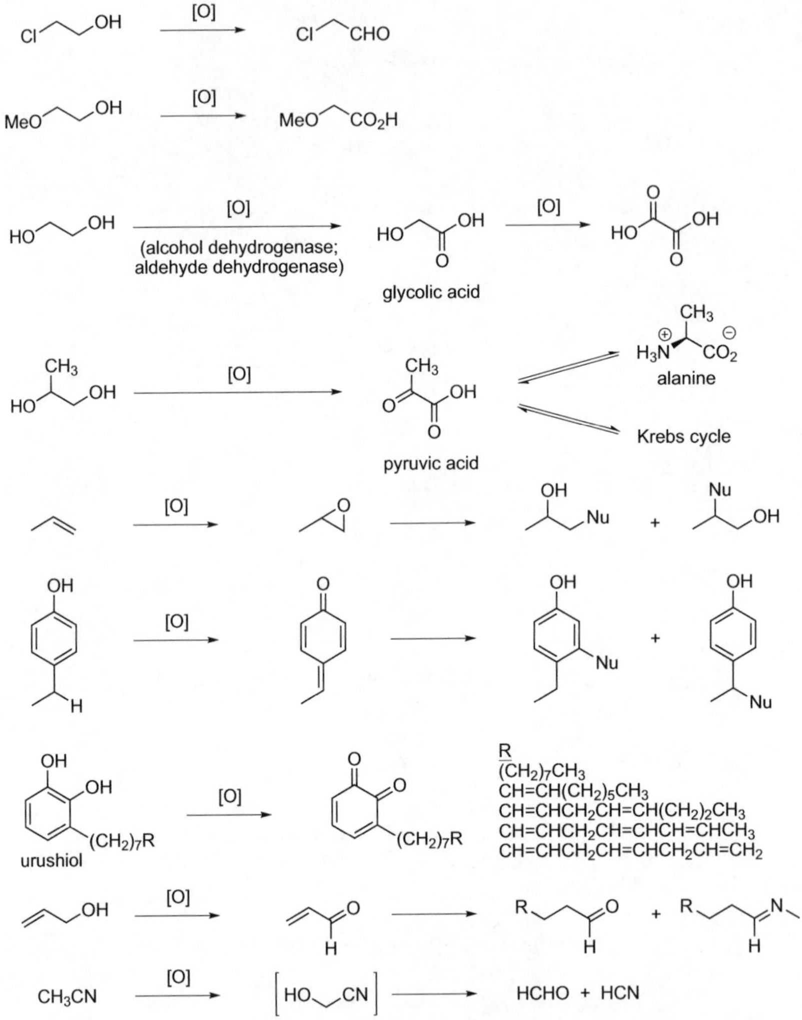

FIGURE 2.2 Examples of bioactivation to produce reactive electrophiles.

Similar metabolism of 1,2-propanediol leads to pyruvic acid, which can enter into the citric acid cycle and generate alanine, making 1,2-propanediol less toxic than ethylene glycol [31]; 1,2-propanediol is GRAS (generally recognized as safe) [32] and is used in formulating APIs into drug products as capsules. Oxidation of phenols can produce toxic quinones, quinonemethides, and similar compounds. *In vivo* oxidation of urushiol (a mixture of five compounds) produces the *o*-quinones, metabolites that may be painfully familiar to those allergic to poison ivy and poison oak [33,34]. Similar oxidative metabolism is involved with the toxicity of benzo[*a*]pyrene and other polycyclic aromatic hydrocarbons [35]. The primary systemic toxicity of acetonitrile is due to *in vivo* oxidation to hydroxymethylcyanide and subsequent breakdown to HCN [36,37]. Some researchers are designing compounds for biodeactivation [38].

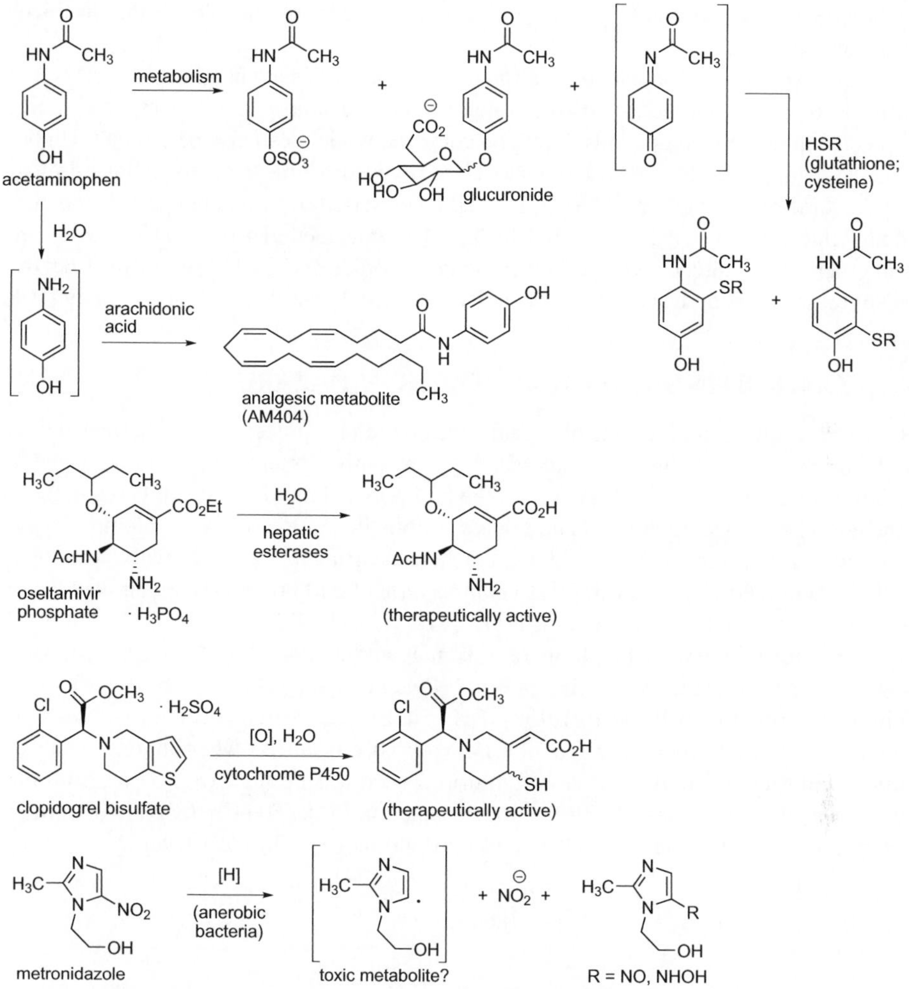

FIGURE 2.3 Metabolism of some prodrugs.

Compounds may be metabolized at multiple sites. For instance, acetaminophen (paracetamol, 4-acetamidophenol) is sulfated, converted to the glucuronide, or hydrolyzed and acylated with arachidonic acid to afford the analgesic species (Figure 2.3) [39,40]. Oxidative metabolism of acetaminophen produces the corresponding iminoquinone, which is toxic to the liver; overdoses of acetaminophen are treated with N-acetyl cysteine, and the mercaptan adds to the quinone ring [41]. Other prodrugs, such as oseltamivir phosphate [42,43] and clopidogrel bisulfate [44], are transformed *in vivo* to the active species by hydrolysis and (in the case of clopidogrel bisulfate) oxidation. Metronidazole, an antibiotic, is reductively converted to the active species by anaerobic bacteria [45].

Although software is available to predict the toxicity of compounds [46,47], compounds must be respected for any unexpected metabolites that may be toxic. Combinations of compounds may also prove toxic; for instance, the combination of melamine and cyanuric acid in pet foods may have been responsible for the deaths of dogs and cats [48].

Highly volatile compounds pose difficulties on scale. When charging to reactors it is difficult to prevent the loss of volatile compounds to the atmosphere or vacuum systems, and containing these compounds during reaction and workup can also be difficult. Highly reactive compounds that are volatile should be avoided for these complications. In that regard, dimethyl sulfate (bp ~ 188 °C) would be preferred over iodomethane (bp 43 °C). Methyl fluorosulfonate (Magic Methyl, bp 92–94 °C) and methyl triflate (bp 94–99 °C) are four orders of magnitude more reactive than methyl iodide or dimethyl sulfate [49], but the use of these compounds has largely been discontinued due to their extreme toxicity [50].

IV. CONTROLLING CHEMICAL REACTION HAZARDS

Runaway reactions are the bane of organic chemists and engineers in the pharmaceutical and fine chemicals industries. Excessively high temperatures promoted some of the runaway reactions shown in Figure 2.4. The build-up and rapid release of gases in these "incidents" damaged equipment and injured people. In some cases swept along with the gases were toxic components, including unreacted starting materials, as with the Bhopal hydrolysis of methyl isocyanate [7]. Decomposition of 2 M borane–THF complex led to an explosion at a Pfizer facility in 2002 [51,52]; $BH_3 \cdot THF$ is now sold as a 1 M solution. (Recommended handling of this reagent includes buying it in cylinders, keeping it pressurized, and storing at 5 °C. Reactions should be run below 40 °C to minimize formation of $B(On\text{-}Bu)_3$ [53]. BH_3 can be generated as needed, as discussed in Chapter 4.) An exothermic reaction unexpectedly occurred when a condensate of thionyl chloride in EtOAc was stored in drums used to ship the $SOCl_2$; zinc from the drums was needed for this three-component reaction [54]. (See Chapter 11 for further safety-related discussions on $SOCl_2$ and phosphorus oxychloride.) Hydrogenation of

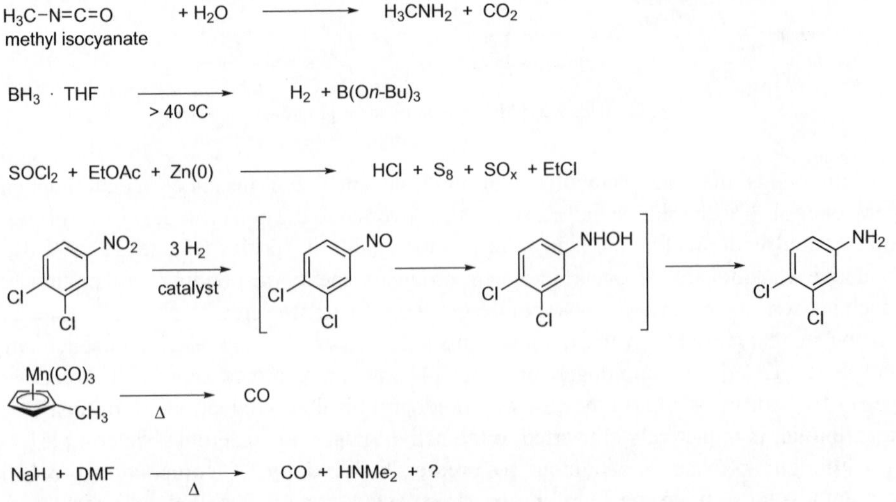

FIGURE 2.4 Some compounds involved in runaway reactions.

3,4-dichloro-nitrobenzene destroyed a batch hydrogenator and a building in 1976; the cause was determined to be the build-up and reaction of the hydroxylamine intermediate at a temperature higher than normal [55]. In processing to make methyl-cyclopentadienyl manganese tricarbonyl a loss of cooling led to an explosion comparable to 1400 lb of TNT [56]. Runaway reactions have occurred between DMF and NaH [57] and between DMF and NaBH$_4$ [58]. DMF is known to disproportionate to carbon monoxide and dimethylamine under acidic or basic conditions [59].

Poor understanding of reaction kinetics and lack of suitable controls often lead to runaway reactions [60]. In an example shown in Figure 2.5, acrylonitrile, cyclohexanol, and H$_2$SO$_4$ were mixed, following a literature reference to make the amide via a Ritter reaction. After two hours at room temperature a violent reaction took place on a 2-mol scale, and most of the reaction mixture was ejected from the flask. The cause of this runaway reaction was identified as a low instantaneous ratio of acid to alcohol; with no temperature control, reaction became faster at higher temperatures, resulting in an autocatalytic exothermic reaction. The optimal conditions developed for safety and good yield were the simultaneous addition of 1.77 equivalents of H$_2$SO$_4$ and a mixture of acrylonitrile and cyclohexanol to 0.25 equivalents of H$_2$SO$_4$ at 45–55 °C. With this charging protocol the acid catalyst was always present in a large excess, even when the reaction mass was relatively dilute at the end of the additions. By adding components at a moderate temperature there was no lag in initiating the reaction, and the reaction temperature would drop if the additions were stopped. When the additions were complete the reaction mixture was held at 60 °C for 3 h to ensure complete reaction. The reaction was poured into ice water and the product was isolated by filtration and washing [61]. (Three aspects of this reaction deserve additional mention. First of all, the presence of a solvent would have made temperature control easier, although the kinetics would have slowed. Secondly, continuous operations may be ideal for conditions using simultaneous or parallel additions. Last of all, when a reaction is run for the first time it is prudent to run it on a small scale.) By understanding an exothermic event and modifying conditions a safe and effective process was developed.

The root cause of runaway reactions is generally the uncontrolled generation and removal of heat from an exothermic reaction. The resulting uncontrolled generation of gases (including volatilized solvents) can expel the contents from a reactor. Along with undesirably high operating temperatures the causes of runaway reactions can include high concentrations of reagents, due to poor mixing and dispersion; high sensitivities to impurities, e.g., metals in vessels; delayed initiations of reactions to form reagents such

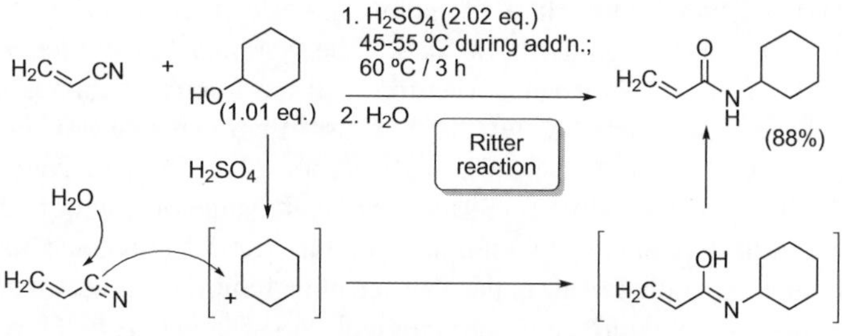

FIGURE 2.5 Moderating an exothermic addition during a Ritter reaction.

TABLE 2.3 Considerations for Designing Safety into a Process

(1) Select safe starting materials and solvents (desk screening)
(2) Develop safe processing conditions (hazard evaluation laboratory)
(3) Validate the safety of steps at each stage of scale-up
(4) Employ strict change control on scale
(5) Design processes that fail safe

as Grignards; and slow undesired reactions and degenerating reaction conditions. If the conditions for a desired and runaway reaction are similar, the likelihood of runaway scenarios increases. When an exothermic reaction is scaled up using batch operations, controlling reaction temperatures through external cooling becomes more difficult; for instance, with a sphere the volume increases as a function of the radius cubed, while the surface area increases as the radius squared [62]. Understanding the chemistry and kinetics is essential for safe operations.

All processes to be scaled up in a pilot plant or manufacturing plant should first be tested in a hazard evaluation laboratory to confirm safe operating conditions. Many reactions are scaled up in kilo laboratories using fume hoods, safety shields and other protective devices; any significant scale-up of a process should be preceded by examination in a hazard evaluation laboratory.

Steps for designing safety into a process are outlined in Table 2.3. It is important to recognize that combinations of compounds can react synergistically, in ways that have not been published or otherwise reported.

Volatile, toxic compounds pose serious considerations for safe operations, in order to reduce the risks of fires and of exposing personnel to toxic materials. The "fire triangle" includes three items needed for a fire: fuel, a source of ignition, and an oxidizer. Compounds may be flammable [63] in air within a range of concentrations defined by the lower and upper flammability limits (LFL and UFL, sometimes termed the lower and upper explosive limits, or LEL and UEL). If the concentration of a flammable mixture in air is below the LFL, the mixture is "lean" and combustion will not occur; similarly, if the concentration of a flammable mixture in air is above the UFL, the mixture is "rich" and combustion will not occur. The flash point (fp) of a liquid is the temperature at which vapors in air can be ignited by a spark. The National Fire Protection Agency (NFPA) has defined three major categories for such hazardous liquids: Class I flammable liquids have flash points below 100 °F, Class II combustible liquids have flash points between 100 and 140 °F, and Class III combustible liquids have flash points above 140 °F. Class IA flammable liquids are the most dangerous, with boiling points below 100 °F and flash points below room temperature [63]. Compounds may ignite at their autoignition temperature in air in the absence of an ignition source; for acetone and diethyl ether these temperatures are approximately 465 °C and 160 °C, respectively [64]. The fact that autoignition temperatures may be significantly higher than the flash

TABLE 2.4 Hazards of Selected Volatile Compounds (1)

Solvent	bp (°C)	Flash Point (°C)	TLV (ppm)	STEL (ppm)
Acetone	56	−20	500	750
MeOH	65	11	200	250
$PhCH_3$	111	4	20	
$CH_3OCH_2CH_2OH$	124	46	0.1	
EtCl	12	−43	100	
$ClCH_2CH_2Cl$	83	18	10	
Ethylene oxide (2)	11	−20	1	

(1) TLV & STEL data: 2009 TLVs and BEIs; ACGIH, Cincinnati, OH, USA.
*(2) Anonymous, Chem. Eng. News **2007**, 85(23), 27.*

points underscores how important it is to work in spark-free, explosion-proof (XP) environments.

Nitrogen gas is used in the laboratory and on scale to preclude fires. Maintaining a positive pressure on a reactor keeps O_2 out of vessels. Slowly passing N_2 through the headspace of a reactor can dilute solvent vapors in the headspace to concentrations below the LFL. Oxidants such as $KMnO_4$ [65] and H_2O_2 [66] generate O_2. Oxidations using or generating O_2 can lead to combustion of the solvent vapors above the reaction mixture, and the headspace of reactors may be swept with N_2 to dilute the O_2 content to less than 6–10% [13,14]. Oxygen sensors can be installed in the headspace of vessels [67].

Permissible exposure limits are calculated based on toxicological data. The threshold limit values (TLVs) are time-weighted averages (TWAs) for exposure to a compound over an 8-h day or a 40-h week. The short-term exposure limit (STEL) is the highest permissible exposure for 15 min. Some selected examples are shown in Table 2.4, and more extensive lists of TWAs and flash points are found in Tables 5.2 and 5.5. Other data on desirable and undesirable solvents are given in Tables 5.3, 5.4, and 5.5.

In a pilot plant or manufacturing facility operators and supervisors do not have the protection afforded by laboratory fume hoods to keep toxic compounds away from them. Protective personnel equipment (PPE) such as self-contained breathing apparatus and suits may be worn [68], but these means of protection

SAFETY — Acetone and methanol are frequently used for cleaning equipment in the laboratory and on scale. These solvents have flash points below room temperature, and should be respected for risk of fires.

can slow operations, lead to clumsy actions, and permit dangerous situations to occur that otherwise would be avoided. For instance, masks and respirators can limit peripheral vision. Due to these limitations and the inherent risks of using dangerous compounds on scale we try to minimize the use of volatile, toxic, and hazardous compounds.

There are now many sources for desktop screening, including *Bretherick's Handbook of Reactive Chemical Hazards* [69], the CHETAH (Chemical Thermodynamic Energy Release Evaluation Program) software program to predict thermochemical properties [70], *Sax's Dangerous Properties of Industrial Materials* [71]; internet searches for {reagent + hazard}; and databases [72]. Amery has reviewed techniques for identifying exotherms and gas evolution, including DSC (differential scanning calorimeter), RSST™ (reactive system screening tool), ARC (accelerated rate calorimeter), the RC1™ (Mettler – Toledo), and QRC (quantitative rate calorimeter) [73]. The NIST WebBook provides information on heats of formation [74]. Estimations of heat of reaction have been shown to be reliable [75]. Reactions should be analyzed for possible hazards, such as exotherms, gas evolution, and safety of handling reagents and products, before running on scale, in the laboratory or pilot plant. When exothermic or explosive events are detected, it is wise to conduct reactions 100 °C below the temperature of the explosive event [11].

By balancing the equation for the reaction to be developed the formation of a gaseous byproduct may be clear, and conditions should be applied to control and possibly moderate gas generation. If gas is generated during a high-temperature reaction through a hot tube that can withstand pressure and suitable seals and check valves are in place, safe processing may be feasible under continuous operations. Under such conditions operating above the boiling point of the solvents in use is possible (see the Newman-Kwart rearrangement in Figure 14.4 and related discussion). In batch operations gas generation is usually controlled by dose-controlled additions at moderate temperatures to ensure rapid reaction and no build-up of reactive intermediates. One example is the Dakin–West reaction, which generates CO_2 as a byproduct (Figure 2.6). Adding alanine (as a solid) in portions over 6 h controlled the rate of CO_2 evolution; 50 kg of alanine generated over 11,000 L of CO_2. (The authors of this excellent, understated paper devoted time to understand the reaction mechanism and minimized the charges of reagents. Both H_2O and three equivalents of Ac_2O are needed for the Dakin–West reaction, and the authors charged just over two equivalents of Ac_2O. The third equivalent of Ac_2O and the H_2O required were generated in situ by reaction of two equivalents of AcOH. The half equivalent of AcOH charged at the beginning of the reaction ensured ready generation of H_2O for hydrolysis and decarboxylation.) The concentrated acetamidobutanone was mixed with malonodinitrile and added to 30% NaOH, and the product pyrrole crystallized from the aqueous mixture, a reactive crystallization. The authors noted that malonodinitrile decomposes autocatalytically at high temperatures, and for safe operations on large scale recommended that malononitrile be added in a dose-controlled manner to solutions of base [76]. The Curtius rearrangement generates stoichiometric amounts of N_2, and uncontrolled release could rupture tanks or send a product stream to a surge tank. Hydrolysis of the diphenylphosphoryl azide (DPPA) reagent or the acyl azide intermediate would generate the toxic and volatile hydrazoic acid (HN_3, bp 37 °C), so the toluene extract of the free base was azeotroped to <0.1% KF. Any HN_3 generated was trapped as triethylammonium azide due to the charge of excess Et_3N, and there was no coolant on condenser to preclude condensation and build-up of HN_3. With a dose-controlled addition of DPPA at a moderate temperature there was no build-up of the potentially acyl azide, hence controlling the evolution of N_2 [77].

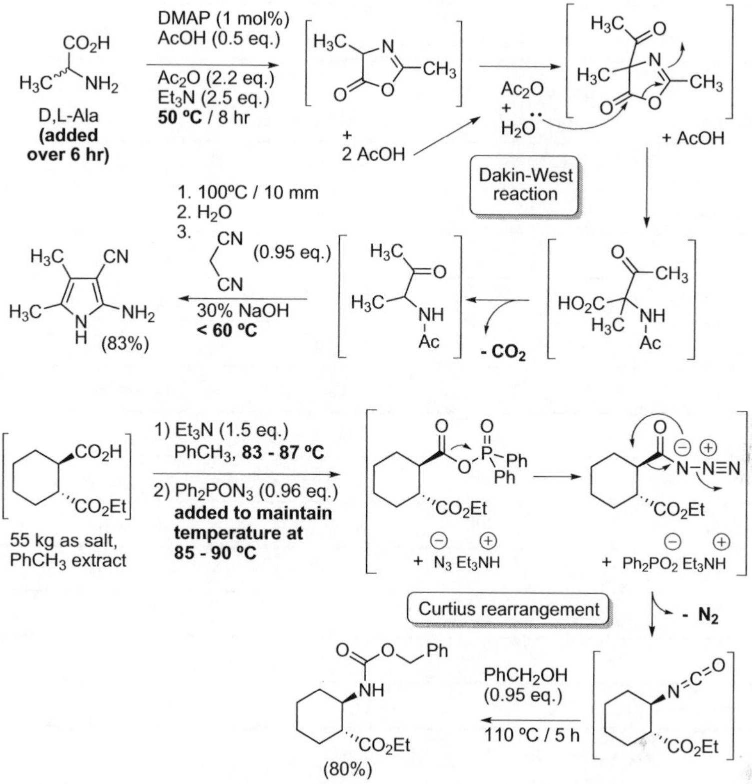

FIGURE 2.6 Safe operating conditions for reactions generating gaseous byproducts.

One can anticipate problems for compounds containing energetic functional groups, as shown in Table 2.5. The values were provided by Rowe in a paper discussing the generation and evaluation of thermal data [78]. Laird suggested that one should be concerned about safety if more than two of the functional groups in the first column were present in a molecule [79]. Low-molecular weight compounds may be especially reactive if only one of these functional groups is present. With new compounds certainty can only be achieved by screening for reactive hazards [80].

Oxygen balance calculations have been used to predict the explosivity of compounds, but results may be misleading. The equation for oxygen balance (Equation 2.1) was derived to develop munitions in World War II, based on molecular formulae. Since water and CO_2 do not fit the profile of highly hazardous compounds, one should not rely too much on oxygen balance data (Figure 2.7). The authors of a paper critiquing oxygen balance data stated that "…oxygen balance does not correlate well with the known hazard potential of compounds…Stoichiometric considerations cannot resolve questions of energy release hazard potential. Thermochemical and kinetic considerations are required for such resolution." [81].

$$OB = -(1600/MW) \times (2x + y/2 - z) \qquad (2.1)$$

where MW = molecular weight, x = # of C, y = # of H, and z = # of O.

TABLE 2.5 Decomposition of Energetic Functional Groups (1)

Compound	Decomposition Energy (kJ/mol)	Compound	Decomposition Energy (kJ/mol)
R_3CNO_2	310–360	Alkene	50–90
R_3CNO (nitroso)	150–290	Epoxide	70–100
$RONO_x$		RCO_3H	240–290
R_2NX		$R_2S{=}O$	40–70
RN_2^+	100–180	RSO_2Cl	50–70
ROOR	230–360	$R_2C{=}NOH$	110–140
ROX		$RN{=}C{=}O$	50–75
Alkyne	120–170	R_2NO (N-oxide)	100–130
RNH–NHR	70–90	$R_2N{-}NO_2$	400–430
RN=NR	100–180	acyl-ONO_2	400–480
RN_3	200–240		
ClO_4^-			
RCN^+O^- (nitrile oxide)			

(1) Rowe, S. M. Org. Process Res. Dev. **2002**, *6, 877.*

Hazard potential ranges, based on oxygen balance calculations

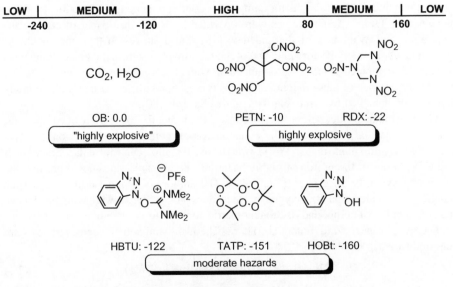

FIGURE 2.7   Oxygen balances calculated for some compounds.

In analyzing for and developing safe conditions three instruments are widely used. Differential scanning calorimetry (DSC) may be carried out on one component, and measures the onset of decomposition (temperature) and the heat of decomposition (by integrating the area under the curve). Using milligrams of materials DSC can also be used to identify exothermic events for raw materials and reaction mixtures. The reactive systems screening tool (RSST™) quickly shows if gases are evolved and at what temperatures exothermic events occur. Adiabatic calorimetry [ARC (accelerated rate calorimeter), or RC1™] provides a quantitative assessment of how much heat is generated (joules/g reaction mass) in a given transformation or event. Once the ability to remove heat from scale-up equipment is known, quantitative data from ARC studies can be used to calculate how rapidly a component can be added to allow the cooling capability of existing equipment to maintain a temperature below a given range. By diluting the reaction mixture with more solvent holding the reaction mixture below a given temperature may be easier, but the reaction may slow unacceptably. The ability of equipment to remove heat on scale can influence process development and equipment selection.

SAFETY — Run reactions below the decomposition temperature of reagents. Literature data may be incomplete, and decomposition of reagents may occur below the melting point. Avoid explosive components, or run reactions in batch operations 100 °C below the onset temperature. Be aware for generation of gases, as over-pressurization of reactors can lead to explosion. Even if a reaction is slightly exothermic in the laboratory, it may be necessary to apply cooling on scale.

A detailed evaluation of energetic compounds and route selection has been described by Kawakubo. DSC analysis confirmed CHETAH calculations that ethyl nitroacetate would have high decomposition energy, and ethyl nitroacetate was found to be sensitive to both impact and friction. Two routes to ethyl nitroacetate were considered (Figure 2.8). Both nitromethane and the dipotassium salt intermediate were assessed as high-risk materials, and the dipotassium salt easily ignited. The route using nitromethane was abandoned in favor of the route generating acetyl nitrate, as the latter did not have risks of explosions [82].

A heat flow study was used to troubleshoot the "simple reaction" shown in Figure 2.9, which worked well on a 100-g scale, but did not proceed as expected on 1-kg scale (in a 22 L flask). Heat flow experiments showed that the temperature rose to 30 °C over 90 min, and then rose to 50 °C over next 30 min. This profile is characteristic of

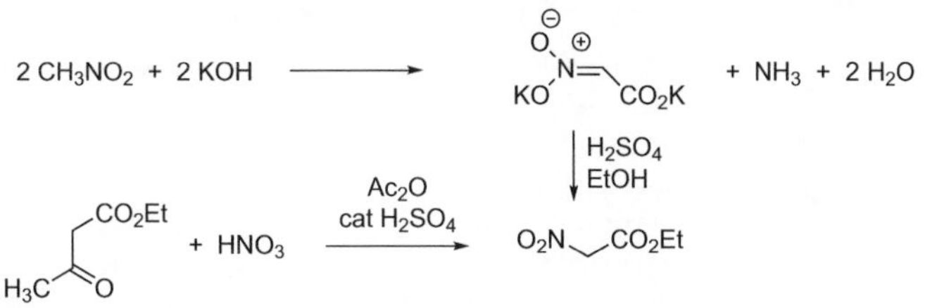

FIGURE 2.8 Routes to ethyl nitroacetate.

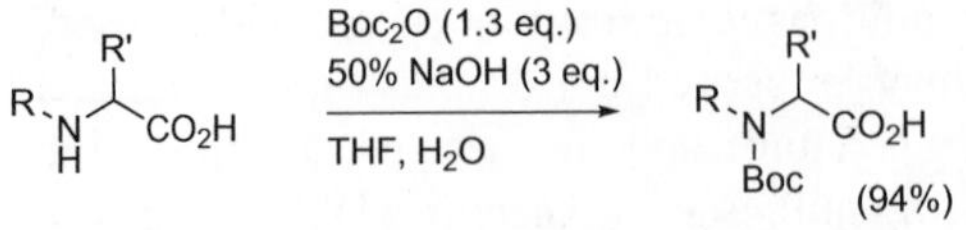

FIGURE 2.9 Preparation of a Boc amino acid.

an autocatalytic reaction; the product was acting as catalyst. For successful scale-up to 1.8 mol, EtOH was added to break the biphasic mixture. Under these conditions it was feasible to develop process with "square tooth" heat flow profile, that is, a process in which the temperature dropped quickly when the addition of a key reagent was ceased [83]. Such dose-controlled additions can be important aspects of safe operations.

The preparation of the Boc amino acid in Figure 2.9 is a good example of using calorimetry to design suitable process controls. Calorimetry can also be used to devise operations to safely initiate the formation of Grignard reagents. For instance a small portion of the Grignard reagent from a prior batch or various activators (Chapter 10) can be added to a suspension of magnesium metal and a small amount of halide, with more halide being added after the exotherm subsides. Dose-controlled additions can also be used to ensure that exothermic reactions do not overwhelm cooling capabilities on scale. Under these conditions it may be wise to conduct the addition at a moderate temperature to encourage rapid reaction as the limiting reagent is dosed into the reactor.

Fail-safe operations should be selected when exothermic reactions are involved. Contingencies should be in place for possible worst-case scenarios, such as an abrupt cessation of stirring, being unable to close an addition valve, unexpected lack of cooling, an extended addition, under-pressurization, over-pressurization, and so on. Processes may need to be re-designed, using Failure Modes and Effects Analysis (FMEA). For example, calorimetry studies for the nitration of the sildenafil intermediate in Figure 2.10 showed that the nitropyrazole decomposed at $100\,^{\circ}C$. Calculations indicated that if the starting material were added on scale to a mixture of HNO_3 and H_2SO_4 the reaction mass would heat to $127\,^{\circ}C$. To handle the exotherm, the starting material and HNO_3 were dissolved separately in H_2SO_4, with 45% of the overall heat evolved being generated in these two separate steps before the reaction occurred. In the pilot plant, only one-third of the dissolved HNO_3 was charged to the header tank at any time, to avoid consequences if a charging valve were stuck open. Complete reaction was verified

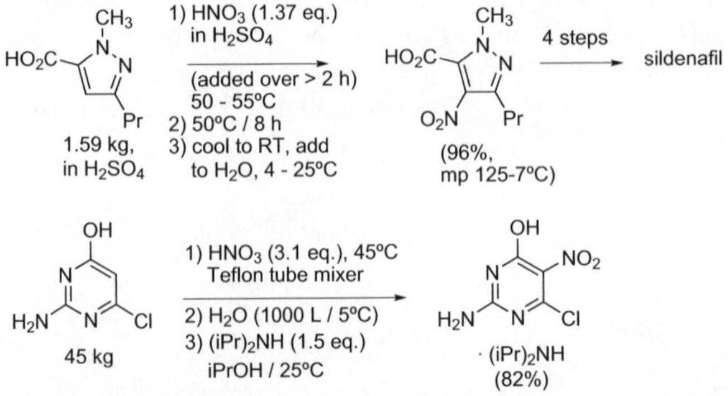

FIGURE 2.10 Conditions for safe nitrations.

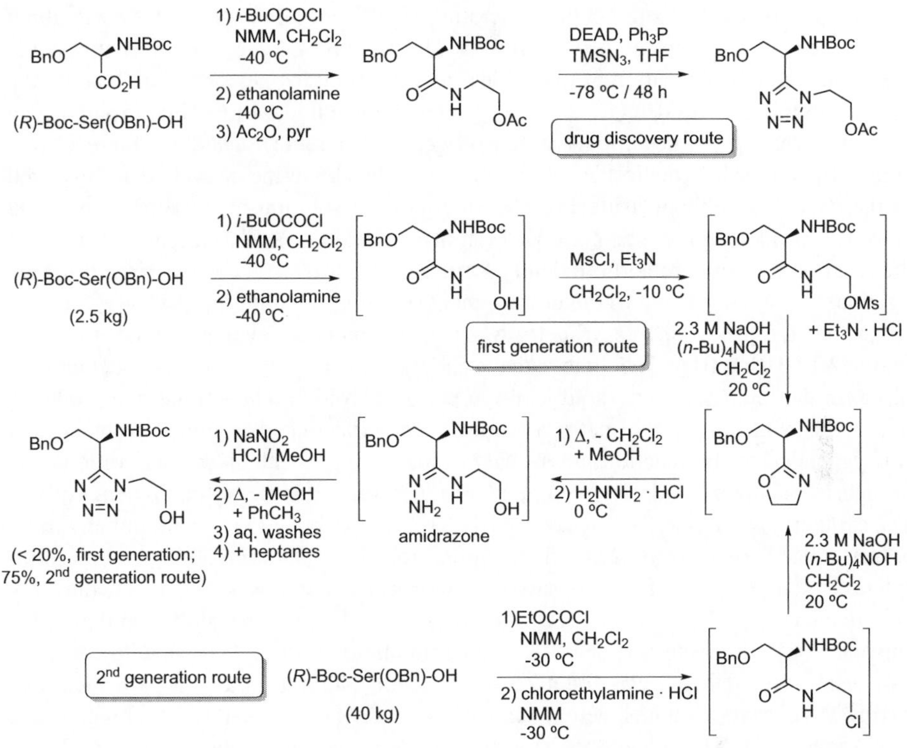

FIGURE 2.11 Safe, high-yielding tetrazole synthesis.

by HPLC before the next portion of dissolved HNO_3 was added [84]. The nitration of the aminopyrimidine in Figure 2.9 was judged too exothermic to be run in a semi-batch mode for operations larger than 22 L, and a continuous nitration process was developed [85]. (DSM has developed safe nitrations for current good manufacturing practices (cGMP) using microreactors [86].)

Reactions forming or involving tetrazoles can be highly exothermic. BMS researchers found that formation of the tetrazolyl acetate in Figure 2.11 was highly exothermic under the conditions used by drug discovery, and decided to pursue other routes for SAFE scale-up. The amidrazone could be converted to the desired tetrazole by nitrosation, thus avoiding issues associated with handling NaN_3 or associated reagents and making the oxazoline a key intermediate. In the first generation route developed by the process chemistry group, the oxazoline was prepared from the mesylate of the hydroxyethylamide using standard conditions (MsCl and Et_3N in CH_2Cl_2), followed by cyclization using phase-transfer catalysis (PTC). SAFETY investigations during the nitrosation reaction showed a delayed exotherm, posing a potential SAFETY issue. Analysis of the reaction components revealed that the cause could be traced back to residual Et_3N from the mesylation: When HCl for the nitration was added to the solution of the amidrazone, residual Et_3N reacted first to form $Et_3N \cdot HCl$, with no significant evolution of heat. After the Et_3N was completely neutralized the reaction mixture became progressively more acidic, generating significant amounts of HONO at about pH 3.2, and exothermic nitration occurred. The BMS researchers found that residual Et$_3$N catalyzed the formation of impurities from the amidrazone, and upon scale-up the yield from this process decreased due to racemization and loss of the Boc group during solvent exchange. In developing a second-generation route to address these issues, chloroethylamine hydrochloride was coupled with the (R)-serine derivative and cyclized to the oxazoline without using Et$_3$N. Furthermore, a solution of the amidrazone and NaNO$_2$ was added *into* methanolic HCl (an inverse addition), allowing good control of the reaction temperature and efficiently forming the tetrazole. Thus calorimetry investigations on two fronts led to the development of a high-yielding process [87].

Other examples where calorimetry was used for process development are shown in Figure 2.12. Using BH$_3$·THF to reduce the carboxylic acid on scale, over-reduction took place. In the multi-kilo run, rapid addition produced a high temperature and produced high levels of the diol. Calorimetry studies were undertaken to determine how much heat was generated in the reaction mass, making it possible to calculate the addition time needed to maintain the desired temperature range for the scale-up equipment chosen. The exotherm was moderated by adding BH$_3$·THF over about 45 min, minimizing over-reduction [88]. A runaway reaction was found for the reaction involving sodium dimsylate at 50–60 °C. (For this reaction sodium dimsylate was found to be the only effective base.) THF (bp 64 °C) was chosen as solvent, in part because volatilizing the solvent would dissipate the heat of anion formation. Adding THF first to the NaH suspension, then DMSO was found to be the safer addition mode, and a 6:1 ratio of THF:DMSO gave a rapid reaction of NaH with DMSO. For safe quenching, EtOH was added, then H$_2$O, and the heptane was removed by azeotroping with EtOH (bp 72 °C, 52% heptane) [89]. In the Mizoroki–Heck reaction LiCl was added to increase the stability of the catalyst, and an amine base was employed to increase reactivity and provide faster reactions. Calorimetry showed that under these conditions an adiabatic temperature rise of 84 °C could occur, overwhelming cooling in the pilot plant. As a result, Et$_3$N was added in portions of 10%, 20%, and 70%. The HBr byproduct was found to neutralize the Et$_3$N, causing a slower reaction. Hence LiOAc was added to neutralize HBr that was generated and to stabilize the catalyst [90]. Calorimetry studies were used to follow reaction rate in a chiral hydrogenation, demonstrating that the amine product inhibited the reaction, probably through complexing to the catalyst. By conducting the reaction in the presence of excess Boc$_2$O product inhibition was eliminated [91]. Calorimetry studies have been used to optimize other hydrogenations [92] and to understand reaction kinetics [93].

V. PERSPECTIVE ON SAFETY

Unfortunately safety usually should be considered more often. Almost everyone knows of someone who narrowly missed an accident, or "dodged a bullet". Many safety considerations are "common-sense," e.g., quenching agents should be non-volatile, don't quench NH$_3$ fumes with bleach. Other considerations rest upon the chemist's knowledge of reactions. For instance, quenching ICl by treatment with Na$_2$S$_2$O$_3$ may seem sensible, but mixing with acetone to increase the contact of ICl with aq. Na$_2$S$_2$O$_3$ would be dangerous: the iodoform reaction of methyl ketones [94] may be exothermic.

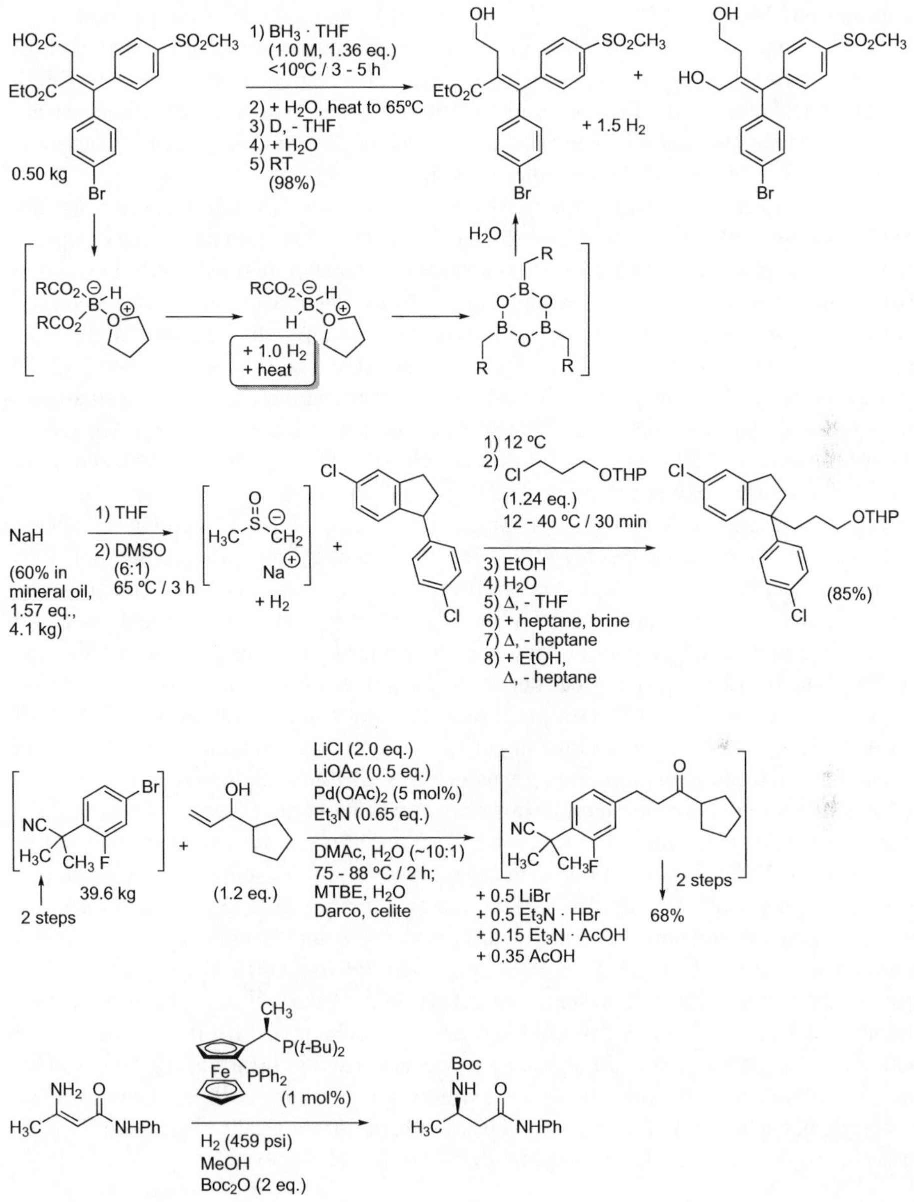

FIGURE 2.12 Some processes developed using calorimetry.

Overlooking the physicochemical characteristics of systems can also easily lead to problems. For instance, quenching a boiling toluene solution with cold water may seem like a quick way to lower the temperature of a reaction mixture – until one realizes that the initial introduction of water will lead to the formation of the toluene–water azeotrope (bp 84 °C), which may flood and overwhelm the condenser, perhaps even ejecting the mixture from the reactor. Using liquid nitrogen may be a quick way to cool something to cryogenic temperatures, but small amounts of liquid nitrogen create a large amount of gases upon warming; any O_2 (bp -183 °C) that had been condensed in the Dewar flask will be enriched as the liquid nitrogen (bp -196 °C) evaporates [95]. Sometimes we are too familiar with our equipment and operations, and we become complacent. Reacting in a hurry to emergencies can create other problems.

Perhaps safety in the kilo lab is most overlooked, since kilo lab operations usually involve scaleups before toxicology and chemical safety hazard examinations are complete. In the laboratory safety considerations [1] include using operating fume hoods, having fire extinguishers and emergency spill equipment readily available, securing gas cylinders, and shielding equipment that is moving, such as vacuum pump belts. We should also take advantage of often-overlooked items for personal safety, such as safety shoes, safety glasses, gloves, and portable safety shields. Safety considerations are present in all aspects of processing, and inherently safe processes increase the chances of success. We should always consider the "inherent SHE," for Safety, Health, and Environment [96], and try to get rid of a hazard, not just control it [97,98].

Safety on a large scale brings other considerations, as the costs of accidents on a large scale can be huge. Static discharge has led to accidents, and process equipment is grounded ("earthed") to prevent dangers from static discharge. Dusts of iron powder, charcoal, flour or grains can be combustible [99]. The importance of thoroughly cleaning process equipment can be overlooked, especially after operations become routine and too familiar. Residues may not be identified if analysts don't know what to look for. Residual metals, such as Mn [100], Cu, and Fe may catalyze the decomposition of peroxides [21]. Residues can become impurities in APIs, and can template the formation of a new polymorph, even prompting premature crystallization. Safe sampling of process streams on scale requires thought. Using a dip cup to withdraw a sample from an open reactor rarely happens, for several reasons. (One might use a dip cup to sample a 10% solution of $NaHCO_3$ in water.) For safety reasons we want to avoid exposing personnel to reactive chemicals, and by keeping the reactive components in the reactor we can control processing to minimize risks. Furthermore, by not opening a reactor we can avoid contaminating the batch, a necessary consideration for cGMP. Displacing a sample from a hot reaction mixture by pressure carries with it the risk that the hot reaction stream could spray personnel if leaks are present, so samples are more likely to be withdrawn by suction. Standard operating procedures (SOPs) for a site may prohibit all transfers of reaction streams above a maximum temperature, such as operations for sampling, quenches, hot phase splits and warm polish filtrations. The increased liability of accidents on scale justifiably heightens concern for safe operations.

SAFETY — Do not take safety for granted. Review conditions frequently to ensure safe operations. Cluttered conditions often lead to accidents. Identify clear paths to the exits in a laboratory, pilot plant, or manufacturing facility. When a dangerous situation occurs, assess the risks of staying to help: you will create a bigger problem if someone has to pull you out. Even if you are a spectator (a JAFO [101]) in a pilot plant or manufacturing facility you can contribute by being alert for unsafe conditions, including leaks. Safety is the responsibility of all chemists and engineers.

REFERENCES

1. *Prudent Practices in the Laboratory.* National Research Council of the National Academies; National Academies Press: Washington, D.C., 2011.
2. McConville, F. X. *The Pilot Plant Real Book*, 2nd ed.; FXM Engineering and Design: Worcester, MA, 2007.
3. http://www.orgsyn.org/hazard.html
4. Pesti, J. *Org. Process Res. Dev.* **2010,** *14*, 483.
5. Hassan, N. *Chem. Eng.* **2011,** *118*(1), 50.
6. Venugopal, B. *Chem. Eng.* **2002,** *109*(6), 54.
7. Heylin, M. *Chem. Eng. News* **1985,** *63*(6), 14; Lepkowski, W. *Chem. Eng. News* **1985,** *63*(6), 16; Worthy, W. *Chem. Eng. News* **1985,** *63*(6), 27; Dagani, R. *Chem. Eng. News* **1985,** *63*(6), 37; Webber, D. *Chem. Eng. News* **1985,** *63*(6), 47; Johnson, J. *Chem. Eng. News* **2011,** *89*(8), 31.
8. http://www.ejnet.org/dioxin
9. Hogue, C. *Chem. Eng. News* **2010,** *88*(46), 30.
10. http://www.hse.gov.uk/chip/phrases.htm
11. Butters, M.; Catterick, D.; Craig, A.; Curzons, A.; Dale, D.; Gillmore, A.; Green, S. P.; Marziano, I.; Sherlock, J.-P.; White, W. *Chem. Rev.* **2006,** *106*, 3002.
12. Conrow, R. E.; Dean, W. D. *Org. Process Res. Dev.* **2008,** *12*, 1285.
13. Slade, J.; Parker, D.; Girgis, M.; Mueller, M.; Vivelo, J.; Liu, H.; Bajwa, J.; Chen, G.-P.; Carosi, J.; Lee, P.; Chaudhary, A.; Wambser, D.; Prasad, K.; Bracken, K.; Dean, K.; Boehnke, H.; Repic, O.; Blacklock, T. J. *Org. Process Res. Dev.* **2006,** *10*, 78.
14. Caron, S.; Dugger, R. W.; Gut Ruggeri, S.; Ragan, J. A.; Brown Ripin, D. H. *Chem. Rev.* **2006,** *106*, 2943.
15. Perepichka, D. F.; Jeeva, S. *Chem. Eng. News* **2010,** *88*(3), 2.
16. Hachoose, R. C. *Chem. Eng.* **2006,** *113*(4), 54.
17. Matlack, A. S. *Introduction to Green Chemistry*, 2nd ed.; CRC Press: Boca Raton, 2010.
18. Hansen, J. *Storms of my Grandchildren.* Bloomsbury: New York, USA, 2009.
19. Wells, A. S.; Coombe, V. T. *Org. Process Res. Dev.* **2006,** *10*, 794.
20. *Chem. Eng. News* **2010,** *88*(29), 27.
21. Barnhart, R. W.; Dale, D. J.; Ironside, M. D.; Vogt, P. F. *Org. Process Res. Dev.* **2009,** *13*, 1388.
22. Drahl, C. *Chem. Eng. News* **2011,** *88*(48), 33.
23. Cavas, T. *Food Chem. Toxicol.* **2001,** (doi:10.1016/j.fct.2011.03.038).
24. Erickson, B. E. *Chem. Eng. News* **2011,** *89*(13), 32.
25. Gropper, S. S.; Smith, J. L.; Groff, J. L. *Advanced Nutrition and Human Metabolism*, 5th ed.; Wadsworth Cengage Learning, 2009.
26. Connors, K. A.; Amidon, G. L.; Stella, V. J. *Chemical Stability of Pharmaceuticals: A Handbook for Pharmacists.* Wiley: New York, 1986; p 63.
27. Chen, Y.-T.; Hsu, C.-I.; Hung, D. Z.; Matsuura, I.; Liao, J.-W. *Food Chem. Toxicol.* **2011,** *49*, 1063.
28. http://echa.europa.eu/doc/consultations/svhc/svhc_axvrep_austria_cmr_2-methoxyethanol.pdf
29. http://www.ncbi.nlm.nih.gov/pubmedhealth/PMH0001778/
30. http://www.calciuminfo.com/calciumquestions/kidneystones.aspx
31. http://en.wikipedia.org/wiki/Propylene_glycol#cite_note-26
32. http://ecfr.gpoaccess.gov/cgi/t/text/text-idx?c=ecfr&sid=786bafc6f6343634fbf79fcdca7061e1&rgn=div5&view=text&node=21:3.0.1.1.14&idno=21#21:3.0.1.1.14.2.1.159; http://edocket.access.gpo.gov/cfr_2009/aprqtr/21cfr184.1666.htm
33. *The Merck Index*, 13th ed.; Merck & Co., Inc.: Whitehouse Station, NJ, 2001; p 1762.
34. Wolf, L. K. *Chem. Eng. News* **2011,** *89*(27), 27.
35. Dai, Q.; Xu, D.; Lim, K.; Harvey, R. G. *J. Org. Chem.* **2007,** *72*, 4856.
36. Patnaik, P. *A Comprehensive Guide to the Hazardous Properties of Physical Substances*, 3rd ed.; Wiley, 2007; p 307.

37. http://toxnet.nlm.nih.gov/cgi-bin/sis/search

38. Boethling, R. S.; Sommer, E.; DiFiore, D. *Chem. Rev.* **2007,** *107*, 2207.

39. http://www.fda.gov/ohrms/dockets/ac/02/briefing/3882B1_13_McNeil-Acetaminophen.htm#_Toc18717570, 2002.

40. Caballero, F. J.; Navarrete, C. M.; Hess, S.; Fiebich, B. L.; Appendino, G.; Macho, A.; Muñoz, E. *Biochem. Pharmacol.* **2007,** *73*, 1013.

41. http://www.liversupport.com/wordpress/2010/10/n-acetyl-cysteine-treats-acute-liver-failure/

42. Johnson, D. S.; Lie, J. J. In *The Art of Drug Synthesis,* Johnson, D. S., Lie, J. J., Eds.; Wiley: New York, 2007; Chapter 7, p 95.

43. Shibasaki, M.; Kanai, M. *Eur. J. Org. Chem.* **2008,** 1839.

44. Pereillo, J.-M.; Maftouh, M.; Andrieu, A.; Uzabiaga, M.-F.; Savi, P.; Pascal, M.; Herbert, J.-M.; Maffrand, J.-P.; Picard, C. *Drug Metab. Dispos* **2002,** *30*, 1288.

45. Lamp, K. C.; Freeman, C. D.; Klutman, N. E.; Lacy, M. K. *Clin. Pharmacokinet* **1999,** *36*, 353; Müller, M. *Biochem. Pharmacol.* **1986,** *35*, 37; Edwards, D. I. *Biochem. Pharmacol.* **1986,** *35*, 53.

46. DEREK (Deductive Estimation of Risk from Existing Knowledge) Long, A.; Combes, R. D. *Toxicol. in Vitro;* **1995,** *9*, 563; http://www.sciencedirect.com/science/article/pii/088723339500040F

47. ToxTree: Raillard, S. P.; Bercu, J.; Baertschi, S. W.; Riley, C. M. *Org. Process Res. Dev.* **2010,** *14*, 1015.

48. Anonymous *Chem. Eng. News* **2007,** *85*(47), 43.

49. Encyclopedia of Reagents for Organic Synthesis.

50. http://www.ncbi.nlm.nih.gov/pubmed/484483

51. am Ende, D. J.; Vogt, P. F. *Org. Process Res. Dev.* **2003,** *7*(6), 1029.

52. BH$_3$·THF was probably used to reduce an imide to the amine: Couturier, M.; Andresen, B. M.; Jorgensen, J. B.; Tucker, J. L.; Busch, F. R.; Brenek, S. J.; Dubé, P.; am Ende, D. J.; Negri, J. T. *Org. Process Res. Dev.* **2002,** *6*, 42.

53. Burkhardt, E. R.; Corella, J. A., II. U.S. Patent 6,048,985, 2000, to Callery; Atkins, W. J., Jr.; Burkhardt, E. R.; Matos, K. *Org. Process Res. Dev.* **2006,** *10*, 1292.

54. Spagnuolo, C.; Wang, S. S. Y. *Chem. Eng. News* **1992,** *70*(22), 2. Wang, S. S. Y.; Kiang, S.; Merkl, W. *Process Safety Prog.* **1994,** *13*(3), 153.

55. Tong, W. R.; Seagreave, R. L.; Wiederhorn, R. *Loss Prevention*; Vol. 11, AIChE: New York, 1977, 71.

56. Johnson, J. *Chem. Eng. News* **2009,** *87*(38), 8.

57. Buckley, J.; Webb, R. L.; Laird, T.; Ward, R. J. *Chem. Eng. News* **1982,** *60*(28), 5.

58. Liu, Y.; Schwartz, J. *J. Org. Chem.* **1993,** *58*, 5005.

59. Perrin, D. D.; Armarego, W. L. F.; Perrin, D. F. *Purification of Laboratory Chemicals*, 2nd ed.; Pergamon: Oxford, UK; 1980; p 224.

60. Etchells, J. C. *Org. Process Res. Dev.* **1997,** *1*, 435.

61. Chang, S.-J. *Org. Process Res. Dev.* **1999,** *3*, 232.

62. Anderson, N. G. *Org. Process Res. Dev.* **2004,** *8*, 260.

63. The term "inflammable" implies that a compound may go "up in flames," not that the compound is not flammable. Materials that are combustible may be referred to as inflammable: Schmidt, M. *Chem Eng.* **2002,** *109*(12), 58.

64. http://www.engineeringtoolbox.com/fuels-ignition-temperatures-d_171.htmls

65. Fray, M. J.; Gillmore, A. T.; Glossop, M. S.; McManus, D. J.; Moses, I. B.; Praquin, C. F. B.; Reeves, K. A.; Thompson, L. R. *Org. Process Res. Dev.* **2010,** *14*, 263.

66. Astbury, G. W. *Org. Process Res. Dev.* **2002,** *6*, 893.

67. For example, see http://www.oceanopticssensors.com/application/organicsolvent.htm

68. Lovasic, S. *Chem. Eng.* **2011,** *118*(5), 51.

69. Urben, P. *Bretherick's Handbook of Reactive Chemical Hazards.* Butterworth-Heinemann: London, 2000.

70. American Society for Testing and Materials: http://www.astm.org/BOOKSTORE/PUBS/DS51F.htm

71. *Sax's Dangerous Properties of Industrial Materials*, 3 volumes, 11th ed.; Wiley, 2004.

72. http://www.ilo.org/legacy/english/protection/safework/cis/products/icsc/dtasht/index.htm

73. Amery, G. *Aldrichimica Acta* **2001**, *34*(2), 61.

74. http://webbook.nist.gov/chemistry/

75. Weisenburger, G. A.; Barnhart, R. W.; Clark, J. D.; Dale, D. J.; Hawksworth, M.; Higginson, P. D.; Kang, Y.; Knoechel, D. J.; Moon, B. S.; Shaw, S. M.; Taber, G. P.; Tickner, D. L. *Org. Process Res. Dev.* **2007**, *11*, 1112.

76. Fischer, R. W.; Misun, M. *Org. Process Res. Dev.* **2001**, *5*, 581.

77. Yue, T.-Y.; McLeod, D. D.; Albertson, K. B.; Beck, S. R.; Deerberg, J.; Fortunak, J. M.; Nugent, W. A.; Radesca, L. A.; Tang, L.; Xiang, C. D. *Org. Process Res. Dev.* **2006**, *10*, 262.

78. Rowe, S. M. *Org. Process Res. Dev.* **2002**, *6*, 877.

79. Laird, T. Contributed talk at AMRI Symposium, May 8, 2003.

80. Dale, D. J. *Org. Process Res. Dev.* **2002**, *6*, 933.

81. Shanley, E. S.; Melham, G. A. *Process Safety Prog.* **1995**, *4*(1), 29.

82. Kawakubo, H. In *Pharmaceutical Process Chemistry*; Shiori, T., Izawa, K., Konoike, T., Eds.; Wiley-VCH, 2011; pp 363–380.

83. Littler, B. Contributed talk at AMRI Chemical Development Symposium, 27–28 September 2007.

84. Dale, D. J.; Dunn, P. J.; Golightly, C.; Hughes, M. L.; Levett, P. C.; Pearce, A. K.; Searle, P. M.; Ward, G.; Wood, A. S. *Org. Process Res. Dev.* **2000**, *4*, 17.

85. De Jong, R. L.; Davidson, J. G.; Dozeman, G. J.; Fiore, P. J.; Kelly, M. E.; Puls, T. P.; Seamans, R. E. *Org. Process Res. Dev.* **2001**, *5*, 216.

86. Short, P. L. *Chem. Eng. News* **2008**, *86*(42), 37.

87. Davulcu, A. H.; McLeod, D. D.; Li, J.; Katipally, K.; Littke, A.; Doubleday, W.; Xu, Z.; McConlogue, C. W.; Lai, C. J.; Gleeson, M.; Schwinden, M.; Parsons, R. L., Jr. *J. Org. Chem.* **2009**, *74*, 4068.

88. Lobben, P. C.; Leung, S. S.-W.; Tummala, S. *Org. Process Res. Dev.* **2004**, *8*, 1072.

89. Dahl, A. C.; Mealy, M. J.; Nielsen, M. A.; Lyngsø, L. O.; Suteu, C. *Org. Process Res. Dev.* **2008**, *12*, 429.

90. Camp, D.; Matthews, C. F.; Neville, S. T.; Rouns, M.; Scott, R. W.; Truong, Y. *Org. Process Res. Dev.* **2006**, *10*, 814.

91. Hansen, K. B.; Rosner, T.; Kubryk, M.; Dormer, P. G.; Armstrong., J. D., III *Org. Lett.* **2005**, *7*, 4935.

92. Littler, B. J.; Looker, A. R.; Blythe, T. A. *Org. Process Res. Dev.* **2010**, *14*, 1512.

93. Zotova, N.; Mathew, S. P.; Iwamura, H.; Blackmond, D. G. In *Process Chemistry in the Pharmaceutical Industry*; Gadamasetti, K., Braish, T., Eds.; 2nd ed.; CRC Press: Boca Raton, Florida, 2008; pp 455–464.

94. Fuson, R. C.; Bull, B. A. *Chem. Rev.* **1934**, *15*, 275.

95. Miller, L. *The Vortex*, November 9, 2001.

96. White, W.; Coombe, V.; Moseley, J. In *Pharmaceutical Process Chemistry*; Shiori, T., Izawa, K., Konoike, T., Eds.; Wiley-VCH, 2011; pp 469–488.

97. Johnson, J. *Chem. Eng. News* **2003**, *81*(5), 23.

98. Edwards, V. H. *Chem. Eng.* **2011**, *118*(4), 44.

99. Hanson, D. J. *Chem. Eng. News* **2011**, *89*(23), 34.

100. Vaino, A. R. *J. Org. Chem.* **2000**, *65*, 4210.

101. Just Another...Friendly...Observer. A JAFO can be recognized by his relaxed stances, such as arms folded across his chest, hands in pockets of his pants, or leaning against a post. Thanks to Dale Trammell, who was a foreman in the Squibb pilot plant, for sharing this term from the Viet Nam war.

Route Selection

"In the end, a 'let the chips fall where they may' attitude paid dividends beyond our best expectation; the lesson learned was that an over-indulgence in protecting groups can work to the detriment of a sound plan…Perhaps the most enduring lesson to be learned from our polycavernoside effort is that a key ingredient for success in complex synthesis is a flexible mindset."

– Prof. James D. White [1]

"A health, SAFETY, and environmental assessment is a critical criterion in our process development….It's got to be dealt with, and it's not taken lightly. If problems are insurmountable, or if there is danger to the environment, a synthesis will be replaced."

– Nick Magnus, Eli Lilly (2007) [2]

"I believe that the ultimate goal for green chemistry is for the term to go away, because it is simply the way chemistry is always done. Green chemistry should just be second nature, the default value."

– Paul Anastas, US Environmental Protection Agency (2011) [3]

Practical Process Research and Development. DOI: 10.1016/B978-0-12-386537-3.00003-4

I. INTRODUCTION

The successful production of chemicals on almost any scale relies on devising chemoselective reaction conditions. Selecting a route is generally the first step; changing the route can affect the selection of reagents, solvents, reaction conditions, workup, crystallization, and more. If the starting materials are inexpensive, then low yields may be tolerated, especially early in the development of an NCE. Although protecting groups add two more steps to a route and are likely to increase the cost of goods (CoGs), protecting groups may be employed to shield reactive centers when chemoselectivity cannot be achieved. Exploiting the reactivity of functional groups can pay big dividends; for instance, Baran's group not only avoided protecting groups, but also exploited the reactivity of functional groups in cascade reactions to synthesize some densely functionalized indole natural products [4]. Enzymes can provide exquisite chemoselectivity, and biocatalysis has been used to "green up" process chemistry [5]. Some remarkable organocatalysis [6] is also being explored, as with the enantiose-lective acetylation of the bis-phenol in Figure 3.1 [7], and the cascade reactions leading to a tetracyclic indole that could be converted to strychnine in only eight steps [8]. Merck's original route to the free base of sitagliptin involved a chiral hydrogenation of an unprotected enamine [9], and this was replaced by the biocatalytic transamination of the ketone precursor to the chiral amine [10]. Merck received US Presidential Green Chemistry awards for both the chiral hydrogenation and the transamination [11]. Researchers at AstraZeneca transformed a methoxy group into an ethyl group in one step using EtMgBr; the original sequence was demethoxylation, triflate formation, and Pd-mediated coupling with BEt_3. Thus the Grignard reaction reduced the number of steps in that route by two [12]. If there is difficulty in reaching chemoselective reaction conditions, starting materials can be changed to ease process development and lower the CoGs. For one preparation of a TGF-beta inhibitor in Figure 3.1 the starting material was 4-aminophenylnitrile [13]; this route avoided a Pd-catalyzed cyanation, a reaction which can require very specific conditions for rugged operations and relatively expensive aryl bromides (see Chapter 10). Innovative chemoselective reactions such as these, often from academic laboratories, can influence researchers to change the route to a compound. Researchers find ways to circumvent those reactions that characteristically have not been rugged, or employ expensive starting materials. Thus route scouting and reagent selection influence each other. Route selection offers many opportunities for creative work and intellectual property (IP).

The route and processes used to manufacture a commercially available product are rarely described: divulging details of optimized processes is like doing years of research for competitors. US patents are required to describe the best examples known at the time of filing with the Patent and Trademark Office [14]; from these examples one can make some sense of desirable routes and operations. Harrington has recently reviewed the patents and journal citations for some top-selling drugs, and proposed the most likely routes [15]. Kleeman and co-workers have summarized a massive number of references for the preparation of drug substances [16]. Companies may provide data on compounds available in bulk through shipping manifestos, and from these it may be possible to elucidate manufacturing routes and devise alternative ones [17]. With few exceptions, manufacturing processes are not divulged.

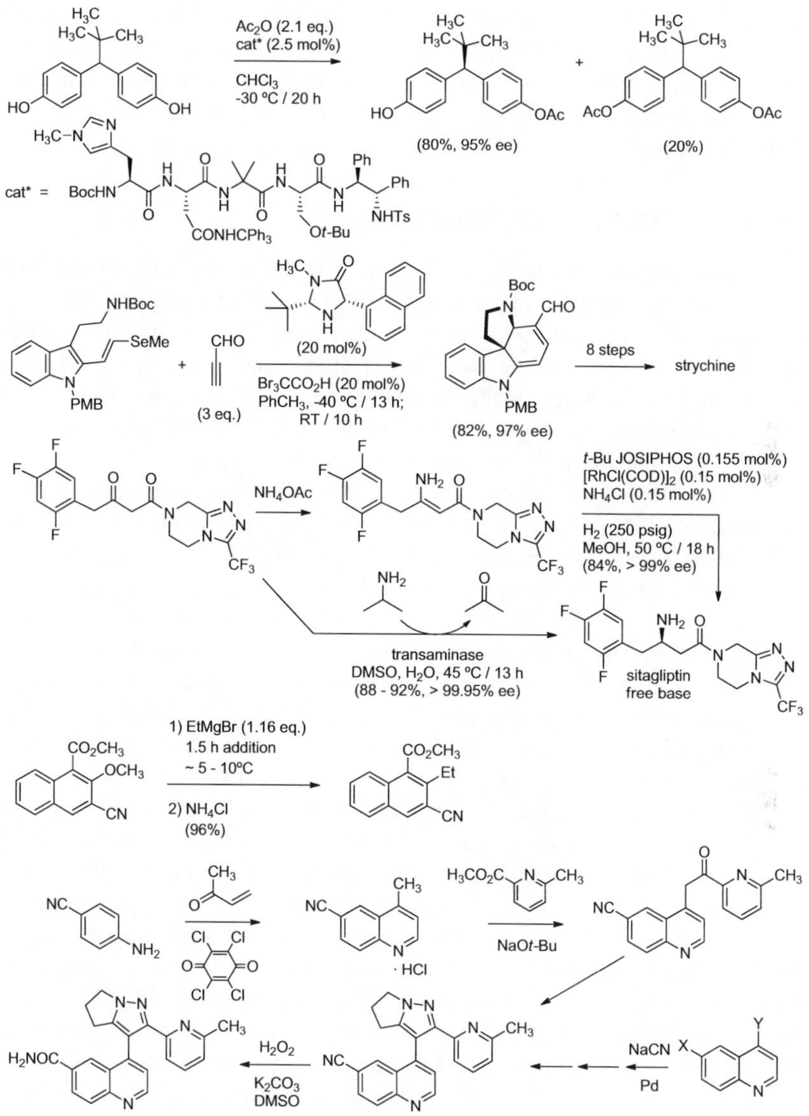

FIGURE 3.1 Inventive transformations and route selection.

In a brilliant review of route selection guidelines, a consortium of authors from three pharmaceutical companies in the UK captured crucial considerations through the SELECT acronym, for SAFETY, Environmental, Legal, Economics, Control, and Throughput [18]. Having SAFETY at the top of the list is no accident. SAFETY considerations were discussed in Chapter 2. Legal considerations generally encompass patenting compounds (composition of matter), patenting processes, and trying to circumvent the areas that have been patented ("freedom to operate"). Control issues are discussed in Chapters 7 and 16. An underlying theme in this book is throughput, or productivity. As mentioned in Chapter 1, economic considerations override the activities within the pharmaceutical and CMO industries; considerations for CoG estimates are included in Section V. Further discussions in this chapter begin with some environmental aspects of route selection, and green chemistry will serve as the framework of the discussions that follow.

II. GREEN CHEMISTRY CONSIDERATIONS

For decades process chemists and engineers have been following many of the tenets of green chemistry; for instance, improving the yields, minimizing the cost of waste disposal, using more concentrated reactions and workups to minimize the volume of solvents that need to be recovered, and raising productivity by using non-hazardous materials, controlling product quality and eliminating process bottlenecks. Green chemistry has incorporated these and other considerations, and the attention-grabbing, positive label "green chemistry" raises the awareness of us all.

Environmental regulations were viewed as punitive. Some over-arching environmental considerations are shown in Table 2.1. Violations of the environment negatively affected corporate images and bank accounts, as discussed in Chapter 2. However, green chemistry is being recognized as profitable. For example, in 1994 Bristol–Myers emitted 500 tons of methyl isobutylketone (MIBK) in Syracuse, primarily from extracting 6-aminopenicillanic acid from fermentation broth, and Bristol–Myers was targeted as one of the biggest polluters in New York State. In 1995–1996 Bristol–Myers introduced a tangential-flow centrifugation and closed-centrifuge extraction process, substituting n-BuOAc for MIBK. A 10% increase in productivity resulted, and operating costs were reduced. The estimated annual profits were $4,900,000, with an estimated payback period of 2.7 years. In 1996 Bristol–Myers received a New York State Governor's Award for Pollution Prevention for this work [19]. In 2001 Blackburn (Baxter International) said "There's a real interplay between environmental and financial motives...and we're pushing that idea across the entire company." [20]. In 2005 McCue (Pfizer) said that "...one can estimate a potential waste coproduced with active pharmaceutical ingredients (APIs) to be in the range of 500 million to 2 billion kg per year. Even at a nominal disposal cost of $1.00 per kg, the potential savings just in waste avoidance is significant and the economic argument for green chemistry becomes compelling.... Green chemistry is an on-the-come innovation that has started to provide the positive economic, environmental, and social benefits these industries will need...." [21]. More details of cost-savings are given with various green chemistry awards.

The Federal Trade Commission wants to regulate what is meant in "green" marketing, including terms such as "eco-friendly," "biodegradable," and "free of [chemical X]." Carbon footprints and carbon offset claims have been touted; the FTC wants to ensure that carbon credits are sold only once [22]. One can conclude that "green" considerations make economic sense to the marketplace. To understand what the label "green" means in a product or a synthesis, one must know what the standard is for comparison.

Green chemistry considerations are increasingly evident, with green chemistry being recognized by the Crystal Faraday Award and the US Presidential Green Chemistry Award, among others. Joining Matlack's compendium on green chemistry [23] are other recent texts, some focused on the pharmaceutical industry [24–28]. Some green chemistry success stories are shown in Figure 3.2. Pfizer workers received an award in 2002 for their work on sertraline. This post-approval process development of sertraline eliminated the use of four solvents and the generation of TiO_2 waste, increased the yield from 20% to 40%, reduced solvent usage from 60,000 gal/ton to 6000 gal/ton, and reduced the annual cost of process waste disposal by more than $100,000. These savings were generated by exploiting the observation that the intermediate imine crystallized from the reaction solvent [29]. (Simulated moving bed chromatography (SMB) has been also used to manufacture sertraline, by resolution of the ketone [30].) Merck received an award in 2005 for the manufacture of aprepitant, through developing four routes over 9 years of process R&D. The current route uses

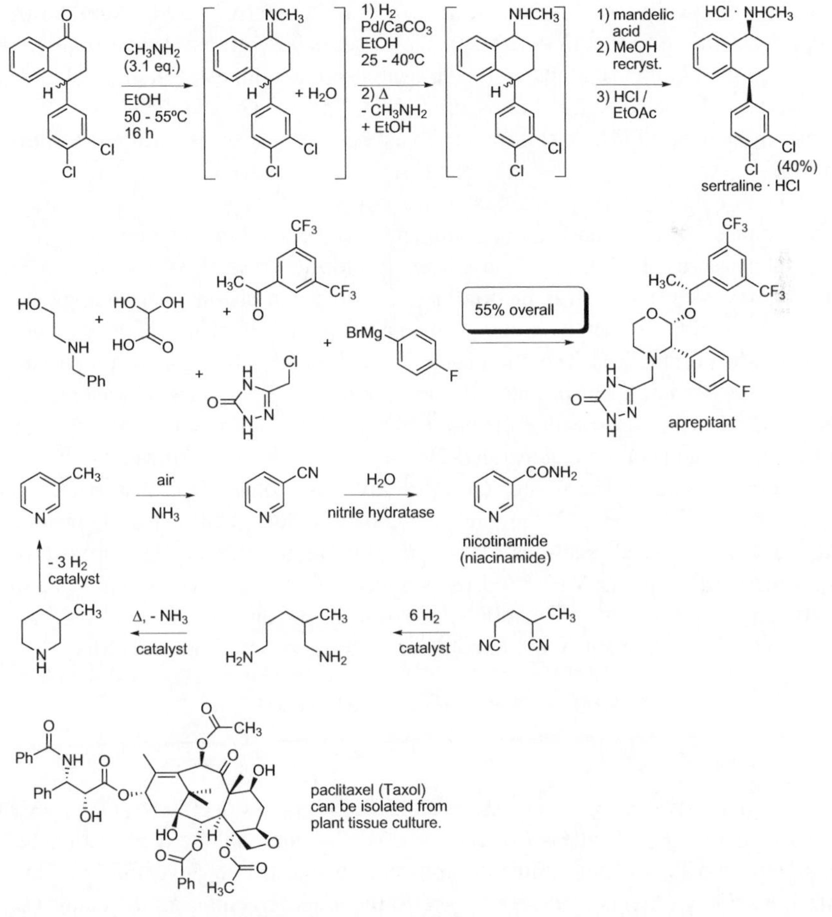

FIGURE 3.2   Green chemistry success stories.

only 20% of raw materials and water of the original route, has only one redox process (a hydrogenolysis), and eliminates 82,000 gal of waste/ton of aprepitant. Aprepitant is made in 55% yield overall, and key to this operation is a crystallization-induced asymmetric transformation (CIAT) [11,31]. Lonza has annually manufactured 5000 metric tons of the vitamin nicotinamide (niacinamide, or niacin) using an immobilized enzyme [32]. All the steps in this green process are catalytic, and all reactions but the biocatalytic step are conducted in the vapor phase. Noteworthy in Lonza's route is the choice of what should be an inexpensive starting material, a side product from reductive dimerization of acrylonitrile to produce adiponitrile, a precursor of nylon-6 [33,34]. Under conditions of traditional organic chemistry the hydration of a nitrile to the primary amide competes with hydrolysis to ammonia and the carboxylic acid, so a nitrile hydratase was chosen and immobilized for increased productivity [32]. Tao and co-workers have converted 3-cyanopyridine to nicotinamide using a recombinant nitrile hydratase at 30 °C [35]. Paclitaxel (Taxol®) is manufactured by plant cell fermentation [36–39], eliminating over 5 years an estimated 35 tons of chemicals, 10 solvents, and six drying steps in converting an intermediate from the leaves and twigs of the European yew into paclitaxel [11,36]. *A priori* a biologically based manufacture of a compound as densely functionalized as paclitaxel is very attractive, especially when compared to the number of steps that have been used to synthesize paclitaxel [40–43]. Many factors must be considered to establish how "green" a biocatalytic process would be for manufacturing [44]. A significant investment may also be required to develop a bio-organic system.

Various principles of green chemistry are shown in Tables 3.1–3.3; the benefits of some principles are well known, and comments are included in some cases.

Several measures of "greenness" have been developed. Trost introduced the concept of atom economy [45], which was incorporated in the green chemistry principles above. Sheldon advanced the E factor for environmental acceptability, which is the weight ratio of waste/product (Table 3.4). The pharmaceutical industry has higher E factors because it has more steps per weight of product. If one multiplies the tonnages of waste times the respective E factors, it appears that the pharmaceutical industry is little better than others regarding the amounts of waste generated. Wastes in the industries shown in Table 3.4 are mainly inorganic salts. Some wastes are more toxic than others, and disposal costs are higher for wastes that are more hazardous. Sheldon has reflected this in his EQ factor [46].

Andraos has analyzed reactions for adhering to green chemistry principles through reaction mass efficiency (RME). Andraos's approach considers yield, atom economy, stoichiometry, and recovery of materials, as outlined in Equations 3.1 and 3.2 [47,48]. RME calculations can quantify the benefits of convergence and other factors.

$$\text{RME} = \varepsilon \times (\text{AE}) \times (1/\text{SF}) \times \text{MRP} \qquad (3.1)$$

$$\text{AE} = 1/(1 + E_{mw}) \qquad (3.2)$$

where ε = yield (as decimal <1), AE = atom economy (as decimal <1), SF = stoichiometric factor (SF > 1 unless 1.0 eq. used), MRP = material recovery parameter (as decimal <1), and E_{mw} = environmental impact factor based on molecular weight.

Atom economy (AE) is inversely related to the environmental impact factor (E_{mw}). Andraos defined reactions with an AE of at least 0.618 as "golden atom economical"

TABLE 3.1 Principles of Green Chemistry (1)

	Principle	Comments
1	Prevention	Better to prevent waste than to reduce waste at the end of the pipeline
2	Atom economy	Incorporate all materials used in processing into final product
3	Less hazardous chemical syntheses	Use and generate less harmful materials
4	Designing safer chemicals	Effective chemicals with minimal toxicity
5	Safer solvents and auxiliaries	Green solvents are discussed in Chapter 5
6	Design for energy efficiency	Try to conduct processes at ambient temperature and pressure. Energy costs are a substantial portion of CoGs for commodity chemicals (2)
7	Use of renewable feedstocks	Renewable and sustainable, not depleting
8	Reduce derivatives	Eliminate protecting groups, as they require additional steps and generate more waste
9	Catalysis	(Reusable) catalysts generate less waste
10	Design for degradation	Rapidly breakdown into harmless byproducts (3)
11	Real-time analysis to prevent pollution	Can prevent formation of impurities
12	Inherently safer chemistry for accident prevention	Minimize potential for accidents, "benign by design"

(1) http://www.chemistry.org/portal/a/c/s/1/acsdisplay.html?DOC=greenchemistryinstitute%5Cgc_principles. html
(2) Energy costs may be 50–75% of the cost of Cl_2 (Reisch, M. Chem. Eng. News 2010, 88 (14), 10).
(3) Botthling, R. S.; Sommer, E.; DiFiore, D. Chem. Rev. 2007, 107, 2207.

reactions, and using over 400 reactions calculated the proportion of reactions within nine categories that reach this standard. As can be seen in Figure 3.3, condensations, rearrangements, and multicomponent reactions (MCRs) are most likely to be very "green." Redox reactions are among the least green reactions, and hence should be avoided when possible. Except for the category of condensations, reagents with lower molecular weights have higher values of AE; hence oxidations with O_2 would be favored over those with higher-molecular reagents such as the Dess–Martin reagent, and reductions with H_2 would be favored over those using $LiAl(O\text{-}t\text{-}Bu)_3H$ [47].

Some aspects of green chemistry and route selection are discussed in more detail in the sections below.

Green chemistry and sustainable operations are interwoven. Earth's resources are finite, as are her abilities to tolerate wastes dumped into landfills and released to air and oceans. Businesses have protested that green considerations are costly and eliminate jobs. But unless we take care of Mother Earth, we may have no economy, and no life worth living.

TABLE 3.2 Twelve More Principles of Green Chemistry (1)

	Principle	Comments
1	Identify and quantify byproducts	Balance the equation; H_2O is the ideal byproduct
2	Report conversions, selectivities, and productivities	Identify opportunities for optimization
3	Establish full mass balance for process	Poor mass balances indicate that operations are not well-understood and appropriate controls may not be in place
4	Measure solvent and catalyst losses in air and aqueous effluents	Reduce impact on environment
5	Investigate basic thermochemistry	SAFETY considerations
6	Anticipate heat and mass transfer limitations	Scale-up considerations for SAFETY and minimizing impurities. Continuous processes may be preferred (Chapter 14)
7	Consult a chemical or process engineer	Seek out the experienced
8	Consider effect of overall process on choice of chemistry	Route selection
9	Help develop and apply sustainability measures	Sustainable measures will ensure operations long-term
10	Quantify and minimize use of utilities	Water and electricity will become more expensive
11	Recognize where SAFETY and waste minimization are incompatible	SAFETY first
12	Monitor, report, and minimize laboratory waste emitted	Sustainability

*(1) Winterton, N. Green Chemistry **December 2001,** G73.*

II.A. Analyzing Waste Streams (First Principle of Green Chemistry)

Following the first principle, it is far better to prevent waste, not just reducing waste at the end of the pipeline. In addition to generating less waste and producing reduced costs for waste disposal, routes with fewer steps require smaller quantities of starting materials, solvents, and reagents, and require less labor.

Lower CoGs and decreased environmental impact result. The greenest route is usually the route with the fewest steps [49,50]. Using RMEs one can calculate the greenest route.

Considering all alternatives may be necessary to determine whether the preferred route is the greenest route. For instance, the Mitsunobu reaction [51] shown in Figure 3.4 generated the byproducts triphenylphosphine oxide (TPPO) and (1,2-diisopropylcarbonyl)

TABLE 3.3 Twelve Principles of Green Engineering (1,2)

	Principle	Comments
1	Design for inherent SAFETY as much as possible	Once operations are SAFE attention can be devoted to other areas
2	Better to prevent waste than to treat or clean up waste	Conserve energy and materials
3	Design separation and purification operations to minimize consumption of energy and materials	Telescoped processes can decrease the time required for operations, and the costs of labor, solvents and reagents, and waste disposal, while increasing yields overall
4	Maximize mass, energy, space, and time efficiency of processes	Chapter 8. Continuous processes may be preferred (Chapter 14)
5	Products, processes, and systems should be "output pulled" rather than "input pushed" through the use of energy and materials	
6	Conserve embedded entropy & complexity for recycle, reuse, or disposal	
7	Target for product durability, not immortality	Immortal products don't decompose in landfills.
8	Design for capacity to meet need and minimize excess	But just-in-time manufacturing may not meet needs if key components are no longer available.
9	Minimize material diversity in multicomponent products to promote disassembly and value retention	Also true for solvent usage
10	Design products, processes, and systems to integrate and interconnect with available energy and materials flows	
11	Design for commercial "afterlife" in products, processes, and systems	
12	Material and energy inputs should be renewable rather than depleting.	

(1) Anastas, P. T.; Zimmerman, J. B. Design through the Twelve Principles of Green Engineering, Environ. Sci. Technol., **2003,** 37 (5), 95.
(2) http://portal.acs.org/portal/acs/corg/content?_nfpb=true&_pageLabel=PP_ARTICLEMAIN&node_id=1415&content_id=WPCP_007505&use_sec=true&sec_url_var=region1

TABLE 3.4 Comparison of Wastes Generated by Segments of the Chemical Industry (1)

Industry Segment	Product Tonnage	E Factor	Tonnage × E Factor
Petroleum	10^6-10^8	<1	$<10^6-10^8$
Fine chemicals	10^4-10^6	1–5	$10^4-5 \times 10^6$
Bulk chemicals	$100-10^6$	5–50	$5 \times 10^2-5 \times 10^6$
Pharmaceuticals	10–1000	25–100	$250-10^5$

*(1) Sheldon, R. A. Green Chem. **2007**, 9, 1273.*

hydrazine, with combined molecular weights about 50% greater than the molecular weight of the product mesylate (a fosinopril sodium intermediate). However, this reaction led to the isolation of the mesylate in high yield, with <0.1% of the *trans*-isomer. Removing the TPPO and hydrazide by filtration (see Chapter 11 for additional tips on removing TPPO) was a *lagniappe* that allowed this process to be effective without chromatography. Implementing the Mitsunobu process saved three steps in manufacturing [52], and thus was very productive overall. The Mitsunobu approach is probably greener than the four-step alternative if one considered all the waste that would have been generated. As another example, resorting to a Mitsunobu reaction avoided generating a troublesome impurity. In the upper route to a phthaloyl amine intermediate a mesylate

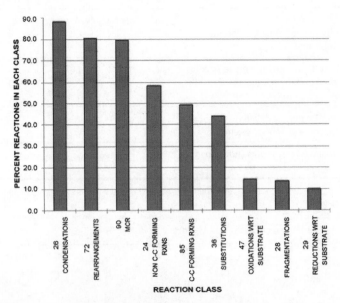

FIGURE 3.3 Proportions of "golden atom economical" reactions. *Reprinted with permission from Andraos, J. Org. Process Res. Dev. **2005**, 9, 404. Copyright 2005 American Chemical Society.*

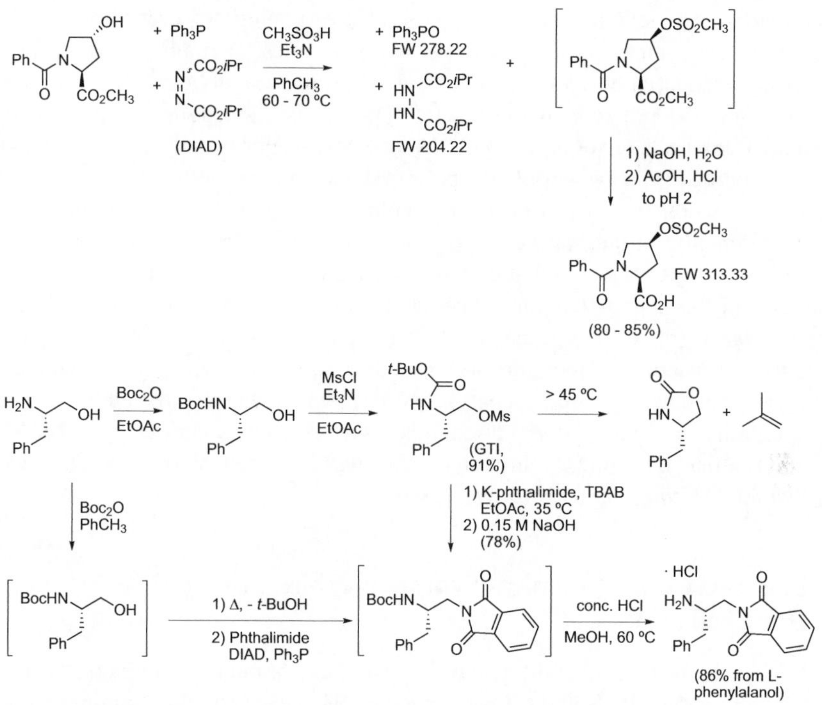

FIGURE 3.4 Benefits from using the "un-green" Mitsunobu reaction.

was first produced and converted to the phthalimide derivative using potassium phthali-mide under PTC conditions. This route was abandoned because the mesylate was found to be mutagenic (a gentoxic impurity, or GTI), and unstable above 45 °C. (The instability of this compound at a relatively low temperature would make it unattractive for outsourcing as an isolated intermediate; beyond the possibility of reduced product quality, the buildup of isobutylene vapors from decomposition could rupture shipping containers.) The preferred route was a one-pot process that did not involve the GTI: L-phenylalanol was reacted with Boc$_2$O in toluene, converted to the phthalimide using Mitsunobu chemistry [51], and deprotected by adding MeOH and concentrated HCl. The Mitsunobu byproducts TPPO and reduced DIAD were dissolved in the mother liquor, and the hydrochloride salt was isolated in 86% yield from this one-pot sequence [53]. In this example SAFETY considerations of not generating a GTI out-weighed the generation of some product waste. The ACS Green Chemistry Institute Pharmaceutical Roundtable highlighted the Mitsunobu reaction as one area where greener alternatives are preferred [54].

II.B. Atom Economy (Second Principle of Green Chemistry)

Atom economy is not the primary consideration for manufacturing in the pharmaceutical industry. Reactivity has greater priority in reagent selection, especially early in devel-opment, because shorter reaction times can lead to greater productivity from equipment (higher space–time yields). Volatile reagents require running at lower temperatures at ambient pressure, which can lead to longer reaction times, or running under pressure, which many instinctively avoid. (As long as the pressure of a reaction mixture is SAFELY below the pressure of the reactor rupture disc, reactions can be run under pressure. See Figure 8.11 for an example using HMDS.) For instance, dimethyl sulfate and methyl iodide [55] are generally the preferred reagents for methylation because they afford rapid reactions and are convenient (Table 3.5). Although methylation of a thio-amide with dimethyl sulfate was very convenient (Figure 3.5), researchers from the Dow Chemical Company considered that dimethyl sulfate was too hazardous for a large-scale commercial process, and they investigated the use of methyl bromide. Subsurface addition of MeBr (bp 4 °C) was begun at 25 °C, and when one-third of the MeBr had been added the reaction temperature was raised to 80 °C. The gas cylinder containing MeBr was warmed to speed up the charging of HBr, which was completed in 5 h. Under these conditions conversions to the thioamidate salt were equivalent using either MeBr or dimethyl sulfate, but certainly dimethyl sulfate was more convenient [56]. *It should be noted that methylating reagents are very toxic.*

II.C. Less Hazardous Chemical Syntheses (Third Principle of Green Chemistry)

SAFETY is a recurring theme with green chemistry, one of many parameters considered in route scouting [57]. PGIs and GTIs are avoided whenever possible to minimize risks from occupational exposure, or these reactive intermediates may be telescoped into the next step. To minimize the impact of purging PGIs from the APIs, reactive reagents are moved from the end of the route. Lilly researchers recommended positioning such reagents at least four steps from the formation of the API [58]. (See discussion around Figure 11.10 and Section IV of Chapter 13.) Reagents that are pyrophoric or explosive may be avoided; for instance, outsourcing should be considered for reactions requiring diazomethane. Pfizer workers analyzed in detail the routes to an API candidate using Heck, Sonogashira. and Suzuki couplings shown in Figure 3.6. They selected the Heck route for scale-up because it avoided thermal hazards associated with acetylenes, and generated less metal waste and less waste overall [59].

II.D. Reduce Use of Protecting Groups (Eighth Principle of Green Chemistry)

The route with the fewest steps may be the greenest and cheapest, and the fewer the steps, the fewer opportunities for errors. Each protecting group requires an additional step for both protection and deprotection. Unless a protecting group confers improved stability or solubility upon a molecule, as with the dimethyleneoxy group used to tether the amide nitrogens of adjacent amino acids [60], it may just take up reactor space and reduce productivity. For rapid development of an NCE the best route may be that with the fewest protecting groups.

Obviously the best protecting group is none at all, but achieving this can be difficult. Protonating amines or ionizing carboxylic acids can sometimes be effective in lieu of

TABLE 3.5 Atom Efficiency of Methylating Reagents (1)

Reagent	FW	bp	Atom Efficiency (%)	Comments/Some Uses
$CH_3OSO_2OCH_3$	126.13	188 °C	12	May be preferred due to reactivity and low volatility
CH_3I	141.94	41 °C	11	May be preferred due to reactivity. May be hard-piped for SAFE handling in manufacturing
CH_3OTs	186.23	144 °C/5 mmHg (mp 27 °C)	8	Low-melting solids may be most conveniently charged as liquids
$CH_3OSO_2CF_3$	164.10	94–99 °C	9	Highly toxic (2)
CH_3OSO_2F (Magic Methyl)	114.1	92–94 °C	13	Highly toxic (2)
CH_3Br	94.94	4 °C	16	Volatile
CH_3Cl	50.49	−24 °C	30	Highly volatile
$CH_3OC(O)OCH_3$ (dimethyl carbonate)	90.08	90 °C	17	Will be used more in future (3,4)
$HCHO, H_2$	32.04	~20 °C	47	Methylation of amines
CH_3OH/cat. H^+	32.04	65 °C	47	Esterification
CH_2N_2	44.04	−23 °C	34	Explosive in presence of selected impurities (5)
TMS-CHN_2	114.22		13	Toxic. Sold as solution in hexanes (6)

(1) Highly toxic. The state of California has listed dimethyl sulfate, iodomethane, and formaldehyde as causing cancer, and has listed methyl bromide and methyl chloride as reproductive toxins: http://www.calepa.ca.gov/Search/default.aspx?q=chemicals+known+to+cause+cancer&cx=016697183436188592925%3Aufud9xiklcq&cof=FORID%3A11#1278

(2) Highly toxic. Four orders of magnitude more reactive than methyl iodide or dimethyl sulfate (Encyclopedia of Reagents for Organic Synthesis). The use of these compounds has largely been discontinued due to their extreme toxicity (http://www.ncbi.nlm.nih.gov/pubmed/484483).

(3) Rekha, V. V.; Ramani, M. V.; Ratnamala, A.; Rupakalpana, V.; Subbaraju, G. V.; Satyanarayana, C.; Rao, C. S. Org. Process Res. Dev. **2009**, 13, 769.

(4) Parrott, A. J.; Bourne, R. A.; Gooden, P. N.; Bevinakatti, H. S.; Poliakoff, M.; Irvine, D. J. Org. Process Res. Dev. **2010**, 14, 1420.

(5) Consider outsourcing to specialty contractors, who may use continuous operations. See Proctor, L. D.; Warr, A. J. Org. Process Res. Dev. **2002**, 6, 884.

(6) Contact with trimethylsilyldiazomethane has led to at least one death. See Barnhart, R. W.; Dale, D. J.; Ironside, M. D.; Vogt, P. F. Org. Process Res. Dev. **2009**, 13, 1388.

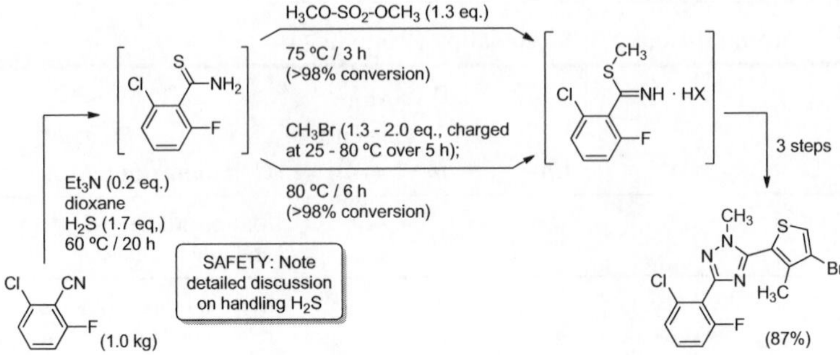

FIGURE 3.5 Conditions for methylation of a thioamide, using dimethyl sulfate or MeBr.

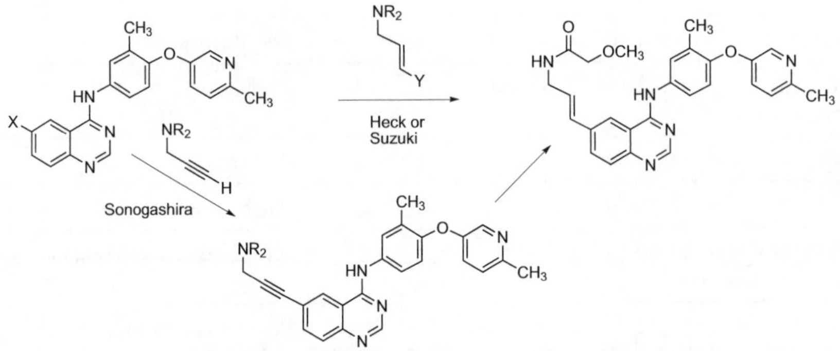

FIGURE 3.6 Heck, Sonogashira, or Suzuki coupling alternatives.

protecting groups. In the example of using a proton as a protecting group in Figure 3.7, adding HCl and using three equivalents of piperazine greatly improved the yield of the mono-alkylated piperazine. In contrast, mono-protection of piperazine with the carbo-benzyloxy (Cbz or Z) group, followed by alkylation and deprotection had afforded only a 2.1% yield from piperazine. For efficient processing the J&J researchers centrifuged the suspension at 60–70 °C. They also found that the product was more efficiently extracted into toluene at an elevated temperature (70 °C) than at room temperature [61]. Piperazines were also protected by trialkylsilyl groups for selective monoacylation. In the example shown in Figure 3.7 the labile TMS group sufficiently protected 2,6-dimethylpiperazine, but bulkier groups were needed for less sterically congested piperazines [62].

The stability of protected molecules may be more important than cost of the protecting groups, however. Feringa's group noted that larger and more stable protecting groups gave higher yields overall (Figure 3.8). Of all the alcohols tried, diphenylpropanediol gave the best yield (64%) of the ketal, in part because the resulting ketal was the most stable. (On scale they found that the yield of the ketal was improved by extracting residual ene-dione starting material into 1 M NaOH as the enolate, avoiding chromatography.) The stable ketal also improved the yield in the subsequent tandem addition to the PGE_1 intermediate [63].

If protecting groups must be used, operations may be more productive if several protecting groups can be removed in one step. When the cefmatilen precursor was

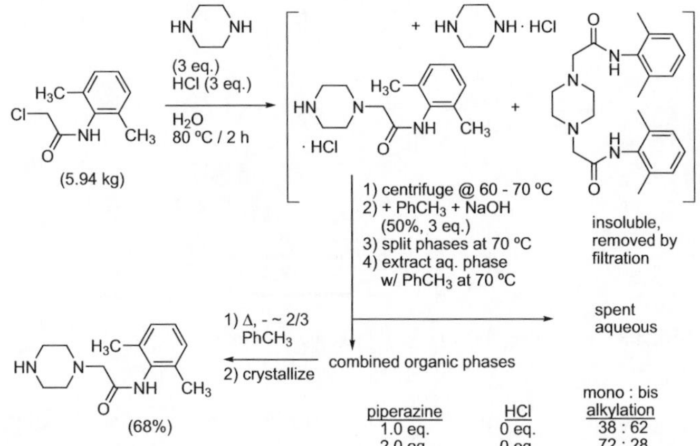

piperazine	HCl	mono : bis alkylation
1.0 eq.	0 eq.	38 : 62
2.0 eq.	0 eq.	72 : 28
3.0 eq.	0 eq.	72 : 28
3.0 eq.	3.0 eq.	93 : 7

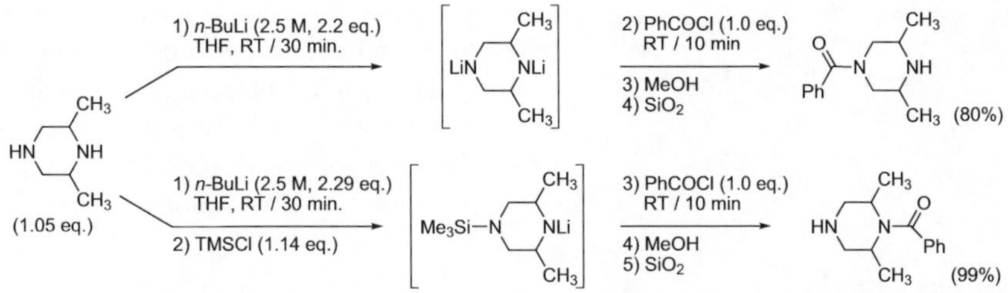

FIGURE 3.7 Using a proton or trimethylsilane as a protecting group.

treated with $AlCl_3$ in the presence of anisole to remove the Boc, *t*-butyl ether, and benzhydryl protecting groups, as expected lower temperatures afforded more selectivity (Figure 3.9). However, when at least 0.2% of the benzhydryl tetrazole impurity was present, the crystal habit was the desired blade form of the API. (Crystal habit often refers to the appearance of minerals, their size and shape. A change in crystal habit does not imply a change in polymorphism. In this case the crystals formed blade-like agglomerates and processing of the API was easier.) As the formation of this impurity

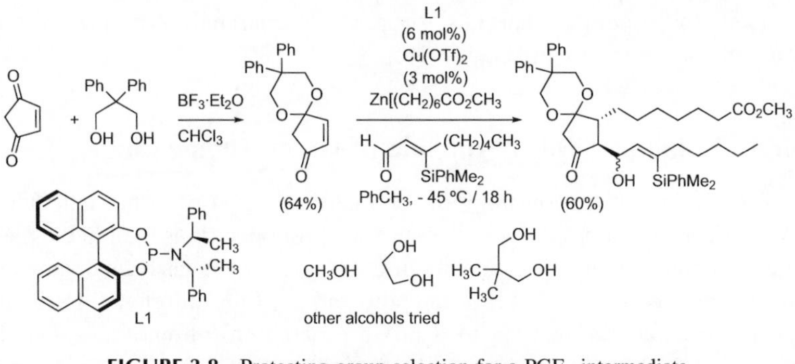

FIGURE 3.8 Protecting group selection for a PGE_1 intermediate.

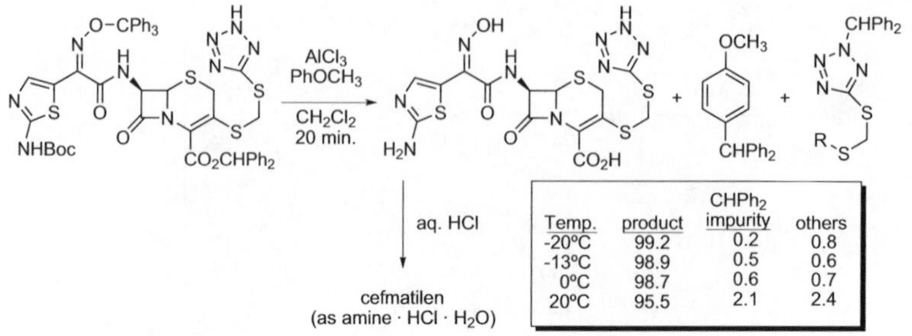

Temp.	product	CHPh2 impurity	others
-20°C	99.2	0.2	0.8
-13°C	98.9	0.5	0.6
0°C	98.7	0.6	0.7
20°C	95.5	2.1	2.4

FIGURE 3.9 Removing three protecting groups in one step.

Common protecting groups for amines are the *t*-butyloxycarbonyl (Boc) and carbobenzoxy (aka carbobenzyloxy, Z, or Cbz). The trifluoroacetyl group has been used to protect amino acids [65]. The chloroacetamide group can be removed by alcoholysis, five times more readily than the acetamide [66]. For cleavage of esters by ammonia, choroacetyl is removed faster than methoxacetyl, which is removed faster than acetyl [67]. These groups, trifluoroacetyl, chloroacetyl, and methoxyacetyl, may offer useful alternatives to protection of amines by Boc and Cbz.

increased the particle size increased. To ensure that the desired crystal habit could be prepared from purified API, the Shionogi researchers examined adding small amounts of polymeric modifiers (cellulose derivatives, poly-vinylpyrrolidone, and poly-vinylalcohol) to crystallizations. The desired blade-like agglomerates were reliably generated by the addition of 0.2% of hydroxypropylcellulose type-M [64], an excipient used in formulations.

II.E. Catalysis (Ninth Principle of Green Chemistry)

With catalytic reactions involving expensive catalysts cost-efficient routes will recycle the catalysts. Heterogeneous catalysts often comprise catalysts adsorbed to insoluble materials such as activated carbon, silica, alumina, or polystyrene. Heterogeneous reagents usually react significantly more slowly, as Halpern noted for quaternary ammonium salts bound to ion exchange resins [68]. For cost-effective preparations ligands with metals and organocatalysts may have to be tethered to insoluble materials, or readily separated from the product by extractions. (See discussions in Chapter 10.)

II.F. Direct Isolations (Third Principle of Green Engineering)

Direct isolations minimize consumption of materials, energy, time and labor (see Section V). Direct isolations can be very convenient for operations in both the plant and the kilo lab. The manufacture of sildenafil citrate (Figure 3.10) has been documented as one of the greenest routes [69,70]. For the preparation of the sulfonyl chloride molten 2-ethoxybenzoic acid was charged to a mixture of chlorosulfonic acid and thionyl chloride; excess $ClSO_3H$ functioned as solvent, and the $SOCl_2$ ensured the preparation

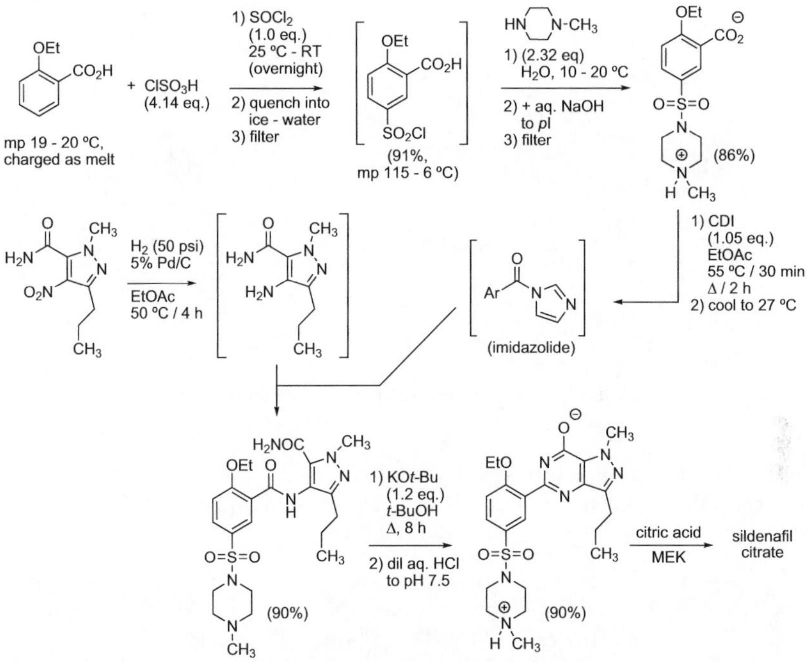

FIGURE 3.10   Manufacture of sildenafil citrate.

of the sulfonyl chloride [70]. The reaction was quenched into water and the intermediate was isolated by filtration. Because the subsequent condensation with N-methyl-
piperazine was conducted in water, there was no need to dry the sulfonyl chloride. By
adding aqueous NaOH to adjust the sulfonamide mixture to the isoelectric point (pI) the
product crystallized in high yield. The latter intermediate was treated with carbonyl
diimidazole (CDI) in EtOAc to form the acyl imidazolide, and condensed with the
aminopyrazole from hydrogenation of the nitro precursor. EtOAc was selected as
solvent for these steps because the product crystallized directly from the reaction and
the starting materials were soluble. The subsequent ring closure was carried out in
t-BuOH, and upon acidifying with dilute aqueous HCl to the pI sildenafil crystallized in
high yield. The five transformations outlined in this paragraph were run at high
concentration, and products were isolated without extractions, crystallized directly
from the reaction mixtures. Only four organic solvents are used in the manufacture of
sildenafil citrate: toluene (used to prepare the amide precursor of the nitropyrazole),
EtOAc, methyl ethyl ketone, and another solvent that replaced t-BuOH, which was
difficult to recover from the aqueous mother liquor. With optimized recovery and
recycle a waste of only 4 L/kg of solvents was expected [71]. Keys to the high
productivity of this route were the efficient isolations of an intermediate and the free
base by adjusting to the pI for direct isolation.

   Direct isolations can also be very convenient for the kilo lab. All reaction products
shown in Figure 3.11 were isolated directly from reactions by filtration, either after
cooling the reaction mixtures or by adding H_2O and then filtering. (The 12-h reactions

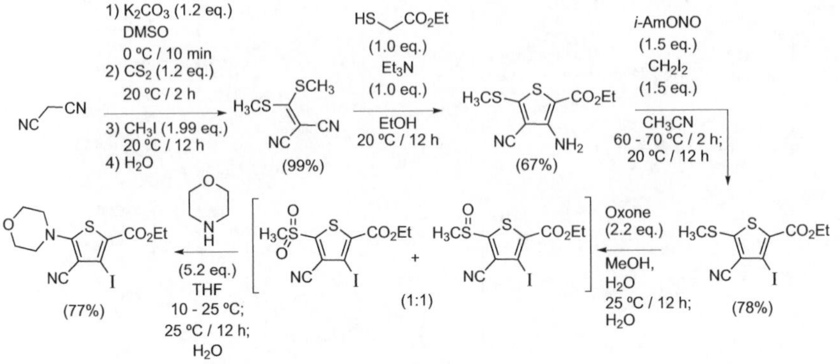

FIGURE 3.11 Direct isolations in the kilo lab.

suggest that the processing times were not optimized.) Fortunately the mixture of sulfoxide and sulfone reacted with the poor nucleophile morpholine [72].

> Select a route with crystalline intermediates, especially with a crystalline penultimate intermediate. Tosylates are often more crystalline than other analogous sulfonates or alkyl halides [73], and *p*-toluenesulfonamides may be more crystalline than other amine derivatives [74]. Compounds with three or more aromatic rings are likely to be crystalline. Crystallization, recrystallization, and reslurrying are reliable techniques to upgrade the quality of intermediates. An excellent approach to control quality of the API is to first control the quality of the penultimate. Simple chemistry at the end of a route, such as ester hydrolysis and salt formation, can simplify routine manufacturing of an API. To conserve efforts throughout drug development, select the optimal penultimate intermediate and "freeze" the preparation and isolation of the NCE. Endeavor to select a penultimate intermediate that can be prepared from multiple transformations, and vary only the approaches to the penultimate, not conversion of the penultimate to the API.

II.G. Convergent Routes, Location of Low-Yielding Steps, and Resolutions

When a synthetic route incorporates convergent subroutes, the total amount of intermediates required to prepare a product can decrease dramatically. This is illustrated in Figure 3.12 with four theoretical assemblies of a fully protected octapeptide made by workers at R. W. Johnson [75]. (The octapeptide was assembled from the carbonyl terminus, an unconventional approach. As seen with the linear route the molecular weight of each intermediate rises from the addition of each element less the molecular weight of MeOH.) Figure 3.13 shows that as the yields per step decrease the amount of intermediates that need to be prepared rise significantly. When subroutes converge closer to the end of the synthesis, the total amount of intermediates which need to be prepared decreases significantly; for example, if the yields per step were 75–80% for each step, using a linear synthesis it would be necessary to prepare about 50 wt% more

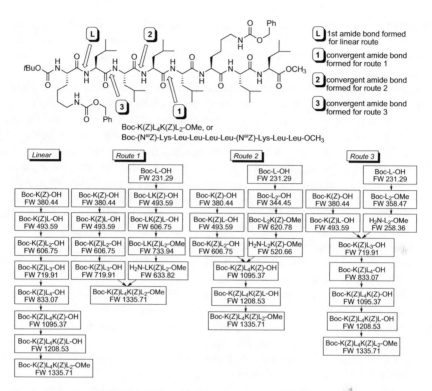

FIGURE 3.12 Model routes to prepare an octapeptide.

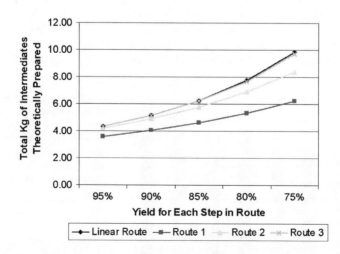

FIGURE 3.13 Total kg of intermediates theoretically required to produce 1 kg of octapeptide in Figure 3.11. Assumptions to model: (1) All derivatized amino acid derivatives were purchased. (2) After each coupling the intermediate ester was hydrolyzed to the carboxylic acid and isolated. (3) All intermediates were stable and no salts were prepared. (4) Yields for each step within a route were identical.

intermediates than using a route convergent at the final step. The total amount of intermediates to be prepared for the early-convergent route is essentially the same as that for the linear route, consistent with the trend that routes convergent at the end of a route require less intermediates to be made.

Using a convergent route may require an additional step, and preparing and assaying an additional intermediate. As shown for the synthesis of the octapeptide in Figure 3.12, the linear route requires one step less than any convergent route. In this case the additional step is necessary due to the multiple functional groups of the amino acids used.

With extended linear syntheses the middle and late intermediates become "increasingly precious." Introducing convergency lessens the value of intermediates and cost of any scale-up errors [76] up to the final step. For manufacturing large amounts of API the risk of failure is generally spread out by campaigning several batches of each step.

Positioning resolutions early in syntheses minimizes wasted resources in preparing the undesired enantiomer. Figure 3.14 shows the trend for the total amount of intermediates required to generate 100 kg of an API, through a hypothetical sequence of six steps and one classical resolution by salt formation. Positioning a resolution at the beginning of this sequence requires purchasing and making more than 1250 kg of intermediates, while positioning the resolution at the end of this sequence requires making and purchasing more than 1960 kg of intermediates. This increased requirement of greater than 50% more intermediates would also require as much of an increase in solvents, reagents, and waste disposal. The amounts of raw materials required will increase if the yields are less than the 90% used in the model. Another conclusion from this model is that placing low-yielding steps early in sequence reduces the raw materials required. (Similarly, placing high-yielding steps at the end of a route may simplify purifications because the amount of waste is less.) Optimizing low-yielding steps first is one of the quickest ways to reduce raw material requirements [77].

For efficient use of materials processes that require resolution steps should be replaced by asymmetric processes. If this replacement is not feasible, ideally conditions can be developed to racemize the undesired isomer for recovery of the desired enantiomer. Crystallization-induced dynamic resolution (CIDR) and crystallization-induced processes asymmetric transformation (CIAT) perform the racemization in situ,

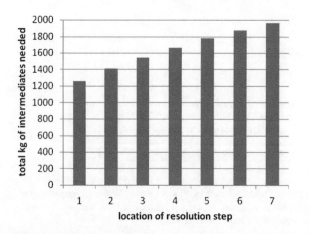

FIGURE 3.14 Benefits of positioning a resolution step early in the route. Assumptions made in the model: a 50% yield for the resolution, and 90% yields for all other steps. For ease of calculations all compounds have the same molecular weight.

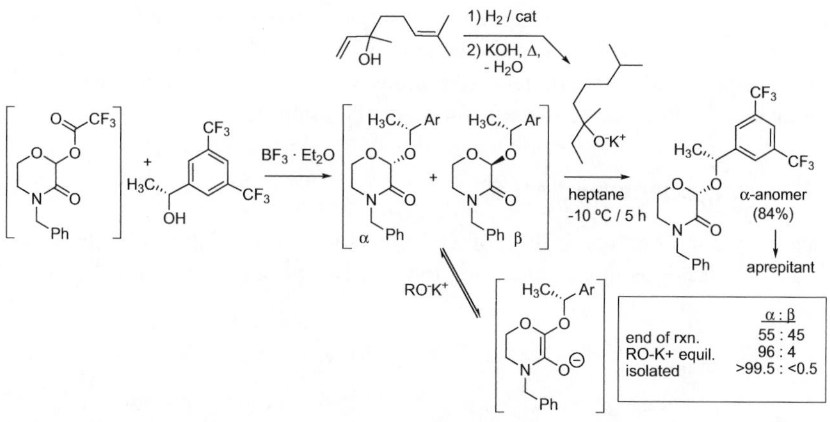

FIGURE 3.15 Manufacture of an aprepitant intermediate through a CIAT process.

theoretically affording 100% yields (Chapter 12). The CIAT step shown in Figure 3.15 efficiently sets the stereochemistry for the second of the three chiral centers in the manufacturing process for aprepitant [31].

SMB is an economical technology used for decades to purify material. For instance, SMB has been used to prepare *o*-xylene, a precursor for phthalate plasticizers [78], and to separate sucrose from molasses [79]. SMB with chiral stationary phases has been shown to be an economical alternative to classical resolutions, and has been used to manufacture sertraline [30], levetiracetam, escitalopram, and others [80]. In a detailed example, Amgen researchers found that a chiral SMB approach to a chiral thiourea was less costly than a standard chemical resolution (Figure 3.16). After extensive optimization, the sequence of Ritter reaction, resolution, and thiourea formation afforded the chiral thiourea in a 15% overall yield; the yield from the chiral SMB approach was 36.5%. Furthermore two steps in the classical resolution route required extended processing times. Using SMB purification required generating a stream of the racemic

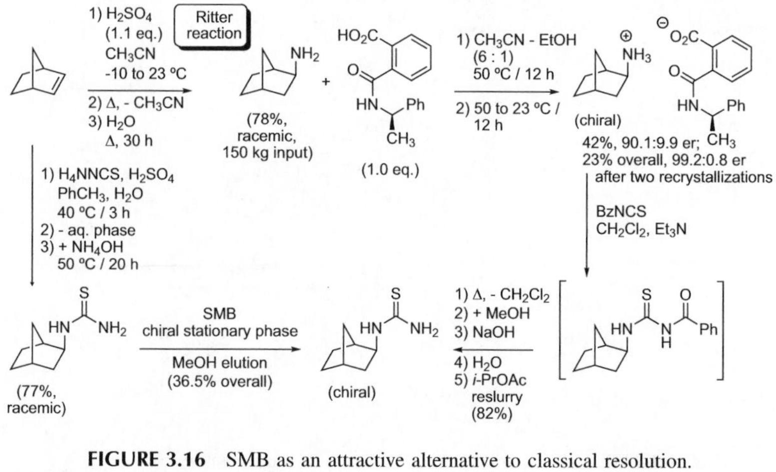

FIGURE 3.16 SMB as an attractive alternative to classical resolution.

thiourea that was essentially free of impurities, in order to increase productivity. MeOH was chosen for elution because it could be readily recovered and recycled, even though other solvents provided more efficient separations. Overall the SMB approach was much more attractive for large-scale manufacturing of the thiourea [81].

II.H. Rearrangements and MCR

Rearrangements and MCRs are very "green," with some of the highest RMEs. Hofmann rearrangements, such as those generating the first two products shown in Figure 3.17,

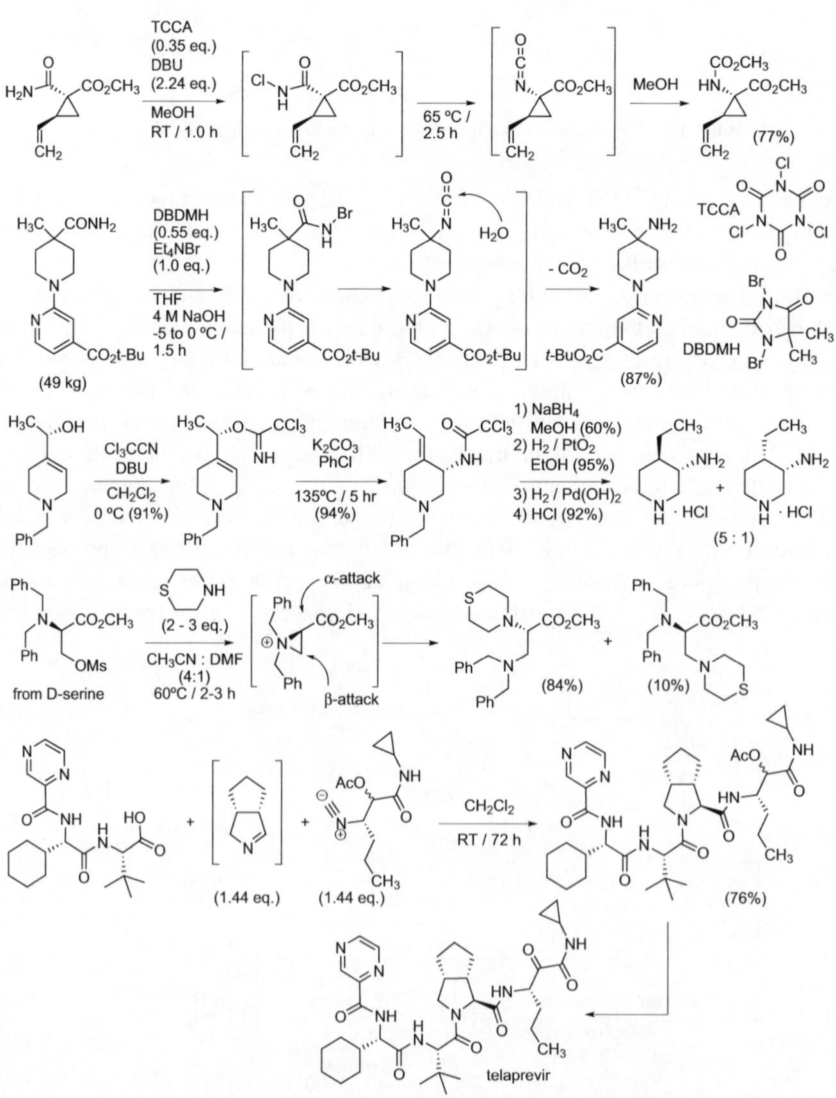

FIGURE 3.17 Rearrangements and a multicomponent reaction.

can efficiently prepare some hindered amines. After chlorination using 1,3,5-tri-chloroisocyanuric acid (TCCA), conversion to the isocyanate took place with heating, and the latter was trapped with MeOH [82]. (TCCA is highly soluble in many organic solvents, and all three chlorines in a molecule can react, making TCCA about one-fourth the cost of NCS [83].) In the presence of tetraethylammonium bromide and aqueous NaOH, bromination with 1,4-dibromo-5,5-dimethylhydantoin (DBDMH) led to the formation of the isocyanate, followed by hydrolysis and decarboxylation [84]. Through the Overman rearrangement shown the chirality was transferred from a distal carbon into the ring of an amino pyrrolidine. (K_2CO_3 was used to preclude acid-catalyzed break-down of the trichloroacetimidate [85].) Similar to a rearrangement, a carbon–nitrogen bond was broken through formation of an aziridinium species in the next example. (The authors noted that with NaN_3 as nucleophile only the product from attack at the α-position was isolated [86]). These rearrangements efficiently afforded some complex molecules, and it should be noted that the success of rearrangements can depend on both electronic and steric factors. Telaprevir was made by a three-component Ugi-type conden-sation of a carboxylic acid, a chiral imine, and an isonitrile, which itself was prepared by a Passerini three-component reaction. Overall this sequence required only 11 steps, vs. 24 in the original procedure, in part due to the use of fewer protecting groups [87]. Even though this route used more than stoichiometric charges of two components, this example shows the promise of using MCRs to make APIs. Rear-rangements and MCRs should be incorporated into syntheses when they are advantageous.

Experimental procedures in a manu-script often prompt questions about the state of development of a process and a drug candidate. For instance, an operation conducted under a very dilute concentration, such as 1 g/100 mL of solvent, may indicate that impurities formed at higher concentrations. Alternatively, the operations may have been con-ducted at these dilute concentra-tions for convenience, and no further process optimization was undertaken because the project was abandoned.

III. BIOTRANSFORMATIONS

Biotransformations provide exciting opportunities for changing synthetic routes, and several books have reviewed this subject recently [88,89]. Previously fermentations were considered appropriate for preparing complex molecules such as antibiotics and steroids, and biocatalysts were used to prepare valuable compounds such as 6-ami-nopenicillanic acid. The stereotypical biocatalytic process was not viewed as productive, being dilute and requiring days for complete reaction. Now practical biocatalytic processes can be run under concentrated conditions at moderate temperatures and in the presence of a high proportion of an organic solvent; these changes are due to high-throughput selection processes and rapid and strategic development of new strains. Biocatalytic routes should not be avoided as potentially expensive. Some products derived from biocatalysis are shown in Figure 3.18. Amyris Biotechnology has developed farnesane, derived from β-farnesene, as a bio-fuel (Figure 3.18) [90], and LS9, another of Kiesling's efforts, is developing biodiesel from bacterial fermentation [91]. Amyris also developed a fermentation to produce

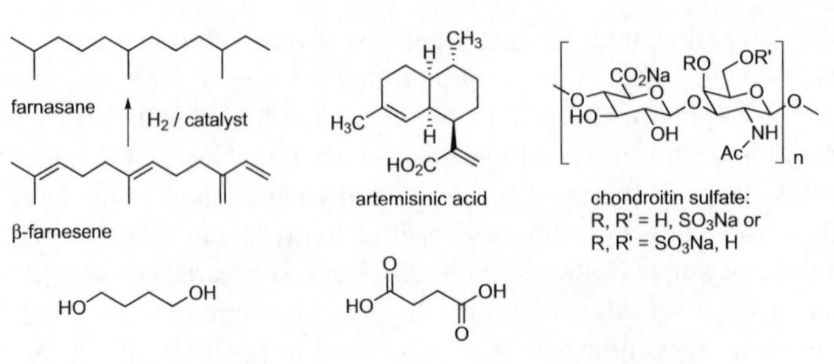

FIGURE 3.18 Some products from biotechnology.

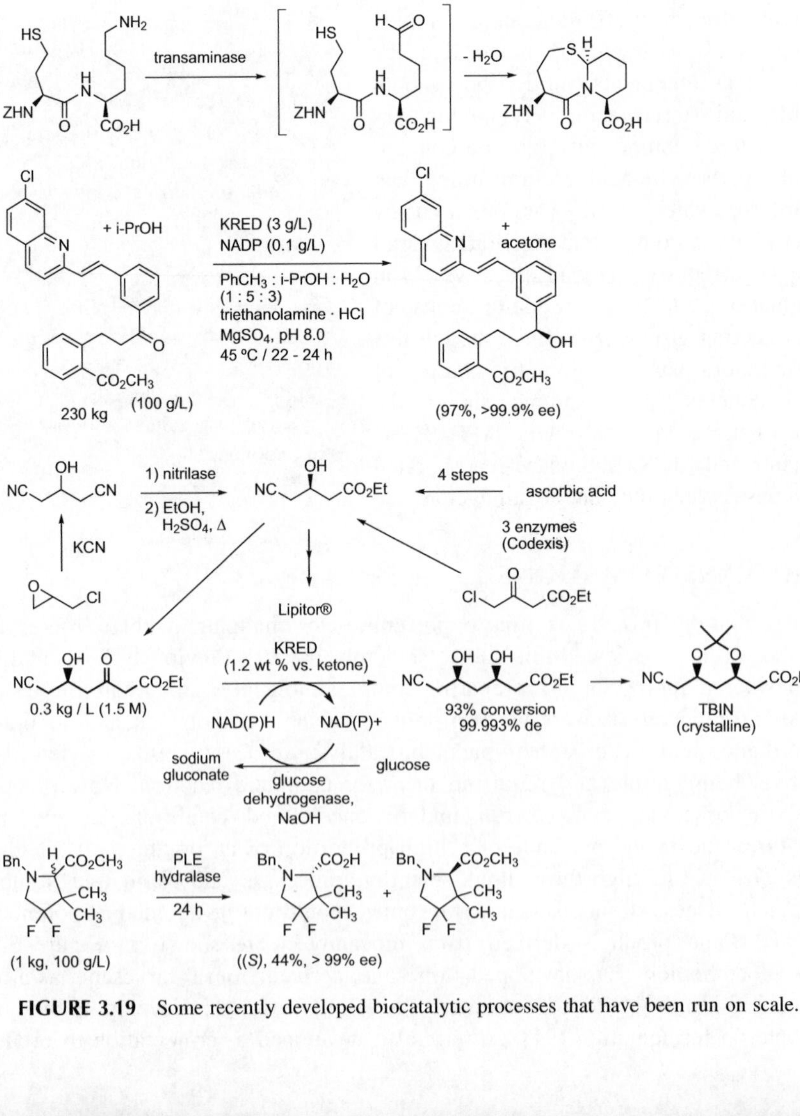

FIGURE 3.19 Some recently developed biocatalytic processes that have been run on scale.

artemisinic acid, a precursor to inexpensive artemisin-based anti-malarial therapies [92,93]. Chondroitin sulfate, a nutraceutical in the US and a prescription drug in Europe used to treat joint pain, has been prepared using microbiological techniques [94], and perhaps this animal-free approach will be used to manufacture this therapeutic agent. BioAmber received a US Presidential Green Chemistry Award for the preparation of succinic acid from corn kernels and corn stover, using a strain of *Escherichia coli* licensed from the Department of Energy. This venture was hailed as the first to commercially produce a commodity chemical more inexpensively than the petroleum-based route [11,95]. Genomatica received a US Presidential Green Chemistry Award for the low-cost preparation of 1,4-butanediol; the recombinant *E. coli* strain can grow on a variety of renewable carbohydrate feedstocks [11,95]. The economical production of the latter two high-volume chemicals illustrates the promise of biotechnology to prepare materials inexpensively.

Biocatalysts have been developed for synthetic purposes beyond the biochemical transformations usually associated with these enzymes. For instance, transaminases, which normally are thought of as speeding the interconversion of α-amino acids and α-keto acids, have been developed for the oxidative deamination of a primary amine (a vanlev precursor) [96,97] (Figure 3.19). The reduction to the montelukast precursor is a slurry-to-slurry conversion, and is run at 100 g/L [98]. Biocatalytic approaches developed for a lipitor precursor shown in Figure 3.19 were developed by Diversa [99], Codexis [100], and Dowpharma [101]. In further work toward lipitor intermediates Codexis researchers were able to prepare the diol precursor to TBIN using a ketoreductase (KRED); essentially none of the undesired diastereomer was produced in this very productive process [102]. The resolution of an unusual α-amino acid using a lipase became practical when the *N*-Boc group was replaced by benzyl [103]. (DSM has developed a recombinant porcine liver esterase (PLE), i.e., an enzyme from a non-animal, BSE-free source [104].) Tao's guidelines for rapid process development using biocatalysts are shown in Table 3.6 [105]. Thus it can be seen that many approaches can be taken to develop an efficient biocatalytic process. Furthermore the time needed for directed evolution is decreasing. Biocatalytic routes offer considerable promise for making quantities of material inexpensively, especially for second-generation routes [32].

TABLE 3.6 Tao's Guidelines for Rapidly Developing Biocatalytic Processes

Quickest to Effect	Change	Example
1	Screen medium	Dissolve substrate in DMF, DMSO, or acetone
2	Modify starting material	Change BOC for benzyl
3	Screen biocatalysts	High-throughput
4	Directed evolution	Can require 6–12 mos

Practical Process Research and Development

IV. SLEUTHING TO DETERMINE ROUTES USED TO MAKE DRUGS

Isotopic ratio mass spectrometry (IRMS) has been used to determine whether routes used to make APIs infringed existing patents [106], for determining the sources of raw materials for illicit street drugs [107], and for other types of forensic chemistry [108]. For example, The Centre for Forensic Science (Queen's University, Belfast) prepared the street drug ecstasy (methylenedioxymethamphetamine, MDMA) by the routes shown in Figure 3.20, and analyzed the batches using IRMS. Using one batch of piperonyl methyl ketone (PMK) as the substrate for the three reductive aminations, five batches of free base from each route were prepared to generate meaningful statistical analyses. Analysis by IRMS showed that the reductive amination using H_2 was most reproducible of the three routes to the free base. (Variability in mass spectrometric assays [109] and in natural isotopic abundances [110] may complicate the analyses for IRMS.)

The preparation of ecstasy in Figure 3.20 deserves some additional comments. Although the yields could probably be improved by an experienced process chemist, the reagents are inexpensive, and one step, the isomerization to isosafrole, is solvent-free. The dihydroxylation to isosafrole glycol requires no metal reagent. (Notably the experimental procedures did not describe any monitoring or quenching of excess H_2O_2.) Conversion of the glycol to PMK is probably driven by the stabilization of a benzylic cation by the electron-rich aromatic ring, hence taking advantage of the inherent nature of the molecules. The experienced process chemist applies such practical thinking in route selection, and recognizes the extreme danger of manufacturing both ethical pharmaceuticals and street drugs (as witnessed by fires in illegal laboratories making methamphetamine).

V. CoG ESTIMATES

A projected CoG can greatly influence the route selection and development of an API [18]. (In accounting parlance COGS is used to indicate the cost of manufacturing and shipping a product.) Customers can use commercial programs to calculate a CoG estimate [111,112].

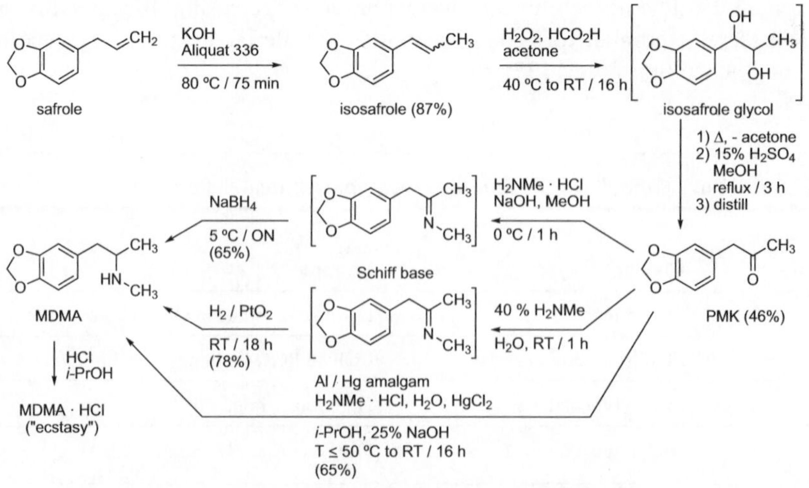

FIGURE 3.20 Preparation of ecstasy by reductive amination.

In 2002 Braish mentioned a "back of the envelope formula" to estimate the CoG for a compound by multiplying by 100 the sum of the number of rings, stereocenters, regiocenters, functional groups and heteroatoms. The resulting value is the cost per kilogram when a process has been optimized and commercialized, and includes the cost of raw materials, licensing, manufacturing, and waste disposal [113]. Many individuals and companies derive CoG estimates using spreadsheet programs developed in-house [114].

Defining all the parameters and assumptions is imperative for reliable CoG estimates (Table 3.7). The scale on which an API is produced is one of the most important parameters in a CoG estimate: the CoG decreases as the scale increases, a trend called "economy of scale" or "economy on scale." The prices of starting materials greatly influence CoG estimates. For early CoG estimates an assumption may be made for the price of starting materials in bulk; in the 1990s it was reasonable to predict that a compound could be obtained on scale at 20–25% of the best catalog price, but in the second decade of the twentieth century bulk prices have been estimated at 3–10% of the best catalog price [44]. Larger companies employ outsourcing specialists to seek out quotes for compounds purchased in bulk [115], thus obtaining firmer data for CoG estimates. By choosing starting materials and reagents that are readily available from multiple vendors, one can develop routes with low CoGs through competitive bidding [116]. These efforts to source amino acids for the 36-amino acid drug Fuzeon [117] reduced the prices of amino acids for other purchasers [118]. (Remarkable reductions in costs can be reached by finding CMOs that make products similar to those needed, and asking for quotes [119]. Just as it may be wise to outsource processing problems to the specialists, purchasing starting materials may save time and money. Trends in outsourcing, particularly to Asia, have been reviewed recently [120,121].) Early CoG estimates may not include the cost of solvents and waste disposal. Labor and overhead costs may not be considered for in-house manufacturing if a company has unused capacity, but a CMO may build into quotations the costs for labor, capitalization (e.g., purchase and amortization of equipment) and operations (e.g., maintenance). Reaction yields and the number of equivalents of reagents also influence the CoGs

TABLE 3.7 Parameters to Consider for CoGs Estimates

Parameters	Examples	
Necessary	Yields, MW, equivalents of intermediates & reagents Scale (amount of product to be made) State basis for estimated costs of purchased intermediates and reagents	
Smaller impact (often disregarded)	Solvent costs and disposal costs Recovery and reuse of solvents cGMP, QC, QA, and analytical R&D costs Documentation costs Cycle time for cleaning and validation	
Labor (ignore if in-house capacity available)	Research	number of full-time equivalents (FTE) is variable
	Pilot plant	$330–420/h (highly variable, less for Asia?)
	Manufacturing	$500–750/h (highly variable, less for Asia?)

Depending on the business model, the price charged by a CMO for the initial kilos of API may be low if the CMO hopes to get future business to manufacture the approved API, or a CMO may charge individually for delivery of the API, identification of impurities, and preparation of impurity standards. A CRO or CMO may levy additional charges if extenuating circumstances are encountered. In approving any contract understanding the details in advance is crucial.

estimate. For accurate comparison of CoGs estimates it is necessary to define the parameters and assumptions.

The isolation of a product contributes heavily to the CoGs. About 40% more time was needed for an extractive workup on scale, as shown in Figure 3.21. Initially the product phosphinic acid was isolated directly from the concentrate after azeotropic removal of the solvent acetonitrile and byproduct hexamethyldisiloxane; cooling the liquid–liquid dispersion of the concentrate led to the crystallization of the phosphinic acid. Despite oil droplets seen under the microscope in early laboratory runs, the product crystallized in 97–98% purity, and operations were simple. This process was successfully demonstrated in the pilot plant (eight runs) and in the manufacturing facility (five runs). Unfortunately there were problems with the subsequent manufacturing campaign: the phosphinic acid was slowly converted to the next fosinopril sodium intermediate, and hence the phosphinic acid was routinely upgraded by recrystallization from MIBK. The process was redeveloped to include an extraction into MIBK, followed by concentration and cooling to crystallize the phosphinic acid [122]. The additional processing for extraction and crystallization increased the time for processing by 40% for a 30 kg batch [123]. Based on the

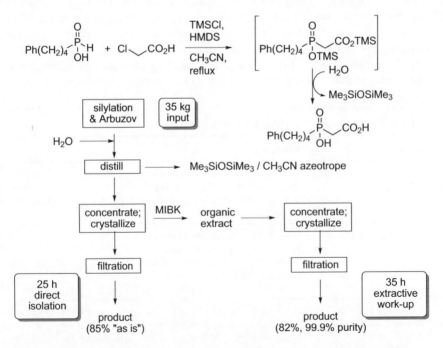

FIGURE 3.21 Crystallizing a product by direct isolation or after an extractive workup.

processing times for these two workups and the generalization that a 10-fold scale-up doubles processing time, the CoGs for these two processes were calculated (Table 3.8).

There are several conclusions that can be drawn from these CoG estimate data, as shown in Table 3.8. First of all, the CoG is increased by about 20% by using the extraction–crystallization process instead of the direct isolation process. The increased cost of waste disposal for the extraction–crystallization process is small compared to the overall CoG. Direct isolation procedures are cheaper by decreasing the labor costs, not the waste disposal costs [123]. Even if a product is made in-house and labor is not considered a part of the CoGs, by running extended processes the opportunity is lost to prepare something else in that equipment (an opportunity cost). Secondly, larger batches are cheaper due to decreased labor costs (economy on scale); in this example by scaling up 10-fold the CoGs dropped to about 60% of the CoGs for the smaller batches. However, the cost of failing a batch is greater with larger batches, and hence campaigns of several batches of intermediates may be run to decrease the potential impact of batch failure.

Simply changing reagents and workups may not markedly decrease the CoGs. As an example, an ester could be made by alkylative esterification (CH_3I, base, DMF) or by Fischer esterification (H_2SO_4, MeOH). For the hypothetical example in Figure 3.22 the product can be crystallized by adding H_2O to the alkylative esterification, or EtOAc and heptanes to the Fischer esterification, and times for isolation are identical. Iodomethane is more costly than H_2SO_4, but the cost of raw materials is very small compared to the cost of the starting material (Table 3.9). The biggest difference is seen with the extended time of a typical Fischer esterification, which raises the labor cost and the CoG; the increased labor cost narrows the difference between the CoGs. (The yield and perhaps the reaction time would improve if reagents such as $SOCl_2$ or trimethyl orthoformate were added to remove H_2O from the equilibrium of the Fischer esterification.) Other factors come into play, such as reliability of processing and SAFETY. For SAFETY reasons on scale most plants would prefer to handle H_2SO_4/MeOH over CH_3I (or dimethyl sulfate)/DMF.

Factors other than the cost of purchasing an expensive reagent can greatly influence the location of this step in route selection. In the hypothetical route shown in Figure 3.23 positioning the reagent (diethylamino)sulfur trifluoride (DAST) at the last step requires the least amount of DAST, as expected. Unfortunately the number of CMOs that can perform fluorinations under cGMP conditions might be rather limited. Positioning DAST at step *a* requires almost twice as much DAST as is needed for step *b*, but the functionality in the starting material for step *a* may be simpler, allowing intermediate A to be out-sourced competitively. Intermediate C could be a good choice for outsourcing if it could be prepared in high yield using DAST, as fewer equivalents of DAST would be required and the contribution to the CoGs might be the least of these four choices. Depending on the route filed with regulatory authorities, locating the fluorination at step *c* could also require that intermediate C be manufactured under cGMP conditions.

VI. SUMMARY AND PERSPECTIVE

The preferred route reliably provides suitable quality product within the timelines (fit for purpose). Hence to prepare material for early GO/NO GO studies the best route may be the one with which people have had the most experience and the most success, and

TABLE 3.8 CoGs Estimates for Scaling Up Direct Isolation and Extraction–Crystallization Processes

	Unit Cost		35 kg Input Batches		350 kg Input Batches	
			Direct	Extraction	Direct	Extraction
Input $C_{10}H_{15}O_2P$	$500	/kg	$17,500	$17,500	$175,000	$175,000
Other raw materials	$62	/kg of $C_{10}H_{15}O_2P$	$2162	$2162	$21,618	$21,618
MIBK	$5	/kg	0	$1835	0	$18,351
hours/batch			25	35	50	70
Labor	$500	/h	$12,500	$17,500	$25,000	$35,000
Volume of waste		L/batch	267	690	2667	6901
Waste disposal	$500	/55 gal drum	$641	$1657	$6406	$16,574
Water			$0	$0	$0	$0
Utilities (1)			$0	$0	$0	$0
Amortization of equipment			$0	$0	$0	$0
QC charge (2)	$6560	/batch	$6560	$6560	$6560	$6560
Total cost		/batch	$39,363	$45,380	$234,585	$254,753
Product output		kg/batch	38.5	37.1	385	371
Output purity vs. standard			98.0%	99.9%	98.0%	99.9%
Product output, corrected		kg/batch	37.7	37.1	377	371
All raw materials		/kg product	**$522**	**$580**	**$522**	**$580**
Labor		/kg product	**$332**	**$472**	**$66**	**$94**
Waste disposal		/kg product	**$17**	**$45**	**$17**	**$45**
QC charge		/kg product	**$174**	**$177**	**$17**	**$18**
CoG		/kg product	**$1044**	**$1274**	**$622**	**$737**

(1) Utility costs were calculated at <1% of the total CoGs and so were ignored.
(2) QC charges can be 30% of the CoGs for an API (Mullin, R. Chem. Eng. News **2009**, 87 (39), 38). For these calculations the QC charge was estimated at 20% of the raw materials, labor, and waste disposal for the direct isolation on a 35 kg scale.

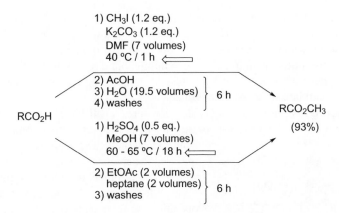

FIGURE 3.22 Alkylative esterification and Fischer esterification.

TABLE 3.9 Comparison of CoGs for Alkylative and Fischer Esterifications (1)

	Alkylative Esterification	Fischer Esterification
RCO$_2$H (100 kg)	$80,000	$80,000
Raw materials	$7106	$2476
Labor for reaction	$500 (1 h)	$9000 (18 h)
Labor for isolation (6 h)	$3000	$3000
Waste disposal	$12,056	$4629
Total (for 100 kg input)	$102,662	$99,105

(1) The cost of the carboxylic acid was defined at $800/kg, and the costs of the other ingredients were taken as 20% of the best prices in the 2009–2010 Aldrich catalog.

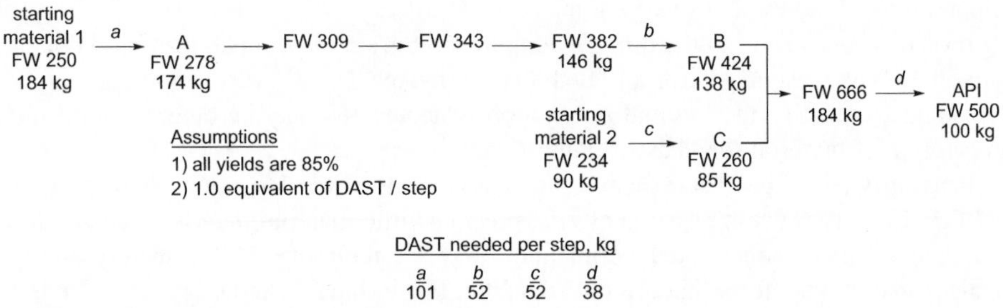

FIGURE 3.23 How the location of DAST step in route changes the amount of DAST needed to make 100 kg of API. The molecular weights in the long sequence increase by about 10% per step, and the loss of 166 Da in the last step is equivalent to hydrolyzing a methyl ester and removing a Boc group. Substituting a fluoride for an alcohol group does not significantly change the molecular weight for a pair of molecules.

relatively low yields may be acceptable if operations are simple [124]. A tabular summary of considerations for early route selection is shown in Table 3.10. For routine manufacturing on scale the objectives are different. By holding in mind the desired characteristics for routine manufacturing one throughout development activities one can accelerate the development of a process that is suitable for scale-up when successful development candidates have been identified.

Route selection is not a major consideration for a small biotech or pharmaceutical company – they may not have money or time for more than proof of concept. Employees from a small biotech or pharmaceutical company may be content to use dangerous reagents that would prompt others to redesign a route. CROs may prefer changing a route to ensure SAFE operations in their facilities, for the sake of for their equipment and reputation. Clients may not want to pay a CRO to conduct SAFETY studies to assure SAFE scale-ups.

> CROs and CMOs have found that educating some clients about SAFE and efficient operations on scale is not always easy.

The optimal route in general should be in place for the preparation of material for Phase 3 clinical trials, and should be considered for Phase 2 trials. Since the cost of re-registering a process is significant, perhaps $500,000–$1,000,000, the optimal process should be defined before registration. "Freezing the last step" may be a good way to prepare material for developing an NCE, allowing future optimization without having to submit material again for toxicology screening. In this approach the penultimate intermediate and chemistry to convert the penultimate to the API candidate are declared the means for making all future batches of the NCE and API. Only the preparation of the penultimate may be changed. If the impurity profiles of batches of the penultimate intermediate can be suitably controlled, the quality of API produced may be similar to or better than that of the batch initially submitted for toxicology screening. Nonetheless sponsors should anticipate that resubmitting material for toxicology screening is likely to be necessary, as impurity profiles change with process development successes. Grabowski has discussed the timing on route optimization [125].

Developing a route that has biologically active ingredients only at the very end of the route makes processing easier, as the need to contain the biologically active ingredients is reduced. Converting a step to GMP status could triple the production cost, so it is important to define the FDA starting materials [126].

Investing funds early in the drug development program to establish the optimal final form may save resources overall. Stability of the NCE is a very important issue. Screening to identify the optimal polymorph (Chapter 13) may be deferred until the preparation of material for Phase 2 clinical trials.

In identifying the best route many approaches have to be tried to reduce theory to best practice. In early process research efforts, precious little material may be available for experiments, and using model compounds may be tempting. Unfortunately subtle changes may have effects that are not so subtle. For instance, DeShong's group found that changing from a hydrogen to a methyl group changed the stereochemistry of the product from a Mitsunobu reaction (Figure 3.24) [127]. To paraphrase R. B. Woodward, the best model compound may be the undesired enantiomer. Route scouting is best undertaken on the desired starting material.

TABLE 3.10 Considerations for Early Route Selection

	Examples/comments
Use established transformations, especially those known to work well on scale	In-house knowledge *Organic Process Research & Development* *Organic Syntheses* *Practical Synthetic Organic Chemistry;* *Reactions, Principles, and Techniques;* Caron, S., Ed.; Wiley; **2011** Biocatalysis using commercial enzymes
Position high-yielding steps at end of route	Decrease burden of purifying product Ester hydrolysis Removing a Boc group
Position low-yielding steps early in route	Decrease amounts of raw materials purchased
Position steps with impurities that must be controlled at low levels 3–4 steps from the end of route (1)	Purification in subsequent steps can help purge impurities (2) Genotoxic impurities, Pd QbD may ease analyses
Outsource intermediates requiring special equipment	Highly energetic chemistry Corrosive conditions, e.g., fluorination Cryogenic temperatures ($-70\ °C$) or extremely high temperatures ($>140\ °C$)
Design convergent routes to decrease cost of materials	Shorter routes usually provide lower CoGs
Select final steps that you can conduct in your pilot plant	May allow better control of timelines
Consider chromatographic purification for early supplies (3,4)	Preparative chromatography, especially SMB, can be best route for routine manufacturing on scale (5)
Crystalline intermediates, especially the penultimate, ease purifications of product	Telescoped operations can avoid inevitable losses from isolation
Avoid reagents, solvents and starting materials that are restricted or unstable	HOBt, Et_2O (shipping restrictions) Piperidine, piperonal (DEA List I Regulated Chemicals) Materials sourced from animals (BSE) Cyanohydrins, aldehydes, acid chlorides
Telescope reactive intermediates into next step, do not isolate	SAFETY: minimize toxicological burden and hazards Genotoxic impurities Explosive compounds
Prepare preliminary COGs ("guestimate") to compare routes	Assume quantitative yields and 1 equivalent of reagents. Ignore costs of solvents and common reagents for workup

(Continued)

TABLE 3.10 Considerations for Early Route Selection—cont'd

	Examples/comments
Use outsourcing to identify vendors for attractive but expensive intermediates	Identify vendors who make products similar to your desired intermediates
"Green Chemistry" considerations	Decrease environmental impact Decrease waste and waste disposal costs Develop simple, effective work-ups

(1) Pierson, D. A.; Olsen, B. A.; Robbins, D. K.; DeVries, K. M.; Varie, D. L. Org. Process Res. Dev. **2009,** 13, 285.
(2) Anderson, N. G.; Coradetti, M. L.; Cronin, J. A.; Davies, M. L.; Gardineer, M. B.; Kotnis, A. S.; Lust, D. A.; Palaniswamy, V. A. Org. Process Res. Dev. **1997,** 1, 315.
(3) Welch, C. J.; Fleitz, F.; Antia, F.; Yehl, P.; Waters, R.; Ikemoto, N.; Armstrong, J. D., III; Mathre, D. J. Org. Process Res. Dev. **2004,** 8, 186.
(4) Welch, C. J.; Henderson, D. W.; Tschaen, D. W.; Miller, R. A. Org. Process Res. Dev. **2009,** 13, 621.
(5) Caille, S.; Boni, J.; Cox, G. B.; Faul, M. M.; Franco, P.; Khattabi, S.; Klingensmith, L. M.; Larrow, J. F.; Lee, J. K.; Martinelli, M. J.; Miller, L. M.; Moniz, G. A.; Sakai, K.; Tedrow, J. S.; Hansen, K. B. Org. Process Res. Dev. **2010,** 14, 133.

Green chemistry innovations are profitable. Responsible chemists and engineers actively address green chemistry issues in developing APIs. Advances will continue in the design and use of enzymes, because of the benefits from mild reaction conditions and avoiding protecting groups. Continuous operations offer many green advantages, including greater margins of SAFETY and improved heat transfer and mass transfer. Telescoping decreases solvent usage and labor costs, and workups offer substantial opportunities to "green up" processes. Direct isolations decrease waste (primarily solvents and inorganic salts) and decrease processing times, but some laboratory time may be necessary to develop that selectively remove impurities. SMB will be exploited more in the future. The ACS Green Chemistry Institute Pharmaceutical Roundtable has pointed out areas where greener alternatives are preferred [54]. Efficient route design will afford the greenest and most economical routes, and green considerations will be incorporated into more plant designs [3].

Process chemists and engineers must be quick to adopt new technologies [57]. In that regard, biocatalysis and membrane technology are being considered more by the

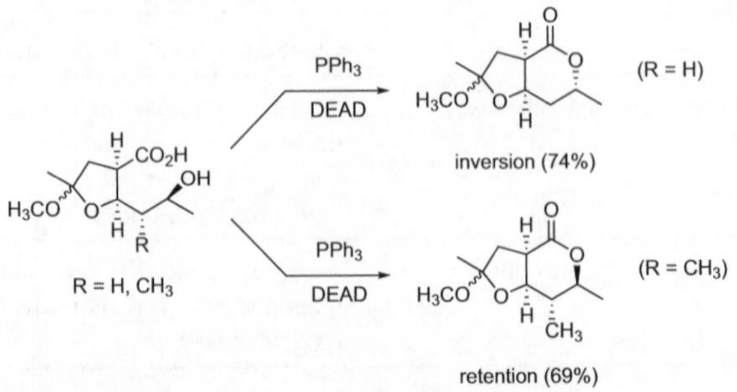

FIGURE 3.24 The "methyl effect" during a Mitsunobu reaction.

pharmaceutical and CMO industries, although they are not new to some outside of pharmaceutical and the CMO industries. Ley's group has reviewed some lesser-known technologies for synthesis and workup [128]. Electro-organic chemistry is very rarely considered in these industries, but where "un-green" oxidation reactions must be used electro-organic chemistry may become practical [129,130].

In general the preferred route will have the fewest intermediates, the fewest steps, and the fewest unit operations, and accomplish more in each step. Hudlicky termed this "brevity of design" [77]. With fewer unit operations there are fewer transfers and the risk of physical losses and contamination is reduced. A route with more steps may be preferred if the costs of the starting materials are significantly lower than the attendant labor costs. In trying to assess the best route from an analogous literature transformation, reviewing the details of operations may be most helpful to identify inefficient operations and those with SAFETY concerns. Often simple manipulations, such as removal of an acetate protecting group, will be reserved for the final or penultimate step. Many of these limitations become clear by reviewing the experimental procedures found in a manuscript or in its supporting information.

Many factors must be weighed to select a route. The Kepner–Tregoe Decision Analysis for route selection may help; these evaluations help to identify key decision points by classifying objectives into "musts" and "wants," and summing the multiplied weightings to evaluate routes [131]. As time changes the weighting factors, routes may also change. For instance, many creative routes to oseltamivir phosphate (Figure 3.25) [132–136] have been compared [137]. The chirality and the all-*trans* substitution on the cyclohexene ring of this densely functionalized molecule are evident. Conditions must be chosen carefully to avoid aromatizing the cyclohexene ring. Not surprisingly, in many routes to this molecule assembling the ether is one of the lowest-yielding steps; installing this group early in the route may minimize the downstream use of solvents and reagents, and thus the amount of waste overall. Early approaches to this molecule used azide chemistry to insert the amine groups onto the carbon skeleton derived from shikimic acid. Low-molecular weight azides may be prone to explosive release of N_2 (see Chapter 2); Roche authors warned readers that the two isomeric azides should not be handled neat above 40 °C nor in solution above

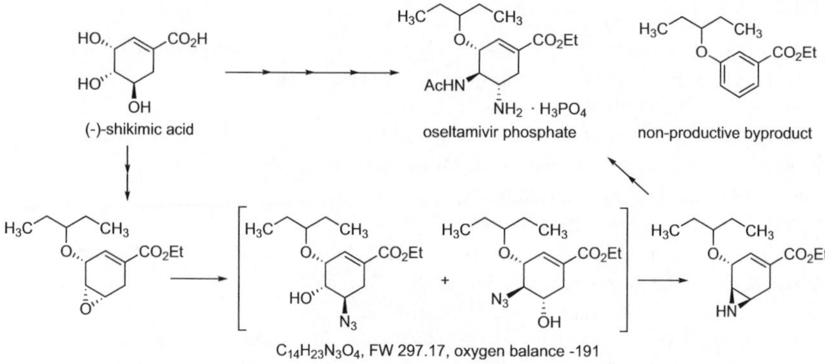

FIGURE 3.25 Preparation of oseltamivir phosphate.

70 °C [133]. Installation of the azide functionality often involved treatment with a combination of NaN_3 and acids such as NH_4Cl; additional care is needed to protect personnel from the volatile, toxic hydrazoic acid (bp 37 °C) formed during processing. (Note: PTC under basic conditions can provide facile displacement by azide, as discussed in Chapter 4. The Sharpless group has also shown that azides can be used under neutral conditions in water to form tetrazoles [138]. In later processes oseltamivir was manufactured by reaction of NaN_3 under neutral conditions [134].) To avoid the hazard of generating alkyl azides the amine functionalities were installed using NH_3 (or ammonia derivatives), t-BuNH$_2$, allylamine [139], L-serine or a pyridine [140]. Cyclohexenes have been prepared through a Diels–Alder reaction using *de novo*-designed enzymes [141], making this approach tantalizing for preparing starting materials and products by fermentation. Many routes to oseltamivir phosphate were examined in part due to the limited availability and variable quality of shikimic acid derived from Chinese star anise. The shortest route (eight steps, 30% overall yield) does not involve shikimic acid, but employs a relatively uncommon protecting group and two oxidations, and may be patented [142]. Almost all of these routes proceed through epoxide and aziridine intermediates, which are subject to additional handling concerns as PGIs or reactive intermediates; reactions involving these intermediates are probably telescoped, both to minimize the risk of exposure and to increase the overall yield. CMOs experienced with energetic chemistry have manufactured oseltamivir intermediates using azides. (–)-Shikimic acid, a biosynthetic precursor to the aromatic amino acids L-phenylalanine, L-tyrosine, and L-tryptophan, eventually became available in ton quantities through fermentation or extraction from star anise [134]. Andraos used green metric calculations to comprehensively examine 12 approaches to oseltamivir phosphate. He pointed out the importance of both defining "readily available starting materials" and disclosing all materials used and energy requirements for processing in experimental details. Andraos concluded that "optimization is a multivariable problem" and that "only through the exploration of several routes [will] the true optimum be found" [137]. Selecting the optimal route involves many considerations, which change with time and the pressures to bring a compound to market.

REFERENCES

1. Blakemore, P. R.; White, J. D. Synthesis of Polycavernoside A. In *Strategies and Tactics in Organic Synthesis*; Vol. 6, Harmata, M., Ed.; Elsevier, 2005; Chapter 6, pp 207–208.
2. *Chem. Eng. News* **2007**, *85*(32), 19.
3. Sanderson, K. *Nature* **2011**, *469*, 18.
4. Baran, P. S.; Maimone, T. J.; Richter, J. M. *Nature* **2007**, *446*, 404.
5. Ritter, S. K. *Chem. Eng. News* **2010**, *88*(43), 45.
6. Miller, S. *J. Acc. Chem. Res.* **2004**, *37*, 601.
7. Lewis, C. A.; Gustafson, J. L.; Chiu, A.; Balsells, J.; Pollard, D.; Murry, J.; Reamer, R. A.; Hansen, K. B.; Miller, S. *J. Am. Chem. Soc.* **2008**, *130*, 16358.
8. Jones, S. B.; Simmons, B.; Mastracchio, A.; MacMillan, D. W. C. *Nature* **2011**, *475*, 183.
9. Hansen, K. B.; Hsiao, Y.; Xu, F.; Rivera, N.; Clausen, A.; Kubryk, M.; Krska, S.; Rosner, T.; Simmons, B.; Balsells, J.; Ikemoto, N.; Sun, Y.; Spindler, F.; Malan, C.; Grabowski, E. J. J.; Armstrong, J. D., III. *J. Am. Chem. Soc.* **2009**, *131*, 8798.

10. Savile, C. K.; Janey, J. M.; Mundorff, E. C.; Moore, J. C.; Tam, S.; Jarvis, W. R.; Colbeck, J. C.; Krebber, A.; Fleitz, F. J.; Brands, J.; Devine, P. N.; Huisman, G. W.; Hughes, G. J. *Science* **2010**, *329*, 305.

11. http://www.epa.gov/gcc/pubs/docs/award_recipients_1996_2011.pdf

12. Parker, J. S.; Smith, N. A.; Welham, M. J.; Moss, W. O. *Org. Process Res. Dev.* **2004**, *8*, 45.

13. Mundla, S. U.S. Patent 2010/0120854, to Lilly.

14. Frederick C. Williams. Personal communication.

15. Harrington, P. J. *Pharmaceutical Process Chemistry for Synthesis; Rethinking the Routes to Scale-Up*; Wiley-VCH, 2011.

16. Kleemann, A.; Engel, J.; Kutscher, B.; Reichert, D. *Pharmaceutical Substances: Syntheses, Patents, Applications*, 5th ed.; Thieme, 2008, http://www.thieme-connect.com

17. http://www.row2technologies.com

18. Butters, M.; Catterick, D.; Craig, A.; Curzons, A.; Dale, D.; Gillmore, A.; Green, S. P.; Marziano, I.; Sherlock, J.-P.; White, W. *Chem. Rev.* **2006**, *106*, 3002.

19. http://www.p2pays.org/ref/09/08147.htm; http://www.dec.ny.gov/public/1096.html?showprintstyles

20. Deutsch, C. H. Together at Last: Cutting Pollution and Making Money. *The New York Times*. Sept. 9, 2001.

21. Cue, B. W., Jr. *Chem. Eng. News* **2005**, *83*(39), 46.

22. Hogue, C. *Chem. Eng. News* **2010**, *88*(43), 42.

23. Matlack, A. S. *Introduction to Green Chemistry*, 2nd ed.; CRC Press: Boca Raton, 2010.

24. Afonso, C. A. M.; Crespo, J. P. S. G., Eds.; *Green Separation Processes*; Wiley: New York, 2005.

25. Tao, J.; Kazlaukas, R. J., Eds.; *Biocatalysis for Green Chemistry and Chemical Process Development*; Wiley-VCH, 2011.

26. Jiménez-González, C.; Constable, D. J. C. *Green Chemistry and Engineering: A Practical Design Approach*; Wiley, 2011.

27. Dunn, P., Wells, A.; Williams, M. T., Eds.; *Green Chemistry in the Pharmaceutical Industry*; Wiley-VCH, 2010.

28. Leitner, W., Jessop, P. G., Li, C. J., Wasserscheid, P.; Stark, A., Eds.; *The Handbook of Green Chemistry – Green Solvents*; Wiley-VCH, 2010.

29. Colberg, J. C.; Pfisterer, D. M.; Taber, G. P. U. S. Patent 6,232,500, 2001 (to Pfizer); Rouhi, A. M. *Chem. Eng. News* **2002**, *80*(16), 30. Taber, G. P.; Pfisterer, D. M.; Colberg, J. C. *Org. Process Res. Dev.* **2004**, *8*, 385.

30. Hawkins, J. M.; Watson, T. J. N. *Angew. Chem. Int. Ed. Eng.* **2004**, *43*, 3224.

31. Nelson, T. D. In *Strategies and Tactics in Organic Synthesis*; Vol. 6, Harmata, M., Ed.; Elsevier: Amsterdam, 2005; p 321.

32. Leresche, J. E.; Meyer, H.-P. *Org. Process Res. Dev.* **2006**, *10*, 572.

33. Chuck, R., *Green Sustainable Chemistry in the Production of Nicotinates*, presented at the First International IUPAC Conference on Green-Sustainable Chemistry, 21.09.2006. http://www.lonza.com/group/en/company/news/publications_of_lonza.-ParSys-0002-ParSysdownloadlist-0038-DownloadFile.pdf/Chuck1.pdf

34. Meyer, H.-P.; Ghisalba, O.; Leresche, J. E. In *Green Catalysis*; Vol. 3, Anastas, P. T., Ed.; Wiley-VCH: Weinheim, 2009; pp 171–212.

35. Li, B.; Su, J.; Tao, J. *Org. Process Res. Dev.* **2011**, *15*, 291.

36. Venkat, K. http://www.iupac.org/symposia/proceedings/phuket97/venkat.html

37. Tabata, H. In *Advances in Biochemical Engineering/Biotechnology*; Vol. 87, 2004; p 1.

38. Mountford, P. G. In *Green Chemistry in the Pharmaceutical Industry*; Dunn, P., Wells, A.; Williams, M. T., Eds.; Wiley-VCH, 2010; p 145.

39. Williams, M. *Chem. Eng. News* **2010**, *88*(45), 8.

40. Holton, R. A.; Kim, H. B.; Somoza, C.; Liang, F.; Biediger, R. J.; Boatman, P. D.; Shindo, M.; Smith, C. C.; Kim, S.; Nadizadeh, H.; Suzuki, Y.; Tao, C.; Vu, P.; Tang, S.; Zhang, P.; Murthi, K. K.; Gentile, L. N.; Liu, J. H. *J. Am. Chem. Soc.* **1994**, *116*, 1599.

41. Nicolaou, K. C.; Ueno, H.; Liu, J.-J.; Nantermet, P. G.; Yang, Z.; Renaud, J.; Paulvannan, K.; Chadha, R. *J. Am. Chem. Soc.* **1995,** *117,* 653; and references therein.

42. Danishefsky, S. J.; Masters, J. J.; Young, W. B.; Link, J. T.; Snyder, L. B.; Magee, T. V.; Jung, D. K.; Isaacs, R. C. A.; Bornmann, W. G.; Alaimo, C. A.; Coburn, C. A.; Di Grandi, M. J. *J. Am. Chem. Soc.* **1996,** *118,* 2843.

43. Wender, P. A.; Badham, N. F.; Conway, S. P.; Floreancig, P. E.; Glass, T. E.; Houze, J. B.; Krauss, N. E.; Lee, D.; Marquess, D. G.; McGrane, P. L.; Meng, W.; Natchus, M. G.; Shuker, A. J.; Sutton, J. C.; Taylor, R. E. *J. Am. Chem. Soc.* **1997,** *119,* 2757.

44. Tufvesson, P.; Lima-Ramos, J.; Nordblad, M.; Woodley, J. M. *Org. Process Res. Dev.* **2011,** *15,* 266.

45. Trost, B. M. *Angew. Chem. Int. Ed. Eng.* **1995,** *34,* 259.

46. Sheldon, R. A. *Green Chem.* **2007,** *9,* 1273.

47. Andraos, J. *Org. Process Res. Dev.* **2005,** *9,* 149; Andraos, J. *Org. Process Res. Dev.* **2005,** *9,* 404; Andraos, J. *Org. Process Res. Dev.* **2006,** *10,* 212.

48. Andraos, J.; Izhakova, J. *Chimica Oggi/Chemistry Today* **2007,** *24*(6), 31.

49. Anderson, N. G. *Org. Process Res. Dev.* **2008,** *12,* 1019.

50. Blaser, H.-U.; Eissen, M.; Fauquex, P. F.; Hungerbühler, K.; Schmidt, E.; Sedelmeier, G.; Studer, M. In *Asymmetric Catalysis on Industrial Scale*; Blaser, H. U.; Schmidt, E., Eds.; Wiley-VCH: Weinheim, 2004; p 91.

51. Kumara Swamy, K. C.; Bhuvan Kumar, N. N.; Balaraman, E.; Pavan Kumar, K. V. P. *Chem. Rev.* **2009,** *109,* 2551.

52. Anderson, N. G.; Lust, D. A.; Colapret, K. A.; Simpson, J. H.; Malley, M. F.; Gougoutas, J. Z. *J. Org. Chem.* **1996,** *61,* 7955.

53. Bellingham, R.; Buswell, A. M.; Choudary, B. M.; Gordon, A. H.; Moore, S. O.; Peterson, M.; Sasse, M.; Shamji, A.; Urquhart, M. W. J. *Org. Process Res. Dev.* **2010,** *14,* 1254.

54. Constable, D. J. C.; Dunn, P. J.; Hayler, J. D.; Humphrey, G. R.; Leazer, J. L., Jr.; Linderman, R. J.; Lorenz, K.; Manley, J.; Pearlman, B. A.; Wells, A.; Zaks, A.; Zhang, T. Y. *Green Chem.* **2007,** *9,* 411.

55. Thaper, R. K.; Kumar, Y.; Kumar, S. M. D.; Misra, S.; Khanna, J. M. *Org. Process Res. Dev.* **1999,** *3,* 476.

56. Hull, J. W., Jr.; Romer, D. R.; Adaway, T. J.; Podhorez, D. E. *Org. Process Res. Dev.* **2009,** *13,* 1125.

57. Davies, I. W.; Welch, C. J. *Science* **2009,** *325,* 701.

58. Pierson, D. A.; Olsen, B. A.; Robbins, D. K.; DeVries, K. M.; Varie, D. L. *Org. Process Res. Dev.* **2009,** *13,* 285.

59. Ripin, D. H. B.; Bourassa, D. E.; Brandt, T.; Castaldi, M. J.; Frost, H. N.; Hawkins, J.; Johnson, P. J.; Massett, S. S.; Neumann, K.; Phillips, J.; Raggon, J. W.; Rose, P. R.; Rutherford, J. L.; Sitter, B.; Stewart, A. M., III; Vetelino, M. G.; Wei, L. *Org. Process Res. Dev.* **2005,** *9,* 440.

60. Wu, X.; Park, P. K.; Danishefsky, S. J. *J. Am. Chem. Soc.* **2011,** *133,* 7700.

61. Guillaume, M.; Cuypers, J.; Vervest, I.; De Smaele, D.; Leurs, S. *Org. Process Res. Dev.* **2003,** *7,* 939.

62. Wang, T.; Zhang, Z.; Meanwell, N. A. *J. Org. Chem.* **2000,** *65,* 4740.

63. Arnold, L. A.; Naasz, R.; Minaard, A. J.; Feringa, B. L. *J. Org. Chem.* **2002,** *67,* 7244.

64. Masui, Y.; Kitaura, Y.; Kobayashi, T.; Goto, Y.; Ando, S.; Okuyama, A.; Takahashi, H. *Org. Process Res. Dev.* **2003,** *7,* 334.

65. Jass, P. A.; Rosso, V. W.; Racha, S.; Soundararajan, N.; Venit, J. J.; Rusowicz, A.; Swaminathan, S.; Livshitz, J.; Delaney, E. J. *Tetrahedron* **2003,** *59,* 9019.

66. Oh, L. M.; Wang, H.; Shilcrat, S. C.; Herrmann, R. E.; Patience, D. B.; Spoors, P. G.; Sisko, J. *Org. Process Res. Dev.* **2007,** *11,* 1032.

67. Greene, T. W.; Wuts, P. G. M. *Protective Groups in Organic Synthesis*, 3rd ed.; Wiley: New York, 1999; p 161.

68. These quaternary ammonium salts were used in phase-transfer reactions: Halpern, M. "PTC Tip of the Month" for June 2011. http://phasetransfer.com

69. Dunn, P. J.; Galvin, S.; Hettenbach, K. *Green Chem.* **2004,** *6,* 43.

70. Dale, D. J.; Dunn, P. J.; Golightly, C.; Hughes, M. L.; Levett, P. C.; Pearce, A. K.; Searle, P. M.; Ward, G.; Wood, A. S. *Org. Process Res. Dev.* **2000,** *4,* 17.

71. Dunn, P. J. In *Process Chemistry in the Pharmaceutical Industry*; Gadamasetti, K., Braish, T., Eds, 2nd ed.; CRC Press: Boca Raton, Florida, 2008; p 267.

72. Huang, Q.; Richardson, P. F.; Sach, N. W.; Zhu, J.; Liu, K. K.-C.; Smith, G. L.; Bowles, D. M. *Org. Process Res. Dev.* **2011,** *15,* 556.

73. Anderson, N. G.; Ary, T. D.; Berg, J. L.; Bernot, P. J.; Chan, Y. Y.; Chen, C.-K.; Davies, M. L.; DiMarco, J. D.; Dennis, R. D.; Deshpande, R. P.; Do, H. D.; Droghini, R.; Early, W. A.; Gougoutas, J. Z.; Grosso, J. A.; Harris, J. C.; Haas, O. W.; Jass, P. A.; Kim, D. H.; Kodersha, G. A.; Kotnis, A. S.; LaJeunesse, J.; Lust, D. A.; Madding, G. D.; Modi, S. P.; Moniot, J. L.; Nguyen, A.; Palaniswamy, V.; Phillipson, D. W.; Simpson, J. H.; Thoraval, D.; Thurston, D. A.; Tse, K.; Polomski, R. E.; Wedding, D. L.; Winter, W. J. *Org. Process Res. Dev.* **1997,** *1,* 300.

74. Sledeski, A. W.; Kubiak, G. G.; O'Brien, M. K.; Powers, M. R.; Powner, T. H.; Truesdale, L. K. *J. Org. Chem.* **2000,** *65,* 8114.

75. Abdel-Magid, A. F.; Cohen, J. H.; Maryanoff, C. A.; Shah, R. D.; Villani, F. J.; Zhang, F. *Tetrahedron Lett.* **1998,** *39,* 3391.

76. Turner, S. *The Design of Organic Synthesis*; Elsevier: New York, 1976; p 71.

77. Hudlicky, T. *Chem. Rev.* **1996,** *96,* 3.

78. Morse, P. M. *Chem. Eng. News* **2011,** *89*(22), 28.

79. Crabb, C. *Chem. Eng.* **2000,** *107*(5), 26.

80. Ruland, Y.; Colin, H.; David, L. *Biopharm. J.* **2011,** *3,* 4. http://submit.biopharmj.ru/ojs/index.php/biopharmj/article/view/85/pdf_125

81. Caille, S.; Boni, J.; Cox, G. B.; Faul, M. M.; Franco, P.; Khattabi, S.; Klingensmith, L. M.; Larrow, J. F.; Lee, J. K.; Martinelli, M. J.; Miller, L. M.; Moniz, G. A.; Sakai, K.; Tedrow, J. S.; Hansen, K. B. *Org. Process Res. Dev.* **2010,** *14,* 133.

82. Crane, Z. D.; Nichols, P. J.; Samakia, T.; Stengel, P. J. *J. Org. Chem.* **2011,** *76,* 277.

83. Tilstam, U.; Weinmann, H. *Org. Process Res. Dev.* **2002,** *6,* 384.

84. McDermott, T. S.; Bhagavatula, L.; Borchardt, T. B.; Engstorm, K. M.; Gandarilla, J.; Kotecki, B. J.; Kruger, A. W.; Rozema, M. J.; Sheikh, A. Y.; Wagaw, S. H.; Wittenberger, S. J. *Org. Process Res. Dev.* **2009,** *13,* 1145.

85. Reilly, M.; Anthony, D. R.; Gallagher, C. *Tetrahedron Lett.* **2003,** *44,* 2927.

86. Couturier, C.; Blanchet, J.; Schlama, T.; Zhu, J. *Org. Lett.* **2006,** *8,* 2183.

87. Znabet, A.; Polak, M. M.; Janssen, E.; de Kanter, F. J. J.; Turner, N. J.; Orru, R. V. A.; Ruijtr, E. *Chem. Commun.* **2010,** *46,* 7918.

88. Whittall, J.; Sutton, P. *Practical Methods for Biocatalysis and Biotransformations*; Wiley, 2010.

89. Polaina, J.; MacCabe, A. P., Eds.; *Industrial Enzymes: Structure, Function and Applications*; Springer, 2010.

90. Ritter, S. K. *Chem. Eng. News* **2011,** *89*(7), 11.

91. Arnaud, C. *Chem. Eng. News* **2010,** *88*(5), 11.

92. http://en.sanofi.com/press/press_releases/2008/ppc_14642.asp

93. Jarvis, L. *Chem. Eng. News* **2008,** *86*(10), 15.

94. Bedini, E.; De Castro, C.; De Rosa, M.; Di Nola, A.; Iadonisi, A.; Restaino, O. F.; Schiraldi, C.; Parrilli, M. *Angew. Chem. Int. Ed.* **2011,** *50,* 6160; Ritter, S; K. *Chem. Eng. News* **2011,** *89*(23), 38.

95. Ritter, S. K., http://pubs.acs.org/cen/news/89/i26/8926notw1c.html

96. Patel, R. N.; Banerjee, A.; Nanduri, V. B.; Goldberg, S. L.; Johnston, R. M.; Tully, T. P.; Szarka, L. J.; Swaminathan, S.; Venit, J. J.; Moniot, J. L.; Anderson, N. G.; Lust, D. A.; Crispino, G.; Srivastava, S. K. US Patent 6,515,170, 2003. (to Bristol–Myers Squibb).

97. Hanson, R. L. Stereoselective Enzymatic Synthesis of Intermediates Used for Antihypertensive, Antiinfective, and Anticancer Compounds. In *Process Chemistry in the Pharmaceutical Industry, Vol. 2:*

Challenges in an Ever Changing Climate; Gadamasetti, K.; Braish, T., Eds.; CRC Press: Boca Raton, 2008; pp 279–294.

98. Liang, J.; Lalonde, J.; Borup, B.; Mitchell, V.; Mundorff, E.; Trinh, N.; Kochrekar, D. A.; Cherat, R. N.; Pai, G. G. *Org. Process Res. Dev.* **2010**, *14*, 193.

99. DeSantis, G.; Zhu, Z.; Greenberg, W. A.; Wong, K.; Chaplin, J.; Hanson, S. R.; Farwell, B.; Nicholson, L. W.; Rand, C. L.; Weiner, D. P.; Robertson, D. E.; Burk, M. J. *J. Am. Chem. Soc.* **2002**, *124*, 9024.

100. 2006 Presidential Green Chemistry Award to Codexis for directed evolution of three enzymes to make hydroxy nitrile. http://www.epa.gov/gcc/pubs/docs/award_recipients_1996_2011.pdf

101. Bergeron, S.; Chaplin, D. A.; Edwards, J. H.; Ellis, B. S. W.; Hill, C. L.; Holt-Tiffin, K.; Knight, J. R.; Mahoney, T.; Osborne, A. P.; Ruecroft, G. *Org. Process Res. Dev.* **2006**, *10*, 661.

102. Huisman, G. W. *Society of Chemical Industry Symposium*, Cambridge, UK, December 6, 2007.

103. Hu, S.; Martinez, C. A.; Kline, B.; Yazbeck, D.; Tao, J.; Kucera, D. J. *Org. Process Res. Dev.* **2006**, *10*, 650.

104. http://www.dsm.com/en_US/downloads/dpp/RSS_PLE.pdf

105. Tao, J.; Zhao, L.; Ran, N. *Org. Process Res. Dev.* **2007**, *11*, 259.

106. Jasper, J. P.; Weaner, L. E.; Hayes, J. M. *Pharma. Technol.* **2007**, *31*, 68.

107. Buchanan, H. S. A.; Daéid, N. N.; Meier-Augenstein, W.; Kemp, H. F.; Kerr, W. J.; Middleditch, M. *Anal. Chem.* **2008**, *80*, 3350.

108. Everts, S. *Chem. Eng. News* **2011**, *89*(26), 32.

109. Arnaud, C. *Chem. Eng. News* **2010**, *88*(38), 24.

110. Jacoby, M. *Chem. Eng. News* **2010**, *88*(51), 35.

111. Schultis, K. in Thayer, A. M. *Chem. Eng. News* **2011**, *89*(5), 13. http://www.rondaxe.com/software

112. http://www.aspentech.com

113. Henry, C. M. *Chem. Eng. News* **2002**, *80*(22), 53.

114. A sample spreadsheet was shown in *Practical Process Research & Development*; 1st ed.; p 49.

115. One resource is the Directory of Worldwide Chemicals Producers. http://www.chemicalinfo.com/dwcp

116. Prices may increase as specifications tighten. (Thanks to Mike Standen for mentioning this.)

117. Bray, B. *Nat. Rev. Drug Disc.* **2003**, *2*, 587; July.

118. Thayer, A. M. *Chem. Eng. News* **2011**, *89*(22), 21.

119. Wang, P.; Ouellette, M. Personal communication.

120. Lu, J.; Shinkai, I. In *Process Chemistry in the Pharmaceutical Industry*; Gadamasetti, K. G.; Braish, T., Eds, 2nd ed.; CRC Press: Boca Raton, Florida, 2008; p 465.

121. Robins, D.; Hannon, S. In *Process Chemistry in the Pharmaceutical Industry*; Gadamasetti, K. G.; Braish, T., Eds, 2nd ed.; CRC Press: Boca Raton, Florida, 2008; p 471.

122. Anderson, N. G.; Ciaramella, B. M.; Feldman, A. F.; Lust, D. A.; Moniot, J. L.; Moran, L.; Polomski, R. E.; Wang, S. S. Y. *Org. Process Res. Dev.* **1997**, *1*, 211.

123. Anderson, N. G. *Org. Process Res. Dev.* **2004**, *8*, 260.

124. Becker, C. L.; Engstrom, K. M.; Kerdesku, F. A.; Tolle, J. C.; Wagaw, S. H.; Wang, W. *Org. Process Res. Dev.* **2008**, *12*, 1114.

125. Grabowski, E. J. J. In *Fundamentals of Early Clinical Drug Development*; Abdel-Magid, A. F.; Caron, S., Eds.; Wiley: Hoboken, NJ, 2006; p 1.

126. ICH Q7A Good Manufacturing Practices for Active Pharmaceutical Ingredients, 2001. http://www.fda.gov/downloads/Drugs/GuidanceComplianceRegulatoryInformation/Guidances/ucm073497.pdf.

127. Anh, C.; DeShong, P. *J. Org. Chem.* **2002**, *67*, 1754.

128. O'Brien, M.; Denton, R.; Ley, S. V. *Synthesis* **2011**, 1157.

129. Fry, A. J. *Synthetic Organic Electrochemistry*, 2nd ed.; Wiley, 1989.

130. Shono, T. *Electroorganic Synthesis*, Academic: San Diego, 1991.

131. Parker, J. S.; Moseley, J. D. *Org. Process Res. Dev.* **2008**, *12*, 1041; Moseley, J. D.; Brown, D.; Firkin, C. R.; Jenkin, S. L.; Patel, B.; Snape, E. W. *Org. Process Res. Dev.* **2008**, *12*, 1044; Parker, J. S.; Bower, J. F.; Murray, P. M; Patel, B.; Talavera, P. *Org. Process Res. Dev.* **2008**, *12*, 1060.

132. Johnson, D. S.; Lie, J. J. In *The Art of Drug Synthesis*; Johnson, D. S.; Lie, J. J., Eds.; Wiley: NewYork, 2007; Chapter 7, p 95.

133. Shibasaki, M.; Kanai, M. *Eur. J. Org. Chem.* **2008,** 1839.

134. Karpf, M. In *Pharmaceutical Process Chemistry*; Shiori, T., Izawa, K.; Konoike, T., Eds.; Wiley-VCH, 2010; p 1.

135. Karpf, M.; Trussardi, R. *Angew. Chem. Int. Ed. Eng.* **2009,** *121*, 5760.

136. Weng, J.; Li, Y.-B.; Wang, R.-B.; Li, F. Q.; Liu, C.; Chan, A. S. C; Lu, G. *J. Org. Chem.* **2010,** 75, 3125.

137. Andraos, J. *Org. Process Res. Dev.* **2009,** *13*, 161.

138. Demko, Z. P.; Sharpless, K. B. *J. Org. Chem.* **2001,** *66*, 7945.

139. Harrington, P. J.; Brown, J. D.; Foderaro, T.; Hughes, R. C. *Org. Process Res. Dev.* **2004,** *8*, 86.

140. Kipassa, N. T.; Okamura, H.; Kina, K.; Hamada, T.; Tetsuo Iwagawa, T. *Org. Lett.* **2008,** *10*, 815.

141. Siegel, J. B.; Zanghellini, A.; Lovick, H. M.; Kiss, G.; Lambert, A. R.; St. Clair, J. L.; Gallaher, J. L.; Hilvert, D.; Gelb, M. H.; Stoddard, B. L.; Houk, K. N.; Michael, F. E.; Baker, D. *Science* **2010,** *329*(5989), 309.

142. Trost, B. M.; Zhang, T. *Angew. Chem. Int. Ed. Eng.* **2008,** *47*, 1.

Reagent Selection

"Ironically, I moved into the area of organocatalysis, i.e., the phosphine-catalyzed reactions, in 1991 by accident…when the transition metal was omitted, the reaction still underwent the same transformation to conjugated electron-deficient 1,3-alkadienes under catalysis by the phosphine ligand(s) alone…. I believe that due to the diversity of metals and reactive organocatalysts, more and more new reactions with new mechanisms will be discovered both by accident and delicate conceptual design."

– Prof. Xiyan Lu [1]

"Liquid smoke is made by channeling smoke from smoldering wood chips through a condenser, which quickly cools the vapors, causing them to liquefy…. Curious about the manufacturing process for this product, we wondered if we could bottle up some for ourselves. To do this we created a small-scale mock-up of the commercial method, involving a kettle grill, a duct fan, a siphon, and an ice-chilled glass coil condenser. In a comparison of homemade and store-bought liquid smoke, homemade was praised for its clean, intense smoky flavor. But we spent an entire day and $50 on materials to make 3 tablespoons of liquid smoke. Commercial liquid smoke is just fine…"

– J. Kenji Alt [2]

I. INTRODUCTION

New reagents with improved chemoselectivity continue to be developed, by accident and by theory. The most useful reagents combine high chemoselectivity with minimal impacts on SAFETY, health, the environment, and cost. Sometimes "old-fashioned"

Practical Process Research and Development. DOI: 10.1016/B978-0-12-386537-3.00004-6

reagents provide an expeditious and economical approach that is more practical than the current, intellectually stimulating reactions and reagents. In other cases considerable time and money can be saved by purchasing a key reagent instead of generating it in-house. The ideal characteristics of reagents (Table 4.1) change with the desired transformation, starting materials, and products.

TABLE 4.1 Some Ideal Characteristics of Reagents

Characteristic	Comments or Examples
Specific for the desired synthetic transformation	Isolation is easier from high-yielding reactions
SAFE: poses no chemical reaction hazard	Liability increases with scale
No shipping restrictions	CH_3Li in Et_2O; anhydrous HOBt
No purchasing restrictions	Time is required to obtain licenses to purchase and handle compounds such as piperonal, which could be used to make illegal drugs
Non-toxic	PPE is time-consuming, cumbersome, inconvenient
BSE-free	Not sourced from animals; for example, DSM sells recombinant porcine liver esterase (PLE)
If highly toxic, non-volatile	Dimethyl sulfate may be preferred over CH_3I
Readily available from multiple vendors	One vendor may hold up supply or raise cost
Inexpensive, with consistent quality batch to batch	Inexpensive, high-quality reagents lower CoGs
Stable, good shelf-life	Greater flexibility in scheduling manufacturing
Readily generated and used if short shelf-life	Greater flexibility in manufacturing operations
Handled without attention to special conditions, such as H_2O, O_2	Greater flexibility in manufacturing operations
Readily transferred into reactors	Greater flexibility in manufacturing operations
Facilitates work-up and isolation of pure product, or has no effect on work-up of the product	Increased productivity
Produces non-toxic byproducts	Lowers disposal costs
Catalytic, and readily recovered and reused	Lowers CoG
Requires no specialized equipment or facilities	Greater flexibility in siting manufacturing, may avoid building a specialized plant, such as one needing a hydrogenation facility

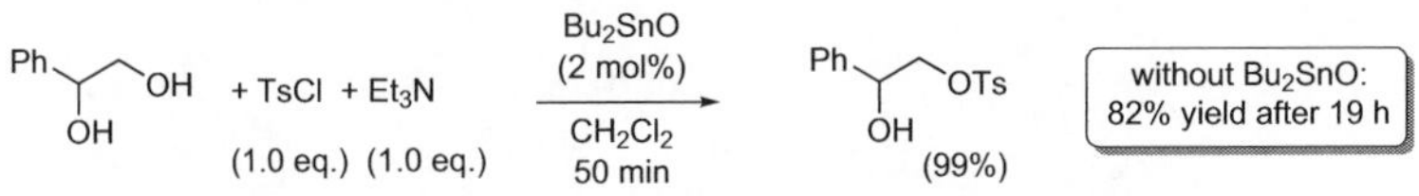

FIGURE 4.1 Tosylation in the presence of $(n\text{-Bu})_2\text{SnO}$.

Caron and co-workers have provided a collection of proven methodology for many applications [3], and this chapter will have minimal overlap. The examples presented in this chapter may discuss the power or selectivity of reagents, ease of workup, occasionally the difficulty of a workup, or other features.

The benefits of a catalyst may outweigh disadvantages. For instance, Lilly researchers found that monotosylation was very rapid in the presence of 2 mol% of $(n\text{-Bu})_2\text{SnO}$, a reagent used in carbohydrate chemistry (Figure 4.1). In the absence of the tin reagent, the yield of monotosylate was 82% after 19 h, with the bis-tosylate being the major impurity [4,5] (Figure 4.1). Although tin is a toxic metal, the benefits of a rapid, high-yielding reaction may make a reaction such as this very attractive, especially if the reaction with the tin reagent was not the last step in the sequence [6].

Balancing the equation of an intended reaction can give great insight into potential impurities and how to work up a reaction. For instance, ammonia generated from DIBAL-H reduction of a nitrile may react with reagents downstream. Greater care will be needed to work up a reaction that generates benzene or uses excess sodium azide. Balancing the equation can be extremely powerful.

The importance of screening to find optimal solvents, reagents, and leaving groups can hardly be overstated. Kurth's group found that the products from bis-alkylation of an ester depended on whether *cis*- or *trans*-1,4-dichloro-2-butene or the *cis*-1,4-dimesyloxy-2-butene was used [7] (Figure 4.2). Boehringer-Ingelheim researchers found that cyclopropanation was heavily favored using lithium bases in apolar solvents, probably through aggregation of anionic intermediates, and identified the azepine as a side product from the *cis*-cyclopropane intermediate [8]. In the *ipso*-displacement of a fluoride from an electron-rich aromatic ring Merck researchers found that potassium hexamethyldisilazide was the superior base, provided that polar aprotic solvents were not used [9]. Parallel screening of reagents and solvents can be extremely useful, especially early in the development of a process.

Polymeric reagents can afford easy workups and be easy to recycle. In the acid-mediated preparation of benzothiazoles water was found to be the best solvent for transformation of an *ortho*-amino thiol and an aldehyde (Figure 4.3). The sulfonic acid resin Dowex 50W was filtered off and could be reused [10]. Other examples of heterogeneous catalysis can be found in the discussion in Section V of Chapter 10.

The following sections highlight biocatalysis, phase-transfer catalysis (PTC), and peptide coupling. Also included are shorter sections on oxidation, reduction, and removing Boc groups. Approaches to optimizing cross-coupling reactions are discussed in Chapter 10.

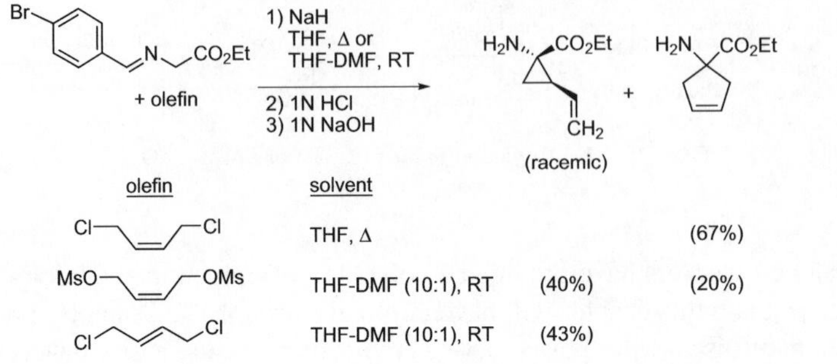

olefin	solvent		
Cl~~~Cl	THF, Δ		(67%)
MsO~~~OMs	THF-DMF (10:1), RT	(40%)	(20%)
Cl~~~Cl	THF-DMF (10:1), RT	(43%)	

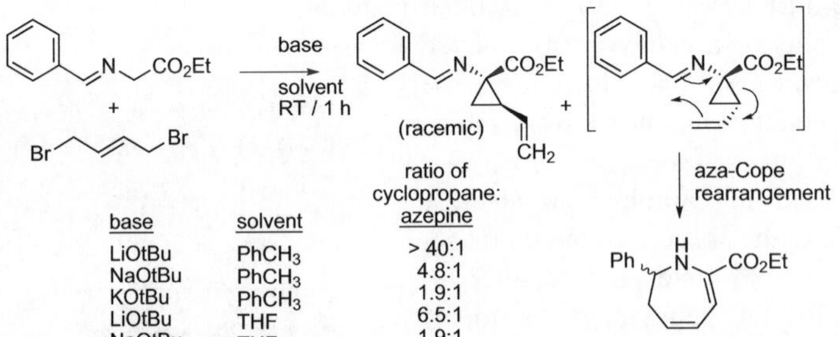

base	solvent	ratio of cyclopropane: azepine
LiOtBu	PhCH₃	> 40:1
NaOtBu	PhCH₃	4.8:1
KOtBu	PhCH₃	1.9:1
LiOtBu	THF	6.5:1
NaOtBu	THF	1.9:1
KOtBu	THF	1.2:1
LiHMDS	hexane / PhCH₃	> 40:1
NaHMDS	PhCH₃	14:1
KHMDS	PhCH₃	3.2:1

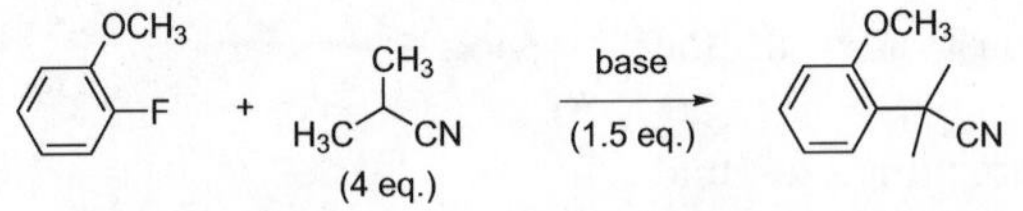

base	solvent	temperature	time (h)	yield
Cs₂CO₃	THF	75°C	24	no reaction
LDA	THF	RT	<1	decomp
tBuOK	THF	70°C	24	decomp
LiHMDS	THF	60°C	23	3%
NaHMDS	THF	60°C	23	49%
KHMDS	THF	60°C	23	95%
KHMDS	PhCH₃	60°C	18	95%
KHMDS	DMSO	75°C	24	no reaction
KHMDS	NMP	75°C	24	1%

FIGURE 4.2 The importance of screening leaving groups, reagents, and solvents.

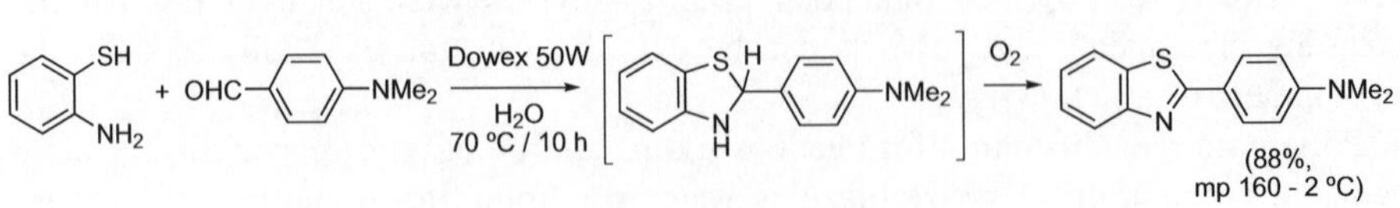

FIGURE 4.3 A reaction with a polymeric reagent.

II. BIOCATALYSIS

Perhaps the most familiar application of enzymes to those in the pharmaceutical industry is the preparation of penicillins. The enzymatic transformations shown in Figure 4.4 are highly selective for these multifunctional, rather reactive molecules. In processing the

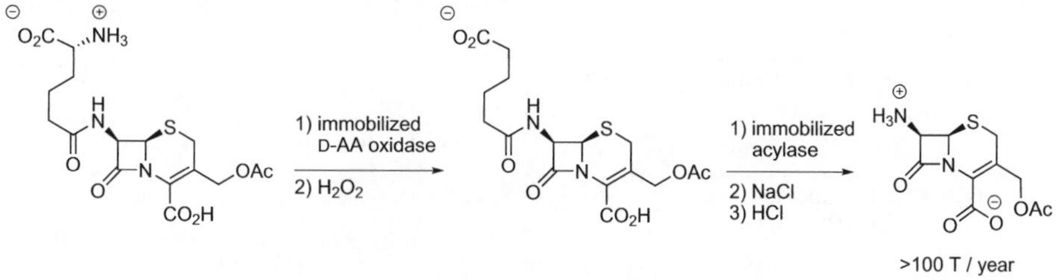

FIGURE 4.4 Enzymatic preparation of penicillins.

fermentation broth was filtered to remove the mycelia, clarified by ultrafiltration, and then purified by chromatography using XAD 16 and IRA 68 resins in all-aqueous processing. The keys for these processes were the use of recombinant strains to enhance enzymatic activity and being able to reuse the enzymes through immobilization. Telescoping eliminated the need for isolations, and organic solvents were avoided by controlling the pH [11]. The fermentation industries are well established, including the preparation of beer, wine, cheese, coffee, tea, and other commodities.

For efficient route scouting and reagent selection biocatalysis offers many opportunities, especially for chiral chemistry. Hydrolysis using lipases and esterases, estimated in 2002 to account for about 60% of biotransformations [12], is often employed for resolutions [13]. Ketoreductases (KREDs) have become increasingly utilized to reduce ketones to alcohols, and in some departments may be the method of choice. Ene reductases reduce polarized double bonds akin to the *beta*-addition of hydride to an enone. Oxidations include the epoxidation of olefins, the oxidation of ArH to ArOH, the oxidation of arenes to the corresponding dihydrocatechols, and Baeyer–Villiger oxidation of ketones to lactones. Alcohol oxidases mediate the oxidation of RCH_2OH to RCHO and allylic alcohols to enones. Laccases can remove the *p*-methoxyphenyl (PMP) group from amines or imines. Biocatalytic halogenations, even with fluoride, have been documented. Nitriles have been hydrated to primary amides by hydratases [14] or hydrolyzed to RCO_2H by nitrilases; these reactions have been used for resolution. Epoxides have been hydrolyzed, often for resolution, or formed from halohydrins. Peptidases can be used to form or hydrolyze amides. Transaminases mediate the R_2CHNH_2–ketone (or aldehyde) interconversion [15], as for sitagliptin. Other reactions with aldehydes include aldol reactions and the addition of cyanide to RCHO. Biocatalytic glycosylation can avoid the use of protecting groups in sugars. Since enzymes frequently catalyze equilibrium reactions driving biocatalytic processes to completion is important, often by removing a byproduct. Hudlicky and Reed have reviewed using biocatalytic systems for the synthesis of complex molecules [16]. Some transformations are summarized in Figure 4.5 [12,17,18]. Faber has reviewed biotransformations [19].

Enzymes have been developed that can operate under what have been considered extreme conditions. For instance, Codexis is developing carbonic anhydrase enzymes that can capture CO_2 by methyl diethanolamine at 90 °C; such enzymes could be used to sequester the byproduct CO_2 from power plants [20]. Ager, Turner, and co-workers have shown that non-specific reductants ($BH_3 \cdot NH_3$, 40 equivalents; $BH_3 \cdot t$-$BuNH_2$, 20 equivalents; or 1 M ammonium formate and 10% Pd/C) can be used in conjunction with

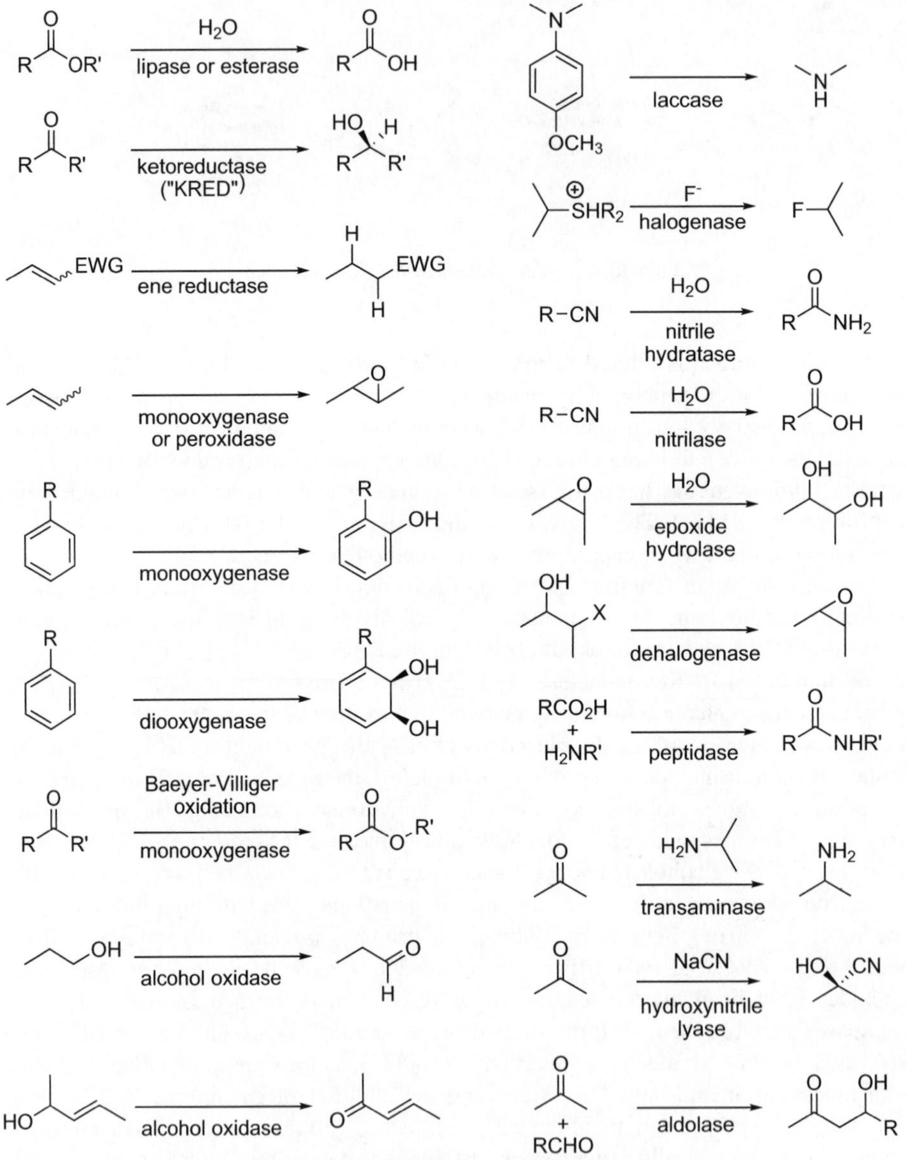

FIGURE 4.5 General biocatalytic transformations.

a specific oxidant (an *S*-selective amino acid oxidase) to resolve racemic amino acids to the (*R*)-enantiomers (Figure 4.6). The reagents were charged in large excess to prevent hydrolysis of the imine; significantly, however, the biocatalysts tolerated such concentrations of reagents that were foreign to the enzyme [21].

Codexis has described in detail some of the biocatalytic processes they have developed. Codexis actively generated amine oxidases to produce biocatalysts for the

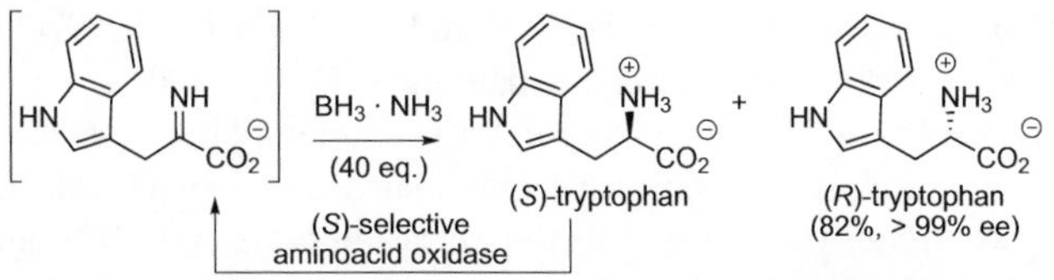

FIGURE 4.6 Biocatalytic resolution to (R)-tryptophan.

chiral bicyclic proline analogs central to boceprevir and telaprevir (Figure 4.7). The enantioselective oxidation of the *meso*-pyrrolidines to the imines proceeds through the reduction of the flavin cofactor (enz-FAD) to the reduced flavin cofactor (enz-FADH₂), and oxidation with O₂ regenerates the flavin cofactor. On scale agitation and sparging

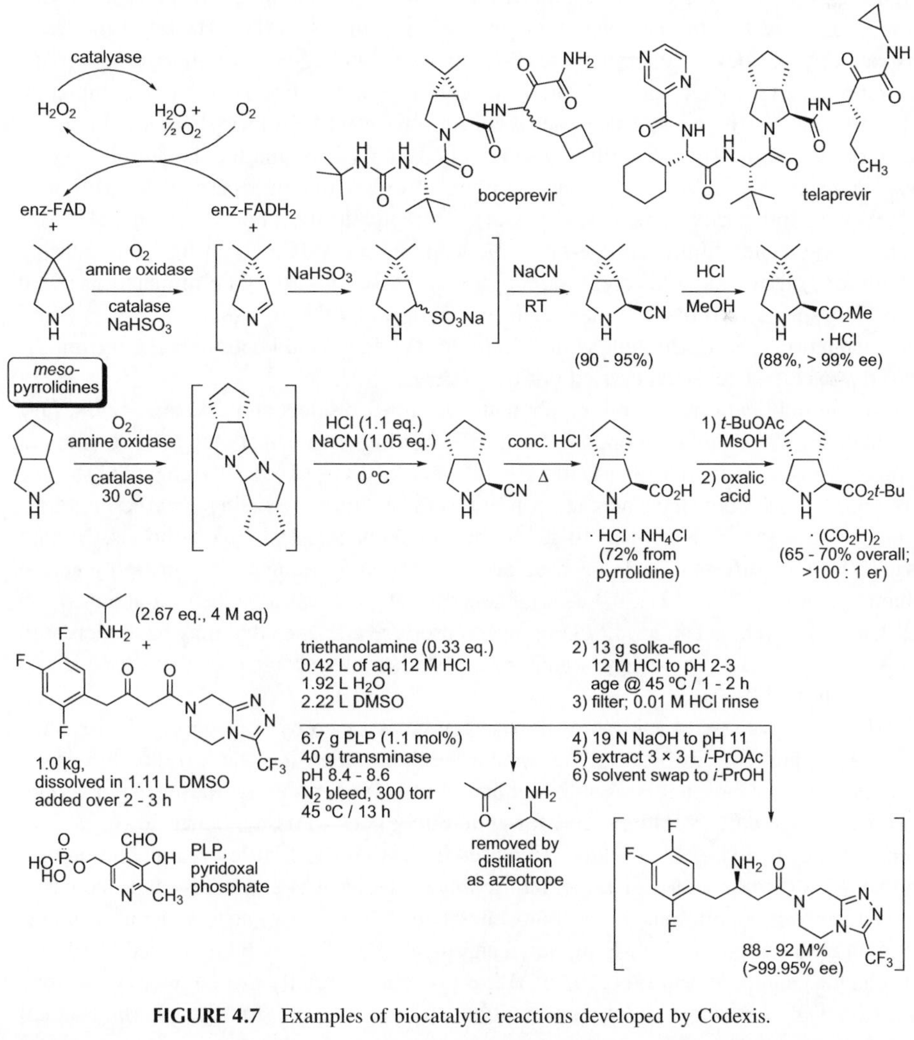

FIGURE 4.7 Examples of biocatalytic reactions developed by Codexis.

with air were optimized to increase the availability of O_2 needed for the catalytic systems. The byproduct of the latter oxidation is H_2O_2, which was reduced by commercially available catalase added to preclude oxidizing either the starting material or the product. The imine for the boceprevir precursor was trapped by $NaHSO_3$, and the bisulfite adduct was readily converted to the chiral aminonitrile. Through the Pinner reaction the methyl ester precursor for boceprevir was prepared in high overall yield. The imine for the telaprevir precursor formed a dimer, which was purified by steam distillation. The addition of HCl and NaCN produced the chiral aminonitrile, which was converted to the *t*-butyl ester as the oxalate salt in high yield. These biocatalytic processes gave yields that were 2.5 times higher than yields that had been reported for those chiral proline analogs [22]. Codexis and Merck have detailed the biocatalytic transamination of the β-ketoamide to prepare the free base of sitagliptin, for which they were recognized by a US Presidential Green Chemistry Award [23]. This process is highly efficient, with an overall concentration of the ketone at 0.5 M and an enzyme charge of 4 wt% of the ketone. The reaction is run in 1:1 DMSO:H_2O at a moderate temperature; Codexis generated enzymes to meet such goals for high productivity. Under reduced pressure with a N_2 bleed the equilibrium reaction is driven to completion by removing the byproduct acetone as an azeotrope with isopropylamine. When the reaction was complete the mixture was acidified to pH 2–3, and the denatured enzyme was removed by filtration. Standard workup afforded the free base of sitagliptin in excellent chiral purity, which was converted directly to the phosphate salt [24]. This biocatalytic route eliminated one synthetic step and a crystallization in manufacturing sitagliptin, improved yields and productivity, decreased waste, and eliminated the need to build a specialized plant to manufacture sitagliptin by the previously approved chiral hydrogenation (see discussion with Figure 15.8). These publications afford glimpses into the excellent research carried out by Codexis.

Recent publications by Codexis show it is feasible to select and develop enzymes for productive processing, e.g., operations at 0.5 M or higher in substrate at above ambient temperatures with up to 50% organic solvents and loadings of <1 wt% enzyme [25–27]. Previously researchers have adapted conditions to best utilize existing enzymes, but now biocatalysts can be developed to meet the demands for desired operations. Mutant organisms with different enzymes have been generated by using error-prone polymerase chain reactions (PCRs) [21]. A bacterial genome can now be sequenced for about $1000 [28], and researchers can modify genomes to produce enzymes that may have increased activity. Codexis has capability to analyze and statistically design organisms for optimal performance [29].

The attractiveness of using biocatalysis can be assessed by considering the benefits, drawbacks, and cost of biocatalytic systems. Benefits may include a route with fewer intermediates, or intermediates with higher yield and higher quality; these changes could result in fewer unit operations and fewer intermediates to assay, hence lower cost of goods. Other benefits may include decreased emissions of volatile organic compounds (solvents), decreased cost of remediating solvent-laden waste streams, and decreased solvent recovery operations. Since many enzymatic systems operate in aqueous mixtures at room temperature or slightly higher, a significant benefit may be increased flexibility in selecting equipment to prepare an API. Some commercially available enzymes may function well for a desired transformation and be suitable for the pharmaceutical industry; for instance, a chiral arachidonic acid derivative has been prepared for clinical trials through a lipoxygenase used in the last step [30]. DSM has prepared recombinant porcine liver esterase (PLE) from an animal-free source [31]. Lonza researchers have posed that biocatalysis may be the ideal choice for a second-generation process to make an NCE [32]. Mahmoudian has summarized the benefits of biocatalysis as a green technology for many fields of chemistry [33].

Drawbacks of biocatalytic systems can include a potentially significant time to develop a biocatalytic process, unless a commercially available biocatalyst can be suitably employed. Ownership of intellectual property and licensing should be clearly addressed before a contract research organization develops a biocatalyst specifically for an intermediate. If only one vendor can supply the biocatalyst, contracts must be carefully composed to ensure reliable delivery at an attractive price. Conditions and compounds that inhibit the biotransformation must be identified and controlled, similar to other catalytic processes. If the biocatalyst must be shipped to the site of API manufacturing, stability of the biocatalyst should be considered.

Recently Tufvesson and co-workers have reviewed in detail the factors for estimating the cost of preparing a biocatalyst. (These authors did not estimate the cost of developing a biocatalyst.) With high fermentation yield the cost of the biocatalyst decreases; not all enzymes may be produced in high titer. To increase the turnover number an enzyme may be purified or even immobilized [34]. The extent of purification has a great impact, with costs increasing as the processing evolves from whole cells to crude enzymes to purified enzymes to immobilized enzymes. As production volume increases the cost of the biocatalyst decreases, as expected through economy on scale [35]. The "greenness" of a biocatalytic process must include the operations and waste both in the process and in generating the biocatalyst. Ultimately the acceptability of the cost of a biocatalyst must be judged relative to the value of the product. The value of increased flexibility in equipment choice and improved ruggedness in quality of a product may not be quantifiable before routine manufacture of an API. Biocatalytic processes may provide clear advantages even early in the development of API candidates.

III. STRONG BASES

In academia n-BuLi and NaH are commonly used to deprotonate weakly acidic molecules. These reagents are rarely selected for use on scale, as they generate the pyrophoric byproducts butane and hydrogen. Some alternatives to n-BuLi and NaH shown in Table 4.2 may be preferred over n-BuLi and NaH for SAFETY reasons. For instance, n-hexyl lithium and n-octyl lithium are not pyrophoric, and can be stored at room temperature. The byproducts from these reagents are more recoverable than butane, and hence are "greener" [36]. The alkali metal salts of hexamethyldisilazane are moderately strong bases and do not generate a highly flammable byproduct upon reaction with a protic compound. PTC reactions can use the inexpensive NaOH, and can provide SAFE and productive alternatives to reactions using strong bases (see Section IV).

KOt-Bu is a moderately strong base, and its use on scale can be hampered by its relatively low solubility in organic solvents. The salts of t-AmOH have much greater solubility (Table 4.3) and should be considered for use on scale in place of KOt-Bu. NaOt-Am and KOt-Am are readily available.

TABLE 4.2 Reagents for Deprotonation (1)

Reagent	Typical Concentration	Comments
nBuLi	10–30 wt% in PhCH$_3$	Pyrophoric, store at 0 °C
nHexLi	25–40% in hexane	Not pyrophoric, store at RT or below
nOctLi	15–40% in hexane	Not pyrophoric, store at RT or below
LDA	1.8 M in THF/heptane	Contains "Mg stabilizer"
LiN[SiMe$_3$]$_2$	1.0 M in THF or hexanes	Also available as a solid
NaN[SiMe$_3$]$_2$	1.0 M in THF 0.6 M in PhCH$_3$	Also available as a solid
KN[SiMe$_3$]$_2$	0.9 M in THF 0.7 M in PhCH$_3$	Also available as a solid
NaOH	30–50% aqueous	Phase-Transfer Catalysis (PTC)

(1) Data from FMC Corporation, New Product SAFETY and Data Sheets, 3/94 and 6/94; Callery Chemical Company: http://www.callery.com; and Aldrich Chemical Company.

Triethylamine is probably the amine most commonly used to scavenge acids generated during reactions. Triethylamine is readily used on scale; however, other amines may provide distinct advantages in some cases. Table 4.4 shows some inexpensive amines useful for scale-up, grouped as primary, secondary, and tertiary amines. It is significant to note that secondary amines are more basic in water than the analogous tertiary amines. This is probably due to the increased stabilization from hydrogen bonding with water that is possible for protonated secondary amines (Figure 3.8) [37]. In solvents in which hydrogen bonding cannot occur, basicities are expected to be

TABLE 4.3 Solubility of Selected Alkoxides in Organic Solvents at 25 °C

Solvent	NaOt-Bu	KOt-Bu	NaOt-Am	KOt-Am
Toluene	6 wt%	2.3 wt%	46 wt%	36 wt%
Cyclohexane	14 wt%	NA	>50 wt%	27.6 wt%
Hexanes	11 wt%	0.3 wt%	>50 wt%	30.0 wt%
t-Butanol	NA	17.8 wt%	NA	NA
Pyridine	NA	19.0 wt%	NA	NA
Diglyme	22 wt%	42.0 wt%	47 wt%	47.0 wt%
Tetrahydrofuran	38 wt%	>25 wt%	>30 wt%	>50 wt%

Data from Callery Chemical technical bulletins, 8/97. See http://www.inorganics.basf.com

TABLE 4.4 Characteristics of Some Amines Useful for Scale-Up (1)

Compound	Boiling Point	pK_a (2)	Comments
NH$_3$	−33 °C	9.25	
Methylamine	−6 °C	10.63	
Ethylamine	17 °C	10.70	
3-Dimethylaminopropylamine	133 °C	~10.6–11	Used to remove excess RCOCl (3)
Diethylamine	56 °C	11.02	
Diisopropylamine	84 °C	11.05	
Piperidine	106 °C	11.12	
Morpholine	129 °C	8.33	
Dimethylisopropylamine	66 °C	10.36	
1-Methylpyrrolidine	81 °C	10.32	
Triethylamine	89 °C	10.75	HCl salt insoluble in CH$_3$CN (4)
1-Methylpiperidine	106–7 °C	10.08	
Diisopropylethylamine	127 °C	10.4	HCl salt soluble in CH$_3$CN (4)
4-Methylmorpholine	115 °C/750 mm	7.13	Peptide coupling (5)
Imidazole	89–91 °C (mp)	7.0	Peptide coupling (6), silylation (7)
Pyridine	115 °C	5.25	
2-Methylpyridine (2-picoline)	128 °C	5.97	
2,6-Dimethylpyridine (2,6-lutidine)	143 °C	6.65	Sulfation (8)
4-Dimethylaminopyridine (DMAP)	mp 112–4 °C	9.7	Acylation catalyst (9,10)
1,4-Diazabicyclo[2.2.2]octane (DABCO)	158 °C	8.6	
1,8-Diazabicyclo[5.4.0]undec-7-ene (DBU)	80 °C/0.6 mm	13.2	
1,5-Diazabicyclo[4.3.0]undec-7-ene (DBN)	95 °C/7.5 mm	~13.2	

(Continued)

Practical Process Research and Development

TABLE 4.4 Characteristics of Some Amines Useful for Scale-Up (1)—cont'd

(1) The Merck Index; 12th ed., Budavari, S., Ed., Merck & Co., Inc.: White House Station, NJ, 1996; Perrin, D. D.; Dempsey, B.; Serjeant, E. P. pK$_a$ Prediction for Organic Acids and Bases; Chapman and Hall: New York, 1981; Serjeant, E. P.; Dempsey, B. Ionization Constants of Organic Acids in Aqueous Solution; Pergamon: New York, 1979; CRC Handbook of Chemistry and Physics; 76th ed., CRC Press: New York, 1995–1996; Lange's Handbook of Chemistry; Dean, J. A., 14th ed., McGraw-Hill: New York, 1992.

(2) pK$_a$ data are for the protonated amines (conjugate acids).

(3) Evans, D. D.; Evans, D. E.; Lewis, G. S.; Palmer, P. J.; Weyell, D. J. J. Chem. Soc., **1963**, 3578.

(4) Anderson, N. G.; Ary, T. D.; Berg, J. A.; Bernot, P. J.; Chan, Y. Y.; Chen, C.-K.; Davies, M. L.; DiMarco, J. D.; Dennis, R. D.; Deshpande, R. P.; Do, H. D.; Droghini, R.; Early, W. A.; Gougoutas, J. Z.; Grosso, J. A.; Harris, J. C.; Haas, O. W.; Jass, P. A.; Kim, D. H.; Kodersha, G. A.; Kotnis, A. S.; LaJeunesse, J.; Lust, D. A.; Madding, G. D.; Modi, S. P.; Moniot, J. L.; Nguyen, A.; Palaniswamy, V.; Phillipson, D. W.; Simpson, J. H.; Thoraval, D.; Thurston, D. A.; Tse, K.; Polomski, R. E.; Wedding, D. L.; Winter, W. J. Org. Process Res. Dev. **1997**, 1, 300.

(5) Anderson, G. W.; Zimmermann, J. G.; Callahan, F. M. J. Am. Chem. Soc. **1966**, 88, 1338.

(6) Stoner, E. J.; Stengel, P. J.; Cooper, A. J. Org. Process Res. Dev. **1999**, 3, 145.

(7) Yasuda, N.; Huffman, M. A.; Ho, G.-J.; Xavier, L. C.; Yang, C.; Emerson, K. M.; Tsay, F.-R.; Li, Y.; Kress, M. H.; Rieger, D. L.; Karady, S.; Sohar, P.; Abramson, N. L.; DeCamp, A. E.; Mathre, D. J.; Douglas, A. W.; Dolling, U.-H.; Grabowski, E. J. J.; Reider, P. J. J. Org. Chem. **1998**, 63, 5438.

(8) Floyd, D. M.; Fritz, A. W.; Pluscec, J.; Weaver, E. R.; Cimarusti, C. M. J. Org. Chem. **1982**, 47, 5160.

(9) Hofle, G.; Steglich, W.; Vorbruggen, H. Angew. Chem. Int. Ed. Eng. **1978**, 17, 569.

(10) Encyclopedia of Reagents for Organic Synthesis; Paquette, L. A. Ed., John Wiley & Sons: New York, 1995.

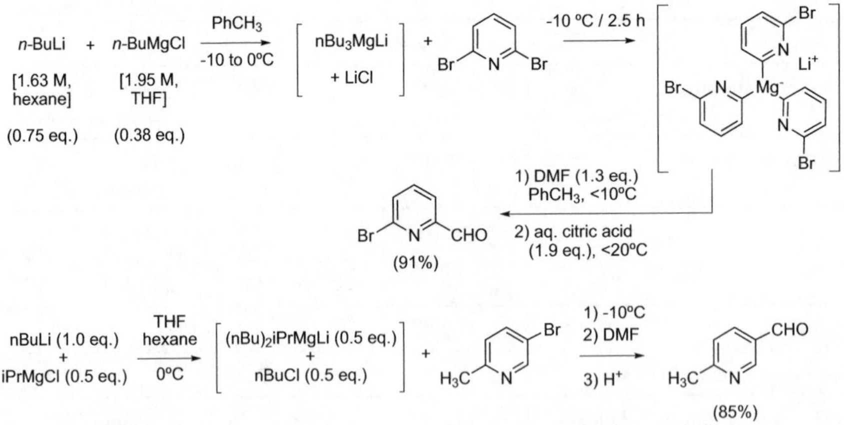

FIGURE 4.8 "Ate" complexes as strong bases not needing cryogenic temperatures.

---

consistent with the inductive effect for substituted amines: the order of basicity in chlorobenzene is $BuNH_2 < Bu_2NH < Bu_3N$ [38].

Complex bases may not require cryogenic temperatures. For instance, a magnesium "ate" complex of 2-bromopyridine reacted at -10 to 0 °C with DMF to afford the aldehyde in high yield (Figure 4.8). Treating 2,6-di-bromopyridine with n-BuLi at this temperature range caused extensive decomposition, and n-BuMgCl or $(n$-Bu$)_2$Mg were much less reactive [39]. The "ate" complex of a 3-bromopyridine was similarly converted in high yield to the corresponding aldehyde [40].

Strong bases may not be necessary. In the first step for the preparation of the iodothiophene in Figure 3.11 K_2CO_3/DMSO was substituted for the NaH/THF used in drug discovery. The only other bases used in this sequence were Et_3N and an excess of the weak base morpholine. Each reaction product was isolated directly from the reaction mixtures, either first by cooling or after adding water. These steps are excellent examples of convenient transformations and direct isolations [41].

> Do not develop a process using a reagent that might be considered too dangerous to be use-tested prior to scale-up.

IV. PHASE-TRANSFER CATALYSIS

In place of basic amines or bases such as NaH, alkyl lithiums, alkoxides, and Grignard reagents, PTC conditions with NaOH or other inorganic bases can often be used. Recoverable organic solvents such as toluene can be used, avoiding reactions that might otherwise be run in high-boiling, polar aprotic solvents such as DMSO; any solvent that does not react under the PTC conditions can be used in principle. Phase-transfer catalysts include quaternary ammonium salts, crown ethers, and ethylene glycol derivatives. PTC using polyethylene glycols (PEGs) has been described by Weber and Gokel [42], and Halpern is the leading authority for PTC [43,44]. The use of PTC reactions and reagents has become rather widespread, and often quaternary ammonium salts and PEGs are not considered as PTC reagents in the literature. TBAB is perhaps most widely used in this regard [45]. PTC applications often are not well referenced in the literature. There are many examples of PTC reactions throughout this book.

Improved selectivity in PTC processes results from four factors. The critical factor most often overlooked is controlling the amount of water in solvating the anion. As Landini's group noted, the basicity of NaOH increased 50,000 times by decreasing hydration number of hydroxide from 11 (15% NaOH) to three (solid) [46]. Secondly, the quaternary cation should be chosen to make a tight or loose ion pair, based on the accessibility of the quaternary center. Thirdly, the polarity of the solvent can make a difference. Last of all, controlling the concentration of the catalyst in the reaction can influence the reaction rate [47].

The PTC reactions shown in Figure 4.9 illustrate the power of PTC. Hydrolysis–esterification processes were some of the first reactions to be applied for PTC, as shown with Lilly's sequence with a β-lactam [48]. The irbesartan precursor was prepared using Aliquat 175. These conditions avoided the use of H_2O-soluble polar solvents, such as DMSO or DMAC, and strong bases, such as NaH or alkoxides. The process was simple, and the product crystallized in good yield [49]. In forming the Ru Pybox catalyst tetrabutylammonium hydroxide was used to facilitate the ring closures. The resulting ruthenium complex was crystallized by adding the reaction mixture into an antisolvent [50]. Merck researchers carried out a solvent-free (neat) alkylation, generating a compound with a quaternary carbon. The product from this reaction was a liquid, facilitating the solvent-free conditions [51]. Terfenadine was produced by reduction of the corresponding ketone under PTC conditions in xylene, offering several advantages

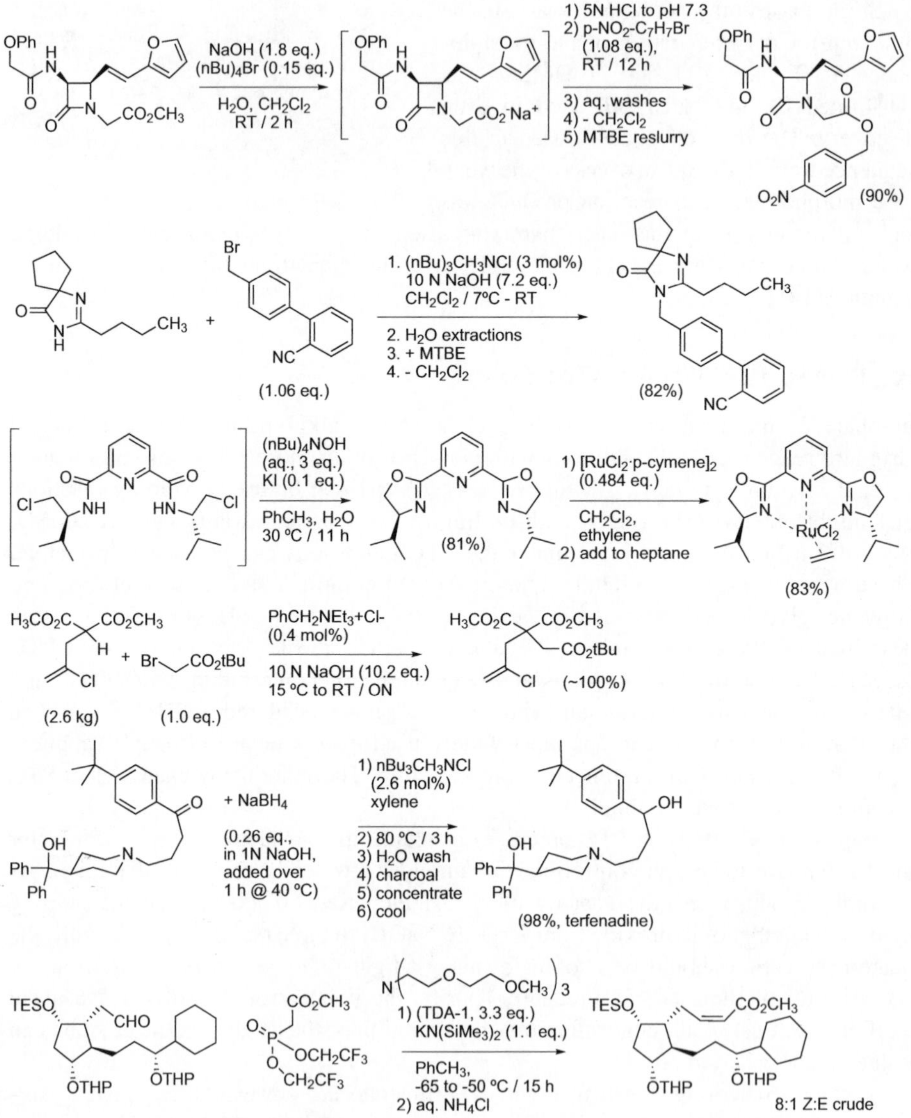

Practical Process Research and Development

FIGURE 4.9 Examples of PTC reactions.

over reductions run in alcoholic solvents. First of all, only 0.26 equivalents of sodium borohydride was required. The latter was charged as a solution in aqueous NaOH, taking advantage of the fact that $NaBH_4$ is more stable in aqueous NaOH than in alcoholic solutions. Secondly, the product crystallized from the reaction solvent (xylene) as the desired polymorph, whereas an undesired polymorph was produced from reductions in EtOH [52]. A Horner–Emmons olefination was carried out using TDA-1 in place of 18-crown-6 to sequester the potassium ion. TDA-1 is an inexpensive ethylene glycol derivative that can be readily removed by extractive workup [53].

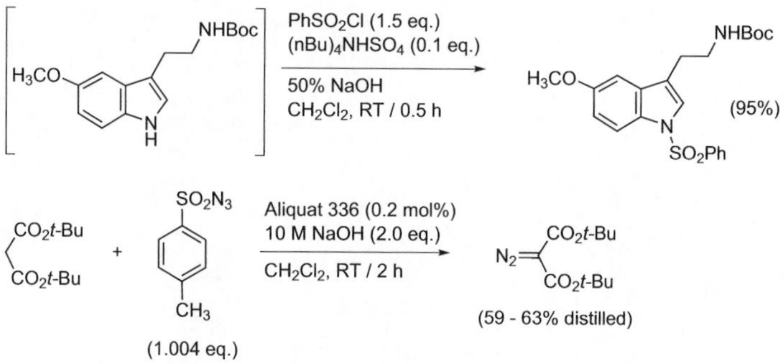

FIGURE 4.10 PTC carried out with water-sensitive functionalities.

Even though strongly basic aqueous phases are usually employed with PTC reactions, hydrolysis of water-sensitive groups can be avoided. Reaction of a sulfonyl chloride with an indole was carried out with 50% NaOH (Figure 4.10). If the sulfonylation had been carried out using conditions such as sodium hydride in THF, scrupulously dry conditions would have been necessary to preclude hydrolysis of the sulfonyl chloride. Perhaps notable is that this high-yielding reaction was not highlighted in the article [54]. Diazo transfer was carried out under PTC conditions to afford the bis-*t*-butyl malonate using 10 M NaOH. If saturated Na_2CO_3 is used this procedure works with ethyl acetoacetate and *t*-butyl acetoacetate as well, and the yields are higher than the example shown. Methyl esters are too readily hydrolyzed to be useful substrates [55]. Over-agitation can hydrolyze components of water-sensitive reactions [56].

Although liquid–liquid PTC conditions are usually employed, solid–liquid PTC conditions are also effective, and may be selected for use with water-sensitive reactions. A 4-bromomethyl pyrrole was converted to the corresponding acetate by solid–liquid PTC in an unoptimized process [57]. Under solid–liquid PTC conditions with potassium carbonate Harrington and co-workers found that a sulfonamide reacted once with a dichloropyrimidine (Figure 4.11). The H_2O formed as a side product from disproportionation of the $KHCO_3$ byproduct was removed by azeotroping with the solvent. With the addition of three equivalents of an ethylene glycol derivative and solid NaOH bosentan was produced in high yield [58]. In a PTC reaction using 18-crown-6, tripotassium phosphate was found to be nearly as effective as KOH, with a 94% conversion after 4 h using K_3PO_4 as compared to 98% conversion with KOH in that same period [59].

PTC reactions can markedly increase the reactivity of cyanide and azide. For the hindered benzyl chloride in Figure 4.12 PTC conditions were analyzed as being more economical and less wasteful than conventional reactions in DMSO. This was an intrinsic reaction; reducing the agitation rate lowered the reaction temperature [60]. In the displacement of fluoride by cyanide, Aliquat

If PEG is not available, an effective substitute may be to use as many equivalents of smaller glycol units to constitute the large PEG [Gokel, G. W.; Goli, D. M.; Schultz, R. A. *J. Org. Chem.* **1983**, *48*, 2837]. Hence perhaps three equivalents of dimethoxyethane would be similar to one equivalent of the toxic 18-crown-6.

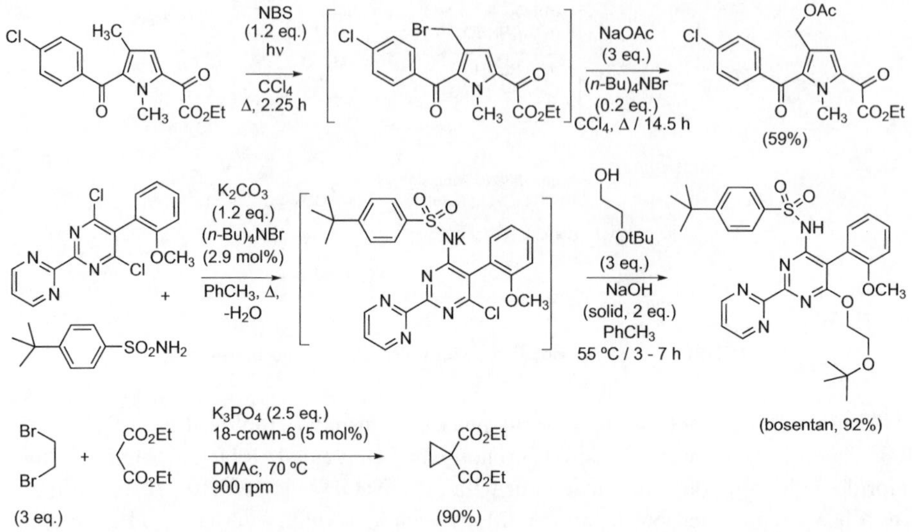

FIGURE 4.11　Examples of solid–liquid PTC.

Aliquat 336™ is methyltricaprylyl-ammonium chloride, where "caprylyl" implies a carbon chain similar to eight linear carbons. The long chains in Aliquat 336™ are about 2:1 octyl:decyl, with average FW 432.

336 (aka Aliquat 128) functioned well; when Aliquat 336 was replaced with $(n$Bu$)_4$NBr, only 82% conversion was seen after 23 h at 25 °C [61]. The use of 18-crown-6 to generate an alkyl azide, as a precursor to the amine, was key; without 18-crown-6 the primary products were the olefins [62]. Under PTC conditions ring closure of a γ-chlorobutyramide to a γ-lactam took place, and displacement with sodium azide generated the racemic α-azidoamide. The chloride byproduct underwent some exchange with the α-bromide, and the α-chloro side product was transformed in situ to the desired azide under PTC conditions. The α-chloro side product did not react with azide under the DMSO conditions examined [63].

Guidelines to check the feasibility of PTC for a transformation. Use 10 mol% TBAB or $(n$-Bu$)_4$NHSO$_4$, three equivalents of 10 M NaOH or 10 M KOH, and toluene. If these conditions show promise additional optimization may be appropriate.

Three factors should be considered in choosing PT catalysts: selecting the optimum catalyst and conditions for conversions; stability of catalyst and side products; and removing the catalyst to acceptably low levels in the products. The choice of catalyst does matter for efficient reactions, as shown in Figure 4.12 with the cyanation of a 2-fluoropyridine. While TBAB often works well to determine the feasibility of PTC conditions, other PT catalysts may give more efficient conversions.

Halpern's guidelines for selecting a quaternary ammonium phase-transfer catalyst [64] are shown in Figure 4.13. Organophilicity and accessibility affect PT-catalyzed

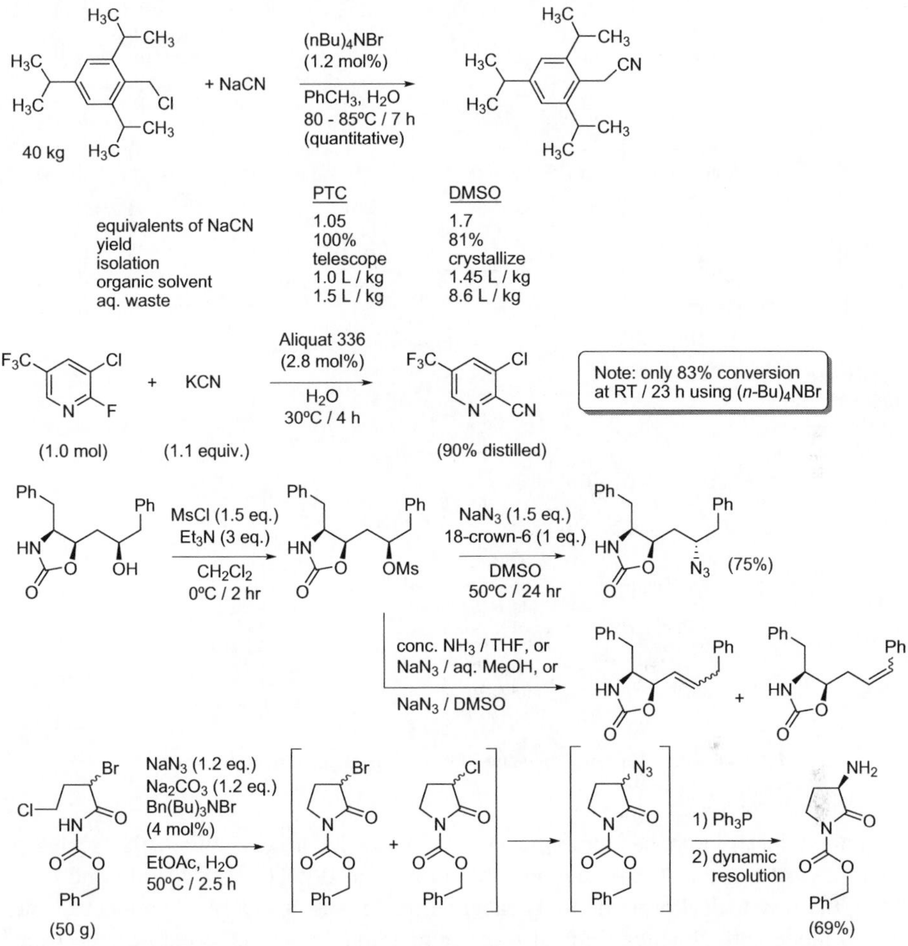

FIGURE 4.12 Activation of cyanide and azide by PTC.

reaction rates. The greater the number of carbon atoms in the quaternary ammonium salt the greater the organophilicity. The less hindered the nitrogen is, the greater the accessibility. Ripin has provided an excellent chapter on basicity and pK_a [65].

Poor stability of phase-transfer catalysts can affect processing, as shown in Figure 4.14. GSK researchers carried out the formation of a chiral amino acid precursor by asymmetric PTC alkylation [66] using a quaternary ammonium salt of cinchonidine. In the laboratory the imine and catalyst were dissolved in dichloro-methane, 45% KOH was added, and the mixture was cooled to 0–5 °C before the benzhydryl bromide was added; this protocol produced the alkylated imine with an enantiomeric ratio of 80:20 (S):(R), with an isolated yield of 55%. Unfortunately on kilogram scale the reaction was slow, producing racemic ester in 65% isolated yield. Decomposition of the catalyst was suspected, as a period of 30–60 min on the large scale was required to cool the mixture before adding the benzhydryl bromide, while

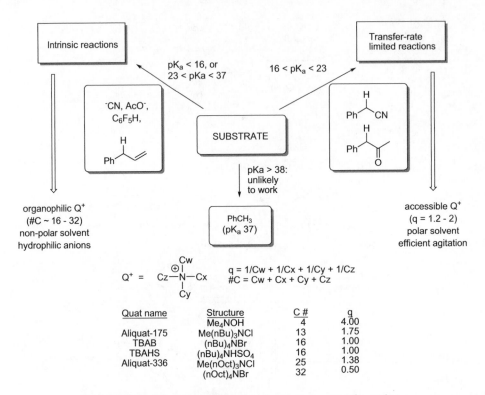

FIGURE 4.13 Halpern's guidelines for selecting a phase-transfer catalyst.

not more than 10 min had been necessary in the laboratory. When the scale-up conditions were use-tested by cooling the mixture to 0–5 °C over 45 min and then adding the benzhydryl bromide, only racemic imine was produced. Treatment of the catalyst alone with aqueous KOH at room temperature gave the tertiary amine from Hofmann elimination, and three other degradents were detected by HPLC–MS. The GSK authors noted that the quaternary ammonium salts from reaction of the two tertiary amine degradents with the benzhydryl bromide catalyzed the alkylation, but unselectively. By adding the aqueous KOH last at 0–5 °C the product was isolated with high enantiomeric excess. Under these conditions the catalyst decomposed to the enol ether when the reaction was complete, suggesting that recovering and recycling the catalyst could be difficult [67].

The degradation of quaternary ammonium salts complicates not only recovery and reuse on scale, but also the analysis and removal of degradents. The Hofmann elimination pathway is the primary degradation for the hindered tetrabutylammonium ion, and dealkylation occurs for benzylic quaternary ammonium salts and less-hindered ammonium salts [68] (Figure 4.14). Streams of tetrabutylammonium hydrogen sulfate have been effectively recovered and recycled in manufacturing pharmaceutical intermediates.

Analysis of residual quaternary ammonium salts has been performed using hydrophilic interaction liquid chromatography (HILIC) as a means of quantifying

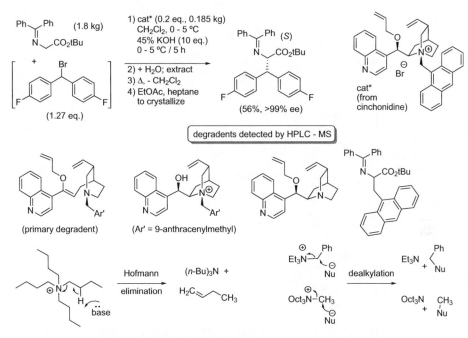

FIGURE 4.14 Examples of degradation of quaternary ammonium salts.

sulfonate esters and sulfate esters [69], and should be suitable for analysis of phase-transfer catalysts. Ion chromatography and capillary electrophoresis may also be suitable.

The pharmaceutical industry may choose PT catalysts based on the ease of removing them from the products. For instance, $(nBu)_3CH_3NCl$ (Aliquat® 175) can be readily extracted into water. Several extractions may be required to remove $(nBu)_4NBr$ (Aliquat® 100) or $(nBu)_4NHSO_4$ from an organic extract [70]. The lipophilic $(nOct)_3CH_3NCl$ (Aliquat® 336) can be left in the organic phase, and the product can be removed by crystallization [71] or by extraction into water [43,72,73].

Quaternary ammonium salts have also been used stoichiometrically or in excess to isolate compounds (Figure 4.15). Valine was converted to its benzyl-trimethylammonium salt for amination with an oxaziridine; valine itself was essentially non-reactive [74]. Merck researchers extracted a water-soluble triazolylaldehyde from DMSO–water into toluene as the Aliquat 336 salt (not characterized); then the triazolylaldehyde was extracted into aqueous acetic acid and isolated from 2-butanol [75]. Squibb researchers isolated crystalline N-sulfated β-lactams in high yields as the tetrabutylammonium salts [76].

There are many other applications of phase-transfer catalysts and phase-transfer reagents. For instance, Sonogashira couplings without transition metals were made possible using water and PEGs under microwave conditions [77]. In copper-free, ligand-free, and amine-free Sonogashira couplings tetrabutylammonium acetate was found to be the preferred reagent, better than cesium carbonate and other bases [78].

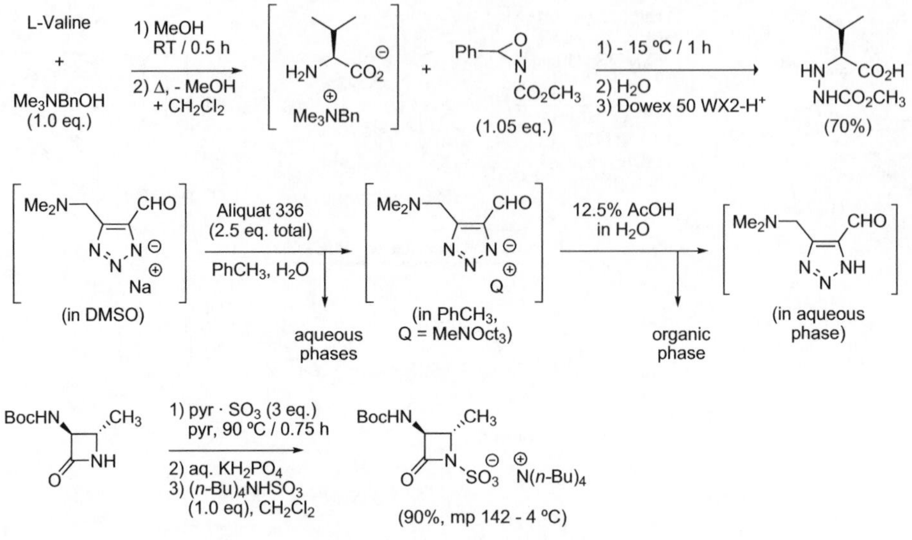

FIGURE 4.15 Quaternary ammonium salts used to isolate compounds.

Noyori's group has developed solvent-free, tungsten-catalyzed, PTC oxidations of alcohols to ketones and olefins to epoxides [79–81]. There are many applications and advantages of PTC.

V. PEPTIDE COUPLING

Peptide coupling reagents are more than fancy dehydrating reagents [82], and there is a wide range of costs. Some commercially available reagents shown in Table 4.5 are ranked by the cost per mol relative to thionyl chloride, along with some exemplary references. (Thionyl chloride was selected as the least expensive reagent, for the formation of acid chlorides. Acid chlorides are rarely used for amide couplings of amino acids, and acid fluorides have some advantages over acid chlorides [83].) HBTU and HATU highly active and may be used for more difficult amide bond formations, as with resins or hindered carboxylic acids [84]. Byproducts from HBTU and BOP-Cl are water-soluble. An amine and a carboxylic acid can be mixed with either carbodiimides or T3P for coupling without prior activation of the carboxylic acid. The urea byproduct from the water-soluble carbodiimides can be readily removed by extractions, due to the basic amine functionality. Dicyclohexylurea will crystallize from DMF and can be removed by filtration. The phenolic byproduct from CDMT couplings can be removed by extractions into a basic aqueous phase. SAFETY precautions should be taken with all these reagents, as they can be severe irritants.

The selection of peptide coupling reagents may change as the development of an NCE progresses. The water-soluble carbodiimides have been used for initial studies for convenience, and then processes have been developed to use reagents that are more atom-economical. On large scale the mixed anhydrides from chloroformates, pivaloyl chloride, or benzoyl chloride are often used. More recently carbonyl diimidazole (CDI)

TABLE 4.5 Relative Costs of Peptide Coupling Reagents

Reagent	Relative cost/mol	Reference
O-(Benzotriazol-1-yl)-N,N,N',N'-tetramethyluronium hexafluorophosphate (HBTU)	1048	(1)
Bis(2-oxo-3-oxazolidinyl)phosphinic chloride (BOP-Cl)	273	
2-Chloro-4,6-dimethoxy-1,3,5-triazine (CDMT)	250	(2, 3)
(Chloromethylene)dimethyliminium chloride (Vilsmeier reagent)	82	(4)
N-(Dimethylaminopropyl)-N'-ethyl-carbodiimide (as the liquid free base or solid hydrochloride: EDAC, EDC, WSC)	70	(5)
Carbonyl diimidazole (CDI)	53	(See text)
Propylphosphonic anhydride (cyclic trimer, T3P)	44	(6)
Diethyl chlorophosphate	19	(7, 8)
Dicyclohexylcarbodiimide (DCC)	10	
Isobutyl chloroformate	5	(9)
Pivaloyl chloride	2	(10)
Thionyl chloride	1	(11)

(1) Sleebs, M. M.; Scanlon, D.; Karas, J.; Maharani, R.; Hughes, A. B. J. Org. Chem. **2011,** 76, 6686.

(2) Garrett, C. E.; Jiang, X.; Prasad, K.; Repic, O. Tetrahedron Lett. **2002,** 43, 4161.

(3) For use of trichlorotriazine for coupling: Rayle, H. L.; Fellmeth, L. Org. Process Res. Dev. **1999,** 3, 172.

(4) Jass, P. A.; Rosso, V. W.; Racha, S.; Soundararajan, N.; Venit, J. J.; Rusowicz, A.; Swaminathan, S.; Livshitz, J.; Delaney, E. J. Tetrahedron **2003,** 59, 9019.

(5) Hansen, K. B.; Balsells, J.; Dreher, S.; Hsiao, Y.; Kubryk, M.; Palucki, M.; Rivera, N.; Steinhuebel, D.; Armstrong, J. D., III; Askin, D.; Grabowski, E. J. J. Org. Process Res. Dev. **2005,** 9, 634.

(6) Patterson, D. E.; Powers, J. D.; LeBlanc, M.; Sharkey, T.; Boehler, E.; Irdam, E.; Osterhout, M. H. Org. Process Res. Dev. **2009,** 13, 900.

(7) Diethyl chlorophosphate is a cholinesterase inhibitor and very toxic: http://toxnet.nlm.nih.gov/cgi-bin/sis/search/a?dbs+hsdb:@term+@DOCNO+6408

(8) Diethyl phosphates can be excellent leaving groups, as shown for a Grignard reaction: Snyder S. A.; Treitler, D. S. Organic Syntheses; Vol. 88, (2011) p 54.

(9) Mickel, S. J.; Sedelmeier, G. H.; Niederer, D.; Daeffler, R.; Osmani, A.; Schreiner, K.; Seeger-Weibel, M.; Bérod, B.; Schaer, K.; Gamboni, R.; Chen, S.; Chen, W.; Jagoe, C. T.; Kinder, F. R., Jr.; Loo, M.; Prasad, K.; Repic, O.; Shieh, W.-C.; Wang, R.-M.; Waykole, L.; Xu, D. D.; Xue, S. Org. Process Res. Dev. **2004,** 8, 92.

(10) Grosso, J. A. US Patent 5,162,543, 1992 (to E. R. Squibb & Sons).

(11) Butini, S.; Gabellieri, E.; Huleatt, P. B.; Campiani, G.; Franceschini, S.; Brindisi, M.; Ros, S.; Coccone, S. S.; Fiorini, I.; Novellino, E.; Giorgi, G.; Gemma, S. J. Org. Chem. **2008,** 73, 8458.

has frequently been employed on a large scale, even for forming amides of hindered carboxylic acids (Figure 4.16) [85–92]. The benefits of CDI are that the byproducts are innocuous and water-soluble, reactions can be monitored readily by the evolution of CO_2, the imidazolides often are fairly stable thermally, and CDI is inexpensive when purchased in bulk. In general imidazolides are less reactive than the

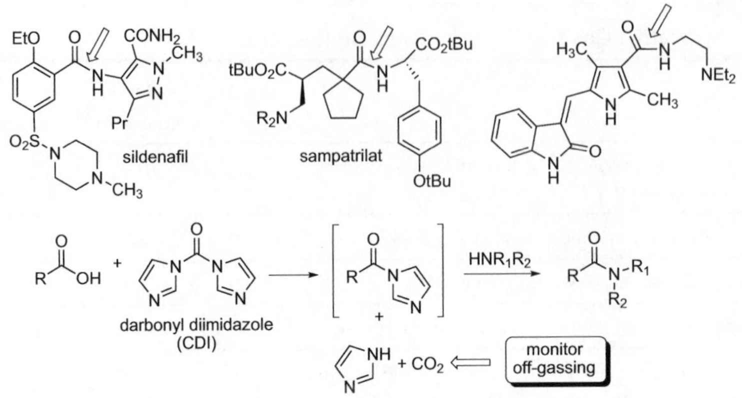

FIGURE 4.16   Some compounds prepared on scale by CDI couplings.

corresponding acid chloride. Imidazolium chloride has been found to catalyst CDI couplings [93], and a small amount of CO_2 has also been shown to be effective [89] (see Figure 17.1).

During peptide coupling, racemization α- to the carbonyl of what was formerly the carboxylic acid group is a concern, usually through oxazolone formation (see discussion with Figure 7.1). N-Hydroxybenzotriazole (HOBt) has classically been used as a catalyst for peptide coupling to minimize racemization and O- to N-acyl transfer (Figure 9.6) through formation of the active ester. Unfortunately HOBt is known to react violently, especially when dry, and the shipping of HOBt is restricted. The nucleus of HOBt is incorporated into some peptide coupling reagents, such as HBTU, HATU, and PYBOP (see Figure 4.17), and these reagents should be treated with care appropriate to HOBt. Pfizer researchers and others have examined substitutes for HOBt [85,86,94]. For the coupling of a hindered imidazolide with the relatively poor nucleophile aniline the activities of the catalysts were found to be (in descending order) HOAt, HOBt·H_2O, 5-nitro-2-hydroxypyridine, and 2-hydroxypyridine. For SAFETY the ranking was (in descending order) 2-hydroxypyridine, 5-nitro-2-hydroxypyridine, HOBt·H_2O, then HOAt. The Pfizer researchers concluded that 5-nitro-2-hydroxypyridine was the preferred catalyst. The inexpensive 2-mercaptobenzothiazole has been shown to be an effective substitute for HOBt in the preparation of peptides by solid-phase peptide synthesis [95], and should be suitable for solution phase syntheses (although it has smelled like rubber tires). Other active esters have been generated or isolated for peptide couplings, such as the pentafluorophenyl ester or the N-hydroxysuccinimide (HOSu) ester. The latter has been shown to undergo a Lossen rearrangement to produce β-alanine derivatives. A lower reaction temperature and less basic conditions during coupling may minimize this type of side product [96,97]. 4-Dimethylaminopyridine (DMAP) is a highly active acylation catalyst [98] that might promote racemization through ketene formation from the acyl pyridinium intermediate. Pyridine has also been used as an acylation catalyst. If a catalyst is needed in place of HOBt, perhaps 2-hydroxypyridine will be effective and economical.

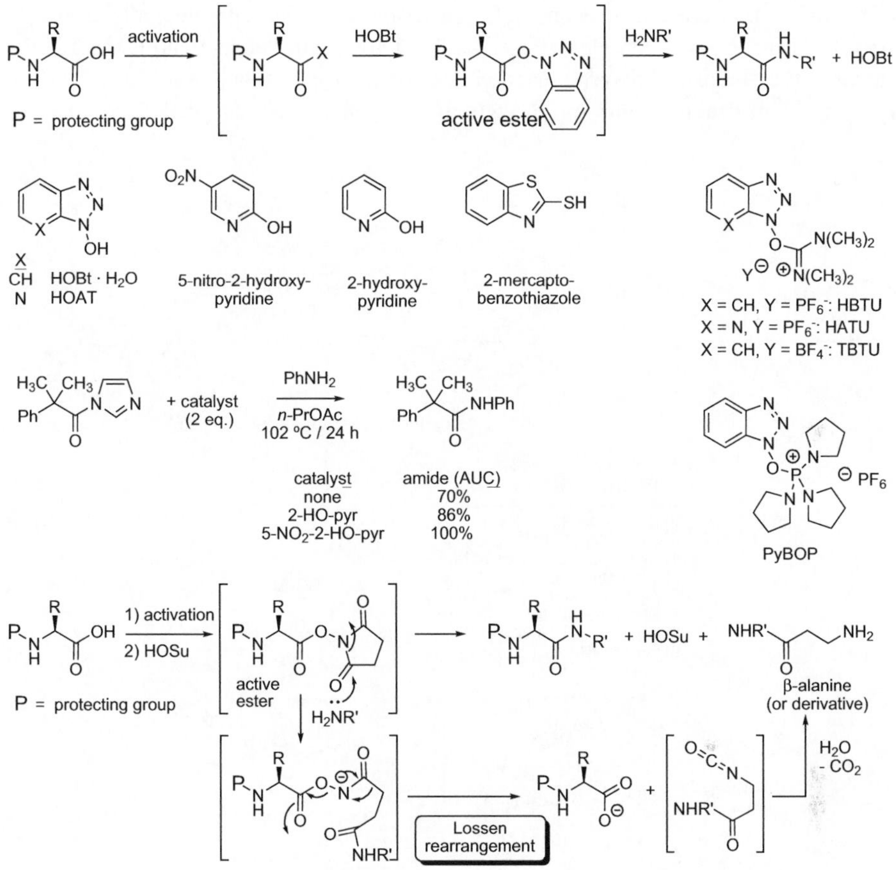

FIGURE 4.17 Alternatives to HOBt.

VI. OXIDATIONS

Oxidations are often troublesome for scale-up. The high reactivity of reagents can lead to hazardous operations, with increased liability for accidents on scale. Shipping of some reagents, such as 90% H_2O_2, *meta*-chloroperoxybenzoic acid, and the Dess–Martin periodinane, is restricted. The toxicity of some reagents can lead to the use of PPE, which is necessary but cumbersome, and may slow operations. Waste disposal costs can be high if toxic byproducts and a large amount waste are produced [99]. Laborious work-ups can ensue, increasing processing time and the risk of poor product recovery. The preferred approach is to design catalyzed oxidations with bulk oxidants that are inexpensive, readily available, less hazardous, and non-toxic. Some useful bulk oxidants are air (O_2 in cylinders poses a greater hazard), H_2O_2 in concentrations no greater than 30%, $NaClO_2$, $NaOCl$ (bleach), urea·H_2O_2, Na_2CO_3·H_2O_2, $NaBO_3$, *N*-methylmorpholine *N*-oxide (NMO), and Oxone®. Oxidations carried out on scale have been reviewed [100].

In Figure 4.18 are shown some selected oxidations. The commercially available pyridine–sulfur trioxide complex is used as the dehydrating reagent and DMSO is the oxidant in the "industrial Swern" reaction described by Parikh and von Doering [101]. Cryogenic temperatures and more vigorous reagents such as oxalyl chloride or

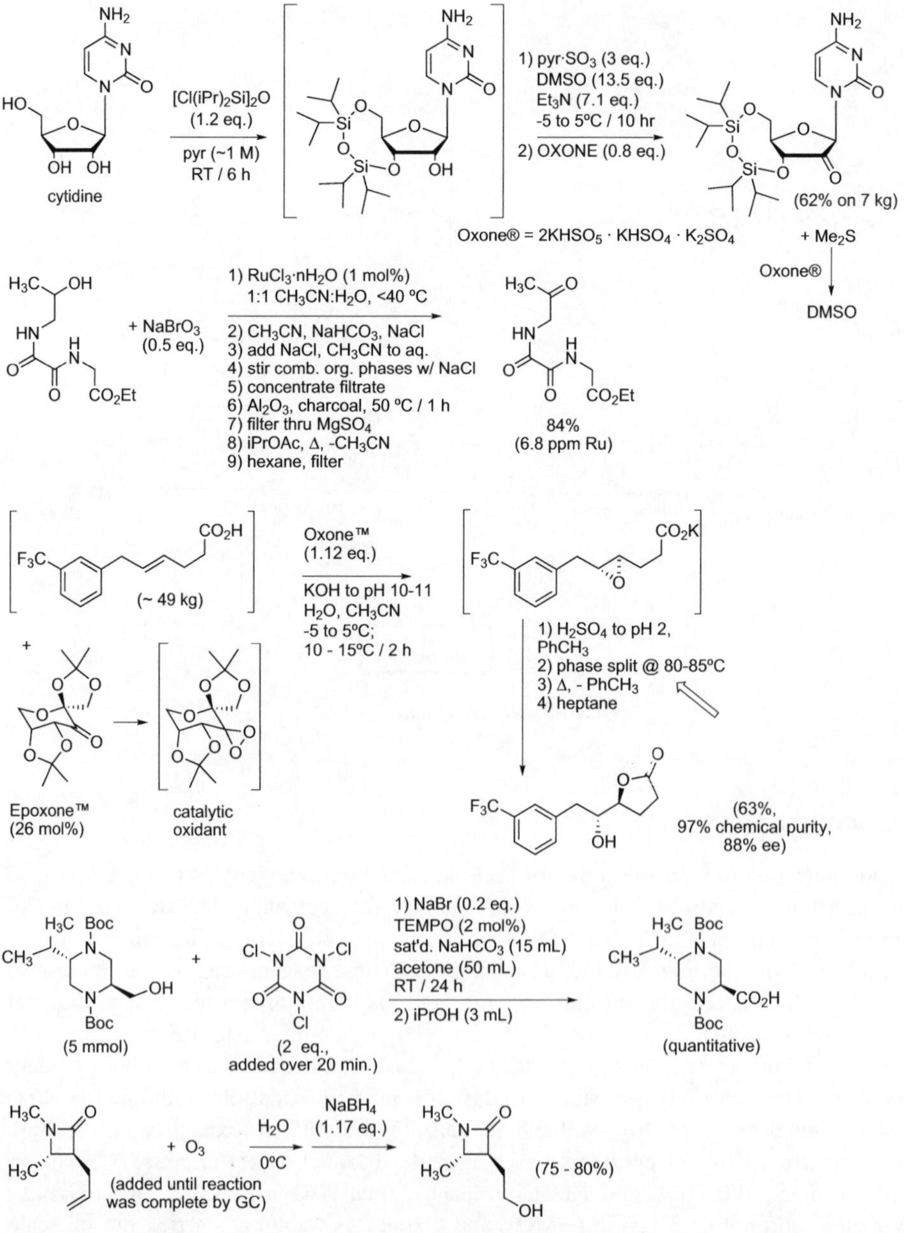

FIGURE 4.18   Selected oxidations.

trifluoroacetic anhydride are not needed. In the process to oxidize the secondary alcohol in the cytidine derivative Oxone® was added during the workup, to convert the malodorous dimethyl sulfide byproduct to DMSO [102–105]. In Merck's ruthenium-catalyzed oxidation the reaction was carried out in acetonitrile–water, and the rather lengthy workup was necessitated because of the high solubility of acetonitrile in water. Notable for this reaction is the extremely low level of residual ruthenium, probably due to the treatment with activated carbon [106]. DSM researchers carried out the chiral epoxidation of an olefin using Shi's reagent [107,108] on scale. A relatively large amount of the precatalyst is used, as it decomposed during the reaction. For facile processing a phase split was carried out at 80–85 °C [109]. The TEMPO-catalyzed oxidation of a primary alcohol to the carboxylic acid using trichloroisocyanuric acid (TCCA) required no metal catalyst. No significant racemization found, and some water was needed to oxidize the intermediate aldehyde to the carboxylic acid [110]. Optimization of reaction conditions might have reduced the number of equivalents of TCCA added, as TCCA has the oxidizing capability of three equivalents of NCS [111]. TEMPO oxidations can be carried out in a variety of solvents, depending on the starting material; efficient agitation of the heterogeneous mixture [112] and control of the pH may be necessary [113]. Ozonolysis [114] of the water-soluble β-lactam was performed on a scale greater than 1 kg scale in water, at 0 °C. Water was chosen as a safer solvent alternative to dichloromethane–MeOH [115]. Ozonolyses have often been carried out at temperatures of −25 °C or colder due to both the low solubility of ozone in most organic solvents is low and concern for the reactivity of the intermediate ozonide. Continuous ozonolysis may be a SAFE, practical alternative [116].

VII. REDUCTIONS

As mentioned in Chapter 5, care must be taken using the borane–THF complex [117,118]. Decomposition of $BH_3 \cdot THF$ in THF generates H_2 and tributyl borate, and above 50 °C decomplexation of $BH_3 \cdot THF$ generates diborane, which is less reactive than BH_3. For SAFE handling of $BH_3 \cdot THF$ in THF, BASF researchers recommended that the reagent be stored at 0–5 °C, and that reactions be conducted below 35 °C [119]. $BH_3 \cdot THF$ in THF is commercially available as 1 M solutions. BH_3 can be generated from reaction of $NaBH_4$ with $BF_3 \cdot Et_2O$, and from the workup of this reaction the stable amine·BH_3 complex could be isolated and recrystallized (Figure 4.19). During the workup methanolysis destroyed unreacted BH_3 first, and at higher temperatures the amine·BH_3 complex was destroyed. The reaction was worked up without H_2O, by quenching with MeOH and removing the $MeOH:B(OMe)_3$ azeotrope (bp 55 °C; $MeOH:B(OMe)_3$ 34:66) [118]. Other research showed that some cleavage of the ketal was found under these conditions that generated BH_3 from $NaBH_4$ and $BF_3 \cdot Et_2O$, and sodium bis-methoxyethoxyaluminum hydride (Red-Al, Vitride) was preferred. No ether was formed with Red-Al treatment, and other byproducts were identified. Before the structure of the ether was known, DoE screening was carried out for reaction temperature, reaction time, and equivalents of Red-Al, and by using 6 experiments ($2^{3-1} + 2$ reactions for the centerpoints) the reduction was rapidly optimized (Figure 4.19) [120].

Some other powerful examples of reductions are shown in Figure 4.20. Reduction of isolated double bonds in food-grade polymers was carried out in water using a very low

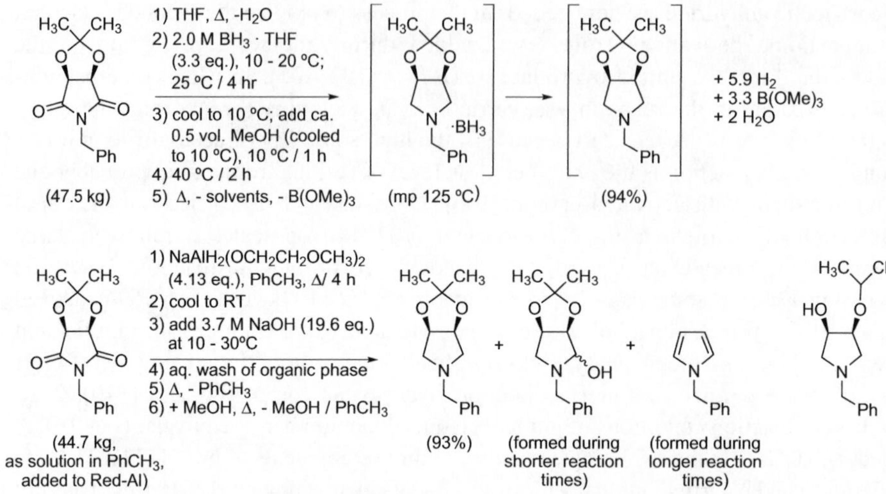

FIGURE 4.19 Reduction of an imide using BH_3 generated *in situ* or Red-Al.

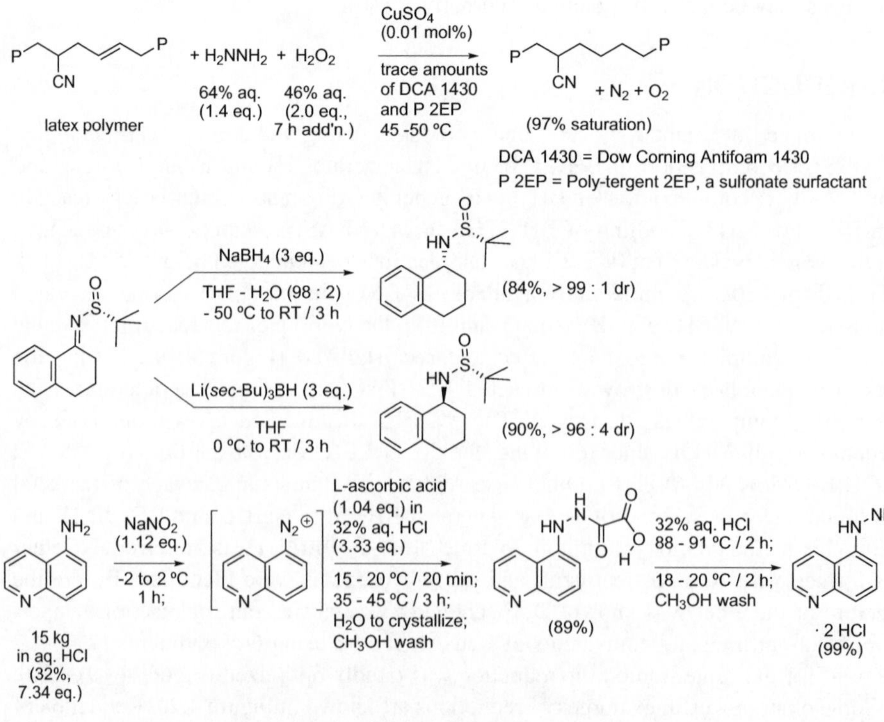

FIGURE 4.20 Miscellaneous reductions.

charge of an inexpensive catalyst. In this reduction H_2 was not used, and the byproducts are very "green." Surfactants were for effective reduction of the starting material, which reacts as a suspension at the interface [121]. This diimide reduction is selective for least-substituted olefins, and holds promise for reactions in solutions if the SAFETY hazards of mixing hydrazine and hydrogen peroxide can be addressed. In the reduction of sulfonyl imines Amgen researchers found that by changing the reductant and solvent they were able to obtain either diastereomer with very high de [122]. Reduction of diazonium salts with ascorbic acid is a very green alternative to reducing diazonium salts with $SnCl_2$. By reducing with ascorbic acid there is no heavy metal waste, and the process is all-aqueous except for the MeOH washes [123].

VIII. TREATMENTS TO REMOVE A BOC GROUP

There are many approaches to removing a Boc group. The one favored by medicinal chemists is usually treatment with trifluoroacetic acid (TFA), because it is volatile and readily removed by rotary evaporation. Many substitutes for the highly corrosive TFA are available for operations on scale, including H_3PO_4/PhCH$_3$ [124], H_2SO_4/PhCH$_3$ [125], p-TsOH/MIBK [8], and 2 M HCl with moderate heat. Isobutylene generated from such acid-catalyzed deprotections can be trapped with alcohols [126]. Under such strongly acidic conditions side products may arise through reaction of the isobutyl carbonium ion that is generated (Figure 4.21). The addition of alcoholic co-solvents can lead to trapping this intermediate as the corresponding ether; BMS researchers found that in order to minimize etherification of the product adding water to the reaction mixture was key [127]. Removing a Boc group from an azetidine was highly effective

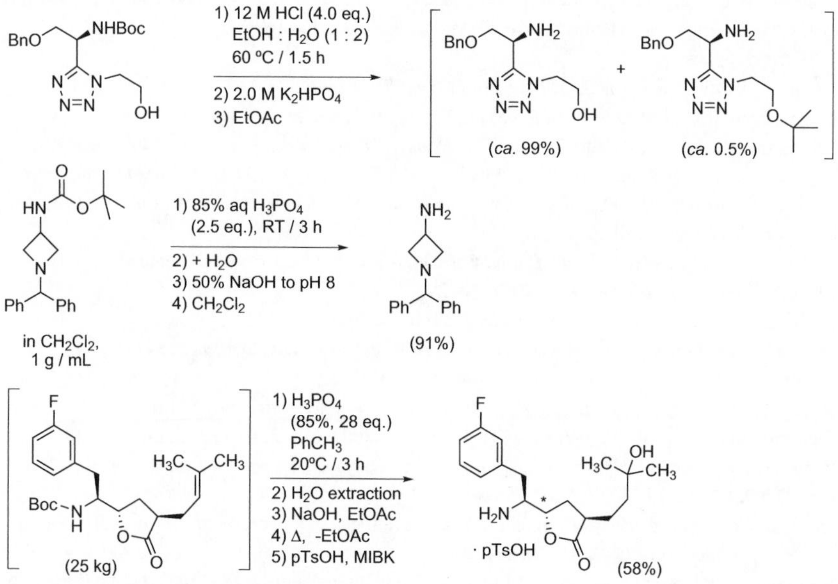

FIGURE 4.21 Reactions to remove Boc groups.

using 85% H_3PO_4; sometimes dichloromethane was added to increase the fluidity of the reaction mixture [128]. (Deprotection of an *N*-Boc azetidine with HCl or TFA led to some dimeric product by attack of the nitrogen on the protonated nitrogen of another azetidine [129]; no mention of such side reactions was made in the Pfizer deprotection shown using 85% H_3PO_4.) Novartis researchers found that 85% H_3PO_4 was effective in removing a Boc group and hydrating an olefin. The two-stage process was sluggish with more dilute H_3PO_4, and some epimerization was seen at the carbon next to the lactone alcoholic oxygen when acids stronger than H_3PO_4 were used [124].

REFERENCES

1. Lu, X. *Synlett* **2011,** 7. preface page.
2. Alt, J. K. *Cooks Illustrated* **2007,** 88(Sept. - Oct.), 30. Reprinted with permission from *Cook's Illustrated* magazine. For a trial issue, call 800-526-8442. Selected articles and recipes, as well as subscription information, are available online at www.cooksillustrated.com. This magazine could be considered a food journal for process chemists, and the editors have compiled recipes into *The Best Recipe*; Boston Common Press: Brookline MA, 1999.
3. Caron, S., Ed.; *Practical Synthetic Organic Chemistry: Reactions, Principles, and Techniques*; Wiley, 2011.
4. Martinelli, M. J.; Vaidyanathan, R.; Pawlak, J. M.; Nayyar, N. K.; Dhokte, U. P.; Doecke, C. W.; Zollars, L. M. H.; Moher, E. D.; Khau, V. V.; Kosmrlj, B. *J. Am. Chem. Soc.* **2002,** *124,* 3578.
5. For the use of catalytic Me_2SnCl_2/K_2CO_3/THF for monobenzoylation of diols, see Maki, T.; Iwasaki, E.; Mausumura, Y. *Tetrahedron Lett.* **1998,** *39,* 5601; Iwasaki, F.; Maki, T.; Onomura, O.; Nakashima, W.; Matsumura, Y. *J. Org. Chem.* **2000,** *65,* 996.
6. For approaches to remove organotin reagents, see Chapter 11. Perhaps Bu_2SnO could be condensed with a lipophilic diol for removal.
7. Park, K.-H.; Kurth, T. M.; Olmstead, M. M.; Kurth, M. J. *Tetrahedron Lett.* **2001,** *42,* 992. SAFETY: NaH and DMF can afford a runaway reaction. See discussion with Figure 2.4.
8. Beaulieu, P. L.; Gillard, J.; Bailey, M. D.; Boucher, C.; Duceppe, J.-S.; Simoneau, B.; Wang, X.-J.; Zhang, L.; Grozinger, K.; Houpis, I.; Farina, V.; Heimroth, H.; Krueger, T.; Schnaubelt, J. *J. Org. Chem.* **2005,** *70,* 5869.
9. Caron, S.; Vazquez, E.; Wojcik, J. M. *J. Am. Chem. Soc.* **2000,** *122,* 712.
10. Mukhopadhyay, C.; Datta, A. *J. Heterocyclic Chem.* **2009,** *46,* 91.
11. Bernasconi, B.; Lee, J.; Roletto, J.; Sogli, L.; Walker, D. *Org. Process Res. Dev.* **2002,** *6,* 152.
12. Tao, J.; Xu, J.-H. Enzymes and Their Synthetic Applications: An Overview In *Biocatalysis for the Pharmaceutical Industry: Discovery, Development, and Manufacturing*; Tao, J., Lin, G.-Q., Liese, A., Eds.; John Wiley, 2009; Chapter 1, pp 1–19.
13. For an example, see Mendiola, J.; García-Cerrada, S.; de Frutos, O.; de la Puente, M. L.; Gu, R. L.; Khau, V. V. *Org. Process Res. Dev.* **2009,** *13,* 292.
14. Li, B.; Su, J.; Tao, J. *Org. Process Res. Dev.* **2011,** *15,* 291.
15. Truppo, M. D.; Rozzell, J. D.; Turner, N. J. *Org. Process Res. Dev.* **2010,** *14,* 234.
16. Hudlicky, T.; Reed, J. W. *Chem. Soc. Rev.* **2009,** *38,* 3117.
17. Meyer, H.-P.; Ghisalba, O.; Leresche, J. E. Biotransformations and the Pharma Industry In *Green Catalysis*, Anastas, P. T., Ed.; Wiley-VCH: Weinheim, 2009, Vol. 3, Chapter 7, pp 171–212.
18. For screening, Codexis has offered many of these biocatalysts in 96 well plates: Thayer, A. *Chem. Eng. News* **2009,** *87*(11), 14.
19. Faber, K. *Biotransformations in Organic Chemistry,* 6th ed.; Springer Verlag: Berlin, 2011.
20. Ondrey, G. *Chem. Eng.* **2010,** *117*(9), 11.
21. Alexandre, F.-R.; Pantaleone, D. P.; Taylor, P. P.; Fotheringham, I. G.; Ager, D. J.; Turner, N. J. *Tetrahedron Lett.* **2002,** *43,* 707.

22. Lalonde, J. J.; Liang, J. In *Asymmetric Catalysis on Industrial Scale: Challenges Approaches and Solutions*; Blaser, H.-U.; Federsel, H.-J., Eds.; 2nd ed.; Wiley-VCH, 2010; p 41.

23. http://www.epa.gov/gcc/pubs/docs/award_recipients_1996_2011.pdf

24. Savile, C. K.; Janey, J. M.; Mundorff, E. C.; Moore, J. C.; Tam, S.; Jarvis, W. R.; Colbeck, J. C.; Krebber, A.; Fleitz, F. J.; Brands, J.; Devine, P. N.; Huisman, G. W.; Hughes, G. J. *Science* **2010**, *329*, 305.

25. Liang, J.; Mundorff, E.; Volari, R.; Jenne, S.; Gilson, L.; Conway, A.; Krebber, A.; Wong, J.; Huisman, G.; Truesdell, S.; Lalonde, J. *Org. Process Res. Dev.* **2010**, *14*, 188.

26. Liang, J.; Lalonde, J.; Borup, B.; Mitchell, V.; Mundorff, E.; Trinh, N.; Kochrekar, D. A.; Cherat, R. N.; Pai, G. G. *Org. Process Res. Dev.* **2010**, *14*, 193.

27. Gooding, O. W.; Voladri, R.; Bautista, A.; Hopkins, T.; Huisman, G.; Jenne, S.; Ma, S.; Mundorff, E. C.; Savile, M. *Org. Process Res. Dev.* **2010**, *14*, 119.

28. Kolata, G. The New York Times, August 29, 2011, http://www.nytimes.com/2011/08/30/science/30microbe.html?_r=1&hp; August 29, 2011.

29. Mundorff, E.; Davis, S. C.; Huisman, G. W.; Krebber, A.; Grate, J. H. U.S. 2009/0118130 A1 (to Codexis).

30. Conrow, R. E.; Harrison, P.; Jackson, M.; Jones, S.; Kronig, C.; Lennon, I. C.; Simmonds, S. *Org. Process Res. Dev.* **2011**, *15*, 301.

31. May, O. http://www.dsm.com/en_US/downloads/dpp/RSS_PLE.pdf

32. Leresche, J. E.; Meyer, H.-P. *Org. Process Res. Dev.* **2006**, *10*, 572.

33. Mahmoudian, M. *Org. Process Res. Dev.* **2011**, *15*, 173.

34. For an example, see Swartz, J. D.; Miller, S. A.; Wright, D. *Org. Process Res. Dev.* **2009**, *13*, 584.

35. Tufvesson, P.; Lima-Ramos, J.; Nordblad, M.; Woodley, J. M. *Org. Process Res. Dev.* **2011**, *15*, 266.

36. Baenziger, M.; Mak, C.-P.; Muehle, H.; Nobs, F.; Prikoszovich, W.; Reber, J.-L.; Sunay, U. *Org. Process Res. Dev.* **1997**, *1*, 395.

37. Perrin, D. D.; Dempsey, B.; Serjeant, E. P. *pK$_a$ Prediction for Organic Acids and Bases*; Chapman and Hall: New York, 1981; p 23.

38. Sykes, P. *A Guidebook to Mechanism in Organic Chemistry*; 6th ed.; Longman: Harlow, Essex, England, 1986; p 67.

39. Mase, T.; Houpis, I. N.; Akao, A.; Dorziotis, I.; Emerson, K.; Hoang, T.; Iida, T.; Itoh, T.; Kamei, K.; Kato, S.; Kato, Y.; Kawasaki, M.; Lang, F.; Lee, J.; Lynch, J.; Maligres, P.; Molina, A.; Nemoto, T.; Okada, S.; Reamer, R.; Song, J. Z.; Tschaen, D.; Wada, T.; Zewge, D.; Volante, R. P.; Reider, P. J.; Tomomoto, K. *J. Org. Chem.* **2001**, *66*, 6775.

40. Kii, S.; Akao, A.; Iida, T.; Mase, T.; Yasuda, N. *Tetrahedron Lett.* **2006**, *47*, 1877.

41. Huang, Q.; Richardson, P. F.; Sach, N. W.; Zhu, J.; Liu, K. K.-C.; Smith, G. L.; Bowles, D. M. *Org. Process Res. Dev.* **2011**, *15*, 556.

42. Weber, W. P.; Gokel, G. W. *Phase Transfer Catalysis in Organic Synthesis*; Springer-Verlag: Berlin, 1977.

43. Starks, C.; Liotta, C.; Halpern, M. *Phase-Transfer Catalysis: Fundamentals, Applications and Industrial Perspectives*; Chapman and Hall: New York, 1994.

44. http://www.phasetransfer.com

45. When TBAB is added to facilitate Heck reactions it may be referred to as the Jeffrey catalyst Jeffrey, T. *Tetrahedron* **1996**, *52*, 10113.

46. Albanese, D.; Landini, D.; Maia, A.; Penso, M. *Ind. Eng. Chem. Res.* **2001**, *40*, 2396.

47. Halpern, M.; Crick, D. *Ind. Phase Transfer Catal.* **2002**, *16*, 1; http://www.ptcorganics.com

48. Doecke, C. W.; Staszak, M. A.; Luke, W. D. *Synthesis*, **1991**, 985.

49. Anderson, N. G.; Deshpande, R. P.; Moniot, J. L. U.S. Patent 6,162,922, 2000 (to Bristol–Myers Squibb).

50. Totleben, M. J.; Prasad, J. S.; Simpson, J. H.; Chan, S. H.; Vanyo, D. J.; Kuehner, D. E.; Deshpande, R.; Kodersha, G. A. *J. Org. Chem.* **2001**, *66*, 1057.

51. Beaulieu, P.; Gillard, J.; Bailey, M.; Beaulieu, C.; Duceppe, J.-S.; Lavallee, P.; Wernic, D. *J. Org. Chem.* **1999**, *64*, 6622.

52. Magni, A. Eur. Patent EP 0, 346, 765, 1989 (to Gruppo LePetit).
53. Hellberg, M. R.; Conrow, R. E.; Sharif, N. A.; McLaughlin, M. A.; Bishop, J. E.; Crider, J. Y.; Dean, W. D.; DeWolf, K. A.; Pierce, D. R.; Sallee, V. L.; Selliah, R. D.; Severns, B. S.; Sproull, S. J.; Williams, G. W.; Zinke, P. W.; Klimko, P. G. *Bioorg. Med. Chem.* **2002**, *10*, 2031.
54. Russell, M. G. N.; Baker, R. J.; Barden, L.; Beer, M. S.; Bristow, L.; Broughton, H. B.; Knowles, M.; McAllister, G.; Patel, S.; Castro, J. L. *J. Med. Chem.* **2001**, *44*, 3881.
55. Ledon, H. J. *Organic Syntheses, Coll.* **1988**, *6*, 414; *Organic Syntheses*, **1979**, *59*, 66.
56. Halpern, M. *Phase Transfer Catal. Commun.* **2000**, *13*, 1.
57. Anderson, N. G.; Carson, J. R. *J. Med. Chem.* **1980**, *23*, 98.
58. Harrington, P. J.; Khatri, H. N.; DeHoff, B. S.; Guinn, M. R.; Boehler, M. A.; Glaser, K. A. *Org. Process Res. Dev.* **2002**, *6*, 120.
59. Qafisheh, N.; Mukhopadhyay, S.; Joshi, A. V.; Sasson, Y.; Chuah, G.-K.; Jaenicke, S. *Ind. Eng. Chem. Res.* **2007**, *46*, 3016.
60. Dozeman, G. J.; Fiore, P. J.; Puls, T. P.; Walker, J. C. *Org. Process Res. Dev.* **1997**, *1*, 137.
61. Dann, N.; Riordan, P. D.; Amin, M. R.; Mellor, M. U.S. Patent 6,921,828 (2005, to Bayer CropScience S.A.).
62. Benedetti, F.; Berti, F.; Norbedon, S. *J. Org. Chem.* **2002**, *67*, 8635.
63. Barrett, R.; Caine, D. M.; Cardwell, K. S.; Cooke, J. W. B.; Lawrence, R. M.; Scott, P.; Sjolin, Å *Tetrahedron: Asymmetry* **2003**, *14*, 3267.
64. Halpern, M. E. In *Phase-Transfer Catalysis*; Halpern, M., Ed.; ACS Symposium Series; American Chemical Society: Washington, DC, 1997; pp 97–107.
65. Ripin, D. H. B. In *Practical Synthetic Organic Chemistry: Reactions, Principles, and Techniques*; Caron, S., Ed.; Wiley, 2011; p 771.
66. Maruoka, K., Ed.; *Asymmetric Phase Transfer Catalysis*; Wiley: New York, 2008.
67. Patterson, D. E.; Xie, S.; Jones, L. A.; Osterhout, M. H.; Henry, C. G.; Roper, T. D. *Org. Process Res. Dev.* **2007**, *11*, 624.
68. Landini, D.; Maia, A.; Rampoldi, A. *J. Org. Chem.* **1986**, *51*, 3187.
69. An, J.; Sun, S.; Bai, L.; Chen, T.; Liu, D. Q.; Kord, A. *J. Pharm. Biomed. Anal.* **2008**, *48*, 1006.
70. For an example using TBAB for a Pd-catalyzed coupling, see Lipton, M. F.; Mauragis, M. A.; Maloney, M. T.; Veley, M. F.; VanderBor, D. W.; Newby, J. J.; Appell, R. B.; Daugs, E. D. *Org. Process Res. Dev.* **2003**, *7*, 385.
71. For a PTC example using 1.4 equivalents of Aliquat 336, see Akao, A.; Hiraga, S.; Iida, T.; Kamatani, A.; Kawasaki, M.; Mase, T.; Nemoto, T.; Satake, N.; Weissman, S. A.; Tschaen, D. M.; Rossen, K.; Petrillo, D.; Reamer, R. A.; Volante, R. P. *Tetrahedron* **2001**, *57*, 8917. Two ionizable groups were present in the starting material.
72. Halpern, M. In *Phase Transfer Catalysis: Mechanisms and Synthesis*; Halpern, M., Ed.; American Chemical Society: Washington, DC, 1997, Chapter 8, p 97.
73. Halpern, M.; Grinstein, R. *Phase-Transfer Catal. Commun.* **1998**, *4*(2), 17.
74. Vidal, J.; Damestoy, S.; Guy, L.; Hannachi, J.-C.; Aubry, A.; Collet, A. *Chem. Eur. J.* **1997**, *3*, 1691; According to Ref. 37 cited therein (Hay, R. W.; Porter, L. J. *J. Chem. Soc.* **1967**, 1261), the pK_a of the amino groups increase from 7.6 for Val-OMe to pK_a 9.7 for valine carboxylate.
75. Journet, M.; Cai, D.; Hughes, D. L.; Kowal, J. J.; Larsen, R. D.; Reider, P. J. *Org. Process Res. Dev.* **2005**, *9*, 490.
76. Floyd, D. M.; Fritz, A. W.; Pluscec, J.; Weaver, E. R.; Cimarusti, C. M. *J. Org. Chem.* **1983**, *47*, 5160.
77. Leadbeater, N. E.; Marco, M. *J. Org. Chem.* **2003**, *68*, 5660.
78. Urganonkar, S.; Verkade, J. G. *J. Org. Chem.* **2004**, *69*, 5752.
79. Sato, K.; Aoki, M.; Ogawa, M.; Hashimoto, T.; Noyori, R. *J. Org. Chem.* **1996**, *61*, 8310.
80. Sato, K.; Aoki, M.; Takagi, J.; Noyori, R. *J. Am. Chem. Soc.* **1997**, *119*, 12386; See also Hronec, M. In *Handbook of Phase Transfer Catalysis*; Sasson, Y.; Neumann, R., Eds.; Chapman and Hall: London, 1997; pp 317–335.
81. Sato, K.; Aoki, M.; Noyori, R. *Science* **1998**, *281*, 1646.

82. For a thorough discussion on peptides, see Lloyd-Williams, P.; Albericio, F.; Giralt, E. *Chemical Approaches to the Synthesis of Peptides and Proteins*; CRC Press: Boca Raton, Florida, 1997.
83. Carpino, L. A.; Sadat-Aalaee, D.; Chao, H. G.; DeSelms, R. H. *J. Am. Chem. Soc.* **1990,** *112*, 9651.
84. Wipf, P.; Uto, Y. *J. Org. Chem.* **2000,** *65*, 1037.
85. Dunn, P. J.; Hoffmann, W.; Kang, Y.; Mitchell, J. C.; Snowden, M. *J. Org. Process Res. Dev.* **2005,** *9*, 956.
86. Bright, R.; Dale, D. J.; Dunn, P. J.; Hussain, F.; Kang, Y.; Mason, C.; Mitchell, J. C.; Snowden, M. J. *Org. Process Res. Dev.* **2004,** *8*, 1054.
87. Dunn, P. J.; Hughes, M. L.; Searle, P. M.; Wood, A. S. *Org. Process Res. Dev.* **2003,** *7*, 244.
88. Dale, D. J.; Dunn, P. J.; Golightly, C.; Hughes, M. L.; Levett, P. C.; Pearce, A. K.; Searle, P. M.; Ward, G.; Wood, A. S. *Org. Process Res. Dev.* **2000,** *4*, 17.
89. Viadyanathan, R.; Kalthod, V. G.; Ngo, D. P.; Manley, J. M.; Lapekas, S. P. *J. Org. Chem.* **2004,** *69*, 2565.
90. Chaudhary, A.; Girgis, M. J.; Prashad, M.; Hu, B.; Har, D.; Repic, O.; Blacklock, T. *J. Org. Process Res. Dev.* **2003,** *7*, 888.
91. Woodman, E. K.; Chaffey, J. G. K.; Hopes, P. A.; Hose, D. R. J.; Gilday, J. P. *Org. Process Res. Dev.* **2009,** *13*, 106.
92. Weisenberger, G. A.; Anderson, D. K.; Clark, J. D.; Edney, A. D.; Karbin, P. S.; Gallagher, D. J.; Knable, C. M.; Pietz, M. A. *Org. Process Res. Dev.* **2009,** *13*, 60.
93. Woodman, E. K.; Chaffey, J. G. K.; Hopes, P. A.; Hose, D. R. J.; Gilday, J. P. *Org. Process Res. Dev.* **2009,** *13*, 106.
94. Lorenz, J. Z.; Busacca, C. A.; Feng, X.; Grinberg, N.; Haddad, N.; Johnson, J.; Kapadia, S.; Lee, H.; Saha, A.; Sarvestani, M.; Spinelli, E. M.; Varsolona, R.; Wei, X.; Zeng, X.; Senanayake, C. H. *J. Org. Chem.* **2010,** *75*, 1155.
95. Evans, D. J. *Specialities Chemicals Magazine*, September 2005, 50.
96. Zalipsky, S. *Chem. Commun* **1998,** *69*.
97. Hlebowicz, E.; Andersen, A. J.; Andersson, L.; Moss, B. A. *J. Peptide Res.* **2005,** *65*, 90.
98. For an interesting review of the uses of DMAP and side reactions, see Höfle, G.; Steglich, W.; Vorbrüggen, H. *Angew. Chem. Int. Ed. Engl.* **1978,** *17*, 569.
99. Decades ago syntheses of steroidal intermediates were sometimes carried out with several oxidations and reductions, often using excesses of chromium-based reagents. For fine examples of Merck's work from an earlier era, see Pines, S. H. *Org. Process Res. Dev.* **2004,** *8*, 708.
100. Caron, S.; Dugger, R. W.; Ruggeri, S. G.; Ragan, J. A.; Ripin, D. H. B. *Chem. Rev.* **2006,** *106*, 2943.
101. Parikh, J. R.; von Doering, W. *J. Am. Chem. Soc.* **1967,** *89*, 5505.
102. McCarthy, J. R.; Matthews, D. P.; Sabol, J. S.; McConnell, J. R.; Donaldson, R. E.; Duguid, R. U.S. Patent 5,760,210 (to Merrell Pharmaceuticals), 1998.
103. Ng, J.; Przybyla, C.; Liu, C.; Yen, J. C.; Muellner, F. W.; Weyker, C. L. *Tetrahedron* **1995,** *51*, 6397.
104. For TFAA/DMSO oxidation at −15 °C, see Appell, R. B.; Duguid, R. *J. Org. Process Res. Dev.* **2000,** *4*, 172.
105. Adding pyridine to minimize acid-catalyzed side reactions: Chen, L.; Lee, S.; Renner, M.; Tian, Q.; Nayyar, N. *Org. Process Res. Dev.* **2006,** *10*, 163.
106. Fleitz, F. J.; Lyle, T. J.; Zheng, N.; Armstrong, J. D., III; Volante, R. P. *Synth. Commun.* **2000,** *30*, 3171.
107. Shu, L.; Shi, Y. *J. Org. Chem.* **2000,** *65*, 8807.
108. Hashimoto, N.; Kanda, A. *Org. Process Res. Dev.* **2002,** *6*, 405.
109. Ager, D. J.; Anderson, K.; Oblinger, E.; Shi, Y.; VanderRoest, J. *Org. Process Res. Dev.* **2007,** *11*, 44.
110. De Luca, L.; Giacomelli, G.; Masala, S.; Porcheddu, A. *J. Org. Chem.* **2003,** *68*, 4999.
111. TCCA may be about one-fourth the cost of NCS per chlorine Tilstam, U.; Weinmann, H. *Org. Process Res. Dev.* **2002,** *6*, 384.
112. Janssen, M. H. A.; Chesa Castellana, J. F.; Jackman, H.; Dunn, P. J.; Sheldon, R. A. *Green Chem.* **2011,** *13*, 905.

113. For oxidizing a secondary alcohol to the ketone the range of pH 8.5–10.5 was found to be optimal Alorati, A. D.; Bio, M. M.; Brands, K. M. J.; Cleator, E.; Davies, A. J.; Wilson, R. D.; Wise, C. S. *Org. Process Res. Dev.* **2007,** *11,* 637.

114. For large-scale application of ozonolysis to drug synthesis, see Van Ornum, S. G.; Champeau, R. M.; Pariza, R. *Chem. Rev.* **2006,** *106,* 2990.

115. Fleck, T. J.; McWhorter, W. W.; DeKam, R. N.; Pearlman, B. A. *J. Org. Chem.* **2003,** *68,* 9612.

116. Allian, A. D.; Richter, S. M.; Kallemeyn, J. M.; Robbins, T. A.; Kishore, V. *Org. Process Res. Dev.* **2011,** *15,* 91.

117. am Ende, D. J.; Vogt, P. F. *Org. Process Res. Dev.* **2003,** *7*(6), 1029.

118. Couturier, M.; Andresen, B. M.; Jorgensen, J. B.; Tucker, J. L.; Busch, F. R.; Brenek, S. J.; Dubé, P.; am Ende, D. J.; Negri, J. T. *Org. Process Res. Dev.* **2002,** *6,* 42.

119. Atkins, W. J., Jr.; Burkhardt, E. R.; Matos, K. *Org. Process Res. Dev.* **2006,** *10,* 1292.

120. Alimardanov, A. R.; Barrila, M. T.; Busch, F. R.; Carey, J. J.; Couturier, M.; Cui, C. *Org. Process Res. Dev.* **2004,** *8,* 834.

121. Parker, D. K.; Roberts, R. F.; Schiessl, H. W. *Rubber Chem. Technol.* **1992,** *65,* 245; Parker, D. K., In *Polymeric Materials Encyclopedia*; Salamone, J. C., Ed.; CRC Press: Boca Raton, Vol. 3, p 2048.

122. Colyer, J. T.; Andersen, N. G.; Tedrow, J. S.; Soukup, T. S.; Faul, M. M. *J. Org. Chem.* **2006,** *71,* 6859.

123. Norris, T.; Bezze, C.; Franz, S. Z.; Stivanello, M. *Org. Process Res. Dev.* **2009,** *13,* 354.

124. Li, B.; Andresen, B.; Brown, M. F.; Buzon, R. A.; Chiu, C. K.-F.; Couturier, M.; Dias, W.; Urban, F. J.; Jasys, V. J.; Kath, J. C.; Kissel, W.; Le, T.; Li, Z. J.; Negri, J.; Poss, C. S.; Tucker, J.; Whritenour, D.; Zandi, K. *Org. Process Res. Dev.* **2005,** *9,* 466.

125. Prashad, M.; Har, D.; Hu, B.; Kim, H.-Y.; Girgis, M. J.; Chaudhary, A.; Repic, O.; Blacklock, T. *J. Org. Process Res. Dev.* **2004,** *8,* 331.

126. Dias, E. L.; Hettenbach, K. M.; am Ende, D. *J. Org. Process Res. Dev.* **2005,** *9,* 39.

127. Davulcu, A. H.; McLeod, D. D.; Li, J.; Katipally, K.; Littke, A.; Doubleday, W.; Xu, Z.; McConlogue, C. W.; Lai, C. J.; Gleeson, M.; Schwinden, M.; Parsons, R. L., Jr. *J. Org. Chem.* **2009,** *74,* 4068.

128. 58% H_3PO_4 has also been used to deprotect *t*-butyl esters and *t*-butyl ethers Li, B.; Berliner, M.; Buzon, R.; Chiu, C. K.-F.; Colgan, S. T.; Kaneko, T.; Keene, N.; Kissel, W.; Le, T.; Leeman, K. R.; Marquez, B.; Morris, R.; Newell, L.; Wunderwald, S.; Witt, M.; Weaver, J.; Zhang, Z.; Zhang, Z. *J. Org. Chem.* **2006,** *71,* 9045.

129. Lynch, J. K.; Holladay, M. W.; Ryther, K. B.; Bai, H.; Hsiao, C.-N.; Morton, H. E.; Dickman, D. A.; Arnold, W.; King, S. A. *Tetrahedron: Asymmetry* **1998,** *9,* 2791.

Solvent Selection

"Solvents in routine GSK development processes consistently average between 85 and 90% of the mass contribution to the overall process mass leading to an API."

– D. J. C. Constable, C. Jimenez-González, and R. K. Henderson [1]

I. INTRODUCTION

Solvents can have huge impacts on processing costs, and on the environment. Many green chemistry principles are reflected in using solvents, such as minimizing waste from excess solvents, minimizing energy to recover and recycle solvents, SAFELY using solvents to preclude accidents, designing safer solvents from sustainable resources, and minimizing pollution from the escape of solvents to the environment. The ACS Green Chemistry Institute Pharmaceutical Roundtable has ranked solvents based on SAFETY, health, the environmental impact on air and water, and the environmental impact of waste [2]. A three-volume set discusses aspects of green chemistry regarding solvents [3].

Solvents may be classified into several groups, sometimes overlapping. Solvents within these groups may display surprising differences in abilities to solubilize materials, and in abilities to accelerate or impede reactions. Reichardt has thoroughly discussed characterization and uses of solvents [4,5]. Categories of solvents include:

- protic, or hydrogen bond-donating solvents (Lewis acids), such as H_2O, EtOH, AcOH, and NH_3

Practical Process Research and Development. DOI: 10.1016/B978-0-12-386537-3.00005-8
                                                  121

- hydrogen bond-accepting solvents (Lewis bases), such as H_2O, Et_3N, EtOAc, acetone, and DMF
- polar aprotic solvents, not hydroxylic, such as DMSO, DMF and DMAc
- chlorocarbon solvents, such as dichloromethane, $CHCl_3$, and CCl_4
- fluorocarbon solvents, such as hexafluoroisopropanol
- hydrocarbon solvents, such as heptanes, iso-octane, and toluene
- ionic liquids
- supercritical gases, such as supercritical CO_2

Solvation is the surrounding of dissolved solutes by solvents, and solvation by H_2O is termed hydration. The solvation number refers to the number of solvent molecules surrounding an ion, and in general the solvation number increases with increased charge and decreased ion size [6]. The reactivity of a species increases with decreasing solvation, as though the solvating molecules shield a reactive species and disperse its charge. One part of a molecule may be preferentially solvated by a solvent; for instance, dipolar aprotic solvents such as DMSO solvate cations, rendering the anionic counterparts more reactive [7]. Crown ethers, often used in phase-transfer catalysis (PTC), similarly complex with cations, rendering anions more reactive [8] (Figure 5.1). Two solvents in a mixture of solvents may solvate different parts of a molecule, making a combination of solvents more effective in dissolving a compound than either solvent alone [9]. A striking example of how decreased solvation affects reactivity was found with NaOH: the basicity of NaOH increased 50,000 times when solid NaOH (three molecules of hydration) was used instead of 15% NaOH (11 molecules of hydration) [10]. (PTC has been said to generate "naked anions" or "bare anions," but a small amount of water may be necessary, especially for solid–liquid PTC reactions [11]. Water content is a key parameter in developing PTC processes, as discussed in Chapter 4.) Solvation is one of the many important aspects for selecting solvents.

The foremost reason to select reaction solvents carefully is to provide SAFE, non-hazardous scale-up conditions for equipment and operators, and to minimize the harm to operators, analysts, and patients. Physicochemical characteristics of solvents, such as polarity, boiling point and water miscibility, influence reaction rates, phase separations, the effectiveness of crystallization, and the removal of volatile components by azeotroping or drying of solids. Other physicochemical characteristics such as viscosity of a mixture affect mass transfer and heat transfer, influencing byproduct formation and physical transfers. The environmental impact of solvents is becoming increasingly important. All of these factors, including the ease of recovering and reusing solvents, can greatly influence the CoG of a product. Reichardt has published extensively on solvents [4,5]. A consortium of researchers described

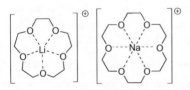

FIGURE 5.1 Complexes of 15-crown-5 and 18-crown-6.

quantitative modeling approaches to solvent selection based on physicochemical data, and showed GSK's ranking of 47 solvents [12]. Caron has extensively tabulated data on solvents and ranked them for utility on scale [13]. A recent chapter has detailed considerations for ranking the "greenness" of solvents, including environmental impact as volatile organic compounds (VOCs), over the life cycle from generation to use to recovery to incineration [14].

Solvents can have a big impact on operations and the CoG of a product. GSK researchers determined that solvents comprised 85–90% of the mass in API development processes, with about 50% of solvent used being recovered and recycled [1]. Solvent recovery and recycling can consume a great deal of operations for routine manufacturing of APIs. High-boiling solvents such as DMF and DMSO are impractical to recover, and chemists and engineers seek alternatives for routine manufacturing.

The best reaction solvent may be the one that crystallizes the product directly from the reaction. Novartis workers developed a process in which a starting material was heated to reflux in ethylene glycol–EtOH–H_2O (2:2:1) leading to what is formally a migration of the aryl group from a ring to a distal nitrogen atom; upon cooling the product from this Dimroth rearrangement crystallized and was isolated in excellent yield [15,16] (Figure 5.2). This solvent mixture was chosen to include H_2O and allow the reaction to occur in a short time frame. (Given the toxicity of ethylene glycol, propylene glycol (1,2-propane diol) may be a good substitute.) As another example of a practical process facilitated by the right choice of solvent, the tosylate salt crystallized directly from deprotection of the Boc-amine [17].

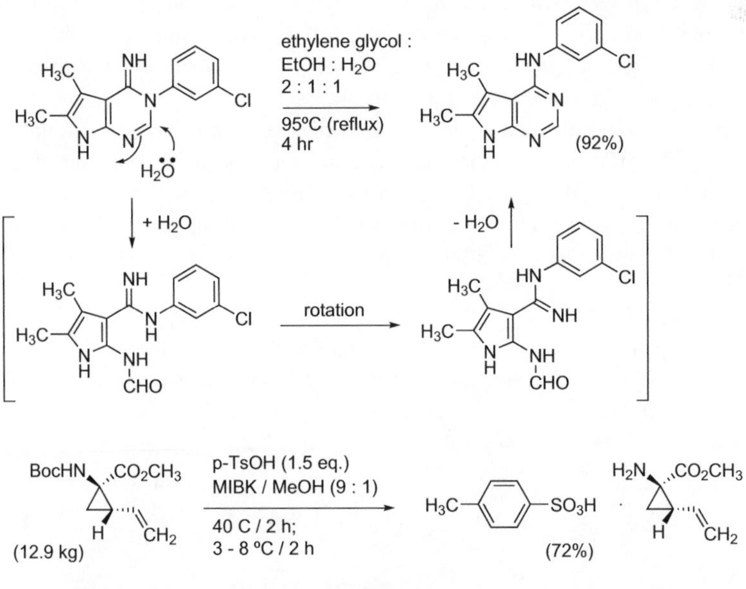

FIGURE 5.2 Crystallization of products directly from reaction mixtures.

A 4-h reaction in the laboratory may be scaled up during two 8-h shifts in a pilot plant. The remaining 12 h is consumed by delivering materials to reactor sites, charging materials, heating and cooling, sampling and analyses, workup and isolation, and the inevitable delays. A reaction time of 2–4 h in the laboratory is convenient for batch processing during one day, still allowing you time to go home at night.

The first key principle of solvent selection for rapid scale-up is that homogeneous reactions are usually much faster and easier to scale-up than heterogeneous reactions, especially suspensions of solids. If heterogeneous conditions are necessary, solvent and reaction conditions should be selected so that the reaction mixture is fluid and easily mixed. (Efficient agitation is extremely important for conventional hydrogenations, which are solid–liquid–gas mixtures.) In many cases the reaction can be driven by conditions in which the product separates out, preferably by crystallization, but not by precipitating or oiling, which can entrap starting materials (see Chapter 12). An example of heterogeneous conditions that proved troublesome on scale-up is shown in Figure 5.3. In the initial scale-up, the dense Na_2CO_3 stayed in the vortex of reactor, and was less available for reaction. PTC conditions created homogeneous conditions for easier scale-up, and increased the reactivity of cyanide. Furthermore, these conditions were always basic, thus avoiding the generation of HCN [18]. GSK workers have similarly discussed the difficulties of implementing on scale a reaction using insoluble Cs_2CO_3 [19].

Heterogeneous conditions can be very advantageous for some processes. An overview of reaction conditions is shown in Table 5.1. Heterogeneous reaction conditions may be employed to facilitate reactions or to minimize degradation of the product during reaction conditions. PTC often employs two immiscible solvents, with the reaction taking place in the organic phase or at the interface. Sometimes solid–liquid PTC conditions are employed, with a base such as K_2CO_3 being suspended in the reaction [20]. Schotten–Baumann conditions often employ two immiscible solvents, with decomposition of the reagents and product being minimized by dissolution in the organic phase, away from the base (for example, see Figure 5.5).

Heterogeneous reaction conditions are sometimes developed with the thought that the starting materials will dissolve as they react. Some reactions are suspensions throughout, as with the CIAT and CIDR transformations of Chapter 12 or the on-water reactions developed by Sharpless and others (Section V). Solvents can be selected to provide some solubility of the components for effective processing, as with the addition of EtOH or

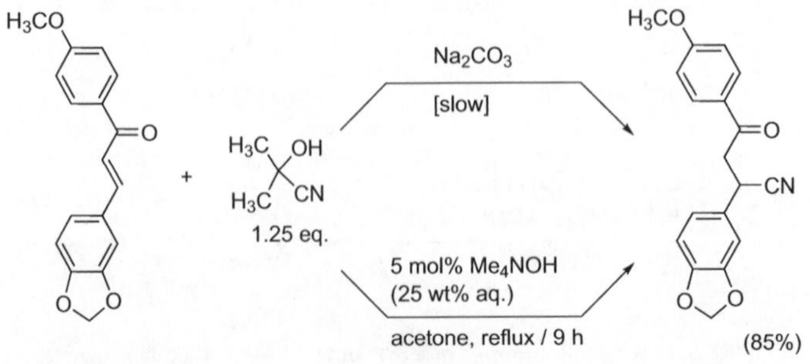

FIGURE 5.3 Successful scale-up by employing a soluble base.

TABLE 5.1 Overview of Reaction Conditions

Initial Conditions				
Starting Material(s)	**Reagent(s)**	**Final Conditions**	**Examples**	**Comments**
Soluble	Soluble	Solution Suspension	Many Direct isolation	Usually rapid reactions May add antisolvent to crystallize
Soluble (usually in organic phase)	Insoluble (solids or dissolved in aqueous phase)	Suspension Suspension Liquid–liquid	Liquid–liquid PTC Solid–liquid PTC Schotten–Baumann reactions Ester hydrolysis	Reactions may occur at interface Separating starting materials and products from reactive phase may improve stability Product may partition into other phase
Insoluble	Soluble	Solution	Hydrolysis of ester in aqueous hydroxide/MeOH	Catalyst may not be necessary
Insoluble	Insoluble	Suspension	On-water reactions CIAT/CIDR	Reactions may be suspensions throughout. Care may be needed for extended reactions to ensure that fines are not produced, which could blind filters

DMSO to reactions in water [21]. In an example from Merck, bromination of the arene with dibromodimethylhydantoin (DBDMH) in H_2SO_4 produced five byproducts totaling 30% of the mass (Figure 5.4); when a small volume of AcOH was added the reaction mixture was agitated more readily and the aryl bromide was produced in high yield [22]. In other work described by Merck the hydrochloride salt of 4-chloro-piperidine was deemed too insoluble for good conversion to the desired iminium salt, and various salts were screened. The triflic acid salt dissolved in 2,2-dimethoxypropane and was converted rapidly to the iminium salt. The HBr salt was only partially soluble in 2,2-dimethoxypropane but readily formed the iminium species. Because the HBr salt precipitated under the reaction conditions it was selected over the more expensive triflate salt, as it allowed purification by filtration and washing. (Ultimately the Merck researchers found that reaction with 2,2-dimethoxypropane could be carried out in the presence of NaCl, thus avoiding isolation of the HBr salt of 4-chloro-piperidine.) [23]. Heterogeneous conditions

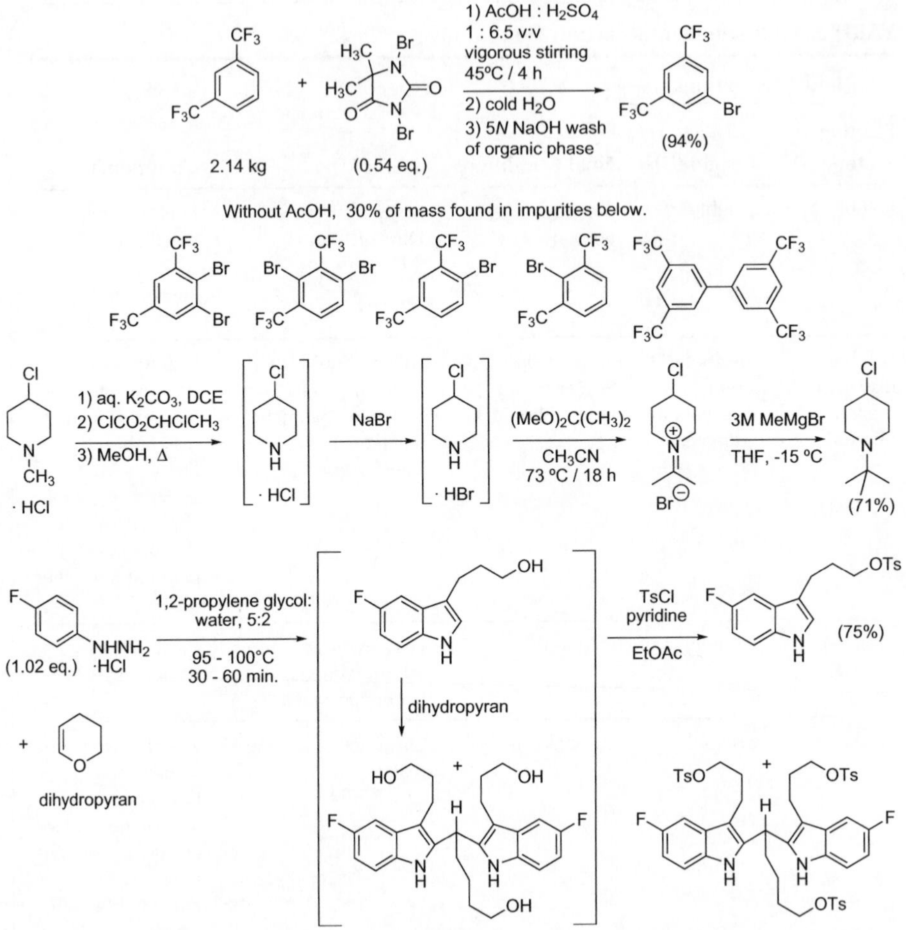

FIGURE 5.4 Some heterogeneous reaction conditions.

can also promote side reactions, as shown with the Fischer indolization in Figure 5.4; under conditions where the starting material was poorly soluble the desired product was slowly formed in solution and reacted further with what was effectively a large excess of reagent to afford the triol byproduct. Unfortunately even small amounts of the triol, when tosylated, greatly reduced the isolated yield of the desired tosylate. A mixture of propylene glycol–H_2O (5:2) was found to solubilize the starting material at 90–95 °C, and to minimize the formation of the triol the dihydropyran was added over 30 min [24]. Thus heterogeneous reaction conditions can provide conditions for ready conversion to the desired product, but extra care may be needed.

Amides are often prepared on scale using Schotten–Baumann conditions, that is, condensing an amine with an acid chloride or anhydride using aqueous alkali to neutralize the acid that is formed. (Without additional base the theoretical yield of the reaction of one equivalent each of an amine with an acid chloride is only 50%.) If no organic co-solvent is used the amide product may precipitate and occlude starting materials [25], so organic

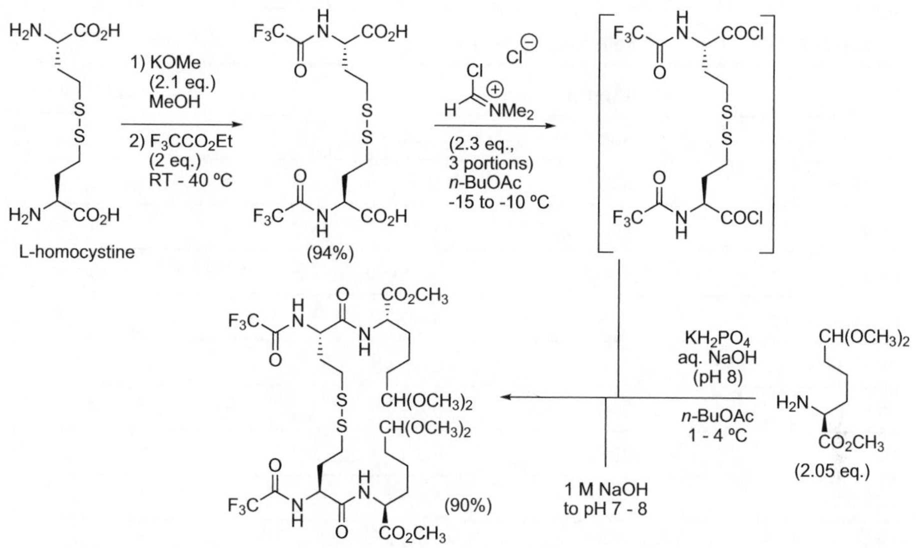

FIGURE 5.5 Acid chloride formation and Schotten–Baumann coupling under heterogeneous conditions.

solvents are often used. Water-immiscible solvents may be employed to decrease the degradation of reagents or products that are readily hydrolyzed. BMS workers developed heterogeneous conditions to control the Schotten–Baumann processing of an acid chloride to amides [26] (Figure 5.5). In these reactions the amine groups were protected by a tri-fluoroacetyl group, which was readily removed by hydrolysis under basic conditions [27]. The carboxylic acid groups were converted to the acid chlorides [28] by the Vilsmeier reagent in n-BuOAc, in which that reagent was relatively insoluble. Although the Vilsmeier reagent was found to be soluble in dichloromethane, these reaction conditions were more exothermic; more care was needed to carry out acid chloride formation in dichloromethane as temperatures above $-10\ °C$ led to racemization through azlactone formation (see Chapter 7). DMF was also unsuitable for acid chloride formation, as racemization was quite high. (DMF and chlorinating reagents can form dimethylcarbamoyl chloride (DMCC), which is an animal carcinogen at ppb levels. The formation and destruction of DMCC is discussed in Chapter 13.) Above pH 8 during the Schotten–Baumann coupling more hydrolysis of the acid chlorides, azlactone formation, and racemization were seen, and below pH 7 the coupling was slowed as more of the amine was protonated. For optimal performance the reaction was buffered to pH 8 and maintained at pH 7–8 during the acid chloride addition by the simultaneous addition of 1 M NaOH. This coupling to make the vanlev intermediate was successfully carried out with inputs of over 100 kg of bis-trifluoroacetyl homocystine.

Solvents commonly used in academia may not be welcomed for industrial applications. Almost any solvent *can* be used in pilot plant and manufacturing settings, but using a hazardous solvent may dramatically slow operations due to SAFETY considerations required, such as measures to prevent leaks of flammable gases and corrosive liquids and wearing cumbersome PPE. Exposure limits for some solvents are shown in Table 5.2; these limits were calculated by the American Conference of Governmental Industrial Hygienists

TABLE 5.2 TLV Limits for Selected Solvents (1,2)

Solvent	TWA (ppm)	Solvent	TWA (ppm)
Acetone	500	MTBE	50
Hexanes	500	THF	50
EtOAc	400	NH_3	25
Heptane	400	Acetonitrile	20
i-Octane	300	n-BuOH	20
i-PrOH	200	Dioxane	20
MeOH	200	MIBK	20
MeOAc	200	$PhCH_3$	20
i-BuOAc	150	AcOH	10
t-BuOH	100	ClPh	10
Cyclohexane	100	1,2-Dichloroethane	10
i-PrOAc	100	DMAc	10
Xylenes	100	DMF	10
CH_2Cl_2	50	Pyridine	1
n-Hexane	50	Et_3N	1
		2-Methoxyethanol (methyl cellosolve)	0.1

(1) 2009 TLVs® and BEIs®, American Conference of Governmental Industrial Hygienists (ACGIH). http:www.acgih.org.
(2) TLV is the Threshold Limit Value, or the concentration of a contaminant in air considered safe for routine daily exposure for nearly all workers. TWA is the Time-Weighted Average for safe exposure over an 8-h day for five days for nearly all workers. The concentration, ppm, is defined as molecules of substance in 10^6 molecules of contaminated air.

(ACGIH) based on toxicological data [29]. A solvent may be avoided for SAFETY reasons due to its low flash point, the temperature at which vapors will ignite in the presence of a source of ignition. Highly flammable solvents and reagents generated in them, such as MeLi in Et_2O, may be restricted to ground transportation. Some solvents rarely used on scale are shown in Table 5.3.

Solvent usage has been a major effort at Pfizer, in terms of solvent selection, minimizing solvent usage, and recovering and reusing solvents. Pfizer researchers have grouped solvents into the three categories shown in Table 5.4.

> Expect the chemical engineer or process equipment owner to demand justification for using a hazardous solvent on scale. And for any operation, for that matter.

TABLE 5.3 Solvents rarely used in the Pharmaceutical Industry

Solvent	Negative Characteristic	Alternative Solvent
Et$_2$O	Flammable	MTBE
(iPr)$_2$O	Peroxide formation	MTBE (1)
HMPA	Toxicity	NMP, DMI (2), DMF, DMAc
Pentane	Flammable	Heptanes, i-octane
Hexane	Electrostatic discharge (3) Neurological toxicity (4)	Heptanes, i-octane
CHCl$_3$	Mutagenicity, environmental, toxicity due to degradant COCl$_2$ (5)	CH$_2$Cl$_2$; 2-MeTHF; PhCH$_3$ w/ CH$_3$CN, n-BuOH or DMF
CCl$_4$, ClCH$_2$CH$_2$Cl	Mutagen, environmental	CH$_2$Cl$_2$
Benzene	Toxicity	Toluene
Ethylene glycol, diethylene glycol (6)	Toxicity	1,2-Propanediol
HOCH$_2$CH$_2$OR (R = Me, Et: cellosolves)	Animal teratogen	1,2-Propanediol
MeOCH$_2$CH$_2$OMe (glyme); dioxane	Animal teratogen suspected carcinogen (7)	THF, 2-Me-THF, (8) DEM (9)
Acetonitrile	Animal teratogen; may generate acetamide, a genotoxin (10)	i-PrOH, acetone–H$_2$O

(1) Laird, T. Org. Process Res. Dev. **2004**, 8, 815.
(2) N,N-dimethyl imidazolidinone (DMI) is more expensive than NMP and is used occasionally.
(3) Straight-chain hydrocarbons with even numbers of carbons are more prone to electrostatic discharge: Ref on static discharge.
(4) Neurological toxicity of hexane, perhaps through formation of hexane-2,5-dione.
(5) Chloroform preserved with amylene was found to be less stable than chloroform preserved with EtOH. Turk, E. Chem. Eng. News **1998**, 76 (9), 6.
(6) Diethylene glycol killed ≥5 in China when substituted for propylene glycol: Tremblay, J.-F. Chem. Eng. News **2006**, 84 (21), 11.
(7) Report on Carcinogens, 11th ed.: http://www.pall.com/pdf/11th_ROC_1_4_Dioxane.pdf.
(8) Aycock, D. F. Org. Process Res. Dev. **2007**, 11, 156.
(9) Boaz, N. W.; Venepalli, B. Org. Process Res. Dev. **2001**, 5, 127.
(10) The possible presence of acetamide in process streams using acetonitrile necessitates controlling and analyzing for acetamide at trace levels. Other solvents may be preferred for preparing APIs and penultimates.

Dichloromethane is a solvent shunned by some, but accepted by others. Due to its low boiling point of 42 °C, containing it and preventing its escape to the environment are difficult, thus causing concerns not only about recovery and reuse but also about environmental impact, especially in Europe. These concerns may cause some CROs and CMOs to shun any process using dichloromethane. GSK has substituted MIBK or

TABLE 5.4 Solvents Preferred for Process Development (1,2)

Preferred	Usable	Undesirable
H$_2$O	Cyclohexane	Pentane
Acetone	Heptane	Hexane(s)
EtOH	Toluene	Diisopropyl ether
i-PrOH	Methyl cyclohexane	Diethyl ether
EtOAc	MTBE	Dichloroethane
i-PrOAc	iso-octane	Dichloromethane
MeOH	Acetonitrile	Chloroform
Methyl ethyl ketone (MEK)	2-Me-THF	DMF
n-BuOH	THF	NMP
t-BuOH	Xylenes	Pyridine
	DMSO	DMAc
	AcOH	Dioxane
	Ethylene glycol	DME
		Benzene
		Carbon tetrachloride

(1) Alfonsi, K.; Colberg, J.; Dunn, P. J.; Fevig, T.; Jennings, S.; Johnson, T. A.; Kleine, H. P.; Knight, C.; Nagy, M. A.; Perry, D. A.; Stefaniak, M. Green Chem. **2008**, 10, 31.
(2) http://www.pfizer.com/responsibility/protecting_environment/green_chemistry_solvent_guide.jsp.

EtOAc for CH$_2$Cl$_2$ [30], perhaps in some extractive workups. Unfortunately there may be no adequate substitute for dichloromethane for some reactions. For instance, in the reduction of an oseltamivir precursor in the presence of a halogenated Lewis acid dichloromethane was the only solvent suitable for reduction, as there was no reaction in MTBE, PhCH$_3$, EtOAc, or THF (Figure 5.4) [31]. In another example in Figure 5.6 the acid-mediated deprotection of an aztreonam penultimate was carried out in dichloromethane; reaction in EtOAc was too slow to be practical [32], and reactions would probably be retarded in other solvents that would be Lewis bases. In circumstances such as these dichloromethane may be chosen, necessitating strict measures to contain that solvent. The only other choice may be to develop another route.

Polarity is a key parameter for solvents. The dielectric constant measures the ability of a solvent to conduct electrical charges. Gutmann donor numbers essentially measure the Lewis basicity of molecules of a solvent [5]. The Hildebrand parameter was used to develop the Hansen Solubility Parameters, which consider van der Waals interactions, dipole interactions, and hydrogen bonding [33]. Reichardt's E_T^N parameter [34] is based on the negative solvatochromism of the $\pi \rightarrow \pi^*$ shifts of solutions of the betaine dye

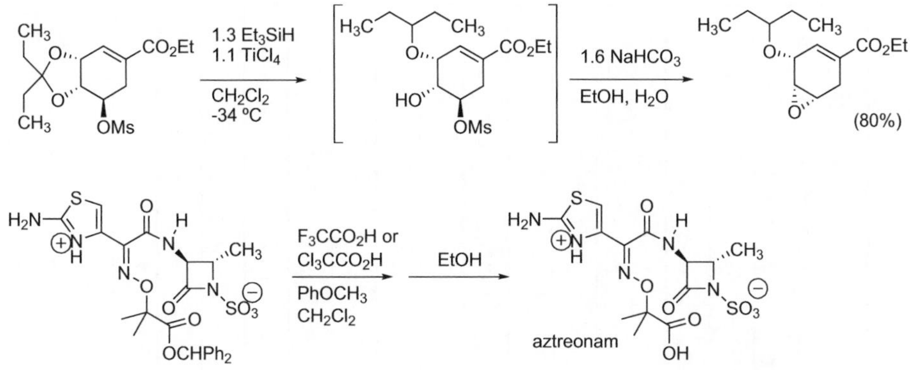

FIGURE 5.6 Two reactions without an acceptable substitute for dichloromethane.

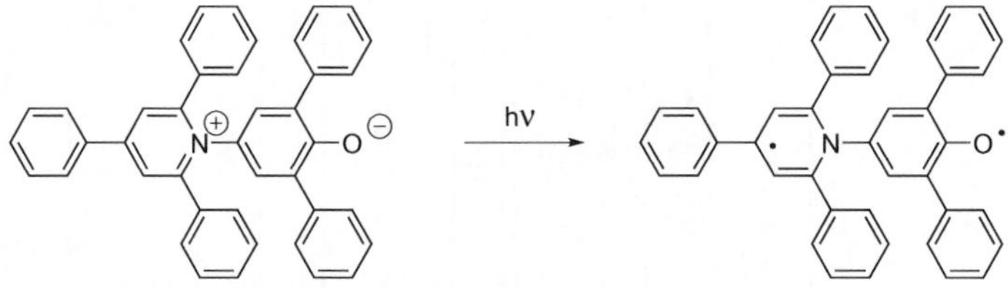

FIGURE 5.7 Betaine dye used to determine solvent polarity.

[2,6-diphenyl-4-(2,4,6-triphenyl-1-pyridinio)-phenoxide] in Figure 5.7. More-polar solvents stabilize the ground energy of the polar dye, producing a greater shift in the position of the $\pi \rightarrow \pi^*$ absorption relative to that found for solutions of the dye in tetramethylsilane. Reichardt has found that the colors of this dye in a solvent are indicative of the polarities of the solvent and solvent combinations used to dissolve it; for instance, red for MeOH, violet for EtOH, green for acetone, blue for isoamyl alcohol, and greenish-yellow for anisole [35]. The E_T^N parameter relates well to solvents used in organic chemistry, and may have merit in calculating solvent substitutions (see Section VI).

Solvents often selected for scale-up are shown in Table 5.5. This table is organized top to bottom by decreasing polarity, based on Reichardt's E_T^N parameter [34]. Two columns describe the characteristics of any azeotrope of water with these solvents; water can be a crucial component of many processes, as discussed in Chapter 6. The last column in Table 5.5 identifies the solvents that are ICH Class 3 (Table 5.7), and so would be recommended over other solvents for the preparation of APIs. McConville has compiled an extensive list of solvents and their characteristics [36].

Boiling points of solvents can be very important in solvent selection. High-boiling solvents, such as xylenes (bp $\sim 140\,°C$), are rarely selected for isolating an API because of the potential difficulty in removing the solvent to acceptable levels. High-boiling, water-soluble solvents (see Section VI) may be more easily removed, by extraction. Solvent selections and practical workups are discussed in Chapter 8.

Some of the solvents in Table 5.5 deserve additional comments, as provided in Table 5.6. One solvent is EtOAc, which can be considerably more reactive than *i*-PrOAc as product-rich extracts. In water EtOAc is hydrolyzed 4.5 times faster than *i*-PrOAc [37]. Peroxides found in laboratory EtOAc oxidized sulfoxides, amines, and ketones (the latter

TABLE 5.5 Characteristics of Solvents Useful for Scale-Up

Solvent	Polarity E_T^N	Melting Point (mp)	Boiling Point (bp)	Flash Point	Solubility in H_2O (wt%)	H_2O Dissolved in (wt%)	bp of H_2O–Solvent Azeotrope	wt% of H_2O Removed by Azeotrope	ICH Solvent Class
Water	1.000	0 °C	100 °C	—	—	—	None	None	
MeOH	0.762	−98 °C	65 °C	11 °C	∞	∞	None	None	
1,2-Propanediol	0.722	−60 °C	188 °C	99 °C	∞	∞	None	None	
EtOH	0.654	−114 °C	78 °C	13 °C	∞	∞	78 °C	4.0	3
AcOH	0.648	17 °C	118 °C	39 °C	∞	∞	77 °C	97	3
i-PrOH	0.546	−90 °C	82 °C	12 °C	∞	∞	80 °C	12.6	3
2-BuOH	0.506	−115 °C	98 °C	26 °C	19.8	65.1	87 °C	73	3
CH$_3$CN	0.460	−48 °C	81 °C	6 °C	∞	∞	76 °C	14.2	
DMSO	0.444	18 °C	189 °C	95 °C	∞	∞	None	None	3
DMF	0.404	−61 °C	152 °C	58 °C	∞	∞	None	None	
t-BuOH	0.389	25 °C	83 °C	11 °C	∞	∞	80 °C	11.8	3
NMP	0.355	−24 °C	204 °C	96 °C	∞	∞	None	None	
Acetone	0.355	−94 °C	56 °C	−20 °C	∞	∞	None	None	3
butanone	0.327	−87 °C	80 °C	−3 °C	26	12	73 °C	11	3

t-AmOH	0.318		−12 °C	102 °C	21 °C	11.0	23.5	87 °C	27.5	
CH$_2$Cl$_2$	0.309		−97 °C	40 °C	–	1.3	0.2	38 °C	1.5	
pyridine	0.302		−42 °C	115 °C	20 °C	8	8	94 °C	43	
MeOAc	0.287		−98 °C	56 °C	−10 °C	24.5	8.2	56 °C	5	3
MIBK	0.269		−80 °C	117 °C	18 °C	1.7	1.9	88 °C	24.3	3
DME	0.231		−58 °C	85 °C	5 °C	8	8	76 °C	10.5	
EtOAc	0.228		−84 °C	77 °C	−4 °C	8.1	3.3	70 °C	8.5	3
i-PrOAc	0.210 (1)		−73 °C	89 °C	2 °C	2.9	1.8	77 °C	10.6	3
THF	0.207		−108 °C	66 °C	−14 °C	8	8	64 °C	5.3	
2-Me-THF	0.179		−136 °C	77 °C	−11 °C	15.1	5.3	71 °C	10.6	
PhCl	0.188		−45 °C	132 °C	28 °C	0.05	0.05	90 °C	28.4	
i-BuOAc	NA		−99 °C	117 °C	18 °C	0.6	1.02	87 °C	16.5	3
1,4-dioxane	0.164		12 °C	101 °C	12 °C	8	8	88 °C	17.6	
MTBE	0.148		−109 °C	55 °C	−28 °C	4.8	1.4	53 °C	4	3
(EtO)$_2$CH$_2$	0.099		−67 °C	88 °C	−6 °C	4.2	1.3	75 °C	10	
PhCH$_3$	0.099		−93 °C	111 °C	4 °C	0.06	0.05	84 °C	13.5	
Xylenes (2)	0.074		(2)	137–144 °C	~27 °C	~0.02	~0.04	~93 °C	~45	
Et$_3$N	0.043		−115 °C	89 °C	−7 °C	5.5	4.6	75 °C	10	
Heptane(s) (3)	0.012		−91 °C	98 °C	−4 °C	0.0004	0.01	79 °C	12.9	3

(Continued)

TABLE 5.5 Characteristics of Solvents Useful for Scale-Up—cont'd

Solvent	Polarity E_T^N	Melting Point (mp)	Boiling Point (bp)	Flash Point	Solubility in H_2O (wt%)	H_2O Dissolved in (wt%)	bp of H_2O–Solvent Azeotrope	wt% of H_2O Removed by Azeotrope	ICH Solvent Class
iso-octane (4)	NA	−107 °C	98–9 °C	−12 °C	0.006		79 °C	11.1	
Cyclohexane	0.006	6 °C	81 °C	−20 °C	0.006	0.01	69 °C	9	

Sources:

1. Horsley, L. H. *Azeotropic Data – III*; Advances in Chemistry Series, 116; ACS: Washington, D.C., 1973.
2. Gmehling, J.; Menke, J.; Fischer, K.; Krafczyk, J. *Azeotropic Data, Parts I and II*; VCH: Weinheim, 1994.
3. Reichardt, C. *Solvents and Solvent Effects in Organic Chemistry*; 2nd ed.; VCH: Weinheim, 1990.
4. *CRC Handbook of Chemistry and Physics*; 78th ed., Lide, D. R., Ed., CRC Press: Boca Raton, FL, 1997–1998.
5. *Industrial Solvents Handbook*; 4th ed., Flick, E. W., Ed., Noyes Data Corporation: Park Ridge, NJ, 1990.
6. McConville, F. X. *The Pilot Plant Real Book*; 2nd ed.; FXM Engineering & Design: Worcester, MA, 2007.
7. Dean, J. A. *Lange's Handbook of Chemistry*; 15th ed.; McGraw-Hill: New York, 1999; p 5.61.
8. http://www.fda.gov/downloads/RegulatoryInformation/Guidances/ucm128282.pdf

(1) Not available. Value listed is for n-PrOAc.
(2) Includes ethyl benzene and isomers of xylene. ETN value shown is for p-xylene. Melting point, flash point, and azeotropes vary slightly with composition.
(3) Data shown are for n-heptane. Heptanes are a fraction from distillation and contain various isomers.
(4) 2,4,4-Trimethyl pentane.

TABLE 5.6 Some Comparisons of Solvents Useful for Scale-up

Solvent(s)	Comments
1,2-Propanediol	GRAS, used in formulation of APIs to make drug products, but not mentioned in ICH solvent lists. May be good substitute for ethylene glycol-based solvents
DMSO	Known to react with acids and bases (1–3)
DMF	Known to react with acids and bases. Disproportionates to give CO and Me_2NH. Reacts with NaH and $NaBH_4$
N-methyl pyrrolidinone (NMP)	ICH Class 2 solvent, less toxic than HMPA
Methyl isobutyl ketone (MIBK)	Good solvent for extractive workups, removes 24.3% H_2O by azeotropic distillation. Underutilized
i-PrOAc (IPAc) vs. EtOAc	In the presence of aqueous base, EtOAc hydrolyzes 4.5 times faster than i-PrOAc. Byproducts AcOH and EtOH may cause difficulties. i-PrOAc often preferred for extractive workups
i-PrOAc	Removes 10.6% H_2O by azeotropic distillation
2-Me-THF vs. THF	2-Me-THF separates from H_2O; THF extracts will separate from H_2O if the ionic strength of the aqueous phase is high enough. 2-Me-THF and THF have similar tendencies to form peroxides, and are generally inhibited with BHT. 2-Me-THF, derived from bagasse, is about twice the cost of THF, and the price may be expected to drop as petroleum feedstocks become scarcer
DEM	May be a good substitute for etheric solvents. Stable to about pH 4 (4)
Solids near room temperature: AcOH, DMSO, t-BuOH, dioxane	These solvents may plug condensers if coolant temperatures are too low

(1) DMSO reacts with acid chlorides at about the same rate as EtOH reacts with acid chlorides. DMSO also reacts with acid anhydrides and MeBr. http://www.gaylordchemical.com/bulletins/bulletin106b/Bulletin106B.pdf
*(2) Heating sodium dimsylate in DMSO led to a runaway reaction at 50–60°C: Dahl, A. C.; Mealy, M. J.; Nielsen, M. A.; Lyngsø, L. O.; Suteu, C. Org. Process Res. Dev. **2008,** 12, 429.*
*(3) Bollyn, M. Org. Process Res. Dev. **2006,** 10, 1299.*
*(4) Boaz, N. W.; Venepalli, B. Org. Process Res. Dev. **2001,** 5, 127.*

were oxidized to Baeyer–Villiger products); probably the oxidant was peracetic acid, generated from air and EtOH via hydrolysis of EtOAc [38]. i-PrOAc is more stable than EtOAc and was preferred for splitting an HCl salt using aqueous NaOH: aqueous $NaHCO_3$ was not effective, and massive hydrolysis occurred when the treatment was carried out with EtOAc and 2 M NaOH [39]. Preparation of a sulfate salt was changed from EtOAc to i-PrOAc because the latter is less readily hydrolyzed under acidic conditions [40]. An acetamide was formed by extraction of a primary amine into EtOAc, and so the product was extracted into dichloromethane [41]. (In the latter case the reaction product was a methoxyacetamide, and under the basic conditions for the Suzuki coupling some hydrolysis of the methoxyacetamide occurred. Purging the acetamide impurity by crystallization probably would have been difficult. Reactions of dichloromethane with amines are discussed with Figure 5.14.) Acetylation of ammonia [42] and n-butyl amine [43] is faster with EtOAc than with i-PrOAc, and the presence of H_2O in such extracts would probably accelerate acetylation of amines. In general fewer impurities are expected from extractions with EtOAc than with i-PrOAc. NMP also should be highlighted; this solvent was thought to be benign, but was recently reclassified as a Class 2 solvent due to issues as a reproductive toxicant [44]. 2-Me-THF has gained popularity for use in the laboratory and on scale, often because it is useful for organometallic reactions. (The utility of this solvent is shown by the availability of 3 M MeLi in 2-Me-THF [45].) 2-Me-THF can be purchased with up to 400 ppm of 3,5-di-t-butyl-4-hydroxytoluene (BHT, "butylated hydroxytoluene," a GRAS preservative) added as a stabilizer, or without BHT; with exposure to air 2-Me-THF forms peroxides slightly faster than THF. 2-Me-THF reacts with HCl slower than THF [46]. Diethoxymethane (DEM) has also been used for commercial solutions of 3 M MeLi [45]. Another solvent deserving additional mention is butanone (methyl ethyl ketone, or MEK), which can form reactive peroxides that initiate polymerization and other reactions [47]. In the presence of O_2 MEK has been used to oxidize Co(II) to Co(III) [48], which is a powerful oxidant. While MEK has a useful boiling point and azeotrope with water, its ability to form peroxides should be respected. MIBK is a high-volume solvent that contributes to ozone formation in the lower atmosphere [49], and has been judged to have a higher potential to contribute to atmospheric ozone than does i-PrOAc [50]. For these reasons some shun MIBK.

SAFETY — Triacetone peroxide is an extremely sensitive, powerful explosive, prepared by reaction of acetone with H_2O_2 and acid. Butanone peroxide can form under similar conditions. Butanone can react with O_2 to oxidize Co(II) to Co(III). In the presence of catalytic transition metals, including Cu, Fe, Mn, and Co, β-dicarbonyl compounds react with tert-butyl hydroperoxide to form mixed peroxides [51]. By extension reactions of ketones with peroxides may be catalyzed by metal contaminants in vessels. If reactions of ketones with peroxides cannot be avoided, reaction conditions should be analyzed in a hazard analysis laboratory to determine SAFE operating conditions.

Solvents are also selected or avoided due to the limits in guidances by regulatory authorities for residual solvents in APIs [52]. These limits have been posed based on toxicological data, which were used to calculate the permitted daily exposure (PDE) for each chemical. As shown in Table 5.7, four categories of solvents have been proposed, based on a daily dosage of 10 g of API [53]. This option is applied to compounds when the dosage of the drug has not been established. Some examples of recommended residual solvent limits in APIs are shown in Table 5.7, based on calculations using Option 1. The International Conference on Harmonization (ICH) has suggested a second option for calculating limits to residual solvents, which sums the amounts of a particular residual solvent present in the API and excipients of the drug product. The sum must be less than the PDE limit for this solvent; in this way the levels of the solvent in the API

TABLE 5.7 ICH Limits for Residual Solvents in APIs, Based on Theoretical Daily Dosage of 10 g (1,2)

Solvent	Limit (ppm)	Solvent	Limit (ppm)
ICH Class 1 Solvents: Carcinogens, Suspected Carcinogens, Environmental Toxins			
Benzene	2	1,2-Dichloroethane	4
CCl_4	4	1,1,1-Trichloroethane	1500
ICH Class 2 Solvents: Non-Genotoxic Animal Carcinogens, and Possible Agents of Irreversible Toxicity, Such as Neurotoxicity or Teratogenicity			
Acetonitrile	410	Dioxane	380
Chlorobenzene	360	Ethylene glycol	620
$CHCl_3$	60	Hexane	290
Cyclohexane	3880	MeOH	3000
CH_2Cl_2	600	NMP	530
DMA	1090	THF	720
1,2-Dimethyoxyethane (DME, glyme)	100	Toluene	890
DMF	880	Xylene	2170
ICH Class 3 Solvents: To Be Minimized By GMP Practices to 5000 ppm			
AcOH	5000	EtOAc	5000
Acetone	5000	Heptane	5000
n-BuOH	5000	i-PrOAc	5000
t-BuOH	5000	i-PrOH	5000
DMSO	5000	MIBK	5000
EtOH	5000	MTBE	5000
ICH Class 4 Solvents: Toxicity and Limits to be Determined			
2-Me-THF (3)		Cyclopentyl methyl ether (3)	

(1) Class 2 and Class 3 solvents listed have been selected as those likely to be used in the pharmaceutical industry. The entire list of limits for residual solvents is available at http://www.fda.gov/RegulatoryInformation/Guidances/ucm128223.htm.

(2) Limits are based on individual PDEs and API theoretical dosages of 10 g/day. If the daily dosages are less than 10 g/day the limits can be calculated accordingly.

(3) Toxicology data recently generated may be used to propose limits for these solvents and inclusion with ICH Class 3 Solvents. Data did not indicate mutagenicity or genotoxicity. Antonucci, V.; Coleman, J.; Ferry, J. B.; Johnson, N.; Mathe, M.; Scott, J. P.; Xu, J. Org. Process Res. Dev. **2011**, 15, 939.

may be higher than an individual limit shown in Table 5.7 [53]. Option 2 is likely to be used when removing a residual solvent from an API is found to be very difficult and the anticipated daily dosage is less than 10 g.

Class 3 solvents have the highest permissible limits for residual solvents in an API. For ease of processing, analysis, and regulatory approval APIs are most conveniently isolated from water and Class 3 solvents. Note that a 1:1 solvate is unlikely to have residual solvent levels that pass the limits posed (for instance, an API with FW 663 solvated with one equivalent of THF contains 9.8% THF).

In early efforts to develop an NCE the cost of solvents is disregarded, since small volumes of solvents are required and the cost of labor is proportionately much higher. As large-scale manufacturing of a compound comes closer to reality the prices of solvents become more critical, especially if solvents contribute greatly to the CoG. Some solvents are ranked by their bulk cost in Table 5.8, relative to the cost of MeOH ($/L). Factors influencing the cost are the grades and water contents of the solvents. As petrochemical resources and energy become more expensive, the costs to purchase, recover and reuse solvents will become increasingly important.

II. USING AZEOTROPES TO SELECT SOLVENTS

An azeotrope is a constant-boiling mixture with a constant mole fraction composition of components. Azeotropes consist of two, three, or more components, and can be homogeneous or heterogeneous (more than one phase) [54]. Abbott researchers have detailed using the water contents and solvent concentrations using azeotropes in chasing H_2O and *i*-PrOAc with *i*-PrOH [55]. Organic chemists probably first experience azeotropes by removing water from a refluxing mixture using a Dean–Stark trap, as with the heterogeneous azeotrope of toluene–water. Most of the azeotropes important to process chemists are minimum-boiling azeotropes, i.e., the boiling point of the mixture is less than the boiling point of any component. (A familiar exception is concentrated aqueous HCl, which is formed as a positive azeotrope.) Extensive data are available from Gmeling's Dortmund Database [56]. All heterogeneous azeotropes are minimum-boiling [57]. Unlike liquids may form azeotropes if the boiling points are relatively similar. For example, toluene azeotropes with H_2O, EtOH, ethylenediamine, and AcOH, but toluene and hexanoic acid do not form an azeotrope. Many organic solvents form azeotropes with water (Table 5.5), and azeotroping out H_2O can conveniently dry product-rich extracts and equipment. For process scientists the value of azeotropes lies primarily in the ability to efficiently remove volatile components of reaction mixtures. Azeotropic removal of volatile components can drive reactions, as shown in Figure 5.8. Azeotropic removal of water formed an imine key to a domino sequence that led to an oseltamivir precursor [58]. Azeotropic removal of water led to the protection of an alcohol as the trityl ether [59]. Azeotropes may facilitate separations and workups. For instance, excess ethylenediamine was removed by azeotropic distillation with toluene in the hydrazinolysis of *S*-serine methyl ester [60] (Figure 5.8); in this case separating the polar hydrazine from the polar, water-soluble product by extraction could be difficult. Hexamethyldisiloxane, a byproduct from acid-catalyzed deprotection of trimethylsilyl compounds, forms an azeotrope

TABLE 5.8 Ranking Solvents by Bulk Cost (1)

Rank	Solvent	Cost Relative to MeOH ($/L)	Rank	Solvent	Cost Relative to MeOH ($/L)
1	MeOH, ACS/reagent	1.00	19	THF, ACS	4.37
2	Acetone, NF	1.57	20	Acetonitrile, ACS	5.03
3	Acetone, ACS	1.68	21	DMSO, ACS	7.24
4	Hexanes, 64% n-hexane, ACS	1.87	22	Toluene, anhydrous	7.64
5	HCl, ACS	1.99	23	NMP 99%	7.89
6	Toluene, ACS	2.05	24	DMF, anhydrous	7.91
7	i-PrOH, ACS	2.33	25	Acetone, anhydrous	8.01
8	EtOAc	2.44	26	Dichloromethane, anhydrous	8.33
9	Cyclohexane	2.50	27	EtOH, 200 proof USP	10.18
10	NaOH solution, 50%	2.84	28	EtOH, 200 proof ACS	10.42
11	Hexanes, 95% n-hexane, ACS	2.86	29	Dioxane	10.46
12	Dichloromethane, ACS	3.08	30	DMSO, anhydrous	11.32
13	i-PrOAc	3.24	31	Acetonitrile, anhydrous	11.48
14	MTBE, ACS	3.30	32	THF, anhydrous	12.33
15	DMF, ACS	3.33	33	2-Me-THF	13.68
16	AcOH, glacial	3.59	34	2-Me-THF, anhydrous	15.84
17	EtOH, denatured	3.72	35	NMP, anhydrous	23.21
18	Heptane	3.75			

(1) Prices were calculated based on 200 L drum quantities. Data from a US CRO/CMO, July 2011.

with esters, alcohols, and trimethylsilanol [61]. For the workup of the TMS-protected product in Figure 12.2 hexamethyldisiloxane was removed as an azeotrope with the solvent acetonitrile [62]. Even if an azeotrope is not rigorously characterized removal of components may be facilitated by a lower boiling point. An azeotrope may also be an economical solvent if the solvent can be recovered and reused, as is the case for the agricultural intermediate in Figure 5.8 [63].

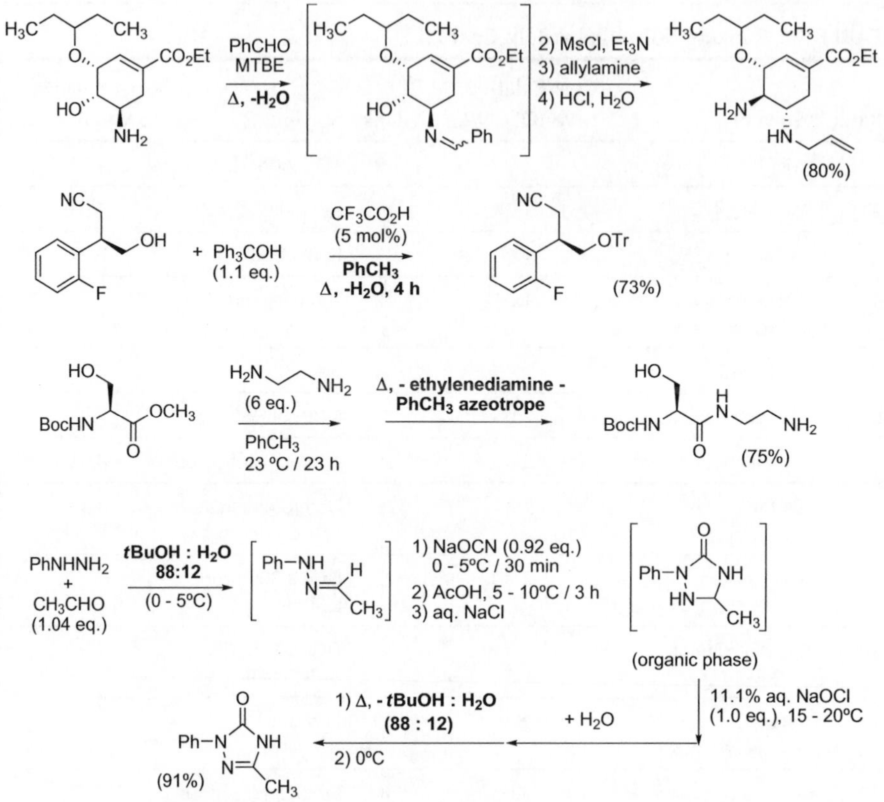

Practical Process Research and Development

FIGURE 5.8 Examples of processing facilitated by azeotropes.

Expect azeotropes to be present in processing. Exploit them in developing processes.

Some useful binary azeotropes of solvents are shown in Table 5.9. When the composition of a binary azeotrope is closer to 1:1 (as is the case for i-PrOH–i-PrOAc, for instance) solvent chasing from either solvent to the other will require less solvent. Some ternary azeotropes (solvent–solvent–water) that may be encountered are shown in Table 5.10 [36]. Some ambiguities exist in the literature around azeotropic data.

Often conducting a distillation at a reduced pressure will reduce the fraction of the minor component in the distillate, as shown in Table 5.11 for the EtOAc–H_2O azeotrope; this is termed "breaking an azeotrope." This behavior is not found for the i-PrOH–H_2O azeotrope. McConville has graphically shown the effects of reduced pressure on the composition of some azeotropes [64].

III. CHOOSING SOLVENTS TO INCREASE REACTION RATES AND MINIMIZE IMPURITIES

Some empirical solvent rules have been formed to predict reaction rates. Hughes and Ingold developed a qualitative model of how solvent polarity affects reaction

TABLE 5.9 Some Binary Azeotropes of Note (1,2)

More-polar Solvent	Heptane (bp 98 °C)		Toluene (bp 111 °C)		Other Solvent		
	Azeotrope bp	% of Polar Solvent	Azeotrope bp	% of Polar Solvent	Second Solvent	Azeotrope bp	% of More-polar Solvent
MeOH (bp 65 °C)	59.1 °C	51.5 wt%	63.8 °C	69 wt%	MTBE	52 °C	15 wt%
EtOH (bp 78 °C)	70.9 °C	49 wt%	76.7 °C	68 wt%	Cyclohexane	65 °C	31 wt%
i-PrOH (bp 82 °C)	76.4 °C	50 wt%	80.6 °C	69 wt%	i-PrOAc	80 °C	52 wt%
2-BuOH (bp 98 °C)	88 °C	37 wt%	95 °C	55 wt%	H_2O	87 °C	73 wt%
t-BuOH (bp 83 °C)	78 °C	62 wt%	None	None	H_2O	80 °C	88 wt%
t-AmOH (bp 102 °C)	92.2	26.5 wt%	100.5 °C	56 wt%	H_2O	87 °C	72 wt%
acetone (bp 56 °C)	56 °C	89.5 wt%	None	None			
CH_3CN (bp 81 °C)	69.4 °C	44 wt%	81.1 °C	78 wt%	EtOH	73 °C	44 wt%
EtOAc (bp 77 °C)	77 °C (2)	6 wt% (2)	None	None	EtOH	72 °C	74 wt%
i-PrOAc (bp 89 °C)	88 °C	67 wt%	None	None	MeOH	65 °C	20 wt%
CH_3CO_2H (bp 118 °C)	95 °C	17 wt%	104 °C	32 wt%			
HCO_2H (bp 100 °C)	78.2 °C	56.5 wt%	86 °C	50 wt%			

(1) Unless otherwise indicated, data are from McConville, F. X. The Pilot Plant Real Book; 2nd ed.; FXM Engineering & Design: Worcester MA, 2007; Section 6.
(2) Lowenthal, H. J. E. A Guide for the Perplexed Organic Experimentalist; 2nd ed.; Wiley: New York, 1990; p 171.

TABLE 5.10 Some Ternary (Solvent–Solvent–Water) Azeotropes that May be Encountered (1)

Solvent A	Solvent B	Azeotrope bp (Atmospheric Pressure)	Azeotrope Composition A:B:water (wt%)
EtOH (bp 78 °C)	EtOAc (bp 77 °C)	70 °C	8:83:9
i-PrOH (bp 82 °C)	i-PrOAc (bp 111 °C)	76 °C	13:76:11
i-PrOH (bp 82 °C)	PhCH$_3$ (bp 111 °C)	76 °C	38:49:13
2-BuOH (bp 88 °C)	Heptane (bp 98 °C)	76 °C	22:67:11

(1) McConville, F. X. The Pilot Plant Real Book; 2nd ed.; FXM Engineering & Design: Worcester MA, 2007; Section 6.

TABLE 5.11 Effect of Reducing Distillation Pressure on EtOAc–H$_2$O and iPrOH–H$_2$O Azeotropes

EtOAc–H$_2$O Azeotrope			iPrOH–H$_2$O Azeotrope		
Pressure (mm)	bp (°C)	Water in Azeotrope (wt%)	Pressure (mm)	bp (°C)	Water in Azeotrope (wt%)
760	70.4	8.5	760		12
250	42.6	6.3	319		12
25	1.9	3.6	153		12
			67		12

Sources:
1. Horsley, L. H. *Azeotropic Data – III;* Advances in Chemistry Series, 116; ACS: Washington, D.C., 1973.
2. Gmehling, J.; Menke, J.; Fischer, K.; Krafczyk, J. *Azeotropic Data, Parts I and II;* VCH: Weinheim, 1994.
3. Reichardt, C. Solvents and Solvent Effects in Organic Chemistry; 2nd ed.; VCH: Weinheim, 1990.
4 CRC Handbook of Chemistry and Physics; 78th ed., Lide, D. R., Ed., CRC Press: Boca Raton, FL, pp 97–98.
5. *Industrial Solvents Handbook;* 4th ed., Flick, E. W., Ed., Noyes Data Corporation: Park Ridge, NJ, 1990.

rates [65]. In general, the effect of increased solvent polarity depends on whether there is higher concentration of charge (charge/volume) in the starting materials or in the transition state. (Sometimes this is described as greater charge localization, and lesser charge localization is sometimes referred to as charge dispersion.) Polar solvents preferentially solvate ions or intermediates with the greatest concentration of charge. If the concentration of charge is greater in the transition state than in the starting materials, a polar solvent stabilizes the transition state and encourages its formation, thus increasing the reaction rate. For example, polar solvents accelerate the quaternization of triethylamine with iodoethane. With greater charge dispersion in the transition state than in the starting materials, a polar solvent would stabilize the starting materials and decrease the reaction rate; examples would be the destruction of a salt and a Finkelstein reaction. (The latter reaction is also driven by the insolubility of the products in the solvent, such as the precipitation of NaCl from acetone in the NaI-mediated exchange of chloride for iodide.) Radical-mediated reactions are little affected by solvent polarity [66]. Some examples are shown in Figure 5.9. The qualitative charge localization/charge dispersion model does not consider other effects of solvents, such as hydrogen bonding, chelation, temperature, and concentration of the reaction [67]. For instance, reaction of azide with iodomethane increased by four orders of magnitude when the reaction was run in formamide instead of MeOH; even though these solvents have very similar polarity, the azide anion in MeOH was solvated by hydrogen bonding, decreasing its reactivity [68]. Changing solvents can also change reaction mechanisms [69].

To illustrate these points further two examples from an API and an API intermediate are shown in Figure 5.10. The oxazoline was treated with thiophenol to afford nelfinavir free base and a regioisomer [70]. By changing to less-polar solvents and incorporating a base the amount of the desired regioisomer increased markedly. (Even though the base Et$_3$N is also non-polar, its influence as a solvent was probably small due to dilution.) Formation of the regioisomer may have proceeded through the epoxide, which would have been formed through the zwitterionic intermediate shown. Polar solvents would be expected to stabilize the zwitterions, thus encouraging the formation of the regioisomer. As another example in Figure 5.10, the

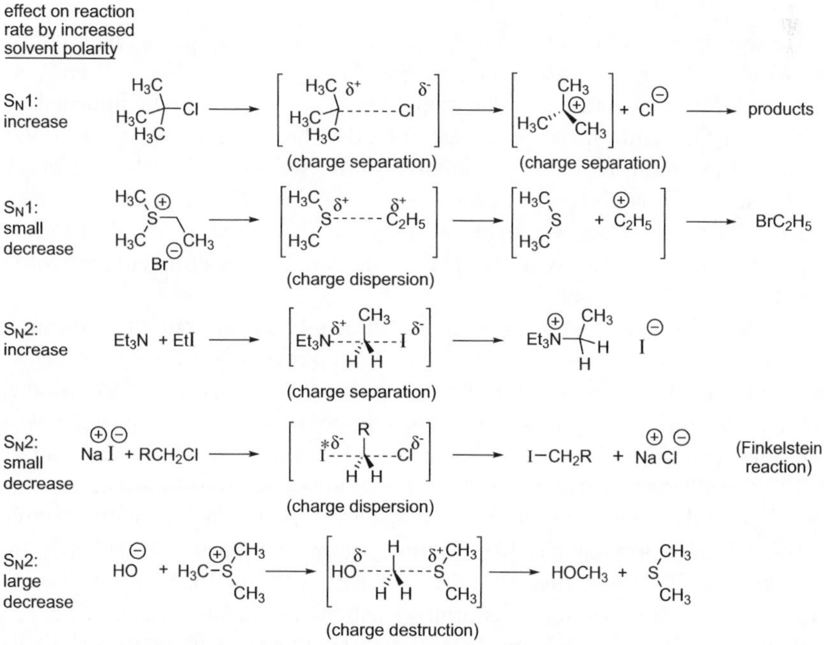

FIGURE 5.9 Choosing solvent polarity to influence reactions.

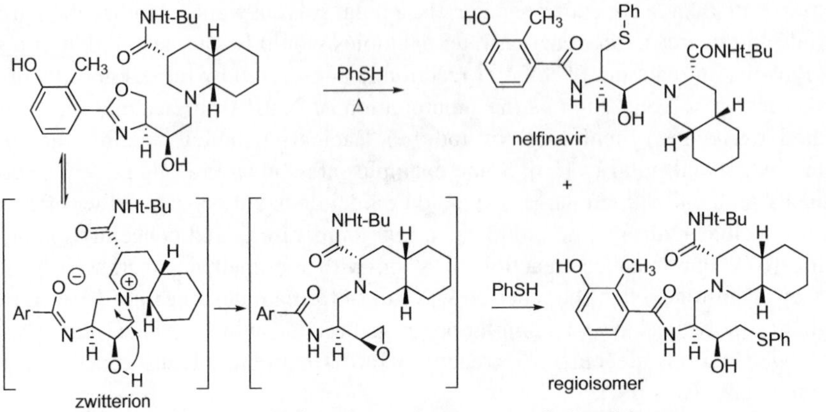

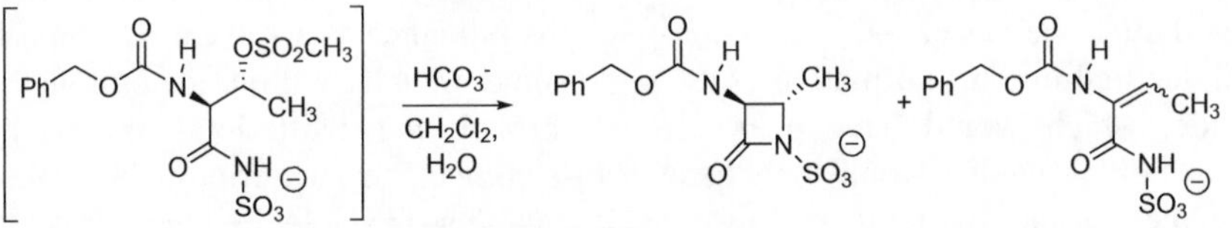

FIGURE 5.10 Byproduct formation influenced by solvent and base.

solvent	base	product : regioisomer
HOCH$_2$CH$_2$OH	none	18 : 82
DMF	none	71 : 29
DMF	Et$_3$N	84 : 16
MIBK	KHCO$_3$	92 : 8

sulfamate was prepared in dichloromethane and treated with aqueous bicarbonate to cyclize to the β-lactam, an aztreonam intermediate [71]. Both the S$_N$2 and the E$_2$ reactions to form the β-lactam and the olefin would be expected to be impeded in the presence of H$_2$O, as the negative charges of both the intramolecular and intermolecular reactions would be dispersed relative to the starting materials. But in more-polar solvents E$_2$ reactions are decelerated more than S$_N$2 reactions, hence the amount of olefin byproduct was expected to decrease relative to the β-lactam [72]. The bigger factor was found to be pH: formation of the elimination byproduct increased as the pH increased [73].

Some additional examples to illustrate the effects of solvent polarity are shown in Figure 5.11. Non-polar solvents have often been preferred over polar solvents for peptide couplings. In the preparation of Cbz-glycyl-L-phenylalanyl-glycine ethyl ester (Figure 5.8) by mixed anhydride coupling, non-polar solvents in general gave the highest yields and lowest levels of racemization [74]. Under the conditions employed, i.e., −15 °C with one equivalent of N-methylmorpholine, activation occurred in as little as 30 seconds. When AstraZeneca workers conducted mixed anhydride couplings using isobutyl chloroformate and N-methylmorpholine at −25 to −10 °C, they found that acetone and DMF were the preferred solvents, with good stability of the mixed anhydrides and no racemization. The primary amide, a frakefamide intermediate, was isolated in 93% yield and 99% purity over two couplings. Although no analyses of diastereomers were disclosed for this intermediate, the authors stated that only traces

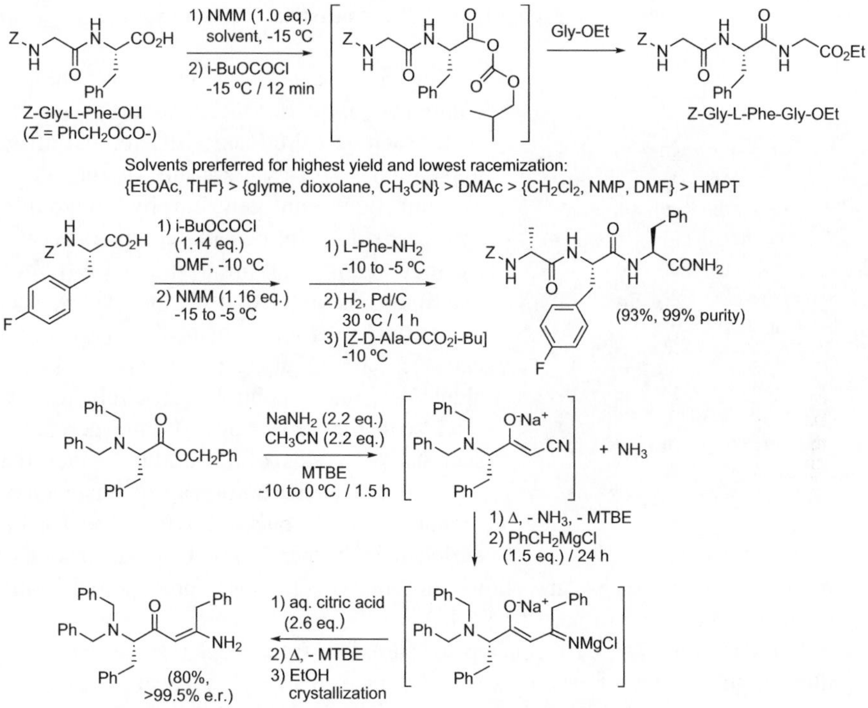

FIGURE 5.11 Solvents preferred in some reactions involving amino acids.

of racemization occurred [75]. Other amino acids, e.g., histidine and cysteine, are more prone to racemize during couplings or in peptides, and for this reason protection of these and other amino acids has been studied extensively [76]. For such peptide couplings using mixed carbonic anhydrides, high temperatures, the presence of bases stronger than *N*-methylmorpholine, and contact with excess base may influence yield and diastereomer formation as much or more than solvent polarity [77]. A less-polar solvent was found to decrease racemization in a reaction involving amino acid derivatives. Using THF, to avoid racemization a reaction temperature of -50 °C was needed; with less-polar solvents, such as MTBE, higher temperatures were possible with minimal racemization [78].

IV. IMPURITIES IN SOLVENTS AND REACTIONS OF SOLVENTS

Water may be the solvent impurity causing the most unforeseen difficulties (see Chapter 6). Many treatments have been developed to remove water from solvents in the laboratory [79,80], and recent papers suggest that molecular sieves may be the most universally useful [81]. "Anhydrous" solvents are more expensive, and reactions using these products must consider the moisture content of the equipment used to transfer the solvents and the reactor. For scale-up solvents and equipment are generally azeotroped dry, or excess reagents are charged to consume any H_2O present.

The best approach of developing reaction conditions that can tolerate some H_2O is not always feasible.

> In the laboratory solvents have sometimes been distilled from $LiAlH_4$ or $Na(0)$—benzophenone. Reaction of these chemicals with H_2O can form a metal oxide crust over excess reagent. When it is time to clean up the laboratory, it can take some thought and time to safely control quenching of excess reagents. With regard to charging such reactive reagents, more is not always better.

Peroxides can form in many solvents commonly used in the laboratory and on scale, such as i-PrOH and EtOAc, and afford unpleasant surprises. Exposure of solvents to air and light can generate hydroperoxides and peroxides. In general compounds with hydrogen atoms that are abstracted in radical reactions are prone to form peroxides, such as tertiary carbons, benzylic and allylic carbons, carbons alpha to an etheric oxygen, aldehydes, and alcohols [82]. Problems can arise with the formation of some peroxides, such as peroxides of diisopropyl ether that crystallize near the bottle cap of this solvent, or during the concentration of a solvent enriched in its peroxides. General classes of solvents that form peroxides are shown in Figure 5.12, and some peroxidizable compounds are shown in Table 5.12 [83].

Besides the obvious SAFETY concerns of peroxides (see Chapter 2), peroxides can also influence the course of reactions. For instance, in the absence of an inhibitor, exposure of aqueous THF (10:1 THF:H_2O) to air led to the formation of ≥ 8 mmol of THF-hydroperoxide, which led to an oxidative byproduct, a phenol, in the Suzuki–Miyaura coupling of a boronic acid (Figure 5.13) [84].

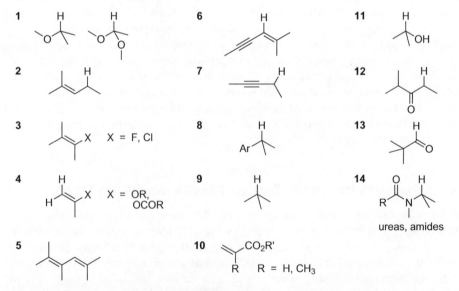

FIGURE 5.12 General structures of functional groups likely to form peroxides, with greatest propensity to form peroxides for those with lowest numbers. (*Kelly, R. J. Chem. Health Saf. 1996, September/October, 28.*)

Frequently check solvents for peroxides. A simple approach is to moisten a peroxide test strip with water and add a drop of solvent. Iodometric titration is a quantitative approach [86].

BHT (about 250 ppm) is often added to commercial shipments of THF and 2-Me-THF as a SAFETY precaution. On scale BHT may be added to condensates of THF and 2-Me-THF for SAFETY. Removing solvent from a reaction using BHT-stabilized THF and 2-Me-THF will afford a concentrate rich in BHT, which may interfere with HPLC and other analyses. A yellow dimer has been identified from oxidation of BHT [86,87].

TABLE 5.12 Classes of Some Peroxidizable Compounds (1)

Compounds that Can Form Explosive Levels of Peroxides Without Concentration		
Diisopropyl ether	Butadiene	Divinyl acetylene
Compounds that can Form Explosive Levels of Peroxides Upon Concentration		
Acetaldehyde	Cyclohexene	i-PrOH
Benzyl alcohol	Decahydronapthalene	2-Me-THF
2-Butanol	Diacetylene	MIBK
Cumene	Dicyclopentadiene	1-Phenylethanol
Cyclohexanol	Dioxane	THF
Cyclohexenol	Glyme (DME)	
Compounds that can Autopolymerize Upon Accumulation of Peroxides		
Acrylic acid	Methyl methacrylate	Vinyl acetylene
Acrylonitrile	Styrene	Vinyl pyridine
Butadiene	Vinyl acetate	

(1) Kelly, R. J. Chem. Health .Saf. **1996,** September/October, 28.

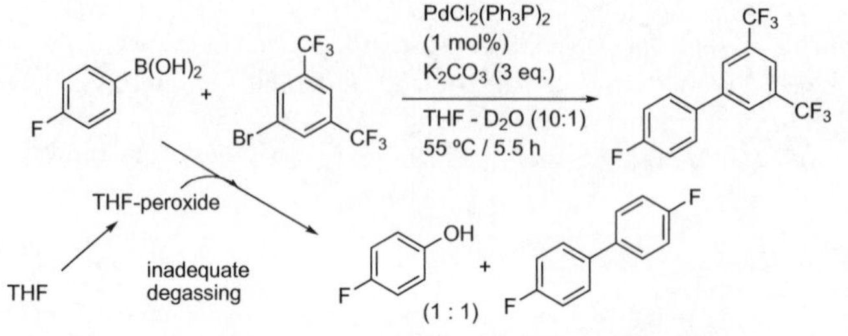

FIGURE 5.13 Formation of an oxidative byproduct in a Suzuki–Miyaura coupling due to the presence of oxygen.

Some impurities often found in solvents are shown in Table 5.13.

Static buildup is a concern on scale, and charging hydrocarbon solvents often proves problematic. Under the 2003 ATEX 137 Directive (European law) it is mandatory to assess for explosion risks on scale. Non-metallic additives, such as Statsafe®, have been developed to reduce the conductivity of solvents and minimize the risk of static discharge [88]. In general the risk of static discharge is greater with hydrocarbon solvents such as heptanes. These additives, proprietary mixtures of polymers and amines, may be considered as impurities in an API if solvents with these additives were used to prepare the API. The risk of static discharge may also be reduced by adding small amounts of a polar solvent, such as i-PrOH.

Solvents can be viewed as simple chemicals that react as shown in sophomore organic chemistry courses. The reactivity of dichloromethane has often been overlooked [89] (Figure 5.14). Bridgehead amines, such as strychnine, quinuclidine, and DABCO, are especially prone to react, followed by tertiary amines with a methyl group and secondary amines [90–92]. AstraZeneca researchers prepared the aminal from proline amide and dichloromethane [93]. Pyrrolidine rapidly formed the aminal, with the second displacement of chloride being much faster than the first [94]. Similarly, pyridines reacted with CH_2Cl_2 to form the bis-pyridinium salt, and the second displacement was faster than the first. DMAP reacted the fastest of the seven pyridines examined [95]. HOBt, a catalysis commonly used in peptide couplings, reacts with dichloromethane [96]. The reactivity of thiols toward dichloromethane may have been overlooked in PTC processes [97]. Grignard reagents have been shown to react with dichloromethane in the presence of anhydrous $FeCl_3$ and other iron salts [98]. A nickel–glycine complex readily reacted twice with dichloromethane to afford a chiral 4-amino glutamic acid [99]. In 1978 Wyeth researchers reported detecting cyanogen chloride in dichloromethane used to isolate metabolites [100]. Dichloromethane was shown to react once with olanzapine, clozapine, and ofloxacin, each of which has an N-methyl-piperazine moiety [101]. Perhaps dichloromethane is the source of many compounds containing symmetrical diamines linked by a methylene carbon. The reactivity of nucleophiles toward dichloromethane should be respected. The low boiling point of dichloromethane makes containing it difficult in order to control the VOC emissions. These are some of the reasons that make dichloromethane unattractive for manufacturing.

> Do not store dichloromethane extracts of amines for prolonged periods. Nucleophilic amines, especially quinuclidines, are prone to react with dichloromethane.

TABLE 5.13 Some Common Solvent Impurities

Solvent	Impurity	Comment
All	H_2O, peroxides	(1)
EtOH, denatured (2,3)	MeOH, 4.7−4.8 wt%	SDA-3A, "IMS" (industrial methylated spirits)
	"Pyridine bases," 0.6 wt%	SDA 6B
	Toluene, 5.1−5.2 wt%	SDA 12A-3
	Acetone, 7.2−7.4 wt%	SDA 23A
	EtOAc, 1.1 wt%	SDA 29-3
	MeOH, 8.9−9.1 wt%	SDA 30A
Acetonitrile	Acetamide	Genotoxin (and approved food additive)
DMF	Me_2NH, CO	Disproportionation is catalyzed by acid and base
CH_2Cl_2	Amylene, cyclohexane	Inhibitor
MIBK	Mesityl oxide (4-methyl-3-pentene-2-one)	From incomplete reduction of diacetone alcohol
EtOAc	Peracetic acid	(4)
ROAc	AcOH	
THF, 2-Me-THF	BHT	Inhibitor ($\sim$250 ppm)
i-Octane	iso-octene	From incomplete hydrogenation

Comments
(1) Peroxides can be found in many solvents: Kelly, R. J. Chem. Health. Saf. **1996**, September/October, 28.
(2) "Ethyl Alcohol Handbook," U.S.I. Chemicals, © 1981 by National Distillers and Chemical Corporation.
(3) Denatured ethanolic solvents are available in 190° and anhydrous conditions.
(4) Hobson, L. A. Chem. Eng. News **2000**, 78(51), 2.

Other common and not-so-common reactions of solvents are shown in Figure 5.15. THF is known to react in the presence of acids, to give both ring-opened and polymeric byproducts. In the laboratory substitution of H_2SO_4 by methanesulfonic acid generated no such byproducts, but on a 20 kg scale ring-opened byproducts were found. Dimethoxyethane proved superior to THF in this case [102]. THF is known to react with acid chlorides and acid bromides [103]. The hydrolysis of THF and 2-Me-THF is similarly slow [46] under aqueous acidic conditions that would be used for extractions. THF and dimethoxyethane were found to polymerize exothermically in the presence of Cl_2 [104]. The borane–THF complex has been implicated in an industrial incident [105] when used in a reaction as 2 M $BH_3 \cdot THF$ in THF [106]. At 10–50 °C reaction of $BH_3 \cdot THF$ in THF generates H_2 and tributyl borate, and above

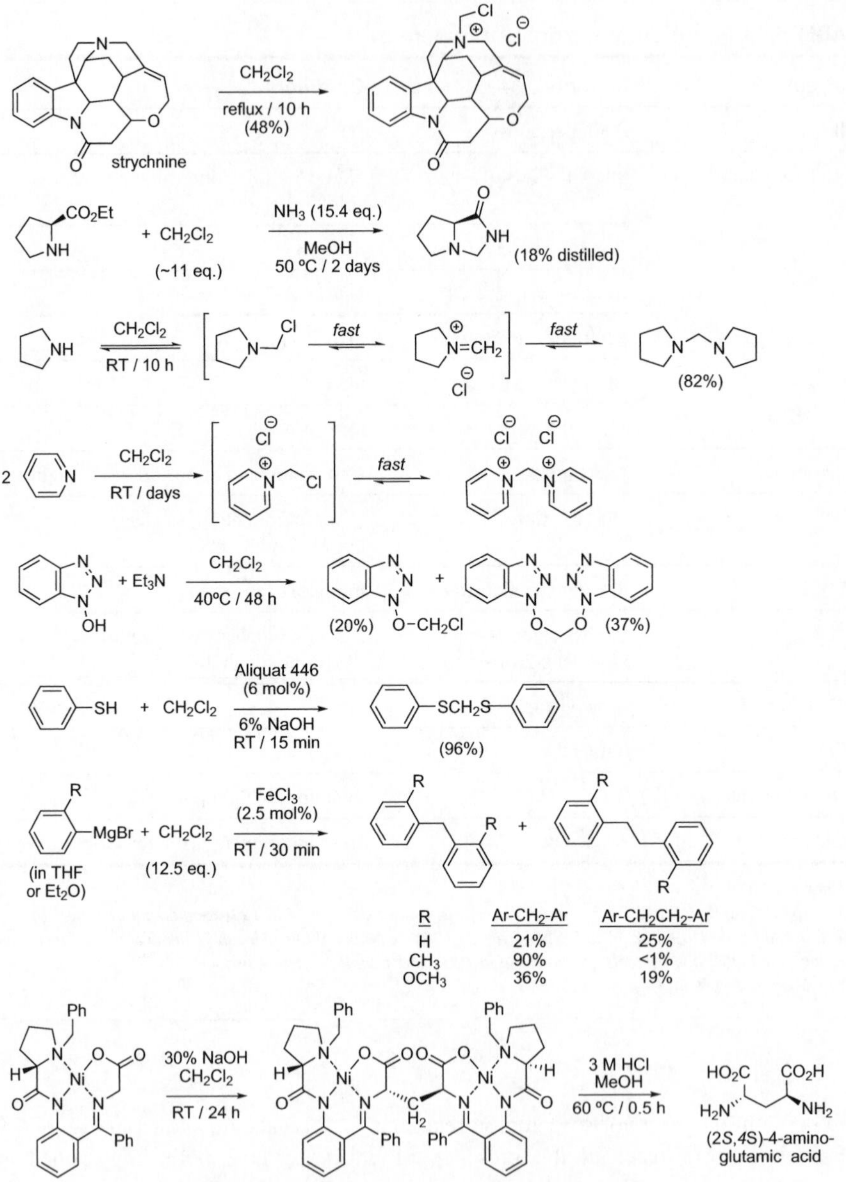

FIGURE 5.14 Some reactions of dichloromethane.

50 °C decomplexation occurs with formation of diborane. For SAFE handling of BH$_3$·THF in THF, BASF researchers recommended that the reagent should be stored at 0–5 °C, and that reactions be conducted below 35 °C [107]. Disproportionation of DMF to CO and dimethylamine is catalyzed by either acid or base, and despite several reports DMF is known to react exothermically with NaH [108] and NaBH$_4$ [109]. NMP and NaH are thermally unstable [110]. DEM is known to hydrolyze at pH

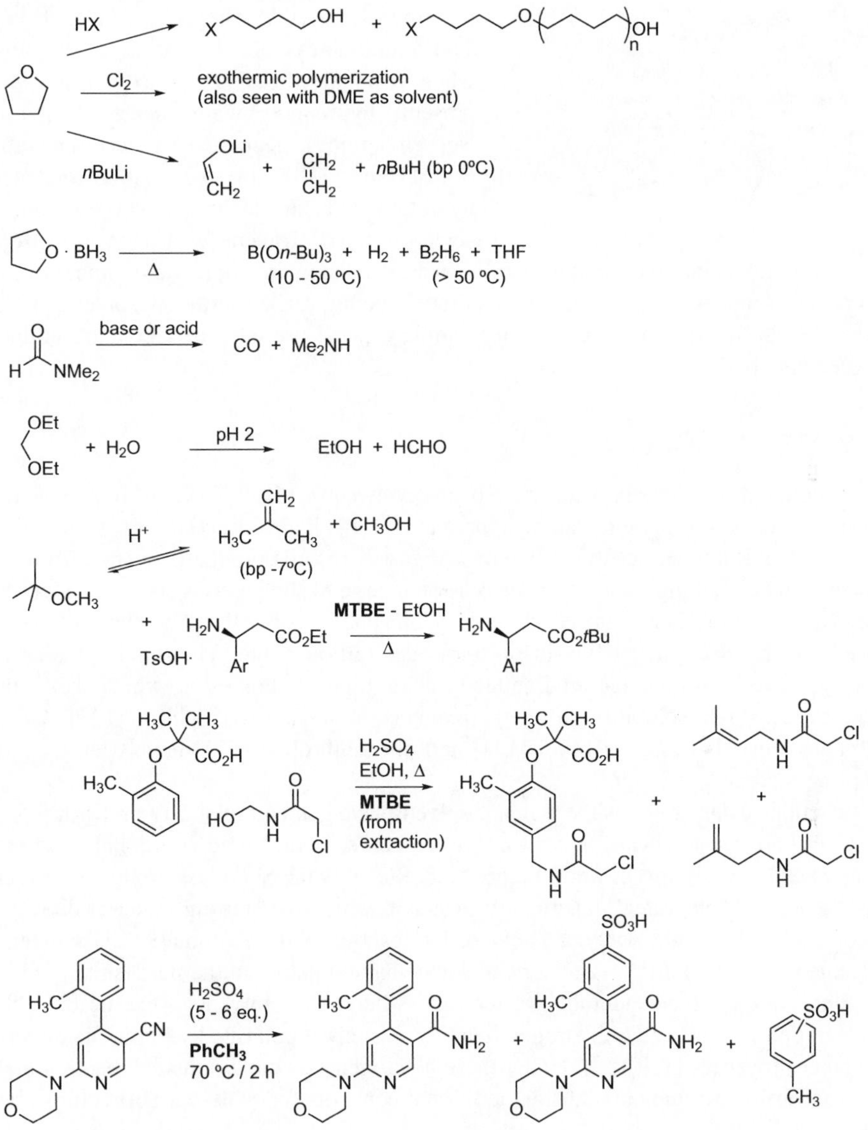

FIGURE 5.15 Some reactions of solvents other than dichloromethane.

2; processing under conditions no more acidic than pH 4 has been recommended [111]. MTBE, which is formed by acid-catalyzed addition of t-BuOH and isobutylene, seems stable under acidic conditions used for extractive workups, but has been shown to react with concentrated HCl at 40 °C, and with H_2SO_4 at elevated temperatures [112]. MTBE has reacted exothermically with thionyl chloride [112] and with Br_2 [113]. In preparing a tosylate salt of an ethyl ester from a refluxing mixture of MTBE–EtOH Pfizer researchers found that the t-butyl ester precipitated, from

> When using THF under anhydrous acidic conditions, expect ring-opened byproducts to form. Other solvents may be preferred for generating salts of amines.

reaction of the ethyl ester with MTBE [114]. The aminomethylation of an ether was slow when residual MTBE from extractions was present; byproducts were identified as the isoprenylamines, generated by decomposition of MTBE [115]. In the H_2SO_4-mediated hydration of a nitrile the product was extensively sulfated. Toluene was added to afford better agitation, and the primary amide was produced in good yield; under these conditions some of the toluene was sulfated, acting as a sacrificial solvent [116]. MIBK has been used to protect primary amines in the presence of secondary amines (Figure 6.9) [117].

V. WATER AS SOLVENT

The benefits of reactions in water have been reviewed [3,118–122]. The simplicity and power of processing in water can be appreciated in the hydroformylation (oxo process) developed by Ruhrchemie/Rhône Poulenc (Figure 5.16): the products separate from the aqueous phase, making the reactor ready for recharge of the gaseous starting materials. The Rh catalyst is held captive in the organic phase, solubilized by the sulfonated ligands, with losses of catalyst in the parts per billion range. Thousands of tons of products have been manufactured annually using biphasic processes, which are facilitated by the high solubility of metal reagents in an aqueous phase [123]. Other sulfonated ligands based on binol [124] and phenanthroline [125] have been used in coupling reactions in water.

Reactions using water as the primary solvent have gained popularity in the past few years, in part because water is viewed as a green solvent. Some additional reactions accelerated by water are shown in Figure 5.17. Recently the Sharpless group has written about reactions "on water," reactions in water in which the starting material does not dissolve [126,127]. On water a Diels–Alder reaction and an aromatic Claisen rearrangement were found to be accelerated somewhat compared to the neat (solvent-free) reactions, but were considerably faster than when other solvents were used [128]. Acceleration may be due to hydrogen bonding, increased polarity, hydrophobic effects, and other properties [129,130]. Under these heterogeneous conditions the reaction may be a suspension throughout, causing additional concerns for scale-up. Difficulties may

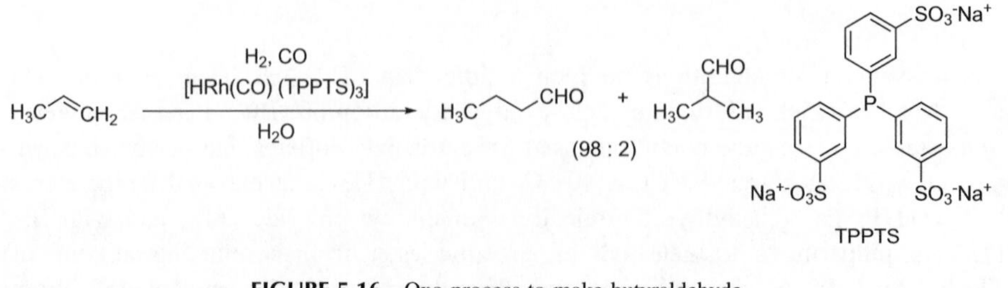

FIGURE 5.16 Oxo process to make butyraldehyde.

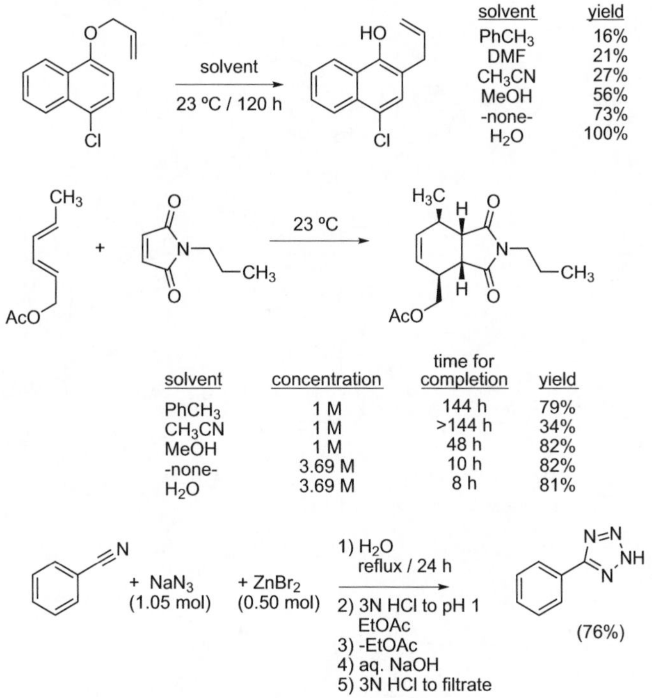

FIGURE 5.17 Some reactions accelerated by water.

be anticipated if the product occludes starting material or other impurities. If extended agitation is needed for conversion on scale, small particles may be produced, complicating filtration and isolation. The Sharpless group has carried out "click" reactions in predominantly aqueous conditions [131] and tetrazole formation under non-acidic conditions in water [132]. H_2O accelerates the Baylis–Hillman reaction [133] and a double elimination to make a 4,6-dienone [134]. Adding small amounts of EtOH or DMSO was shown to accelerate some reactions in water, probably due to their ability to act as surfactants [21].

Various surfactants have been added to water to allow reaction, through the formation of microemulsions or micelles. Kobayashi's group has explored surfactants for reactions such as the aldol reactions shown in Figure 5.18; under these conditions the TMS group is protected from hydrolysis [135]. Polyethylene glycol has been used for Suzuki couplings in water [136]; the PEG could be acting as a surfactant or as a PTC reagent to solubilize metal species. Lipshutz's group has studied reactions using the surfactant PTS [137,138]. PTS is commercially available as a 2.5 wt% solution in water, and another vitamin E–PEG surfactant, TPGS-1000, is used in formulations to increase the bioavailability of APIs [139]. Lipshutz's group has demonstrated the utility of the surfactant approach for facile ring-closing metathesis [140,141], Sonogashira [142], Heck [143], Suzuki [144], Negishi [145], and amination [146] reactions. The metathesis reaction shown in Figure 5.18 is one example of the utility of the PTS surfactant. Since

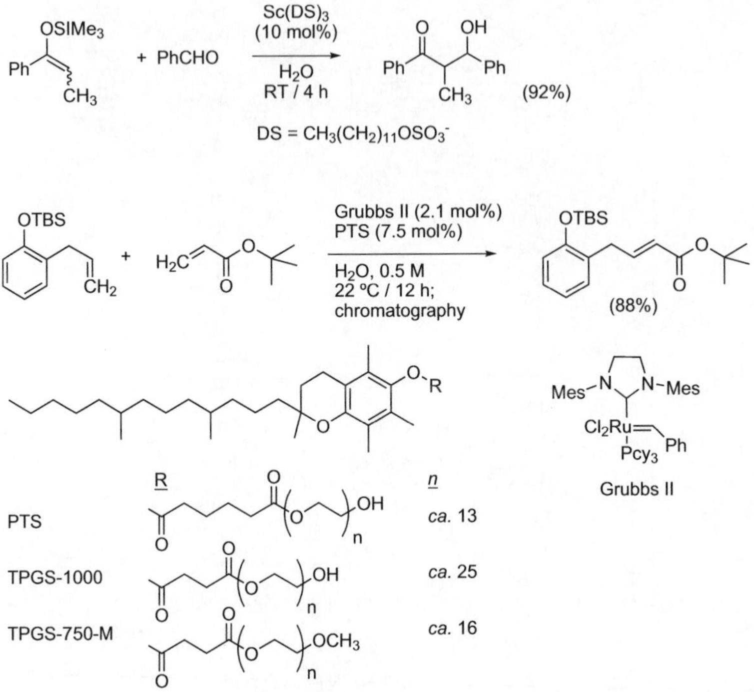

FIGURE 5.18   Reactions in water mediated by micelle formation.

this material has been generally recognized as safe (GRAS), even used as a beverage component [137], there is no need to treat residual amounts of this PTS reagent in APIs as especially toxic. After running reactions with PTS in water, products can be extracted into an organic solvent, and the aqueous phase containing the PTS has been reused for a subsequent reaction [141,143]. More recently the Lipshutz group has described the use of a second-generation amphile for various cross-coupling reactions [147]. For his work with surfactants Lipshutz received a US Presidential Green Chemistry Award in 2011 [139]. The approach of the Lipshutz group and of micelles in general holds considerable promise for manufacturing APIs, especially if the reaction products can be separated by extraction and the micelles can be reused.

One of the best applications of water is to facilitate reactions of polar materials, without resorting to protecting groups. Enzymes often can tolerate diverse functionalities in molecules, and many enzymes function best in media that are at least partially aqueous. Some examples of non-enzymatic transformations of polar molecules in water are shown in Figure 5.19. 2-Hydroxynicotinic acid was readily dissolved in water, and operations using Br_2 were inconvenient for scale-up. A solution of NaBr in bleach (NaOCl) was chosen for bromination; when that solution was added over 1 min a significant amount of the bis-bromo impurity was formed, probably due to micromixing. When the reagent mixture was added over 14 min, a good yield of the monobrominated acid was provided on a 22 L scale [148]. A Suzuki coupling on

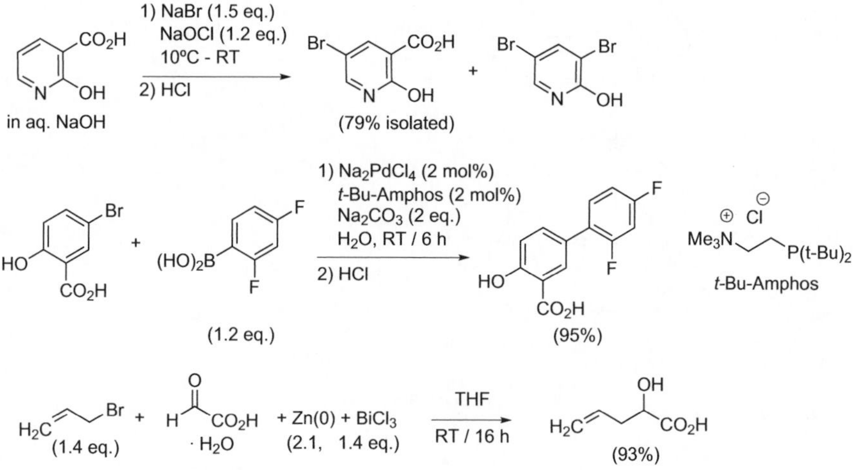

FIGURE 5.19 Reactions conveniently carried out in water without protecting groups.

a hydroxy acid was carried out under mild (basic) conditions in water [149]. A carbon–carbon bond was conveniently formed in water using glyoxal [150]. These examples show the power of matching characteristics of solvent (water in this case) and starting materials.

Water is neither a panacea nor the ideal solvent, even though it is non-flammable and inexpensive [151]. Costs to treat aqueous streams for conscientious disposal and recovery and/or reuse could be significant. These treatments may include stripping out volatile solvents, adsorption to activated carbon, and bioremediation before returning the water to municipal water treatment facilities. Furthermore, the benefit of operating in water may be eliminated if organic solvents are used in workup.

VI. SOLVENT SUBSTITUTIONS

Recently a number of chemicals have been used as inexpensive, "green" solvents, such as 2-methyl-tetrahydrofuran (from bagasse) [46], and DEM (from EtOH and formaldehyde) [111] (Figure 5.20). 2-Methyl-tetrahydrofuran has been used for reactions with organometallic reagents, extractions [46], and PTC reactions [152]. The utility of 2-Me-THF and DEM is demonstrated by the commercial availability of 3 M solutions of MeLi in either of these solvents [153]. 1,3-Propane diol has been prepared biologically, and used for polymers [154]. 1,2-Propane diol can be prepared from glycerol using vapor-phase dehydration and reduction [155], and has been used as a substitute for methyl cellosolve [156]. 1,2-Propane diol is used in formulating drug substances into drug products, and is available in food grade. Glycerol has also been used as a solvent for aza-Michael reactions [157], as a solvent and reagent for transfer hydrogenations [158], and as a solvent for reductions of carbonyls [159]. ("Glycerin" is often reserved to name a commercially available aqueous solution of glycerol, 70–80% pure [160]). Glycerol and the propane-diols may offer an additional advantage if a lipophilic product forms a separate phase.

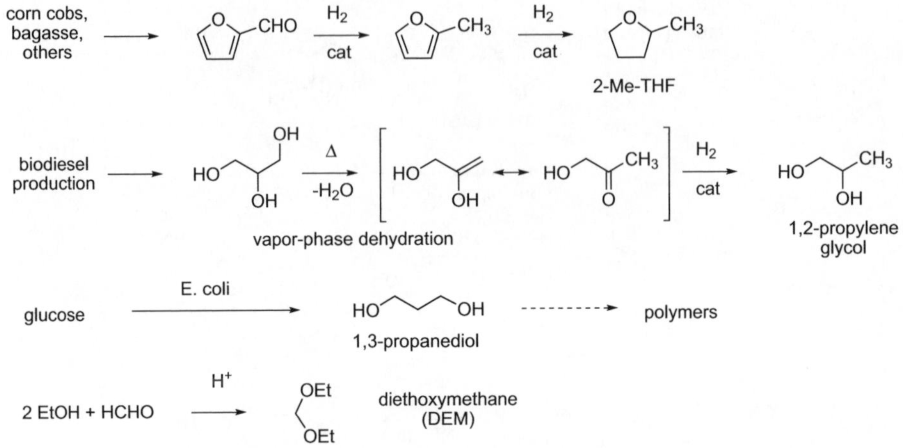

FIGURE 5.20   Preparation of some "green" solvents.

Glycerol can be derived from both plant sources and production of biodiesel, so the price of glycerol and 1,2-propane diol may drop as the production of biodiesel rises.

When high-boiling solvents are needed for reactions, separating the solvents from the product becomes an issue. DW-therm, a mix of triethoxyalkylsilanes with bp 240 °C, was used for a thermal cyclization, and the solvent was recovered by distillation; the other high-boiling solvents tried (DMSO, 1,3,5-tri-isopropyl benzene, mineral oil, and tetramethylethylene sulfone) were not satisfactory [161]. High-boiling, water-soluble solvents may be conveniently removed by extraction into water. Propylene glycol can be regarded as reasonable solvent even though no limits for this solvent have been set by the ICH. The other solvents in Figure 5.21 are derivatives of ethylene glycol. Polyethylene glycol has low toxicity, and is used in laxatives [162], but smaller derivatives of ethylene oxide, such as 1,4-dioxane, are known to be toxic. Ethylene glycol and its metabolites glycolic acid and oxalic acid are toxic to the central nervous system, the heart, and the kidneys [163]. Diethylene glycol has some physical

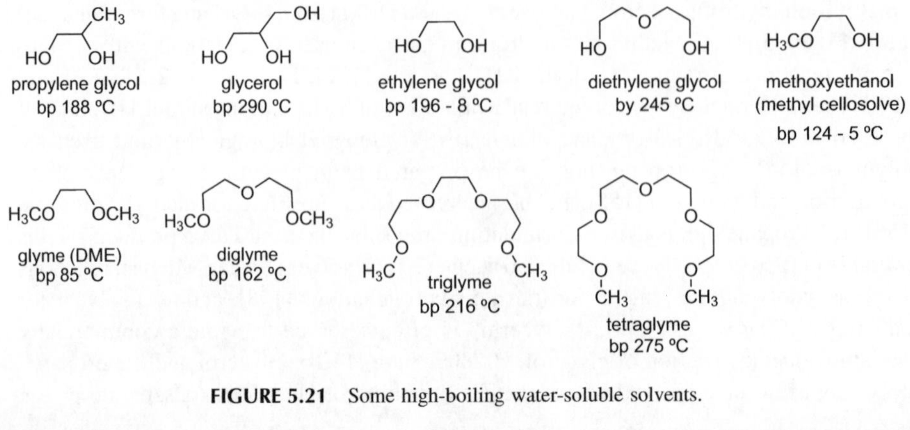

FIGURE 5.21   Some high-boiling water-soluble solvents.

properties similar to propylene glycol, but at least five people died when in China when the less-expensive diethylene glycol was substituted for propylene glycol [164]; 40 people died in Panama from Chinese-made cough syrup formulated with diethylene glycol as an inexpensive substitute for glycerin [165]. The use and manufacture of methoxyethanol (ethylene glycol monomethyl ether, or methyl cellosolve) have been banned in Canada [156], and its use is restricted in Europe [166]. (Methoxyacetic acid is the most toxic metabolite of methoxyethanol. The solvents most widely substituted for methoxyethanol were 1-methoxy-2-propanol (PGME) and 1-butoxy-2-ethanol (EGBE) [167]. Methoxyethanol and ethoxyethanol are toxic for reproduction [168]). The EPA has focused on the health effects of glycol ethers [169] by issuing a rule requiring that people planning to manufacture, import, or process any of 14 glymes for "significant new use" must contact the EPA 90 days in advance. Included in this list of glymes are DME, diglyme, triglyme, and tetraglyme (all as dimethyl ethers) [170]. Hence it may be wise to avoid using ethylene glycol-derived solvents for making APIs and penultimate intermediates.

SAFETY — Diglyme has reacted violently when heated with metallic sodium or metallic aluminum. NaOH-mediated decomposition of diethylene glycol at 200 °C caused an industrial accident. Calculations have indicated that dehydration of 1,2-glycols can be exothermic, and capping the alcohols as methyl ethers may change the thermodynamics little [*Chem. Eng. News* **2010**, *88*(28), 4]. High-temperature reactions in solvents derived from ethylene glycol and propylene glycol should be examined in a hazard evaluation laboratory before reactions are run.

For finding mixtures of solvents as alternatives to a particular solvent Reichardt has mentioned that solvent polarity can be assessed by the color of solutions of the betaine dye shown in Figure 5.7 [171]. Calculations using Reichardt's E_T^N values may also lead to suitable solvent combinations. The amide in Figure 5.22 was initially crystallized from absolute EtOH, at ~60 L/kg. Subsequently the deCODE researchers improved the productivity by recrystallizing this amide from 1:1 acetone–water, at about 32 L/kg [172]. The tetrahydrate of zwitterionic aztreonam initially was crystallized from

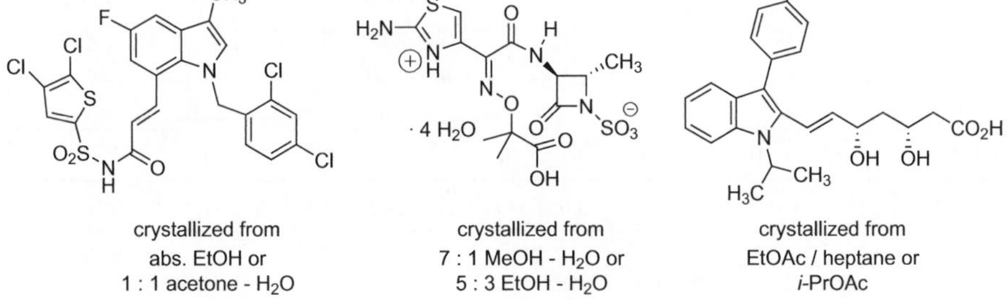

crystallized from
abs. EtOH or
1 : 1 acetone - H₂O

crystallized from
7 : 1 MeOH - H₂O or
5 : 3 EtOH - H₂O

crystallized from
EtOAc / heptane or
i-PrOAc

FIGURE 5.22   Some compounds crystallized by alternative solvents.

7:1 MeOH:water. Later the solvent combination of 5:3 EtOH–water was developed [173], because EtOH (Class 3 solvent) is less toxic than MeOH (Class 2 solvent), and because the API showed better stability in the presence of the less nucleophilic EtOH [174]. In the original manufacturing process fluvastatin was extracted into EtOAc and crystallized with the addition of heptanes; in the optimized manufacturing process i-PrOAc alone was used for both extraction and crystallization, eliminating one solvent and simplifying solvent recovery [175]. Using Reichardt's E_T^N values as shown in Equation 5.1 the approximate polarity of a solvent mixture can be calculated. For the recrystallization of deCODE's amide the E_T^N value of EtOH is equal to the calculated E_T^N of a mixture of 4.6:5.4 water–acetone, which is very close to the 1:1 water–acetone combination used by deCODE (Table 5.14). The calculated E_T^N for 7:1 MeOH:H$_2$O is the same as the calculated E_T^N for 5:3 EtOH:H$_2$O. Only a small amount of heptanes in EtOAc produces a mixture with a calculated E_T^N equal to that of i-PrOAc. (To calculate a solvent mixture to provide a target E_T^N it is necessary to introduce a second equation for the two unknowns, the two proportions of solvent. The second equation can be derived from defining that the sum of the two solvents must equal a given number, such as 10. The formulas for these calculations are shown in Equations 5.2 and 5.3.) Reichardt has noted non-linear effects for combinations of solvents that are dissimilar [34], but for combinations of solvents that are relatively similar this mathematical approach may be useful [176].

$$\text{Calculated } E_T^N = [(V_A \times E_T^N(A)) + (V_B \times E_T^N(B))]/(V_A + V_B) \tag{5.1}$$

TABLE 5.14 Polarities Calculated for Solvent Mixtures

Solvent (or Solvent Mixture)	E_T^N	Substitute Solvent Mixture	Calculated E_T^N
H$_2$O	1.000		
MeOH	0.762	6.9:3.1 EtOH–H$_2$O	0.762
EtOH	0.654	4.6:5.4 acetone–H$_2$O	0.654
Acetone	0.355		
CH$_2$Cl$_2$	0.309		
H$_2$O–saturated CH$_2$Cl$_2$ (0.2:99.8)	0.310	H$_2$O–saturated MIBK (1.9:98.1) H$_2$O–saturated EtOAc (3.3:96.7) H$_2$O–saturated 2-Me-THF (5.3:94.7)	0.283 0.253 0.223
MIBK	0.269		
EtOAc	0.228		
i-PrOAc	0.210	0.8:9.2 heptanes–EtOAc	0.210
Heptanes	0.012		
7:1 MeOH–H$_2$O	0.792	5:3 EtOH–H$_2$O	0.783

$$V_A = 10[(E_T^N(\text{target}) - E_T^N(A))]/[(E_T^N(B) - E_T^N(A))] \qquad (5.2)$$

$$V_B = 10 - V_A \qquad (5.3)$$

where V_x = volume (parts) of solvent X and $E_T^N(X) = E_T^N$ value for solvent X, and

$$E_T^N(B) > E_T^N(\text{target}) > E_T^N(A)$$

The insolubility of fluorocarbons in both water and conventional organic solvents has been exploited for purifications and synthesis [177]. Chromatography over fluorous reverse-phase silica gel [178] could prove valuable to purify impurities in fluorinated APIs. As perfluorinated hydrocarbons are expensive compared to conventional solvents, "fluorous synthesis" has not been used on scale. Benzotrifluoride (PhCF$_3$, bp 102 °C, fp 12 °C, density 1.19 g/mL) has been posed as a substitute for dichloromethane; this solvent reacts with strong reducing agents [179]. It has rarely been used on scale.

Ionic liquids have been posed as green solvents, mainly because their high boiling points essentially preclude evaporative loss [180]. Results from toxicology studies have raised questions regarding the "greenness" of these compounds as solvents [181,182]; perhaps they could be used several steps away from the final step to prepare an API.

Supercritical CO_2 (scCO$_2$) has solubility similar to hexane, and has afforded some interesting applications. H_2 is more soluble in scCO$_2$ than in solvents conventionally used, and the promise of scCO$_2$ has been demonstrated with a continuous asymmetric hydrogenation using an immobilized catalyst [183]. Trace amounts of ruthenium were removed from an API using scCO$_2$, wherein the residual Ru was adsorbed to the walls of the autoclave [184]. scCO$_2$ chromatography is rapid and effective for both analytical and preparative separations [185]. Exposure of crystalline solids to CO_2 has led to the formation of different polymorphs [186]. Since scCO$_2$ has been used to decaffeinate coffee beans, a relatively inexpensive commodity, it seems logical that this solvent could be used more often. Operations in the first multi-purpose commercial plant using scCO$_2$ were begun in 2002 [187]. The benefits of scCO$_2$ and other supercritical fluids have been reviewed [188,189]. The primary drawback of using scCO$_2$ appears to be the expense of equipment to control the condensation and release of the CO_2.

VII. SOLVENT-FREE PROCESSES

In solvent-free (neat) processes a small excess of a liquid reagent may function as a solvent, and the products are often non-crystalline [190–192]. Since solvents were not added in any significant amount to increase the mass of the reaction, quenching such reactions may readily produce high temperatures. In an alternative route to oseltamivir, Harrington described a neat acetylation of a hindered amine; quenching the excess Ac$_2$O required 3 h on a scale of 157 g [193] (Figure 5.23). Chlorosulfonylation of o-ethoxybenzoic acid produced the sildenafil intermediate in the

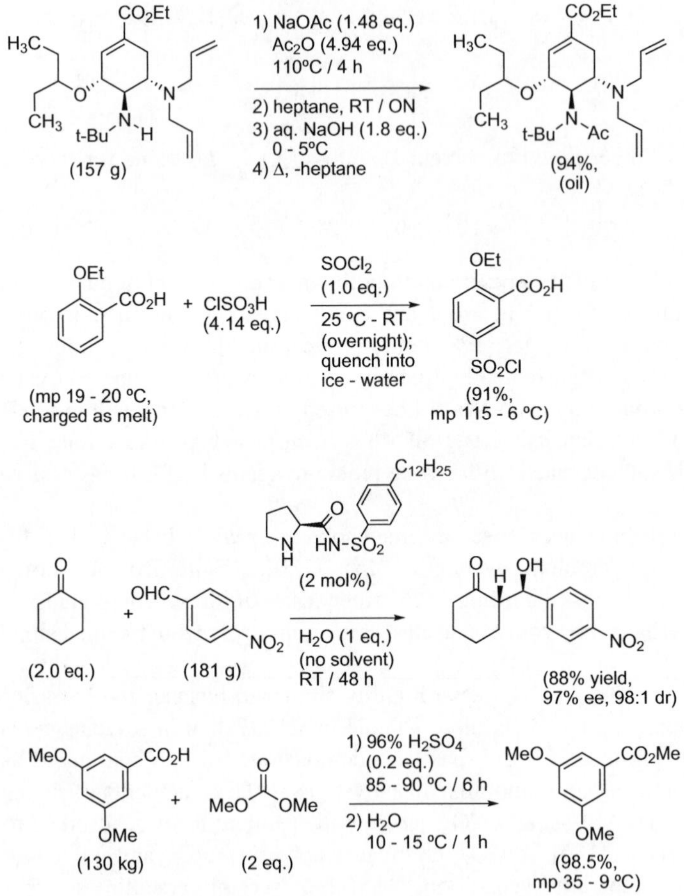

FIGURE 5.23   Some solvent-free (neat) reactions.

commercial process; for ease of transfer the starting material was melted [194]. Using an organocatalyst, Carter's group showed that a high yield from an enantioselective aldol reaction was possible without solvents. Two interesting points in the latter reaction are that one equivalent of H_2O was added to improve the yield and selectivity, and the majority of the product (55% yield) was isolated by simple filtration [195]. Rao demonstrated that carboxylic acids can be esterified productively using dimethyl carbonate with H_2SO_4 as a catalyst; this work improves over other methods requiring stoichiometric DBU or high temperatures [196]. Solvent-free processes can be very effective, especially in increasing the reaction rate for processes that would be slow otherwise.

VIII. SUMMARY AND PERSPECTIVE

Solvents are chosen for many reasons, and the primary reason should be to ensure SAFE operations. Many considerations for solvents are summarized in Table 5.15. A detailed

TABLE 5.15 Perspective on Solvent Selection for Successful Scale-Up

Aspect	Comment
SAFETY	Avoid solvents that are toxic or highly flammable
Promote high-yielding reactions	Compatible with desired chemistry Can isolate product in good yield
Convenient (minimize processing operations)	Isolate product from reaction solvent? Operate at high concentrations Chase solvents to swap (1) Ready phase splits for extractive workup Low [solvent] in API
H_2O-miscibility	Azeotroping ability Control amount of H_2O present during reactions
Cost of bulk and recoverability	More important at end of development cycle
Environmental	Ethics, and cost of recovery and non-compliance
Long-term availability	Cost of petroleum may influence the cost of petroleum-derived feedstocks; more solvents may be derived from syn gas, plants or animals
Acceptability for human use	Demonstrated absence of BSE may be required if source is animals, e.g., glycerol
Water as solvent	May be treated as a biohazard and subjected to microbial treatment. Further treatment to remove dissolved solids and VOCs may be necessary. Cost of treatment and recovery may be high
Neat reactions	Selected applications, but limited heat capacity can make temperature control difficult.
Ionic liquids	Recoverability may lead to manufacturing applications for intermediates. Environmental toxicity is a concern

*(1) By distilling solvents at constant volume Abbott researchers reduced the volume of i-PrOH needed to manufacture erythromycin A oxime by 30%: Li, Y.-E.; Yang, Y.; Kalthod, V.; Tyler, S. M. Org. Process Res. Dev. **2009**, 13, 73.*

analysis of THF is provided in the following tips box, and these types of assessments can be applied to all solvents. Physicochemical characteristics of the components under examination may override theoretical considerations; for instance, the solubility of one component may drive a reaction more than the polarity of a solvent. The latter may be true for a Finkelstein reaction, which is driven to completion by the very small solubility of the byproduct NaCl in acetone as compared to the high solubility of the reagent NaI. As another example, a solvent may be chosen if an azeotrope is discovered to remove a troublesome impurity. Many screening experiments may be needed to determine the ideal solvent for a process.

THF is usually not the first solvent selected for preparing an API, for multiple reasons. THF is used for many organometallic processes, such as Grignard reactions. As an ether it has a considerable tendency to form peroxides, a SAFETY consideration. BHT is usually added to commercial shipments of THF and during operations may be added to THF condensates. The H_2O–THF azeotrope removes only 5.3% H_2O, so a fair amount of THF could be required to dry equipment on scale. The relatively low boiling point of 65 °C limits its use for high-temperature reactions, unless one is willing to run reactions under pressure (a SAFETY consideration). For extractive workups THF is often diluted with water to afford a mixture that can be extracted with a water-immiscible solvent, which increases the V_{max} for operations, or salts are added to effect a phase split, increasing the amount of inorganic waste. The high solubility of THF in water complicates recovery from aqueous phases. As an ICH Class 2 solvent (limit 720 ppm) THF is generally avoided in generating an API. Under anhydrous conditions THF is ring-opened, limiting its use in preparing salts. THF is about twice the cost of MTBE, which can be a suitable substitute. (The cost of MTBE may rise if the demand for it as a gasoline additive falls.)

REFERENCES

1. Constable, D. J. C.; Jimenez-González, C.; Henderson, R. K. *Org. Process Res. Dev.* **2007,** *11*, 133.
2. http://portal.acs.org/portal/PublicWebSite/greenchemistry/industriainnovation/roundtable/CNBP. 027158 Over 60 solvents were ranked. The solvent categories included acids, alcohols, aromatics, bases, dipolar aprotics, esters, ethers, halogenated, hydrocarbons, and ketones. A few of the solvents ranked, such as Et$_2$O, are used in academia more than in industrial process R&D.
3. Leitner, W., Jessop, P. G., Li, C.-J., Wasserscheid, P., Stark, A.; Anastas, P. T., Eds.*; Handbook of Green Chemistry – Green Solvents*, Vol. 5, Wiley, 2010.
4. Reichardt, C. *Org. Process Res. Dev.* **2007,** *11*, 105.
5. Reichardt, C. *Solvents and Solvent Effects in Organic Chemistry*, 3rd ed.; Wiley-VCH, 2003.
6. Ref. 5; p 27.
7. Ref. 5; p 69.
8. Weber, W. P.; Gokel, G. W. *Phase Transfer Catalysis in Organic Synthesis*; Springer Verlag: New York, 1977; p 11.
9. Ref. 5, pp 37–38.
10. Albanese, D.; Landini, D.; Maia, A.; Penso, M. *Ind. Eng. Chem. Res.* **2001,** *40*, 2396.
11. Weber, W. P.; Gokel, G. W. *Phase Transfer Catalysis in Organic Synthesis*; Springer Verlag: New York, 1977; pp 14–15.
12. Gani, R.; Jiménez-González, C.; ten Kate, A.; Crafts, P. A.; Jones, M.; Powell, L.; Atherton, J. H.; Cordiner, J. L. *Chem. Eng.* **2006,** *113*(3), 30.
13. Caron, S. In *Practical Synthetic Organic Chemistry*; Caron, S., Ed.; Wiley, 2011; p 805.
14. Slater, C. S.; Savelski, M. J.; Carole, W. A.; Constable, D. J. C. In *Green Chemistry in the Pharmaceutical Industry*; Dunn, P., Wells, A., Williams, M. T., Eds.; Wiley-VCH, 2010; p 49.
15. Fischer, R. W.; Misun, M. *Org. Process Res. Dev.* **2001,** *5*, 581.
16. For another water-mediated Dimroth rearrangement, see Jensen, M. R.; Hoerrner, R. S.; Li, W.; Nelson, D. P.; Javadi, G. J.; Dormer, P. G.; Cai, D.; Larsen, R. D. *J. Org. Chem.* **2005,** *70*, 6034.
17. Beaulieu, P. L.; Gillard, J.; Bailey, M. D.; Boucher, C.; Duceppe, J.-S.; Simoneau, B.; Wang, X.-J.; Zhang, L.; Grozinger, K.; Houpis, I.; Farina, V.; Heimroth, H.; Krueger, T.; Schnaubelt, J. *J. Org. Chem.* **2005,** *70*, 5869.
18. Ellis, J. E.; Davis, E. M.; Dozeman, G. J.; Lenoir, E. A.; Belmont, D. T.; Brower, P. L. *Org. Process Res. Dev.* **2001,** *5*, 226.

19. Rassias, G.; Hermitage, S. A.; Sanganee, M. J.; Kincey, P. M.; Smith, N. M.; Andrews, I. P.; Borrett, G. T.; Slater, G. R. *Org. Process Res. Dev.* **2009,** *13,* 774.

20. For an example, see Harrington, P. J.; Khatri, H. N.; DeHoff, B. S.; Guinn, M. R.; Boehler, M. A.; Glaser, K. A. *Org. Process Res. Dev.* **2002,** *6,* 120.

21. Breslow, R.; Groves, K.; Mayer, M. U. *Org. Lett.* **1999,** *1,* 117.

22. Leazer, J. L., Jr.; Cvetovich, R.; Tsay, F.-R.; Dolling, U.; Vickery, T.; Bachert, D. *J. Org. Chem.* **2003,** *68,* 3695.

23. Amato, J. S.; Chung, J. Y. L.; Cvetovich, R. J.; Gong, X.; McLaughlin, M.; Reamer, R. A. *J. Org. Chem.* **2005,** *70,* 1930.

24. Anderson, N. G.; Ary, T. D.; Berg, J. L.; Bernot, P. J.; Chan, Y. Y.; Chen, C.-K.; Davies, M. L.; DiMarco, J. D.; Dennis, R. D.; Deshpande, R. P.; Do, H. D.; Droghini, R.; Early, W. A.; Gougoutas, J. Z.; Grosso, J. A.; Harris, J. C.; Haas, O. W.; Jass, P. A.; Kim, D. H.; Kodersha, G. A.; Kotnis, A. S.; LaJeunesse, J.; Lust, D. A.; Madding, G. D.; Modi, S. P.; Moniot, J. L.; Nguyen, A.; Palaniswamy, V.; Phillipson, D. W.; Simpson, J. H.; Thoraval, D.; Thurston, D. A.; Tse, K.; Polomski, R. E.; Wedding, D. L.; Winter, W. J. *Org. Process Res. Dev.* **1997,** *1,* 300.

25. Furniss, B. S.; Hannaford, A. J.; Smith, P. W. G.; Tatchell, A. R. *Vogel's Textbook of Practical Organic Chemistry,* 4th ed.; Addison Wesley Longman Ltd.: Essex, 1989; p 917.

26. Jass, P. A.; Rosso, V. W.; Racha, S.; Soundararajan, N.; Venit, J. J.; Rusowicz, A.; Swaminathan, S.; Livshitz, J.; Delaney, E. *J. Tetrahedron* **2003,** *59,* 9019.

27. Greene, T. W.; Wuts, P. G. M. *Protective Groups in Organic Synthesis,* 3rd ed.; Wiley: New York, 1999; p 557.

28. Amino acids protected on the nitrogen by Cbz or Fmoc were converted to the corresponding acid chlorides and coupled with amines without racemization (dichloromethane, 25 °C). Thus the acid chloride approach to making chiral amides has demonstrated considerable utility. Butini, S.; Gabellieri, E.; Huleatt, P. B.; Campiani, G.; Franceschini, S.; Brindisi, M.; Ros, S.; Coccone, S. S.; Fiorini, I.; Novellino, E.; Giorgi, G.; Gemma, S. *J. Org. Chem.* **2008,** *73,* 8458.

29. http://www.acgih.org/home.htm

30. http://www.genengnews.com/articles/chitem.aspx?aid=2392

31. Federspiel, M.; Fischer, R.; Hennig, M.; Mair, H.-J.; Oberhauser, T.; Rimmler, G.; Albiez, T.; Bruhin, J.; Estermann, H.; Gandert, C.; Göckel, V.; Götzö, S.; Hoffmann, U.; Huberr, G.; Janatsch, G.; Lauper, S.; Röckel-Stäbler, O.; Trussardi, R.; Zwahlen, A. G. *Org. Process Res. Dev.* **1999,** *3,* 266.

32. Anderson, N. G.; Anderson, C. F. U. S. Patent 4,826,973, 1989 (to E. R. Squibb & Sons).

33. Jenkins, S. *Chem. Eng.* **2011,** *118*(1), 22.

34. Reichardt, C. *Pure Appl. Chem.* **2004,** *76,* 1903.

35. Ref. 5; p 364.

36. McConville, F. *The Pilot Plant Real Book,* 2nd ed.; FXM Engineering & Design: Worcester MA, 2007; http://www.pprbook.com

37. Pines, S. Personal communication.

38. Hobson, L. A. *Chem. Eng. News* **2000,** *78*(51), 2.

39. Deshpande, M. H.; Cain, M. H.; Patel, S. R.; Singam, P. R.; Brown, D.; Gupta, A.; Barkalow, J.; Callen, G.; Patel, K.; Koops, R.; Chorghade, M.; Foote, H.; Pariza, R. *Org. Process Res. Dev.* **1998,** *2,* 351.

40. Bradley, P. A.; de Koning, P. D.; Johnson, P. S.; Lecouturier, Y. C.; McManus, D. J.; Robin, A.; Underwood, T. J. *Org. Process Res. Dev.* **2009,** *13,* 848.

41. Ripin, D. H. B.; Bourassa, D. E.; Brandt, T.; Castaldi, M. J.; Frost, H. N.; Hawkins, J.; Johnson, P. J.; Massett, S. S.; Neumann, K.; Phillips, J.; Raggon, J. W.; Rose, P. R.; Rutherford, J. L.; Sitter, B.; Stewart, A. M., III; Vetelino, M. G.; Wei, L. *Org. Process Res. Dev.* **2005,** *9,* 440; See discussion on isolating compound **1** from Suzuki coupling.

42. French, H. E.; Wrightsman, G. G. *J. Am. Chem. Soc.* **1938,** *60,* 50.

43. Baltzly, R.; Berger, I. M.; Rothstein, A. A. *J. Am. Chem. Soc.* **1950,** *72,* 4149.

44. Reisch, M. *Chem. Eng. News* **2008,** *86*(29), 32.

45. Chemetall: http://www.pschem.com/pdfs/MeTHFAGreenAlttoTHF.pdf
46. Aycock, D. F. *Org. Process Res. Dev.* **2007,** *11,* 156.
47. Canadian Centre for Occupational Health and Safety. http://www.ccohs.ca/oshanswers/chemicals/organic/organic_peroxide.html
48. Heiba, E. I.; Dessau, R. M.; Koehl, W. J., Jr. *J. Am. Chem. Soc.* **1969,** *91,* 6830.
49. http://www.gpo.gov/fdsys/pkg/FR-1999-02-23/pdf/99-4320.pdf
50. Murlis, J. In *Volatile Organic Compounds in the Atmosphere*; Hester, R. E.; Harrison, R. M., Eds.; Royal Society of Chemistry, 1995 pp 130–131.
51. Terent'ev, A. O.; Borisov, D. A.; Yaremenko, I. A.; Chernyshev, V. V.; Nikishin, G. I. *J. Org. Chem.* **2010,** *75,* 5065.
52. http://www.fda.gov/RegulatoryInformation/Guidances/ucm128223.htm
53. http://www.ema.europa.eu/docs/en_GB/document_library/Scientific_guideline/2009/09/WC500002674.pdf
54. Luyben, W. A. *Distillation Design and Control Using Aspen™ Simulation*; AIChE, Wiley Interscience: New York, 2007; p 12.
55. Li, Y.-E.; Yang, Y.; Kalthod, V.; Tyler, S. M. *Org. Process Res. Dev.* **2009,** *13,* 73.
56. http://www.ddbst.com/en/products/Products.php
57. *CRC Handbook of Chemistry and Physics,* 88th ed.; Lide, D. R., Ed.; CRC Press: Boca Raton, 2007–2008; pp 6-171–6-172.
58. Karpf, M.; Trussardi, R. *J. Org. Chem.* **2001,** *66,* 2044.
59. Prashad, M.; Har, D.; Chen, L.; Kim, H.-Y.; Repic, O.; Blacklock, T. J. *J. Org. Chem.* **2002,** *67,* 6612.
60. Amedio, J. C.; Bernard, P. J.; Fountain, M.; Van Wagenen, G., Jr. *Synth. Commun.* **1997,** *29,* 2377.
61. Kantor, S. W. *J. Am. Chem. Soc.* **1953,** *75*(11), 2712; Flaningam, O. L.; Williams, D. E. U.S. Patent 5,478,493, 1995 (to Dow); Morgan, D. L.; Williams, D. E. U.S. Patent 5,834,416, 1998 (to Dow).
62. Anderson, N. G.; Ciaramella, B. M.; Feldman, A. F.; Lust, D. A.; Moniot, J. L.; Moran, L.; Polomski, R. E.; Wang, S. S. Y. *Org. Process Res. Dev.* **1997,** *1,* 211.
63. Bailey, A. R.; Halfon, M.; Sortore, E. W. U.S. Patent 5,440,045, 1995 (to FMC).
64. McConville, F. *The Pilot Plant Real Book*, 2nd ed.; FXM Engineering & Design: Worcester MA, 2007; pp 6–26.
65. Ref. 5; pp 137–145.
66. Ref. 5; pp 179–185.
67. Ref. 5; pp 186–188.
68. Sykes, P. *A Guidebook to Mechanism in Organic Chemistry*, 6th ed.; Addison Wesley Longman: Essex, UK; p 81.
69. Ref. 5; p 143.
70. Inaba, T.; Birchler, A.; Yamada, Y.; Sagawa, S.; Yokota, K.; Ando, K.; Uchida, I. *J. Org. Chem.* **1998,** *63,* 7582.
71. Floyd, D. M.; Fritz, A. W.; Cimarusti, C. M. U.S. Patent 4,386,034, 1983 (to E. R. Squibb).
72. Ref. 5; pp 144–145.
73. Anderson, N. G. Unpublished.
74. Anderson, G. W.; Zimmerman, J. E.; Callahan, F. M. *J. Am. Chem. Soc.* **1967,** *89,* 5012.
75. Franzén, H. M.; Bessidskaia, G.; Abedi, V.; Nilsson, A.; Nilsson, M.; Olsson, L. *Org. Process Res. Dev.* **2002,** *6,* 788.
76. Lloyd-Williams, P.; Albericio, F.; Giralt, E. *Chemical Approaches to the Synthesis of Peptides and Proteins*; CRC Press: Boca Raton, 1997; see especially. pp 118–119.
77. Davulcu, A. H.; McLeod, D. D.; Li, J.; Katipally, K.; Littke, A.; Doubleday, W.; Xu, Z.; McConlogue, C. W.; Lai, C. J.; Gleeson, M.; Schwinden, M.; Parsons, R. L., Jr. *J. Org. Chem.* **2009,** *74,* 4068.
78. Haight, A. R.; Stuk, T. L.; Menzia, J. A.; Robbins, T. A. *Tetrahedron Lett.* **1997,** *38,* 4191.
79. Ref. 5; pp 414–415.

80. Armarego, W. L. F.; Chai, C. Purification of Laboratory Chemicals, 6th ed.; Butterworth-Heinemann, 2009.

81. Williams, D. B. G.; Lawton, M. *J. Org. Chem.* **2010,** *75*, 8351.

82. Smith, M. B.; March, J. *March's Advanced Organic Chemistry*, 6th ed.; Wiley, 2007; pp 943–949, 967–971.

83. Kelly, R. J. *Chem. Health Saf.* **1996,** September/October, 28.

84. Butters, M.; Harvey, J. N.; Jover, J.; Lennox, A. J. J.; Lloyd-Jones, G. C.; Murray, P. M. *Angew. Chem. Int. Ed.* **2010,** *49*, 5156; see especially footnote 25 therein.

85. US Peroxide: http://www.h2o2.com/technical-library/analytical-methods/default.aspx?pid=70&name=Iodometric-Titration

86. Cooke, J. W. B.; Bright, R.; Coleman, M. J.; Jenkins, K. P. *Org. Process Res. Dev.* **2001,** *5*, 383.

87. Williams, J. M. In *The Art of Process Chemistry*; Yasuda, N., Ed.; Wiley-VCH, 2010; p 94.

88. http://www.innospecinc.com/assets/_files/documents/may_08/cm__1210589634_Statsafe.pdf

89. Mills, J. E.; Maryanoff, C. A.; Cosgrove, R. A.; Scott, A.; McComsey, D. F. *Org. Prep. Proc. Int.* **1984,** *16*, 97.

90. Nevstad, G. O.; Songstad, J. *Acta Chem. Scan. B* **1984,** *38*, 469.

91. Hansen, S. M.; Nordholm, L. *J. Chromatography A* **1981,** *204*, 97.

92. Faul, M. M.; Kobierski, M. E.; Kopach, M. E. *J. Org. Chem.* **2003,** *68*, 5739.

93. Federsel, H.-J.; Konberg, E.; Lilljequist, L.; Swahn, B.-M. *J. Org. Chem.* **1990,** *55*, 2254.

94. Mills, J. E.; Maryanoff, C. A.; McComsey, D. F.; Stanzione, R. C.; Scott, L. *J. Org. Chem.* **1987,** *52*, 1857.

95. Rudine, A. B.; Walter, M. G.; Wamser, C. C. *J. Org. Chem.* **2010,** *75*, 4292.

96. Ji, J.; Zhang, D.; Ye, Y.; Xing, Q. *Tetrahedron Lett.* **1998,** *39*, 6515.

97. Herriot, A.; Picker, D. *Synthesis* **1975,** 447.

98. Qian, X.; Kozak, C. M. *Synlett* **2011,** *6*, 852.

99. Taylor, S. M.; Yamada, T.; Ueki, H.; Soloshonok, V. A. *Tetrahedron Lett.* **2004,** *45*, 9159.

100. Franklin, R. A.; Heatherington, K.; Morrison, B. J.; Sherron, P.; Ward, T. J. *Analyst* **1978,** *103*, 660.

101. Mohammadi, A.; Amini, M.; Hamedani, M. P.; Torkabadi, H. H.; Walker, R. B. *Asian J. Chem.* **2008,** *20*, 5573.

102. Haight, A. R.; Stuk, T. L.; Allen, M. S.; Bhagavatula, L.; Fitzgerald, M.; Hannick, S. M.; Kerdesky, F. A. J.; Menzia, J. A.; Parekh, S. I.; Robbins, T. A.; Scarpetti, D.; Tien, J.-H. *J. Org. Process Res. Dev.* **1999,** *3*, 94.

103. Goldsmith, D. J.; Kennedy, E.; Campbell, R. G. *J. Org. Chem.* **1975,** *40*(24), 357.

104. See Ref. 7a in Conrow, R. E.; Dean, W. D.; Zinke, P. W.; Deason, M. E.; Sproull, S. J.; Dantanarayana, A. P.; DuPriest, M. T. *Org. Process Res. Dev.* **1999,** *3*, 114.

105. Am Ende, D. J.; Vogt, P. F. *Org. Process Res. Dev.* **2003,** *7*, 1029.

106. Couturier, M.; Andresen, B. M.; Jorgensen, J. B.; Tucker, J. L.; Busch, F. R.; Brenek, S. J.; Dubé, P.; Am Ende, D. J.; Negri, J. T. *Org. Process Res. Dev.* **2002,** *6*, 42.

107. Atkins, W. J., Jr.; Burkhardt, E. R.; Matos, K. *Org. Process Res. Dev.* **2006,** *10*, 1292.

108. Buckley, W.; Webb, R. L.; Laird, T.; Ward, R. J. *Chem. Eng. News* **1982,** *60*(28), 5.

109. Liu, Y.; Schwartz, J. *J. Org. Chem.* **1993,** *58*, 5005.

110. Hutton, J.; Jones, A. J.; Lee, S. A.; Martin, D. M. G.; Meyrick, B. R.; Patel, I.; Peardon, R. F.; Powell, L. *Org. Process Res. Dev.* **1997,** *1*, 61.

111. Boaz, N. W.; Venepalli, B. *Org. Process Res. Dev.* **2001,** *5*, 127.

112. Grimm, J. S.; Maryanoff, C. A.; Patel, M.; Palmer, D. C.; Sorgi, K. L.; Stefanik, S.; Webster, R. R. H.; Zhang, X. *Org. Process Res. Dev.* **2002,** *6*, 938; see experimental findings and footnote 12 therein.

113. http://www.crhf.org.uk/incident09.html

114. Clark, J. D.; Weisenburger, G. A.; Anderson, D. K.; Colson, P.-J.; Edney, A. D.; Gallagher, D. J.; Kleine, H. P.; Knable, C. M.; Lantz, M. K.; Moore, C. M. V.; Murphy, J. B.; Rogers, T. E.; Ruminski, P. G.; Shah, A. S.; Storer, N.; Wise, B. M. *Org. Process Res. Dev.* **2004,** *8*, 51.

115. Oh, L. M.; Wang, H.; Shilcrat, S. C.; Herrmann, R. E.; Patience, D. B.; Spoors, P. G.; Sisko, J. *Org. Process Res. Dev.* **2007**, *11*, 1032.

116. Harrington, P. J.; Johnston, D.; Moorlag, H. G.; Wong, J.-W.; Hodges, L. M.; Harris, L.; McEwen, G. K.; Smallwood, B. *Org. Process Res. Dev.* **2006**, *10*, 1157.

117. Laduron, F.; Tamborowski, V.; Moens, L.; Horváth, A.; De Smaele, D.; Leurs, S. *Org. Process Res. Dev.* **2005**, *9*, 102.

118. Li, C.-J.; Chan, T.-H. *Comprehensive Organic Reactions in Aqueous Media*, 2nd ed.; Wiley-VCH: Hoboken, NJ, 2007.

119. Sheldon, R. A.; Arends, I.; Hanefeld, U. *Green Chemistry and Catalysis*; Wiley-VCH: Weinheim, Germany, 2007.

120. Lindström, U. M., Ed.*; Organic Reactions in Water: Principles, Strategies, and Applications*; Blackwell Publishing: Oxford, U.K., 2007.

121. Hailes, H. C. *Org. Process Res. Dev.* **2007**, *11*, 114.

122. Moulay, S. *Khimiya/Chemistry* **2009**, *18*, 1, http://khimiya.org/pdfs/KHIMIYA_18_3_ONLINE.pdf

123. Wiebus, E.; Cornils, B. In *Organic Reactions in Water: Principles, Strategies, and Applications*; Lindström, U. M., Ed.; Blackwell Publishing: Oxford, U.K., 2007; pp 362–405.

124. Wüllner, G.; Jänsch, H.; Kannenberg, S.; Schubert, F.; Boche, G. *Chem. Commun.* **1998**, 1509.

125. ten Brink, G.-J.; Arends, I. W. C. E.; Papadogianakis, G.; Sheldon, R. A. *Chem. Commun.* **1998**, 2359.

126. Narayan, S.; Muldoon, J.; Finn, M. G.; Fokin, V. V.; Kolb, H. C.; Sharpless, K. B. *Angew. Chem. Int. Ed. Eng.* **2005**, *44*, 3275.

127. Chanda, A.; Fokin, V. V. *Chem. Rev.* **2009**, *109*, 725.

128. Narayan, S.; Fokin, V. V.; Sharpless, K. B. In *Organic Reactions in Water: Principles, Strategies, and Applications*; Lindström, U. M., Ed.; Blackwell Publishing: Oxford, U.K., 2007; pp 350–365.

129. Breslow, R. *Acc. Chem. Res.* **1991**, *24*(6), 159.

130. Skouta, R.; Wei, S.; Breslow, R. *J. Am. Chem. Soc.* **2009**, *131*, 15604.

131. Rostovtsev, V. V.; Green, L. G.; Fokin, V. V.; Sharpless, K. B. *Angew. Chem. Int. Ed.* **2002**, *41*, 2596.

132. Demko, Z. P.; Sharpless, K. B. *J. Org. Chem.* **2001**, *66*, 7945.

133. Yu, C.; Liu, B.; Hu, L. *J. Org. Chem.* **2001**, *66*, 5413.

134. Ohta, T.; Zhang, H.; Torihara, Y.; Furukawa, I. *Org. Process Res. Dev.* **1997**, *1*, 420.

135. Ogawa, C.; Kobayashi, S. In *Process Chemistry in the Pharmaceutical Industry*; Gadamasetti; Braish, Eds, 2nd ed.; CRC Press: Boca Raton, Florida, 2008; pp 249–265.

136. Liu, L.; Zhang, Y.; Wang, Y. *J. Org. Chem.* **2005**, *70*, 6122.

137. Lipshutz, B. H.; Ghorai, S. *Aldrichimica Acta* **2008**, *41*, 59.

138. Abela, A. R.; Huang, S.; Moser, R.; Lipshutz, B. H. *Chimica Oggi* **2010**, *26*(5), 50.

139. Ritter, S. K. *Chem. Eng. News* **2011**, *89*(26), 11, http://pubs.acs.org/cen/news/89/i26/8926notw1b.html

140. Lipshutz, B. H.; Aguinaldo, G. T.; Ghorai, S.; Voigtritter, K. *Org. Lett.* **2008**, *10*, 1325.

141. Lipshutz, B. H.; Ghorai, S. *Org. Lett.* **2009**, *11*, 705.

142. Lipshutz, B. H.; Chung, D. W.; Rich, B. *Org. Lett.* **2008**, *10*, 3793.

143. Lipshutz, B. H.; Taft, B. R. *Org. Lett.* **2008**, *10*, 1329.

144. Lipshutz, B. H.; Petersen, T. B.; Abela, A. R. *Org. Lett.* **2008**, *10*, 1333.

145. Krasovskiy, A.; Duplais, C.; Lipshutz, B. H. *J. Am. Chem. Soc.* **2009**, *131*, 15592.

146. Lipshutz, B. H.; Chung, D. W.; Rich, B. *Adv. Synth. Catal.* **2009**, *351*, 1717.

147. Lipshutz, B. H.; Ghorai, S.; Abela, A. R.; Moser, R.; Nishikata, T.; Duplais, C.; Krasovskiy, A.; Gaston, R. D.; Gadwood, R. C. *J. Org. Chem.* **2011**, *76*, 4379.

148. Haché, B.; Duceppe, J.-S.; Beaulieu, P. L. *Synthesis* **2002**, 528.

149. DeVasher, R. B.; Moore, L. R.; Shaughnessy, K. H. *J. Org. Chem.* **2004**, *69*, 7919.

150. Wada, M.; Honna, M.; Kuramoto, Y.; Miyoshi, N. *Bull. Chem. Soc. Jpn.* **1997**, *70*, 2265; For conversion of this reaction product to a water-soluble tin reagent, see Clive, D. L. J.; Wang, J. *J. Org. Chem.* **2002**, *67*, 1192.

151. Blackmond, D. G.; Armstrong, A.; Coombe, V.; Wells, A. *Angew. Chem. Int. Ed.* **2007**, *46*, 3798.

152. Ripin, D. H. B.; Vetelino, M. *Synlett* **2003,** *15,* 2353.
153. Chemetall: http://www.pschem.com/pdfs/MeTHFAGreenAlttoTHF.pdf
154. Tullo, A. *Chem. Eng. News* **2007,** *85*(25), 36.
155. D'Aquino, R.; Ondrey, G. *Chem. Eng.* **2007,** *114*(9), 31.
156. *Chem. Eng. News* **2006,** *84*(50), 26.
157. Gu, Y.; Barrault, J.; Jérôme, F. *Adv. Synth. Catal.* **2008,** *350,* 2007.
158. Tavor, D.; Popov, S.; Dlugy, C.; Wolfson, A. *Org. Commun.* **2010,** *3,* 70.
159. Wolfson, A.; Dlugy, C. *Org. Commun.* **2009,** *2,* 34.
160. Pagliaro, M.; Rossi, M. *The Future of Glycerol,* 2nd ed.; Royal Society of Chemistry: London, 2010; p 1.
161. Shieh, W.-C.; Chen, G.-P.; Xue, S.; McKenna, J.; Jiang, X.; Prasad, K.; Repic, O.; Straub, C.; Sharma, S. K. *Org. Process Res. Dev.* **2007,** *11,* 711.
162. http://en.wikipedia.org/wiki/Polyethylene_glycol
163. http://en.wikipedia.org/wiki/Ethylene_glycol
164. Tremblay, J.-F. *Chem. Eng. News* **2006,** *84*(21), 11.
165. Tremblay, J.-F. *Chem. Eng. News* **2007,** *85*(23), 8.
166. Hogue, C. *Chem. Eng. News* **2010,** *88*(51), 31.
167. http://echa.europa.eu/doc/consultations/svhc/svhc_axvrep_austria_cmr_2-methoxyethanol.pdf
168. http://echa.europa.eu/doc/consultations/svhc/svhc_axvrep_austria_cmr_2-ethoxyethanol.pdf
169. Hogue, C. *Chem. Eng. News* **2011,** *89*(29), 24.
170. http://www.epa.gov/oppt/existingchemicals/pubs/glymes.html
171. Ref. 5; p 364.
172. Zegar, S.; Tokar, C.; Enache, L. A.; Rajagopol, V.; Zeller, W.; O'Connell, M.; Singh, J.; Muellner, F. W.; Zembower, D. E. *Org. Process Res. Dev.* **2007,** *11,* 747.
173. Anderson, N. G. Unpublished.
174. Florey, K. Aztreonam. In *Analytical Profiles of Drug Substances,* Vol. 17.; Florey, K., Ed.; Academic Press: San Diego, 1988; pp 1–38.
175. Fuenfschilling, P. C.; Hoehn, P.; Mutz, J.-P. *Org. Process Res. Dev.* **2007,** *11,* 13.
176. Reichardt, C. Personal communication, June 2009.
177. Studer, A.; Hadida, S.; Ferritto, R.; Kim, S.-Y.; Jeger, P.; Wipf, P.; Curran, D. P. *Science* **1997,** *275,* 873; Zhang, W.; Curran, D. P. *Tetrahedron* **2006,** *62,* 1183.
178. Curran, D. P.; Luo, Z. J. *Am. Chem. Soc.* **1999,** *121,* 9069.
179. Ogawa, A.; Curran, D. P. *J. Org. Chem.* **1997,** *62,* 450.
180. Martins, M. A. P.; Frizzo, C. P.; Moreira, D. N.; Zanatta, N.; Bonacorso, H. G. *Chem. Rev.* **2008,** *108,* 2015.
181. Ranke, J.; Stolte, S.; Störmann, R.; Arning, J.; Jastorff, B. *Chem. Rev.* **2007,** *107,* 2183.
182. Wells, A. S.; Coombe, V. T. *Org. Proc. Res. Dev.* **2006,** *10,* 794.
183. Stephenson, P.; Kondor, B.; Licence, P.; Scovell, K.; Ross, S. K.; Poliakoff, M. *Adv. Synth. Catal.* **2006,** *348,* 1605.
184. Gallou, F.; Saim, S.; Koenig, K. J.; Bochniak, D.; Horhota, S. T.; Yee, N. K.; Senanayake, C. H. *Org. Process Res. Dev.* **2006,** *10,* 937.
185. Welch, C. J.; Henderson, D. W.; Tschaen, D. M.; Miller, R. A. *Org. Process Res. Dev.* **2009,** *13,* 621.
186. Tian, J.; Dalgamo, S. J.; Atwood, J. L. *J. Am. Chem. Soc.* **2011,** *133,* 1399.
187. Licence, P.; Ke, J.; Sokolova, M.; Ross, S. K.; Poliakoff, M. *Green Chem.* **2003,** *5,* 99.
188. Akien, G.; Poliakoff, M. *Green Chem.* **2009,** *11,* 1083.
189. Jessop, P. G.; Ikariya, T.; Noyori, R. *Chem. Rev.* **1999,** *99,* 475.
190. Metzger, J. O. *Angew. Chem. Int. Ed.* **1998,** *37,* 2975.
191. Tanaka, K.; Toda, F. *Chem. Rev.* **2000,** *100,* 1025.
192. Martins, M. A. P.; Frizzo, C. P.; Moreira, D. N.; Buriol, L.; Machado, P. *Chem. Rev.* **2009,** *109,* 4140.
193. Harrington, P. J.; Brown, J. D.; Foderaro, T.; Hughes, R. C. *Org. Process Res. Dev.* **2004,** *8,* 86.

194. Dale, D. J.; Dunn, P. J.; Golightly, C.; Hughes, M. L.; Levett, P. C.; Pearce, A. K.; Searle, P. M.; Ward, G.; Wood, A. S. *Org. Process Res. Dev.* **2000,** *4*, 17; Dunn, P. J.; Galvin, S.; Hettenbach, K. *Green Chem.* **2004,** *6,* 43.
195. Yang, H.; Carter, R. G. *Org. Lett.* **2008,** *10*, 4649.
196. Rekha, V. V.; Ramani, M. V.; Ratanamala, A.; Rupakalpana, V.; Subbaraju, G. V.; Satyanarayana, C.; Rao, C. S. *Org. Process Res. Dev.* **2009,** *13*, 769.

Effects of Water

"Since, in particular, when moisture-sensitive organometallic reagents or Lewis acids are involved, synthetic chemists go to great lengths to avoid traces of water in the reaction mixture, it can only be speculated how many new H_2O-accelerated processes remain to be (re-)discovered, and how many cases of adventitious water effects have played a significant but yet unrecognized role in influencing the course of the reaction."

Seth Ribe and Peter Wipf [1]

I. INTRODUCTION

Often the influence of water is overlooked until a problem arises. The process chemist may first encounter water as part of a reagent, solvent or co-solvent (Chapter 5), or new intermediates made in the laboratory. Investigations may show that water is a harmful or beneficial impurity. Water can effectively modify organometallic reagents, and the role of water in organic transformations was detailed in an excellent review [1]. Optimized amounts of water may be required for enzymatic reactions conducted in primarily non-aqueous conditions [2–5]. Hygroscopic solids complicate processing (Chapter 8). Water may also be necessary to crystallize a desired hydrate, or to crystallize a polymorph or

Practical Process Research and Development. DOI: 10.1016/B978-0-12-386537-3.00006-X solvate that is not hydrated (Chapters 12 and 13). The importance of cleaning and removing water from stationary equipment is often overlooked (Chapter 16). This chapter will examine some other areas in which water contents need to be considered to control processing.

The key questions surrounding water are:
1. Is water helpful or detrimental to the process?
2. If water is helpful, how little is needed for reliable processing?
3. If water is detrimental, how much can the process tolerate?
4. Where is the best point in operations to analyze for H_2O?
5. What is the best way to sample for H_2O content, especially in equipment containing water-immiscible solvents?

The presence of H_2O has generated some unexpected products (Figure 6.1). In the presence of MeOH hexamethylbenzene reacted with Selectfluor® to give the benzyl

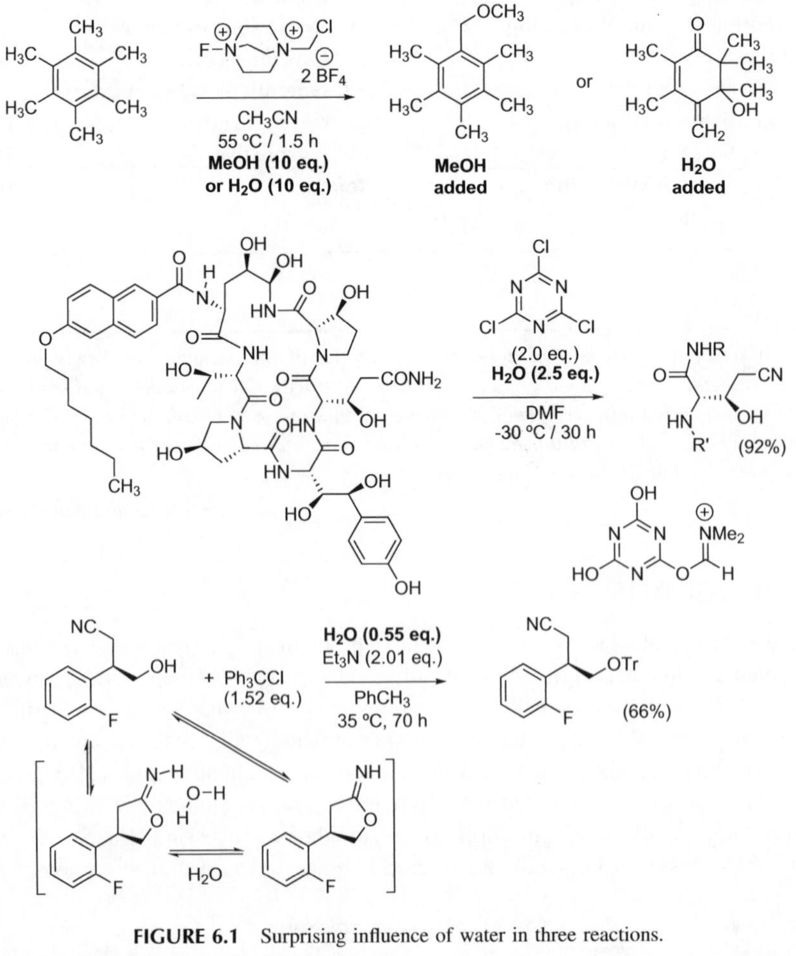

FIGURE 6.1 Surprising influence of water in three reactions.

methyl ether, but in acetonitrile containing H_2O the primary product was the unsaturated ketone from rearrangement. In H_2O hexamethylbenzene did not react with Selectfluor® [6]. Dehydration of the primary amide in the complex cyclic peptide to the nitrile was rapidly carried out using cyanuric chloride at room temperature, but the product quickly decomposed under these conditions. When water-wet amide was charged with two equivalents of cyanuric chloride at $-30\,°C$ the reaction was complete after 30 h with minimal decomposition; adding H_2O to a reaction that is formally a dehydration is counter-intuitive. The reaction was optimized with 2.0 equivalents of cyanuric chloride and 2.5 equivalents of H_2O. The formation of less-reactive species from reaction of cyanuric chloride with water was implicated by the stability of the product under the reaction conditions [7]. Novartis researchers found that under basic conditions the formation of the trityl ether was incomplete and varied from batch to batch with the content of residual water. They developed a process that was catalyzed by H_2O, suggesting that H_2O disrupted an equilibrium of the starting hydroxynitrile with the five-membered imino ether. The role of the nitrile group in suppressing tritylation was indicated by the successful tritylation of 2-phenyl-1-propanol under anhydrous conditions. (Other investigations showed that tritylation could be completed in 4 h using 5 mol% of TFA [8].) These and many other reactions indicate the usefulness of monitoring water contents in reactions.

Water can hydrolyze many reagents and starting materials (Figure 6.2), sometimes very rapidly and exothermically. Other reactions are slower, depending on reaction conditions that the chemist strives to control. For instance, a Schotten–Baumann reaction to generate an amide is usually conducted at pH 8–10 and 0–10 °C (Figure 6.3) [9]. Under more acidic

$$2\ Na(0) + 2\ H_2O \longrightarrow H_2 + 2\ NaOH$$
$$LiAlH_4 + 4\ H_2O \longrightarrow 4\ H_2 + Al(OH)_3 + LiOH$$
$$RLi + H_2O \longrightarrow RH + LiOH$$
$$RMgX + H_2O \longrightarrow RH + XMgOH$$
$$RCOCl + H_2O \longrightarrow RCO_2H + HCl$$
$$(RCO)_2O + H_2O \longrightarrow 2\ RCO_2H$$
$$AlCl_3 + 3\ H_2O \longrightarrow Al(OH)_3 + 3\ HCl$$

FIGURE 6.2 Some reactions quenched by water.

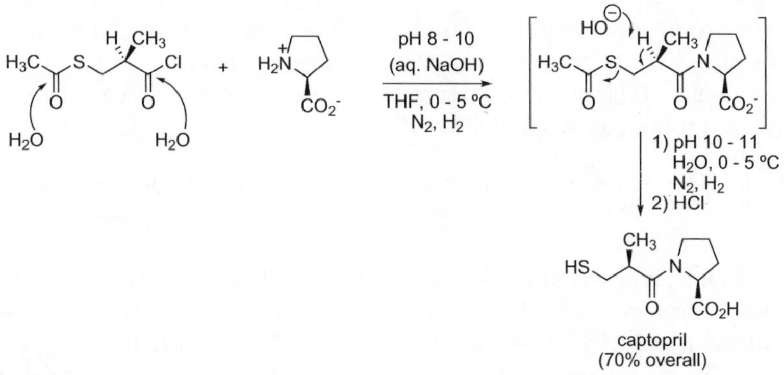

FIGURE 6.3 Use of Schotten–Baumann reaction conditions to prepare a captopril precursor.

conditions the protonated amine will not react, leading to hydrolysis of the acid chloride, and under more acidic conditions hydroxide reacts with the acid chloride. Under more basic conditions and at higher temperatures with extended reaction times it may be difficult to minimize byproduct formation. The chemist uses experience to develop conditions to minimize *in situ* hydrolysis of valuable reagents and to control other processing.

To maintain a constant pH for a reaction, buffers are often used in the laboratory. For instance, a Schotten–Baumann reaction may be run with a phosphate buffer, a mixture of $NaHCO_3$ and Na_2CO_3, or with excess $NaHCO_3$ alone. Unfortunately buffers generally produce a great deal of inorganic waste. Scale-up operations often use a pH autoburette (a unit comprising a pH meter with a titrator to automatically deliver a solution to maintain a pH range, also called a "pH-stat"). Under these circumstances pH is controlled without a buffer, and reagents more basic than Na_2CO_3 can be used effectively with adequate mixing and dilution. For instance, in the Schotten–Baumann reaction in Figure 6.3 aqueous NaOH has been used to control the reaction pH. Such pH-stats are used on scale and are available for laboratory investigations.

II. DETECTING AND QUANTITATING WATER

Chemists apply the assays available to determine how much water is present in samples. Gravimetric analysis, or determining the loss of weight on drying, is slow and often not specific for water. Calibrated ^1H NMR has been used to quantitate H_2O in solids [10], and IR and near-IR can also be used. The Karl Fischer titration (or KF titration or simply KF) is the classic analytical method used to detect water, and is convenient to use. The basis of the Karl Fischer titration is the reaction of water with iodine and sulfur dioxide (Equation 6.1). In the early development of this analytical technique, the solution containing water was titrated with a solution of I_2 in benzene or MeOH until the I_2 color remained, providing a sharp, reproducible endpoint to determine the water content [11]. Today of course benzene is avoided as a solvent for the laboratory and scale-up, and less toxic solvents or solvent mixtures are used. The accurate and rapid coulometric assay, which can detect down to 10 μg of water, is generally preferred; I_2 is generated electrolytically at the cell anode, and the amount of water is determined by the current required for electrolytic oxidation of HI [12]. Coulometric assays are especially useful for investigations in early development, when reactions are frequently run on small scale with starting materials in short supply. Typical units for the concentration of water are % weight/volume, % weight/weight (wt%), or ppm.

$$I_2 + SO_2 + H_2O + 3 \text{ pyridine} + CH_3OH \rightarrow 2 \text{ pyridine} \cdot HI + \text{Pyridine} \cdot HOSO_2OCH_3$$
$$(6.1)$$

Some compounds posing difficulties in Karl Fischer titrations are shown in Table 6.1. Compounds that can be oxidized by I_2 will interfere with assays for water. Compounds that dehydrate readily by intermolecular or intramolecular reactions, e.g., methyl ketones, must be assayed for water content using specialized reagents. Insoluble analytes, e.g., K_2CO_3, react slowly or not at all.

TABLE 6.1 Compounds Known to React with the Karl Fischer Reagent (1,2)

Ascorbic acid	CuX_2	$RCOCH_3$
H_2NNH_2 & $R_2NNR'_2$	FeX_3	Triphenylphosphine
RSH	MOH	Diisopropyl azodicarboxylate
$MnSO_3$	$B(OH)_3$	
$Na_2S_2O_3$	$Na_2B_4O_7$	
$MnCO_3$	Metal oxides	
$MnHCO_3$	$SnCl_2$	

(1) Mitchell, J., Jr.; Smith, D. M. Aquametry; Interscience: New York, 1948, p 22.
(2) R = alkyl, H; M = metal; X = anion.

Water that is bound to surfaces or solutes may not be as "available" as that of distilled water, hence the water activity (fugacity) of the system is lowered [13]. For instance, the water activity of an enzyme–ionic liquid mixture was very non-linear with respect to KF measurements and varied with the ionic liquid [14]. Water activity, not water content, determines the hydration of crystals [15]. By calculating the water activities [16] for solvents to crystallize a hydrate researchers were able to calculate solvent ratios for different solvents to crystallize the desired hydrates [17–19]. Three-component phase diagrams of water, co-solvent, and solids may be helpful in understanding such crystallizations [20]. Water activity has been used to determine the quality, stability, and SAFETY of foods [13,21], and water activity can be determined in solids, liquids, and air. Small and portable instruments with hand-held probes can be used to measure water activity. Levels of water activity are expressed as % of relative humidity (RH) or as a_w units, values ≤ 1.00, with 1.00 a_w being equal to 100% RH [22]. Through calculations, water activity considerations will probably be used more frequently in the pharmaceutical industry to control the crystallization and drying of APIs, especially materials that can crystallize as multiple hydrates.

III. REMOVING WATER FROM ROUTINE ORGANIC PROCESSING

Water can enter a reaction through starting materials, reagents, solvents, byproducts of a reaction, and the atmosphere, and may be present in wet equipment. Routine laboratory operations, such as suction filtration, can draw water into equipment or a process stream [23] (Table 6.2). Since water is nearly ubiquitous, its presence and effects should always be considered, especially for reactive species such as organometallic reagents (SAFETY).

SAFETY — Palladium on carbon and other catalysts are usually provided as 50% wet solids for SAFETY reasons. Dry catalysts can require extreme handling precautions. Excessively suctioning air through a catalyst cake after hydrogenation in a water-miscible solvent may cause a fire.

TABLE 6.2 Entry of Water into Reactions and Means of Removing Water

Source of Water	Means of Avoiding or Removing Water	Comments
Moisture in air	Apply positive pressure of dry N_2 or argon to exclude air	Ar is denser than N_2, useful in the laboratory Ar is more expensive than N_2, rarely used on scale
Solvents	Buy dry solvent Remove by azeotroping Consume with excess reagents	Dry solvents are more expensive Requires additional step, but reliable Convenient if reagents are inexpensive and byproducts are easy to remove
Reagents	Remove by azeotroping Consume with other reagents Set specifications for acceptable material	Requires additional step, but reliable Convenient if reagents are inexpensive and byproducts are easy to remove
Byproduct formation	Remove by azeotroping Adsorb to molecular sieves	Can help to drive reaction May be used if process streams are not stable at temperatures for azeotroping or if water is generated in a solvent unsuitable for azeotroping [24]; treatment of sieves is necessary for recycling
Aqueous extractions	Brine wash Azeotrope dry	Washing with brine may partition NaCl into the organic phase. Brine washes may be unnecessary. Half-saturated brine may be as effective, and produce less waste
Equipment	Clean thoroughly Rinse with H_2O-miscible solvents and flush, then flush with H_2O-immiscible solvent Azeotrope dry Monitor moisture before charging starting materials and reagents, and possibly after charging	Required for reliable processing Low points in transfer lines or condenser piping and lines that have been blanked off ("dead legs") may allow solvents to pool and accumulate water. The presence of H_2O in an equipment train may not be expected, and completely removing H_2O may require some thought and time

III.A. Entry of Water Through Processing Air

At 10 °C the amount of water in breathing air can be as high as 17 g/m^3, depending on the relative humidity. The amount of water that condenses in the reactors and on key intermediates and products will be influenced by the relative humidity, by the temperature of equipment and process streams, by the flow of the atmosphere past the equipment and process streams, and by the hygroscopicity of the various components.

Instituting a positive pressure of a dry gas in processing equipment is the easiest means to inhibit intrusion of water from the atmosphere when water-sensitive reactions are being run. Argon is denser than N_2 (1.784 g/L vs. 1.254 g/L and 1.247 g/L for N_2 and air, respectively [25]) and is sometimes used in the laboratory as an additional precaution to preclude entry of atmospheric water and oxygen. For scale-up operations N_2 is routinely used because it is less expensive than argon. A highly detailed description of how to handle air-- sensitive reagents in the laboratory has been published by the Aldrich Chemical Company [26] and others [27,28]. SAFETY concerns for handling pyrophoric compounds probably cannot be overstated.

> All operations that involve suction or a partial vacuum can draw gases and liquids (especially notable for SAFETY are O_2 and H_2O) into equipment and process streams. Some examples include filling a syringe with a pyrophoric liquid from a SURE-SEAL® bottle, transferring a reagent into a pilot plant reactor using suction, or breaking the vacuum on a dryer. Thinking ahead can avoid difficulties.

III.B. Entry of Water Through Solvents

Solvents are often hygroscopic, and methods for water removal in the laboratory are often specific for a given solvent [29]. Treating solvents with Na_2SO_4 or $MgSO_4$ is generally an ineffective means of removing water. For solvents with dielectric constants less than 15, water and other contaminants can be readily removed on a laboratory scale by filtration through Al_2O_3 (activity I) or SiO_2 (activity I) [29–31]; storage over molecular sieves has been found very effective [29]. Today purchased solvents are usually dry enough (and pure enough) for most routine laboratory operations. Although purchasing an "anhydrous" solvent may be convenient, the cost and availability of "anhydrous" solvents may be limiting; furthermore, contact with water-wet equipment may negate the dryness of an "anhydrous" solvent. The most practical approach is to design processes that can tolerate water.

> Monitor the moisture content of solvents frequently and at critical processing points. A bottle or drum of solvent that has been opened for some time may contain appreciably more H_2O than a recently opened container.

When operations are run in stationary process equipment the most practical way to dry both solvents and equipment is to rinse with dry solvents, or to azeotrope out H_2O before continuing operations (Chapter 5). Acetone cannot be dried without introducing byproducts, as it undergoes intermolecular aldol reaction–Claisen dehydration, producing UV-absorbing impurities (Figure 6.4) [32]. Mesityl oxide (precursor to MIBK) is generated, which is prone to Michael additions by "soft" nucleophiles such as mercaptans (see the theory of hard and soft acids and bases [33]); other possible impurities may arise through retro-Michael/Michael reactions (note that the cyclic ketone is a precursor to TEMPO and N-Ac-TEMPO).

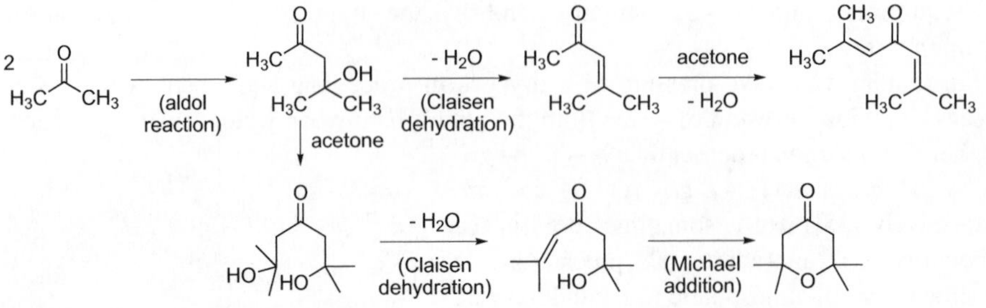

FIGURE 6.4 Possible products from dehydrating acetone.

III.C. Entry of Water Through Reagents

Water is an inherent part of some stable crystalline compounds, both salts and non-salts. Salts more often are found as hydrates, due to the tendency of H_2O to bind and stabilize ionic species [34]. Trying to remove water from hydrated species and then preventing reabsorption of water can be difficult and time-consuming, and such attempts can decompose the reagent. (The instability of "anhydrous" $(nBu)_4NF$ has been noted [35], along with different reactions mediated by the presence of H_2O in "anhydrous" $(nBu)_4NF$ [36]. DeShong et al. have described anhydrous $(nBu)_4NPh_3SiF_2$ as a source of anhydrous fluoride [37]. A low-temperature preparation of anhydrous $(nBu)_4NF$ has been described, and decomposition was noted in the presence of hydroxylic solvents [38]. Anhydrous $(nBu)_4NF$ may not be needed for all applications [39].) A simpler approach to removing H_2O is to substitute a non-hydrated reagent, as shown in Table 6.3; often a crystalline anhydrate can be sourced, but changing the counterion (gegenion)

TABLE 6.3 Some Common Hydrated Reagents and Possible Substitutes

Hydrated Reagent	Possible Non-hydrated Substitute	Hydrated Reagent	Possible Non-hydrated Substitute
$LiOH \cdot H_2O$		$NaH_2PO_4 \cdot H_2O$	NaH_2PO_4
$NaOAc \cdot 3H_2O$	KOAc	$Na_2HPO_4 \cdot 7H_2O$	Na_2HPO_4
$K_2CO_3 \cdot 1.5H_2O$	K_2CO_3	$K_2HPO_4 \cdot 3H_2O$	K_2HPO_4
NaOH pellets (hygroscopic)	K_3PO_4 or 50% NaOH under PTC conditions	$Na_3PO_4 \cdot 12H_2O$	K_3PO_4
KOH, 85% (10–15% H_2O)	K_3PO_4 or 50% NaOH under PTC conditions	$NaMnO_4 \cdot H_2O$	$KMnO_4$
$CaCl_2 \cdot 2 H_2O$	$CaCl_2$ or $Ca(acac)_2$	$CeCl_3 \cdot 7H_2O$	$CeCl_3$
$Fe(C_2O_4) \cdot 2H_2O$	$Fe(acac)_2$	Citric acid $\cdot H_2O$	Citric acid
$Zn(OAc)_2 \cdot 2H_2O$	$Zn(OAc)_2$	$pTsOH \cdot H_2O$	
$SnCl_2 \cdot 2H_2O$	$SnCl_2$	$(nBu)_4NF \cdot 3H_2O$	$(nBu)_4NPh_3SiF_2$

may be necessary. The presence or absence of a small amount of water can significantly change the solubility of a reagent.

III.D. Entry of Water From Process Streams

Grignard reactions are notoriously sensitive to water, and two approaches to controlling H_2O are shown in Figure 6.5. For homologation of the aryl Grignard with formaldehyde, Delrin® acetal was employed as shavings [40]; this homopolymer is effectively anhydrous, in contrast to formalin ($\sim$50% H_2O) or paraformaldehyde (5–9% H_2O). (Some difficulties in cracking paraformaldehyde to produce gaseous HCHO have been described [41].) Lilly researchers conducted a detailed study on initiating the formation of a Grignard reagent and subsequent reaction with a morpholine amide [23] (as a substitute for a Weinreb amide). The free base from splitting the L-(−)-DTTA salt was extracted into toluene in the presence of THF, and the extract was dried by azeotropic distillation to 173 ppm H_2O. Magnesium metal was activated with small amounts of both iodine and $(iBu)_2AlH$ at 60 °C, and formation of the Grignard reagent was initiated at 60 °C; at this temperature the reaction was found to tolerate greater amounts of H_2O, up to 300 ppm. Then the extract of the free base was gradually added to the Grignard preparation. These conditions were robust enough to implement in the pilot plant.

For reliable processing it is important to note when water can react with not only the reagents, as in Figure 6.5, but also with the starting material. In the condensation of the atorvastatin precursor with 4-fluorobenzaldehyde via a Stetter reaction [42] residual water led to the formation of the des-fluoro impurity in essentially a 1:1

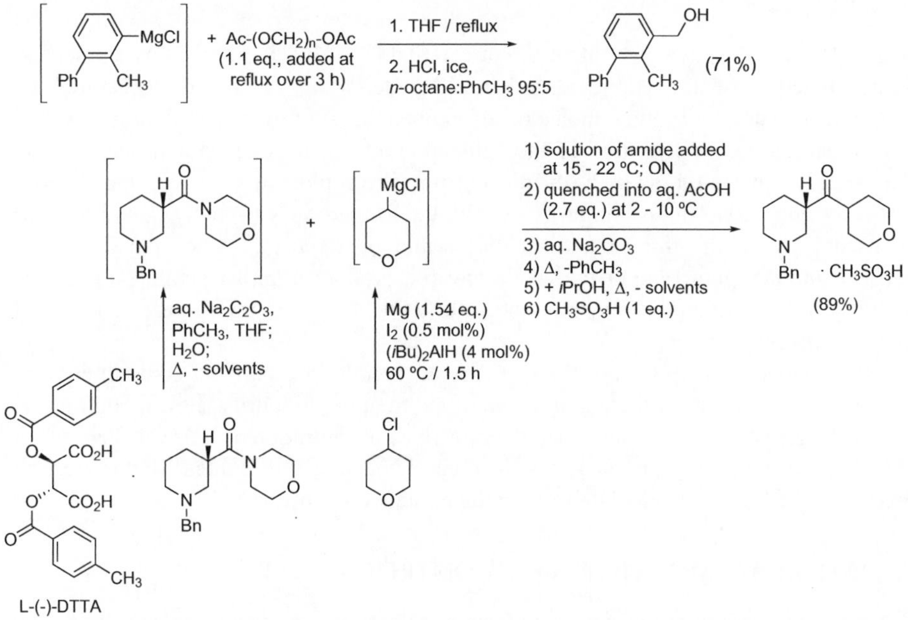

FIGURE 6.5 Forming Grignard reagents and subsequent reactions.

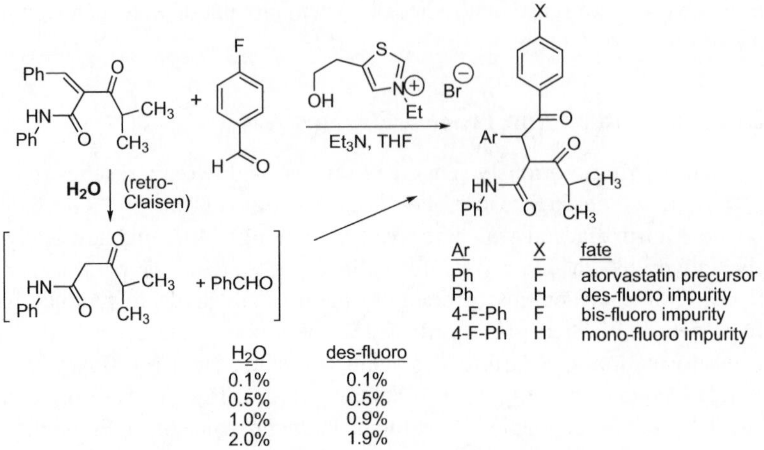

FIGURE 6.6   Tolerance of a starting material to water.

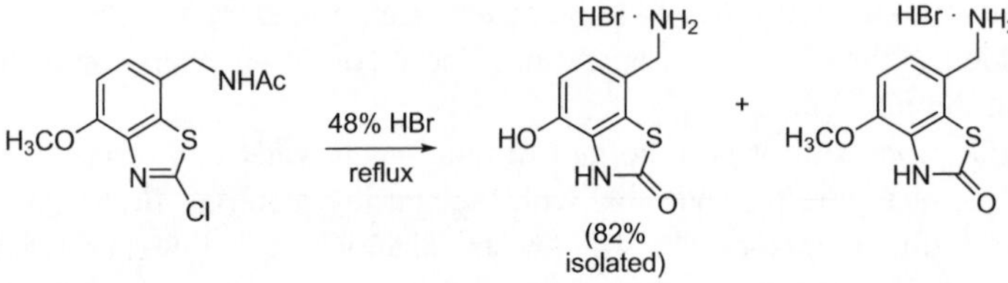

FIGURE 6.7   Minimizing an impurity in an HBr-mediated hydrolysis by controlling H_2O in the starting material.

correlation with the water content (Figure 6.6) [43]. This impurity was formed by Stetter reaction of the enone with benzaldehyde, which had been generated by addition of water to another molecule of enone; the starting material probably had been prepared by condensation of the ketoamide and benzaldehyde and dehydration. (Apparently the bis-fluoro impurity did not pose a problem with processing. Mathematically the mono-fluoro impurity would be formed in extremely low levels as a concurrence of the three reactions that would be required.) Since the difference between the des-fluoro impurity and the desired product is rather small, purging the impurity by crystallization may have been difficult [44], thus prompting strict control of water in this process.

HBr-mediated hydrolyses of the methoxy acetamide in Figure 6.7 sometimes were incomplete, affording significant amounts of the methoxy impurity shown. High levels of this impurity were generated when the effective concentration of HBr was lower than 46%. Key to reliable processing was to control the moisture content of the starting material, which had been isolated following an aqueous quench [45].

IV. WHERE TO MONITOR AND CONTROL WATER

If controlling the content of water is essential to processing, water contents should be monitored at critical stages where inadequate amounts of water would cause SAFETY

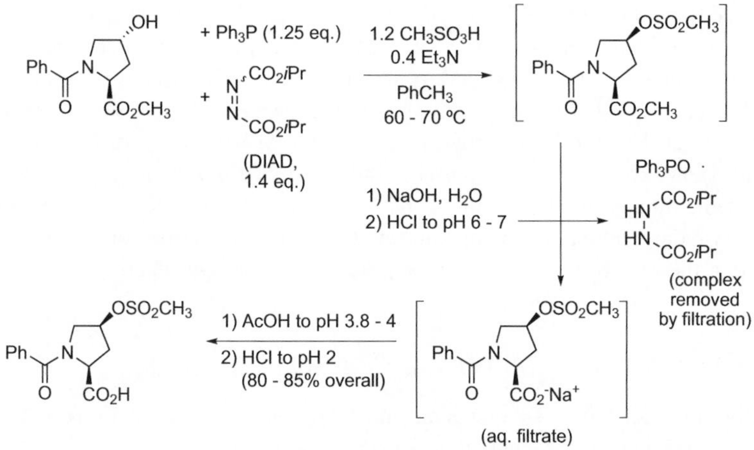

FIGURE 6.8 A Mitsunobu reaction using triphenylphosphine and DIAD.

issues, extend processing, consume valuable materials, or necessitate recharging starting materials or reagents. Knowing which components are hygroscopic is prudent. If an external analysis is required process streams should be sampled only when it is SAFE to withdraw a reaction aliquot (Chapter 8). A detailed analysis for IPC may illustrate some important considerations.

In the Mitsunobu reaction to make the sulfonate in Figure 6.8 [46], excessive water was known to consume the reagents triphenylphosphine and diisopropyl azodicarboxylate (DIAD), leading to incomplete reactions. Controlling the amount of residual water was necessary both to avoid extending operations by recharging Ph_3P and DIAD and to minimize additional waste from the byproducts triphenylphosphine oxide and (1,2-diisopropylcarbonyl)hydrazine. The primary entry for water was the starting ester, and specifications were set to control that amount of water at no more than 0.1%. Specifications were also set to control the amount of water present in Et_3N and methanesulfonic acid. Small excesses of Ph_3P and DIAD (1.25 eq. and 1.4 eq., respectively) were employed so that processing was rarely interrupted to recharge additional amounts of these reagents. (In manufacturing this intermediate on large scale, e.g., 275 kg of ester, azeotropic removal of H_2O would have required a significant amount of time simply to heat and cool the reactor. Charging small excesses of these relatively inexpensive reagents was more productive.) The water contributed to the reaction mass by the solvent was very low since toluene was used, which can dissolve only 0.05% water. In an early iteration of this process the amount of water in the reactor was determined by Karl Fischer titration after toluene, Et_3N, and methanesulfonic acid had been charged; this charging sequence avoided charging Ph_3P and DIAD to overly wet reactor contents. In the optimized sequence Et_3N was the last component added, in order to minimize formation of side products. In the process

To minimize potential issues with water in processing, conduct reactions in solvents that absorb only small amounts of H_2O. For example, run a dehydration reaction in $PhCH_3$ instead of EtOAc, or substitute 2-methyl-THF for THF if an extractive workup follows the reaction.

reintroduced to manufacturing the only KF titration as an IPC was taken on an aliquot of the toluene withdrawn from the reactor before the other components were added, and the IPC limit was set at less than 0.06%. Any water present in the reactor at $>0.05\%$ might not have been detected before processing continued, because H_2O present at greater than the saturation level would have settled to the bottom of the reactor or beaded onto the equipment. Hence the latter IPC essentially indicated whether the toluene in the reactor had been contaminated with H_2O and a chemical, e.g., MeOH from prior cleaning, that could increase the amount of H_2O in solution. Controlling the processing in this fashion required that the reactor be dry before the reaction solvent was charged.

V. OPERATIONS TO REMOVE OR CONSUME WATER

Using anhydrous solvents and reagents may not be necessary, if the reaction can be treated inexpensively with excess reagents that consume the water or if the H_2O formed can be removed by azeotropic distillation. For instance, SO_3 (oleum: fuming sulfuric acid) was added to a nitration, reducing the levels of the regioisomers; when 6% water was added to the reaction the isolated yield was 76%, with an isomer ratio of 90:5:5 (Figure 6.9) [47]. Contamination of commercial NaOEt by NaOH (due to the presence of adventitious H_2O) partially hydrolyzed the esters in the alkylation of a phenol, so the NaOEt solution was pretreated with EtOAc to convert the NaOH to innocuous NaOAc [48]. Bromination using HBr and DMSO [49] was accelerated by azeotroping out the byproduct H_2O [50]. (As with the Swern reaction, the byproduct of this oxidation is the odoriferous dimethyl sulfide. Solutions to scrub Me_2S are found in Table 8.2.) In one approach to generating an imine, H_2O formed was removed by azeotropic distillation with MIBK, as shown for the in situ protection of a primary amine to effect alkylation of a secondary amine [51]. In another process H_2O formed from imine formation was consumed by treating with trimethylorthoformate and the HCl present [52]. On a 1 kg scale the azeotropic removal of H_2O from imine formation of an aniline was stalled, and adding a small excess of triethylphosphite was found to be effective. The byproduct diethyl phosphinite inhibited the subsequent Pd-catalyzed indolization, and was removed by washing a solution of the imine (cyclohexane–Et_3N 9:1) with water [53]. H_2O is the byproduct from Knoevenagel reactions, and unfortunately the malonitrile adduct in Figure 6.9 was both hydrolytically and thermally labile. Treatments with inorganic reagents that remove H_2O were ineffective. By mixing HMDS and AcOH (very exothermic addition!) ammonium acetate and TMSOAc were formed. The latter consumed the H_2O generated, and the NH_4OAc provided the catalyst needed for the Knoevenagel reaction [54].

> DMSO has been shown to react with strong acids and acid chlorides [55] and strong bases [56]. For SAFE operations such reactivity should be considered.

VI. OPERATIONS WHERE WATER CAN BE CRUCIAL

Sometimes the presence of water is essential. The solvent most often used for classical resolutions is EtOH containing 5% water [57], probably because so many salts crystallize as hydrates (see Chapter 12). Water may also dramatically increase a reaction rate

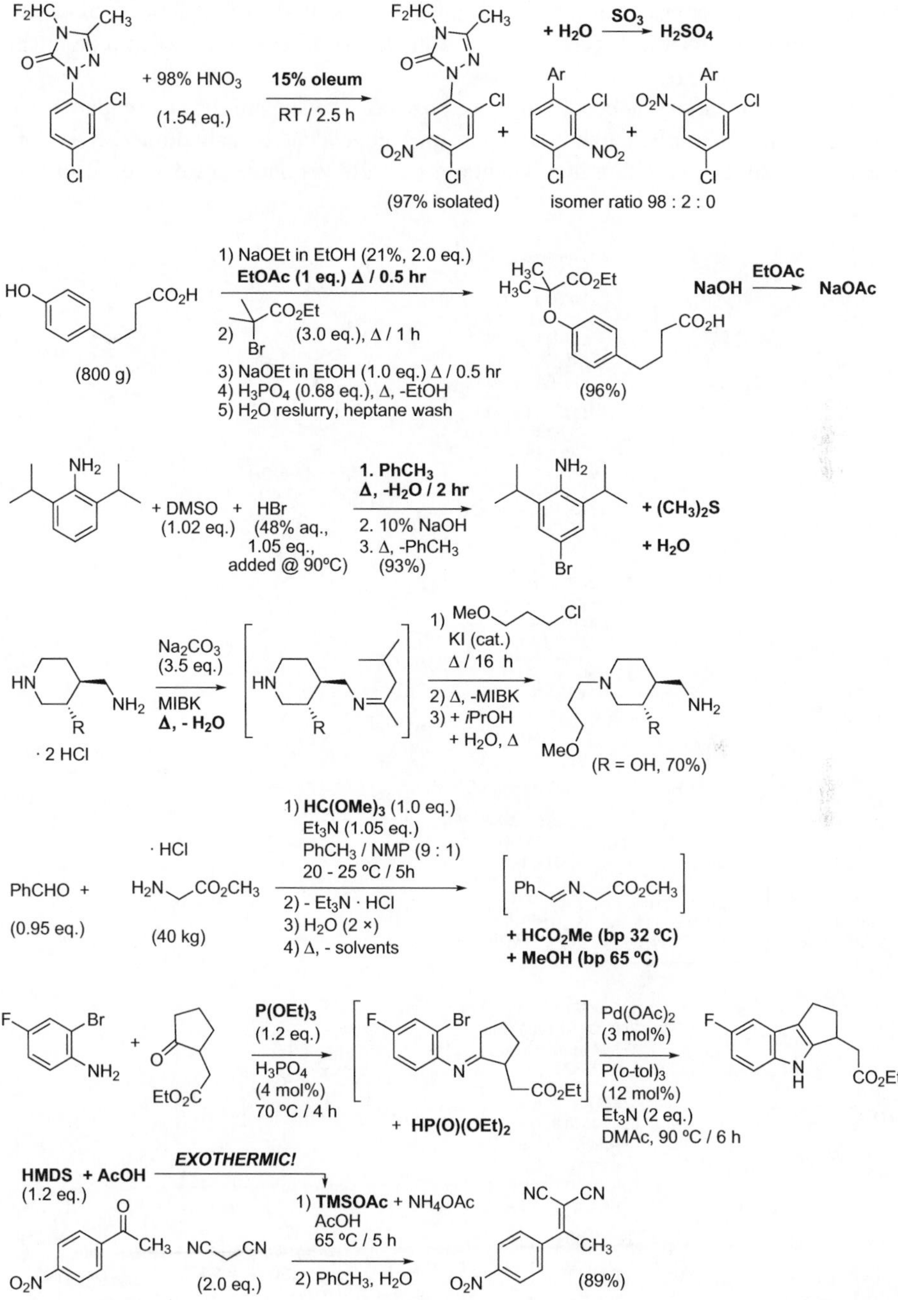

FIGURE 6.9 Processes driven by consuming or removing water.

or moderate the reactivity of a reagent. Small amounts of H_2O in reactions, as in the cross-coupling reactions in Figure 10.3, have made the difference between a practical process and a stalled reaction.

Some other reactions that benefit from the presence of water are shown in Figure 6.10. Researchers at Schering–Plough found that one equivalent of anhydrous LiOH was necessary for alkylation with a high enantiomeric ratio, so one equivalent of H_2O was

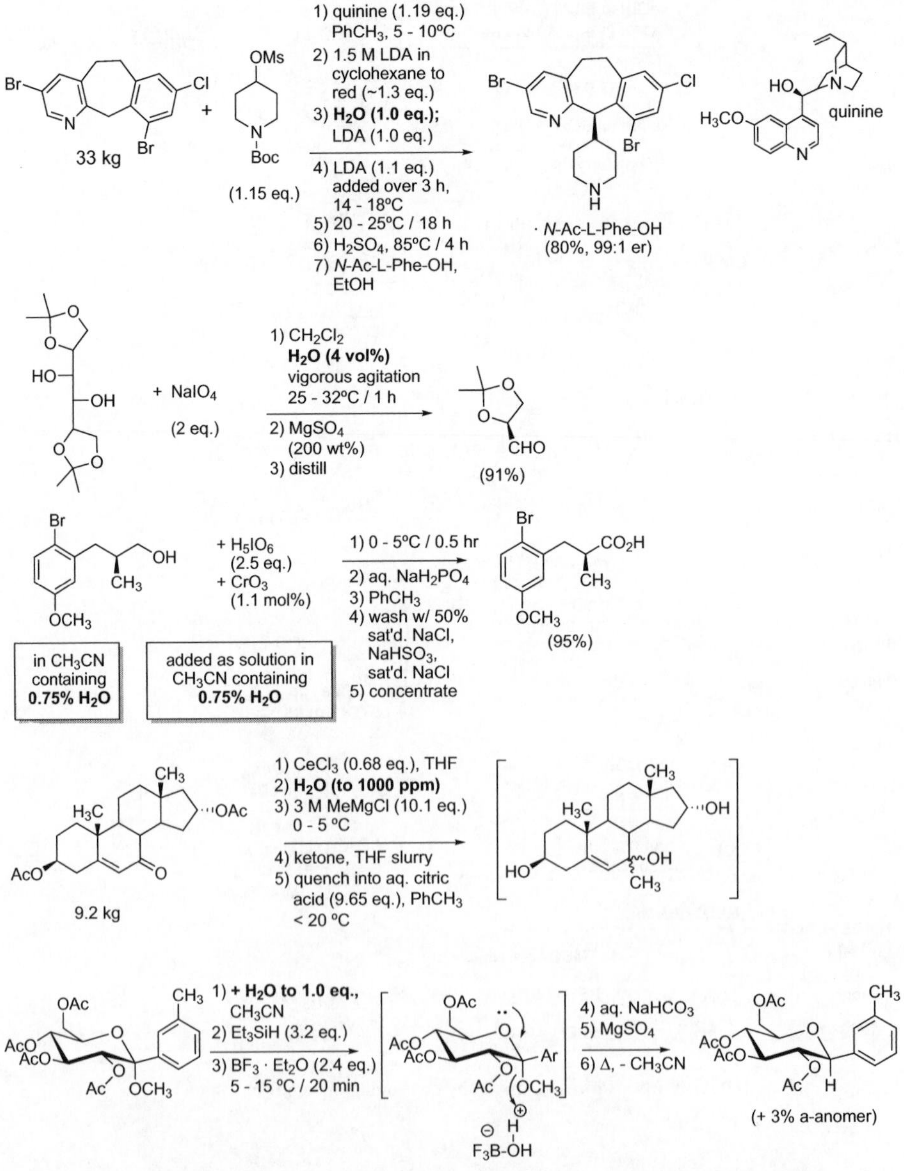

FIGURE 6.10 Some reactions where controlled amounts of water are helpful or necessary.

added, followed by one equivalent of LDA. If no water was added the yield was only 50% (78:22 er) [58]. Water was found necessary for the periodate oxidation shown [59,60], and moderated the reactivity of Merck's catalytic CrO_3 oxidation [61]. For the $CeCl_3$-mediated Grignard reaction in Figure 6.10 a small amount of water was found to be crucial to obtain the crystal form of $CeCl_3$ with enhanced reactivity in the presence of the Grignard reagent [62]. Similarly $Mg(OTf)_2 \cdot 4H_2O$ was an effective catalyst in a Michael addition to a nitrostyrene, while anhydrous $Mg(OTf)_2$ was ineffective; by adding H_2O to the latter the reaction rates were increased to those using $Mg(OTf)_2 \cdot 4H_2O$. Curiously the Michael addition was optimally conducted in the presence of molecular sieves [63]. BMS researchers found that the $BF_3 \cdot Et_2O$-catalyzed reduction of 1-C-aryl methyl glucosides with Et_3SiH proceeded poorly unless one equivalent of H_2O was added; they proposed that reaction of H_2O with $BF_3 \cdot Et_2O$ generated a strong Bronsted acid that facilitated the reduction [64]. A trace of water was found to dramatically raise the yield in a PTC reaction using water-sensitive compounds [65], and controlling water contents is essential for successful PTC reactions (Chapter 4) [66].

> Water can speed heterogeneous reactions with inorganic reagents. A small amount of H_2O may activate the reagent, perhaps by cleaning the surface of the crystals, or may increase the solubility of the reagent in the reaction mixture. Too much H_2O can create a muddy dispersion that complicates mass transfer. The optimal amount of H_2O for a process must be determined experimentally.

A small amount of water can dramatically influence the formation of a desired, non-hydrated crystal. In the preparation of fosinopril sodium more than 0.2% H_2O was found necessary to crystallize the stable, non-hydrated polymorph (Figure 6.11) [67]; the inexpensive, organic solvent-soluble sodium 2-ethyl-hexanoate was used as the sodium carrier for ease of accurate delivery of sodium ion and isolation of product. The addition of one equivalent of H_2O raised the yield of a salt from a CIDR process, by inhibiting the formation of impurities that generated H_2O as a byproduct [68].

> Screen crystallization solvents for H_2O, as a small amount of H_2O in solvents or reagents can affect not only the formation of hydrates but also polymorphs. For example, a small amount of water in benzenesulfonic acid produced an unexpected polymorph of the benzenesulfonate salt as an anhydrate [69].

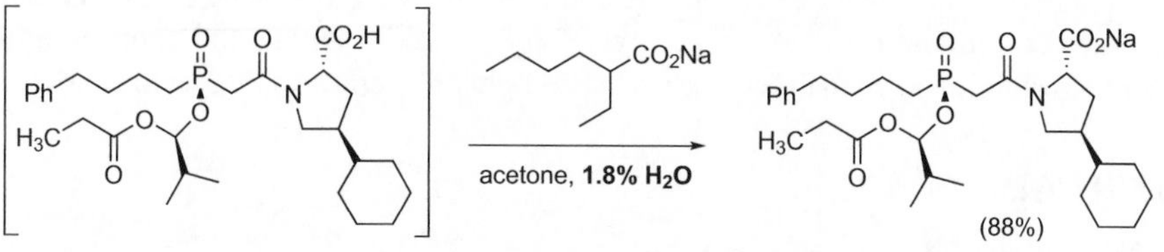

FIGURE 6.11 Control of water to generate desired, non-hydrated crystal forms.

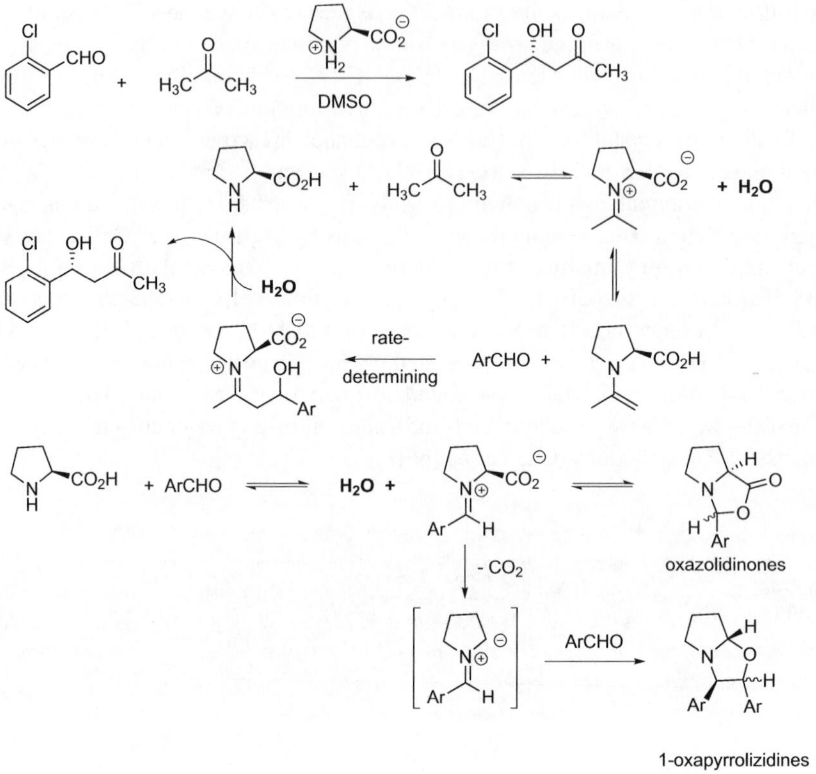

FIGURE 6.12   Effects of water in proline-catalyzed aldol reactions.

VII. PERSPECTIVE

By controlling the amount of water in processes one can exert considerable control on operations, and finding a suitable range for the charge of water may be necessary. For instance, Blackmond's group found in a proline-mediated aldol reaction that excess water slowed the reaction, as expected for equilibrium conditions. (As shown in Figure 6.12, H_2O plays a role twice in the catalytic cycle.) On the other hand, adding water suppressed the formation of both the non-productive oxazolidinones and the 1-oxapyrrolizidines, which consumed two equivalents of aldehyde per proline [70]. (In a solvent-free reaction Carter's group found that optimal conditions for an aldol reaction involving a proline derivative required one equivalent of water [71]). The Blackmond group wrote that their work clarifies "the complex and often opposing roles that water may play both on and off the catalytic cycle" [70]. Understanding and controlling the amount of water involved in processes may shed light on operations that seem capricious.

REFERENCES

1. Ribe, S.; Wipf, P. *Chem. Commun.*, **2001**, 299.
2. Schmitke, J. L.; Wescott, C. R.; Klibanov, A. M. *J. Am. Chem. Soc.* **1996**, *118*, 3360.

3. Carrea, G.; Riva, S. *Angew. Chem. Int. Ed.* **2000**, *39*, 2226.

4. Wells, A. *Org. Process Res. Dev.* **2006**, *10*, 678.

5. Barnwell, N.; Cheema, L.; Cherryman, J.; Dubiez, J.; Howells, G.; Wells, A. *Org. Process Res. Dev.* **2006**, *10*, 644.

6. Kralj, P.; Zupan, M.; Stavber, S. *J. Org. Chem.* **2006**, *71*, 3880.

7. Journet, M.; Cai, D.; DiMichele, L. M.; Hughes, D. L.; Larsen, R. D.; Verhoeven, T. R.; Reider, P. J. *J. Org. Chem.* **1999**, *64*, 2411.

8. Prashad, M.; Har, D.; Chen, L.; Kim, H.-Y.; Repic, O.; Blacklock, T. J. *J. Org. Chem.* **2002**, *67*, 6612.

9. Estaban, J. M. T.; Vidal, C. M.; Marine, J. R.; Diaz, J. M. Spanish Patent ES 556,990, 1987 (assigned to Sueros, Antibioticos y Laboratorio de Vacunoterapia, Sociedad Ltds.); see *Chem. Abstr.* **1988**, *109*, 23391m.

10. Thanks to Trevor Laird for mentioning this point.

11. Mitchell, J., Jr.; Smith, D. M. *Aquametry*, 2nd ed.; John Wiley & Sons: New York, 1980; pp 68–105.

12. Mettler Toledo offers a line of coulometric titrators: http://us.mt.com.

13. Blandamer, M. J.; Engberts, J. B. F. N.; Gleeson, P. T.; Reis, J. C. R. *Chem. Soc. Rev.* **2005**, *34*, 440.

14. Kaftzik, N.; Wasserscheid, P.; Kragl, U. *Org. Process Res. Dev.* **2002**, *6*, 553.

15. Black, S. N.; Phillips, A.; Scott, C. I. *Org. Process Res. Dev.* **2009**, *13*, 78.

16. For example, AspenTech BatchSep: http://www.aspentech.com

17. Variankaval, N.; Lee, C.; Xu, J.; Calabria, R.; Tsou, N.; Ball, R. *Org. Process Res. Dev.* **2007**, *11*, 229.

18. Magano, J.; Acciacca, A.; Akin, A.; Collman, B. M.; Conway, B.; Waldo, M.; Chen, M. H.; Mennen, K. E. *Org. Process Res. Dev.* **2009**, *13*, 555.

19. For another approach to calculating solvent substitutions, see section VI in Chapter 5.

20. Li, Y.; Chow, P. S.; Tan, R. B. H.; Black, S. N. *Org. Process Res. Dev.* **2008**, *12*, 264.

21. http://www.martin.chaplin.btinternet.co.uk/activity.html

22. http://www.novasina.com

23. Kopach, M. E.; Singh, U. K.; Kobierski, M. E.; Trankle, W. G.; Murray, M. M.; Pietz, M. A.; Forst, M. B.; Stephenson, G. A. *Org. Process Res. Dev.* **2009**, *13*, 209; Recovery and recycle of Mg(0) was hampered, probably by absorption of atmospheric water to the metal.

24. For example, H_2O was generated in the formation of an imine in NMP. Clark, J. D.; Weisenburger, G. A.; Anderson, D. K.; Colson, P.-J.; Edney, A. D.; Gallagher, D. J.; Kleine, H. P.; Knable, C. M.; Lantz, M. K.; Moore, C. M. V.; Murphy, J. B.; Rogers, T. E.; Ruminski, P. G.; Shah, A. S.; Storer, N.; Wise, B. M. *Org. Process Res. Dev.* **2004**, *8*, 51

25. Weast, R. C.; Selby, S. M., Eds.; *Handbook of Chemistry and Physics*, 47th ed.; The Chemical Rubber Company: Cleveland, Ohio, 1966; pp B-155, B-201, F-8.

26. "Handling Air-Sensitive Reagents," Technical Information Bulletin Number AL-134, Aldrich Chemical Company, revised 3/97. This bulletin has been shipped with most of the air-sensitive reagents sold by Aldrich. http://www.sigmaaldrich.com/etc/medialib/docs/Aldrich/Bulletin/al_techbull_al134.Par.0001. File.tmp/al_techbull_al134.pdf.

27. Schwindeman, J. A.; Woltermann, C. J.; Letchford, R. J. *J. Chem. Health Saf.* **2002**, *9*, 6.

28. http://www.ehs.uci.edu/programs/sop_library/Pyrophoric/Pyrophorics_combined_works.html; http://www.youtube.com/watch?v=RaMXwNBAbxc; http://chem-courses.ucsd.edu/CoursePages/Uglabs/143A_Weizman/EHS/Episode1720p.html; http://chem-courses.ucsd.edu/CoursePages/Uglabs/143A_Weizman/EHS/Episode2%20720p.html; http://chem-courses.ucsd.edu/CoursePages/Uglabs/143A_Weizman/EHS/Episode3720p.html; http://www.yale.edu/ehs/onlinetraining/OrganoLithium/OrganoLithium.htm

29. Williams, D. B. G.; Lawton, M. *J. Org. Chem.* **2010**, *75*, 8351.

30. Reichardt, C. *Solvents and Solvent Effects in Organic Chemistry*, 2nd ed.; VCH: Weinheim, 1990; p 416.

31. Passing a solvent through SiO_2 or Al_2O_3 may remove stabilizers such as BHT. Thanks to Trevor Laird for mentioning this.

32. Burfield, D. R.; Smithers, R. H. *J. Org. Chem.* **1978**, *43*, 3966, http://curlyarrow.blogspot.com/2010/04/anhydrous-solvents-part-3-acetone-and.html.

33. Pearson, R. G. *Chemical Hardness – Applications From Molecules to Solids*; Wiley-VCH: Weinheim, 1997.
34. Stahly, G. P. *Cryst. Growth Des.* **2007,** *7*, 1007; see Ref. 74 therein.
35. Sharma, R. K.; Fry, J. L. *J. Org. Chem.* **1983,** *48*, 2112.
36. Zdrojewski, T.; Jonczyk, A. *J. Org. Chem.* **1998,** *63*, 452.
37. Pilcher, A. S.; Ammon, H. L.; DeShong, P. *J. Am. Chem. Soc.* **1995,** *117*, 5166; Handy, C. J.; Lam, Y.-F.; DeShong, P. *J. Org. Chem.* **2000,** *65*, 3542.
38. Sun, H.; DiMagno, S. G. *J. Am. Chem. Soc.* **2005,** *127*, 2050.
39. Kuduk, S. D.; DiPardo, R. M.; Bock, M. G. *Org. Lett.* **2005,** *7*, 577.
40. DelMar, E. G.; Kwiatkowski, C. T. U.S. Patent 4,740,637, 1988 (to FMC).
41. Norris, T.; Brown Ripin, D. H.; Ahlijanian, P.; Andresen, B. M.; Barrila, M. T.; Colon-Cruz, R.; Couturier, M.; Hawkins, J. M.; Loubkina, I. V.; Rutherford, J.; Stickley, K.; Wei, L.; Vollinga, R.; de Pater, R.; Maas, P.; de Lange, B.; Callant, D.; Konings, J.; Andrien, J.; Versleijen, J.; Hulshof, J.; Daia, E.; Johnson, N.; Sung, D. W. L. *Org. Process Res. Dev.* **2005,** *9*, 432.
42. Li, J. J. *Name Reactions: A Collection of Detailed Reaction Mechanisms*, 3rd ed.; Springer, 2006 p 567.
43. Zeller, J. R., "*Atorvastatin: A Case Study in the Development of a Commercial Process,*" invited lecture, 3rd International Conference on Process Development Chemistry, Amelia Island, Florida, March 25–28, 1997.
44. For an example, see Brands, K. M. J.; Krska, S. W.; Rosner, T.; Conrad, K. M.; Corley, E. G.; Kaba, M.; Larsen, R. D.; Reamer, R. A.; Sun, Y.; Tsay, F.-R. *Org. Process Res. Dev.* **2006,** *10*, 109.
45. Giles, M. E.; Thomson, C.; Eyley, S. C.; Cole, A. J.; Goodwin, C. J.; Hurved, P. A.; Morlin, A. J. G.; Tornos, J.; Atkinson, S.; Just, C.; Dean, J. C.; Singleton, J. T.; Longton, A. J.; Woodland, I.; Teasdale, A.; Gregertsen, B.; Else, H.; Athwal, M. S.; Tatterton, S.; Knott, J. M.; Thompson, N.; Smith, S. J. *Org. Process Res. Dev.* **2004,** *8*, 628.
46. Anderson, N. G.; Lust, D. A.; Colapret, K. A.; Simpson, J. H.; Malley, M. F.; Gougoutas, J. Z. *J. Org. Chem.* **1996,** *61*, 7955.
47. Hoare, J. H.; Halfon, M. U.S. Patent 5,011,933, 1991 (to FMC).
48. Braden, T. M.; Coffey, S. D.; Doecke, C. W.; LeTourneau, M. E.; Martinelli, M. J.; Meyer, C. L.; Miller, R. D.; Pawlak, J. M.; Pedersen, S. W.; Schmid, C. R.; Shaw, B. W.; Staszak, M. A.; Vicenzi, J. T. *Org. Process Res. Dev.* **2007,** *11*, 431.
49. Inagaki, M.; Matsumoto, S.; Tsuri, T. *J. Org Chem.* **2003,** *68*, 1128.
50. Theriot, K. J. U.S. Patent 5,907,063, 1999 (to Albemarle).
51. Laduron, F.; Tamborowski, V.; Moens, L.; Horváth, A.; De Smaele, D.; Leurs, S. *Org. Process Res. Dev.* **2005,** *9*, 102.
52. Beaulieu, P. L.; Gillard, J.; Bailey, M. D.; Boucher, C.; Duceppe, J.-S.; Simoneau, B.; Wang, X.-J.; Zhang, L.; Grozinger, K.; Houpis, I.; Farina, V.; Heimroth, H.; Krueger, T.; Schnaubelt, J. *J. Org. Chem.* **2005,** *70*, 5869.
53. Chen, C.-Y. In *The Art of Process Chemistry.* Yasuda, N., Ed.; Wiley- VCH, 2010; pp 117–142.
54. Barnes, D. M.; Haight, A. R.; Hameury, T.; McLaughlin, M. A.; Mei, J.; Tedrow, J. S.; Toma, J. D. R. *Tetrahedron* **2006,** *62*, 11311.
55. *Prudent Practices in the Laboratory*; National Research Council of the National Academies; National Academies Press: Washington, D.C, 2011; p 136.
56. SAFE generation and use of dimsyl anion: Dahl, A. C.; Mealy, M. J.; Nielsen, M. A.; Lyngsø, L. O.; Suteu, C. *Org. Process Res. Dev.* **2008,** *12*, 429.
57. Jaques, J.; Collet, A.; Wilen, S. H. *Enantiomers, Racemates, and Resolutions*; John Wiley & Sons: New York, 1981; pp 381–95.
58. Kuo, S. C.; Chen, F.; Hou, D.; Kim-Meade, A.; Bernard, C.; Liu, J.; Levy, S.; Wu, G. G. *J. Org. Chem.* **2003,** *68*, 4984; Wu, G.; Huang, M. *Chem. Rev.* **2006,** *106*, 2596.
59. Jackson, D. Y. *Synthesis* **1988,** 337.
60. Schmid, C. R.; Bryant, J. D.; Dowlatzedah, M.; Phillips, J. L.; Prather, D. E.; Schantz, R. D.; Sear, N. L.; Vianco, C. S. *J. Org. Chem.* **1991,** *56*, 4056.

61. Zhao, M.; Li, J.; Song, Z.; Desmond, R.; Tschaen, D. M.; Grabowski, E. J. J.; Reider, P. J. *Tetrahedron Lett.* **1998,** *39,* 5323.

62. Humphrey, G. R.; Miller, R. A.; Li, W. U.S. Patent 6,187,925 B1, 2001 (to Merck).

63. Barnes, D. M.; Ji, J.; Fickes, M. G.; Fitzgerald, M. A.; King, S. A.; Morton, H. E.; Plagge, F. A.; Preskill, M.; Wagaw, S. H.; Wittenberger, S. J.; Zhang, J. *J. Am. Chem. Soc.* **2002,** *124,* 13097.

64. Deshpande, P. P.; Ellsworth, B. A.; Buono, F. G.; Pullockaran, A.; Singh, J.; Kissick, T. P.; Huang, M.-H.; Lobinger, H.; Denzel, Th.; Mueller, R. H. *J. Org. Chem.* **2007,** *72,* 9746.

65. Mueller, W.-D.; Naumann, P. U.S. Patent 6,790,981, 2004 (to Clariant).

66. For example, the basicity of NaOH increased 50,000× by the decreasing hydration number of the anion from 11 (15% NaOH) to 3 (solid): Albanese, D.; Landini, D.; Maia, A.; Penso, M. *Ind. Eng. Chem. Res.* **2001,** *40,* 2396.

67. Grosso, J. A. U.S. Patent 5,162,543, 1992 (to E. R. Squibb & Sons).

68. Bravo, F.; Cimarosti, Z.; Tinazzi, F.; Smith, G. E.; Castoldi, D.; Provera, S.; Westerduin, P. *Org. Process Res. Dev.* **2010,** *14,* 1162.

69. Gross, T. D.; Schaab, K.; Ouellette, M.; Zook, S.; Reddy, J. P.; Shurtleff, A.; Sacaan, A. I.; Alebic-Kolbah, T.; Bozigian, H. *Org. Process Res. Dev.* **2007,** *11,* 365.

70. Zotova, N.; Franzke, A.; Armstrong, A.; Blackmond, D. G. *J. Am. Chem. Soc.* **2007,** *129,* 15100.

71. Yang, H.; Carter, R. G. *Org. Lett.* **2008,** *10,* 4649.

Chapter 7

In-Process Assays, In-Process Controls, and Specifications

"So returning to the boiled linseed oil, I told my companions at table that in a prescription book published about 1942 I had found the advice to introduce into the oil, toward the end of the boiling, two slices of onion, without any comment on the purpose of this curious additive. I had spoken about it in 1949 with Signor Giacomasso Olindo, my predecessor and teacher, who was then more than seventy and had been making varnishes for fifty years, and he, smiling benevolently behind his thick white mustache, had explained to me that in actual fact, when he was young and had boiled the oil personally, thermometers had not yet come into use: one judged the temperature of the batch by spitting into it, or, more efficiently, by immersing a slice of onion in the oil on the point of a skewer; when the onion began to fry, the boiling was finished. Evidently, with the passage of years, what had been a crude measuring operation had lost its significance and was transformed into a mysterious and magical practice."

– Primo Levi [1]

I. INTRODUCTION

In-process controls (IPCs) are essential for processes for three reasons: to make high-quality products, to conduct operations productively, and to facilitate troubleshooting. For instance, the managers of manufacturing facilities rely upon IPCs to ensure that

Practical Process Research and Development. DOI: 10.1016/B978-0-12-386537-3.00007-1

processes run effectively, efficiently, and consistently. Processes that require frequent troubleshooting or reworking of products are not productive. Regarding product quality, IPCs are crucial for process validation, and processes must be validated for permission to sell drugs for use in humans. IPCs are established to guide operations to prepare APIs and drug products that meet the critical quality attributes (CQAs) accepted by the FDA [2]. The FDA is concerned primarily about quality (such as potency and impurities) of APIs and drug products and not about yields and productivity; however, variable yields may indicate that a process is not controlled, which could result in batches with additional, potentially toxic impurities. IPCs are filed with the FDA as part of the Chemistry, Manufacturing & Controls (CMC) section of New Drug Applications (NDAs) [3] and are part of the current good manufacturing practices (cGMPs) that must be followed to prepare material for use in humans [2]. (Investigational New Drug (IND) applications do not require IPCs, but focus instead on the characterization of the NCE [4].) After the NDA has been approved IPCs must be followed to ensure that the operations are in compliance with the process filed with the FDA.

IPCs include actions in three areas. First of all, monitoring and adjusting processing operations to meet guidelines, such as confirming the weights of reactants delivered to a reactor, and controlling parameters such as agitation rate, temperature, and pH. These important actions seem routine and second-nature. Secondly, confirming that process operations have gone as expected through in-process assays, using both qualitative and quantitative assessments. Last of all, redirecting operations as needed according to the results from the in-process assays. Because IPCs divert time and attention away from other productive actions, process scientists minimize the number of IPCs that must be run or approved by those not working on the manufacturing floor. Nonetheless no IPC should be overlooked if it may be critical to the success of operations.

Without IPCs there is only hope that processing will proceed as planned.

In-process assays are investigated in the laboratory, and used to establish IPCs. In-process assays can determine an acceptable operating "space" in the Quality by Design (QbD) approach favored by the FDA [5]; IPCs can help keep processing within acceptable ranges. Some common in-process assays are shown in Table 7.1. An in-process assay can be used for *each* operation in processing.

Accuracy and reproducibility are essential for good analyses of both the progress of operations and the quality of isolated products. Accuracy is similar to shooting an arrow at a target and having it hit the bullseye; a good archer can reproducibly place arrows in the same region, but if only a small amount of the target is visible the archer may not hit the bullseye. This analogy is valid for assays performed in the pharmaceutical and fine chemicals industries: the more that is known about the chemicals and analytical procedures, the better the accuracy and reproducibility of the results. And the better the operations can be directed for routine production of high-quality product. Analytical Research & Development (AR&D) departments and later Quality Control (QC) departments may develop assays to quantitatively assess the purity of reaction streams and isolated compounds, but initially organic chemists usually settle for reproducibility.

TABLE 7.1 Some Common In-Process Assays

Analytical Technique	Typical Applications	
	Time of Processing	Example/Comments
HPLC	Before key addition	Confirm HPLC system suitability
	During and after key addition	Minimize impurity formation
	End of reaction	Optimize levels of impurities
	During extractive workups	Track removal of impurities
	Before and after concentration	Monitor growth of impurities
	During aging of crystal slurry	Determine when crystallization has effectively stopped, and minimize co-crystallization of impurities
	Washing wet cakes	Track removal of impurities
	After drying	Quantitate impurities present
GC	End of reaction	Determine completion of reaction
	Solvent chasing	Confirm low level of residual solvent
KF titration	Before charging reagents	Confirm reactor is dry
	Solvent chasing	Confirm extract has been azeotroped dry
	Before crystallization	Suitable water content to crystallize hydrate
pH meter	During reaction	Optimize reaction conditions
	During extraction	Control extraction conditions
Conductivity meters	During extractions	Suitable removal of salts from extract
	Washing wet cake with water	Removing salts from product
Loss on drying (LOD)	Drying product cake	Removing product from dryer

Data generated through in-process assays and IPCs can prove extremely valuable. In-process assays can provide a window to understand the formation of impurities, especially using kinetics [6,7]. By reviewing data from in-process assays it may become apparent which assays are critical, and what limits are suitable as IPCs. In troubleshooting (Chapter 17) the historical data amassed from in-process assays and IPCs can help identify the troublesome chemical inputs or troublesome areas of processing.

In assessing data it can be important to consider the types of "purity" assessment. HPLC and GC integrators routinely generate data for peak areas as a function of the total peak areas for a given run. Such data are expressed as area under the curve (AUC, also termed area% or homogeneity index or HI); these data do not automatically consider the responses of individual compounds (for UV detectors, the extinction coefficient). Using a standard of high quality the purity (or potency, wt/wt% or wt%) of a sample can be calculated as shown in Equation 7.1:

$$\text{Purity of sample(wt\%)} = \frac{(\text{area counts of sample}/[\text{sample}])}{(\text{area counts of standard}/[\text{standard}])} \times 100 \qquad (7.1)$$

Such purity assessments can be important in several ways. First of all, they permit a quantitative assessment of the quality of input materials; for instance, an intermediate might contain NaCl or a deleterious impurity that would not be detected by routine assays. Secondly, such purity values can confirm the quality of a product, or indicate the presence of an impurity such as water that cannot be detected by HPLC. Third, purity values allow an accurate assessment of yields, and can identify areas for process improvement. A yield that has been corrected for the potency of the product may be designated as "X M%." Yields may also be corrected for the quality of the inputs, for even more accurate comparisons. (In the case of "quantitative" yields, it is likely that not all impurities [8] and the accuracy of laboratory techniques [9] have been considered.) "As is" yields (sometimes termed mass % yields) are calculated without correction for product purity; such yields should be designated as "X% (as is)" but usually the qualifier in parentheses is omitted. Occasionally yields are expressed as "X wt%," calculated relative to the weight of an input; these yields are useful for operators to judge reproducibility of batches in campaigns, and for bookkeepers.

A high-quality batch can be selected as a standard within the process development laboratory, with no need for extensive documentation; for consistent comparisons the same lot should be used each time. To simplify the calculations, weigh similar amounts of standard and sample into separate volumetric flasks of the same size, and dilute to the volumes.

In the 1980s the isomer ratios in APIs could be determined by the ratio of the corresponding infrared peaks, which were basically limit tests (e.g., the ratio of enantiomers A to enantiomers B must be between 40:60 and 60:40). Today we could measure these four compounds specifically by chiral HPLC, chiral GC, or chiral supercritical fluid chromatography (SFC). In the early days of HPLC the areas under the curve (AUCs) were computed by measuring peaks on a chromatogram with a ruler. Today the AUCs are computed by an integrator using a choice of data-gathering options to separate tailing main peaks from other analytes. Capabilities for specific assays have improved greatly.

The following sections discuss benefits of in-process assays, examples of IPCs, specifications, and process analytical technology (PAT).

For rapid comparison of results, use raw data. Converting to another format takes time and may be less descriptive. For instance, a product to impurity ratio of 92.3:4.1 (as taken directly from the HPLC chromatogram) allows the calculation that the remaining peaks accounted for 3.6% of the total area counts; the corresponding product to impurity ratio of 22.5:1 doesn't readily provide that information. As another example, if one considers ratios of mono-brominated product:dibrominated byproduct:starting material, a ratio of (13.2:1.0:0.9) seems much better than the ratio of (10.4:1.0:0.9). But when these values are normalized to 100, the numbers become (87:7:6) and (85:8:7), a difference within the error of running two ^{1}H NMRs. The convention of enantiomeric excess (ee), proposed in 1971 when the extent of racemization was determined by polarimetry, is outmoded [10]. Today the diastereomeric ratio (dr) may be measured by NMR, and normalizing the data to 100 also seems like a waste of time; the use of NMR should be mentioned if the dr data were generated by NMR. The use of ee and diastereomeric excess (de) may provide cosmetic value for some people, but the terms enantiomeric ratio (er) and dr are more direct.

II. UNDERSTANDING CRITICAL DETAILS BEHIND IN-PROCESS ASSAYS

The best assays have the fewest manipulations prior to assaying, are specific for the target molecules or physical characteristics of a sample, and are readily reproducible. Delving into some analyses from several decades ago, as extreme examples, may best illustrate of the importance of understanding the limitations of analyses.

In 1967 Lederle workers wrote a manuscript on peptide coupling through mixed anhydrides, an influential paper that has been cited over 100 times [11]. N-Protected amino acids are activated using a chloroformate in the presence of a tertiary amine, and with the addition of a primary or secondary amine the amide bond forms rapidly (Figure 7.1). Tertiary amines with a methyl group, such as N-methylmorpholine, form the active ester more rapidly than more hindered amines, implicating the initial reaction of the tertiary amine with the chloroformate. With increasing basicity of the tertiary amine more racemization occurs, probably through an oxazolone. The optimal temperature for coupling was determined to be −15 °C, and the use of polar solvents increased the amount of racemization. This work was analyzed by a racemization test through preparing N-carbobenzoxycarbonyl-glycyl-L-phenylalanyl-glycine ethyl ester (Z-Gly-L-Phe-Gly-OEt), as shown in Figure 7.1. The assay for racemization involved recrystallizing the *isolated* tripeptide as a 2% solution in absolute EtOH at 0 °C; under these conditions the racemic product crystallizes first. [As discussed in Chapter 12 this is a classic example of a true racemate, with the racemate showing a higher melting point (mp 132–133 °C) and crystallizing before the more-soluble single enantiomer (mp 116.5–119.5 °C).] Using this method, which was developed in 1952 [12], several crops of racemic crystals were collected, and then crystals of the desired L-enantiomer were collected. Racemization was calculated by comparing the overall yield from crops of racemic crystals to the theoretical yield of the desired enantiomer. Dozens, if not hundreds, of crystallizations were carried out to determine the trends found by the

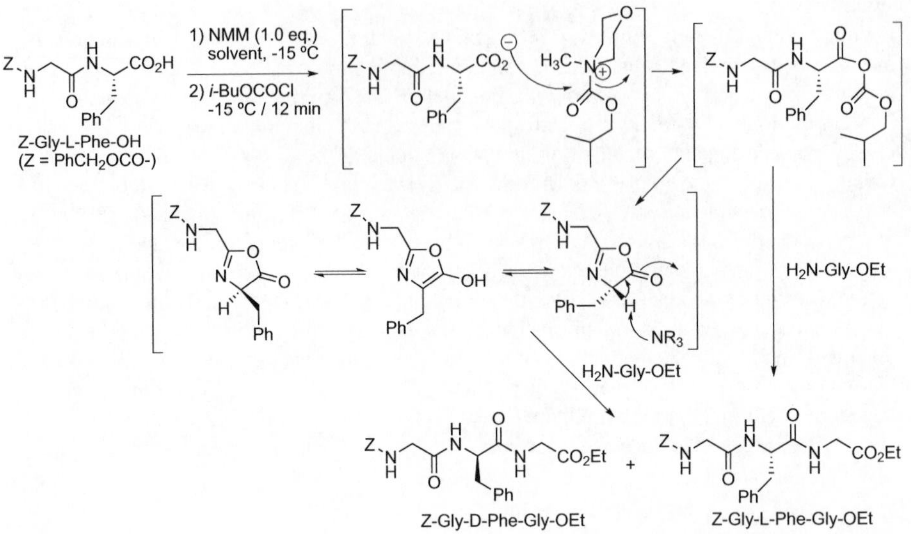

FIGURE 7.1 Mechanism of peptide coupling and racemization using the mixed carbonic anhydride method.

authors. The operations for peptide coupling and racemization were quite reproducible: for three identical coupling experiments the yield of the isolated L-enantiomer was 73% + 3%, with 7.4% + 0.8% racemate isolated [12]. Nonetheless the operations to analyze racemization were tediously repetitive, with many opportunities for small variations in aging times and other factors that could have influenced the outcomes. Today, without isolating the product, we could directly determine the amount of racemization through HPLC analysis using a chiral column, comparing the amount of the separated D-isomer to the desired L-isomer. Other choices include supercritical fluid chromatography (SFC) with a chiral stationary phase, or capillary electrophoresis with a chiral medium. Gas chromatography using a chiral GC column could be employed; derivatizing the tripeptide to increase its volatility would probably be necessary, necessitating additional care in the sample preparation.

Assays that are specific for the desired compound are key to rapid, effective in-process analyses. For instance, the Lederle workers in the above example noted that melting points and optical rotations were not adequate to determine the amount of racemization, as these values may be altered by other impurities present in the samples [12]. Another example, again from the early 1980s, is shown for residual NaCl in captopril (Figure 7.2). At that time NaCl was analyzed by a residue on ignition (ROI) assay. A small amount of captopril was accurately weighed in a covered platinum crucible, which was then fired at high temperature to combust and volatilize the organic

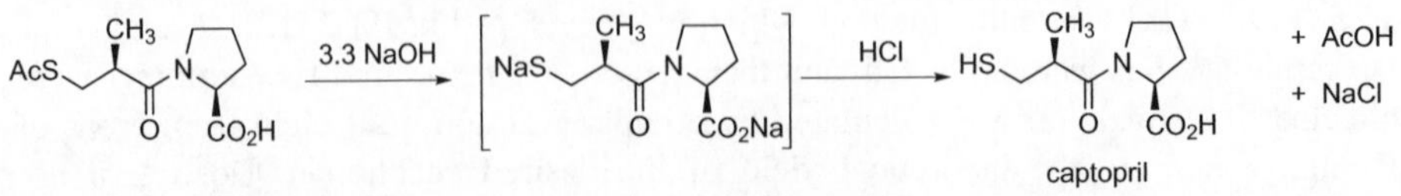

FIGURE 7.2 Generation of NaCl in the preparation of captopril.

contents. The hot crucible was cooled in a desiccator to preclude atmospheric moisture from condensing onto the crucible, and weighed. The percentage of weight remaining was the ROI. This general assay is suitable for non-combustible impurities such as glass chips and bits of metal screens, and obviously is not specific for NaCl. At 0.2 wt% NaCl in captopril the accuracy of the method was ±0.1 wt%, which did not afford much assurance for accuracy at low levels of residual NaCl. Today we could investigate ion-liquid chromatography for chloride, or possibly a chloride-specific electrode. The mercaptan of captopril could interfere with the analysis of residual chloride, by coordinating with a key analyte or by being co-oxidized under conditions where oxidation of chloride is part of the assay. Thus depending on the basis of the assay, analysis of chloride in captopril might not be straightforward.

The two examples discussed in detail above, the crystallization assay for racemization and the ROI assay, do not qualify as in-process assays: those assays were carried out on isolated products. In-process assays are used to confirm and/or redirect operations before the product is isolated, or (in the case of an IPC used during drying) before the product is removed from the dryer. A broad range of in-process assays is available to the process scientist, as shown in Table 7.2. Although some of these assays may seem somewhat dated, such as iodometric titration for the presence of oxidants [13] or titration of solutions of *n*-BuLi or Grignard reagents [14,15], they may prove very useful.

> How meaningful are the data? Is a reproducibility of ±0.1 wt% sufficient to address the processing issues at hand? What are the limit of quantitation (LOQ) and the limit of detection (LOD)? How firm are values below the LOD? A value stated relative to a limit test, such as NMT 200 ppm, may have limited utility in problem-solving. Understanding how meaningful the data are can simplify efforts.

III. BENEFITS OF IN-PROCESS ASSAYS IN OPTIMIZING PROCESSES

In-process assays during additions can provide data that are key to optimizing a process. Abbott researchers studied in detail the cyclization of a keto ester to the diketone (as the enolate). NaO*t*-Bu and THF were the optimal base and solvent, as shown in Figure 7.3. Lower reaction temperatures and excess base decreased the yield, as did adding the starting material to a solution of the base. When solutions of the starting material and the base were added simultaneously to the reactor, in-process HPLC assays showed that the level of enolate formed was consistently high [16]. Such a reaction with parallel additions to control stoichiometry is probably well suited to continuous operations. The chlorination of a hydroxypyrimidine gave a high yield when the reaction was run on a small scale with a rapid addition of the limiting reagent, phosphorus oxychloride; when the addition time was extended a pseudo-dimer formed and the yield fell off. In-process assay showed that dimerization took place at the very beginning of the $POCl_3$ addition, leading to the reaction as shown in Figure 7.3. In the hope of effecting complete phosphorylation without further reaction of the phosphorochloridate, the $POCl_3$ was added below 40 °C, and then the mixture was heated to 60–70 °C to complete chlorination [17]. (Three other

TABLE 7.2 In-Process Assays Useful in Chemical Process R&D

Analytical Method	Ease of Quantitation	Used on Scale	Usefulness/Comments
^{1}H NMR	+	+	Early development
^{13}C NMR			Used under duress
^{31}P NMR and ^{19}F NMR	+	+	Early development
IR (including in-process probes)	+++	++	Very useful for PAT and continuous operations
HPLC: UV detector (including diode array)	+++	+++	Use solvent gradient to elute "all" impurities
Redox detector	++	++	Baseline noise may limit usage
Thermal electron det.	+	+	Replaces RI & ELSD
SFC HPLC	++	+	Chiral resolutions
GC	+++	+++	Solvent displacement
TLC	+	++	Good for troubleshooting, as it "shows everything." Can be quantitative (1)
MS	++	++	ESI allows fast assays (2)
AA (ICP-AA)	++	++	Removing metals from streams (3)
KF titration	+++	++	Quick and reproducible
Chloridometer	++	+	Effective for Cl^-, Br^-, and I^-
Conductivity	++	++	Portable
Titration	+	++	Quick, reliable
Test strips	++	++	Portable
pH meter	++	++	Portable (4)
Cl-specific electrode	+	++	Applicable for resins too (5)
LOD	++	+	Determine end of drying
XRPD	+	+	Polymorphism (6)
Raman (in-line)	++	+	Polymorph transformation (7)
Near-IR	++	++	Determine end of drying
Refractive index	+	+	Solvent displacement
UV−visible (8)	+	+	Often not selective enough; impurities such as solvents can change results
Specific rotation	+	+	
Melting point	+	+	

TABLE 7.2 In-Process Assays Useful in Chemical Process R&D—cont'd

Analytical Method	Ease of Quantitation	Used on Scale	Usefulness/Comments
DSC	++	+	More quantifiable than mp; could be used as LOD IPC
Hygrometer	++	++	Portable; dew-point hygrometer can monitor vapors for LOD (9)
Particle size (FBRM)	++	++	Determine optimal filtration time
Capillary electrophoresis	++	++	For ionic materials (10)
Ion chromatography	++	+	For ionic materials

(1) Johnsson, R.; Träff, G.; Sundén, M.; Ellervik, U. J. Chromatogr. A, **2007**, 1164, 298.

(2) Electrospray ionization (ESI) was used for rapid screening in the Sharpless laboratories for compounds to accelerate the dihydroxylation of olefins: Dupau, P.; Epple, R.; Thomas, A. A.; Fokin, V. V.; Sharpless, K. B. Adv. Synth. Catal. **2002**, 344(3 + 4), 421.

(3) Atomic absorption or inductively coupled plasma atomic absorption. ICP-AA is usually run by dedicated personnel, and requesting their services in advance may be necessary.

(4) For details on pH measurement and control, see McMillan, G.; Baril, R. Chem. Eng. **2010**, 117(8), 32.

(5) Raillard, S. P.; Ji, G.; Mann, A. D.; Baer, T. A. Org. Process Res. Dev. **1999**, 3, 177.

(6) XRPD could detect 3 wt% of Form II in bulk ritonavir: Bauer, J.; Spanton, S.; Henry, R.; Quick, J.; Dziki, W. Porter, W.; Morris, J. Pharm. Res. **2001**, 18, 859.

(7) Hu, Y.; Liang, J. K.; Myerson, A. S.; Taylor, L. S. Ind. Eng. Chem. Res. **2005**, 44, 1233.

(8) Hunter Laboratories offers equipment for convenient assaying of colors: http://www.hunterlab.com.

(9) Cypes, S. H.; Wenslow, R. M.; Thomas, S. M.; Chen, A. M.; Dorwart, J. G.; Corte, J. R.; Kaba, M. Org. Process Res. Dev. **2004**, 8, 576.

(10) May be useful for monitoring resolutions. Schneider, I. Today's Chemist at Work, **September 1999**, 44; Frost, N. W.; Jing, M.; Bowser, M. T. Anal. Chem. **2010**, 82, 4682.

processing considerations deserve to be mentioned for this chlorination. First of all, when Et$_3$N was used as base for the chlorination a viscous oil formed on the walls of the reactor. In-process assay of the mobile solution showed no starting material, as the salt byproducts had entrapped the starting material. The substitution of (i-Pr)$_2$EtN gave a fluid dispersion, which was readily scaled up. Secondly, inverse addition of the starting material to the POCl$_3$ was not employed because the hydroxypyrimidine was poorly soluble in toluene, and dose-controlled additions of suspensions can be troublesome. Third, a source of chloride (e.g., Et$_4$NCl) or components to generate chloride (e.g., DMF) could have been added to suppress formation of the pseudo-dimer at the start of the POCl$_3$ addition, but this tact was avoided because employing another reagent would have meant purchasing, qualifying, handling, and analyzing for another component. But perhaps adding 5 mol% H$_2$O to the reaction mixture initially would have been simple and effective in suppressing dimerization.)

IV. IN-PROCESS CONTROLS

IPCs can include adhering to simple physical guidelines, such as accurate delivery of weights and volumes; reaching a color endpoint, reaching and maintaining critical

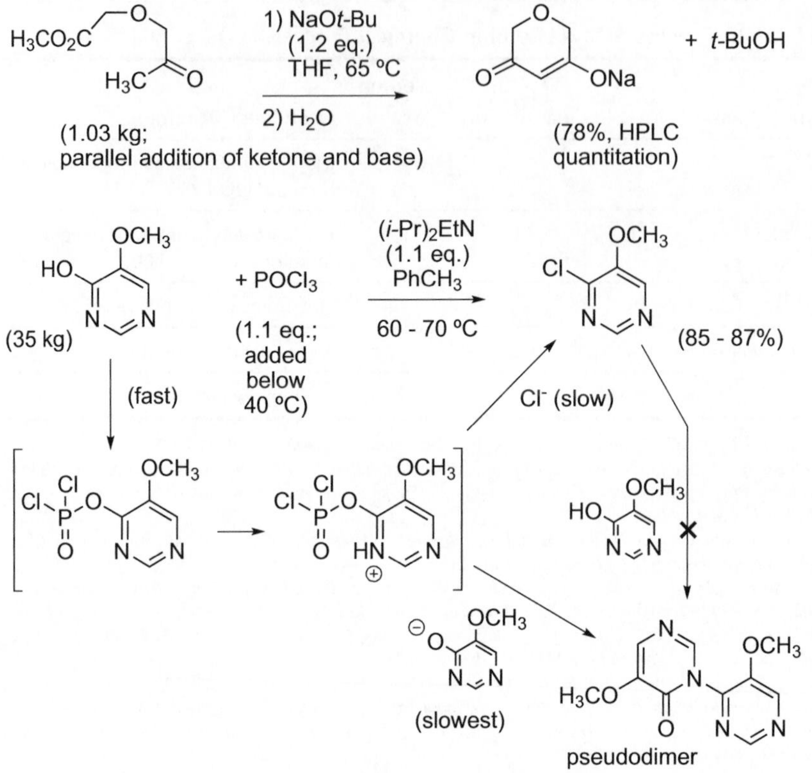

FIGURE 7.3 Use of in-process assays during additions to optimize processing.

temperatures, addition times, or pH [18]; concentrating to the appropriate volume, and so on. The acylation of an oxazolidinone with a mixed anhydride was preceded by the addition of LDA to just produce a dark red-brown color, which was consistent with complete deprotonation (Figure 7.4) [19]; since sight glasses on reactors usually fog up, fiber optic probes may ease detection of a color endpoint. IPCs for pH control were critical for both the enzyme-mediated resolution and the subsequent iodination. (An interesting side note in the latter process is that the sodium salt of the resolved carboxylic acid was generated; the ammonium salt formed from controlling pH was unstable during concentration over 12 h in the plant, due to the disproportionation of the ammonium salt to the free carboxylic acid plus ammonia, with loss of ammonia during concentration. Under these circumstances the free acid co-crystallized with the ammonium salt [18].)

In-process assay considerations include (1) obtaining a representative sample of the process stream, (2) preparing the sample for accurate analysis, (3) conducting the assay, and (4) analyzing the data for reproducibility and accuracy. Obtaining a representative sample is straightforward from a room-temperature solution, but if a sample separates into different phases upon cooling it may be necessary to assay each phase individually to determine how to best prepare an in-process assay. Sampling cryogenic reactions may

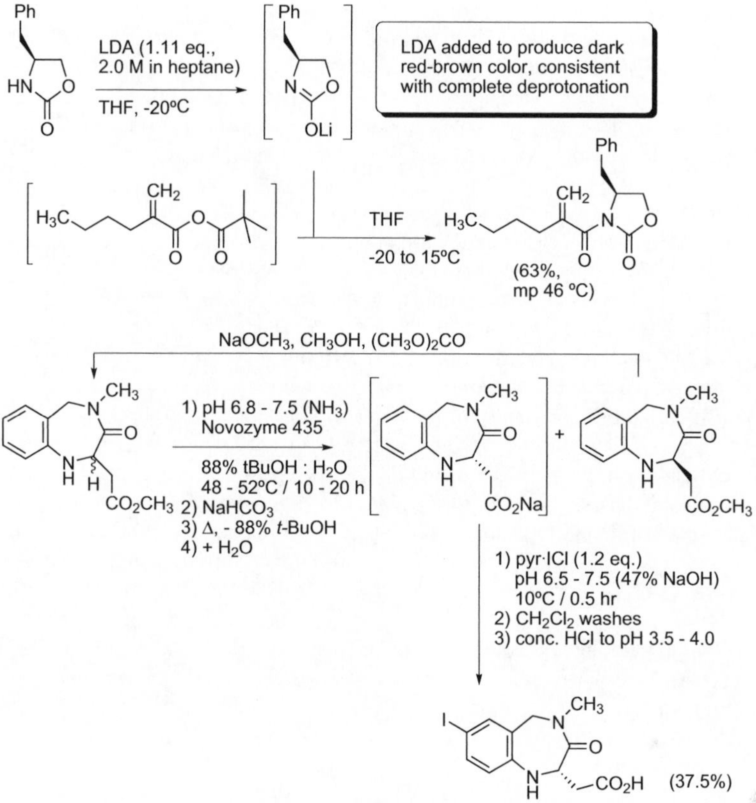

FIGURE 7.4 Color and pH as IPCs.

require additional care if the content of the samples could change upon warming as the sample is withdrawn from the reactor. Impurities can be created as artifacts, due to sampling, sample preparation, or analysis. For instance, when soft drinks were heated at 100 °C for 30 min benzene was formed from the sodium benzoate added as a preservative [20].

Sampling a process stream in a pilot plant or manufacturing plant has additional considerations. Standard operating procedures (SOPs) at a site may require that hot reaction streams be transferred below a maximum temperature. If a process stream must be cooled for SAFE sampling, care must be taken to ensure that the sample accurately represents the contents of the reactor. A reactor is rarely opened for direct sampling, to minimize risks of both exposing personnel to reactive streams and contaminating the batch, so samples are usually withdrawn by suction into approved sampling equipment. If the reaction must be warmed to continue processing after the in-process assay, productivity is affected by the time required for cooling to the sampling temperature, holding for running (and approving) the in-process assay, and warming back to the reaction temperature. These considerations make in-line assays attractive.

For rapid development and routine use of in-process assays
- Make in-process assays convenient, simple, reliable, and easy to describe.
- In-process assays should have broad tolerance for operator technique.
- Try to devise in-process assays that can be used in the laboratory and on scale.
- When possible, bias analyses in favor of key impurities, not the product, unless the product decomposes to impurities that are not detected.
- If a sample is heterogeneous, assay each phase separately to determine which phase contains more information. If one phase contains no analytes, using the heterogeneous mixture for sample preparation and analysis may be possible.
- For convenient laboratory phase separations, use syringes with disposable filters, disposable pipettes as separatory funnels, and mini centrifuges.
- Keep sample preparation consistent. Dilute promptly, with the same order of addition, to hold constant any side reactions such as reaction with air.
- Be consistent with the volumes for dilutions, to hold constant both the amount of any byproduct generated through second-order side reactions during sample preparation and analysis, and the identification and quantitation of small peaks.
- Use volumetric flasks, microliter syringes and micro-pipettors for accurate dilution.
- Be aware of saturating the detector and integrator. If one peak is present at an amount much greater than the calculation limit of the integrator, the area percentage of the smaller peaks will register artificially high.
- Break down in-process assays into component operations for ready troubleshooting.
- Know sample stability; for example, rerun the first vial in the autoinjector queue.
- Keep retains of isolated products for future assay development, and as informal bulk stability studies.

In-line (also called "on-line") assays provide "windows" for opaque reactors without withdrawing samples and sample preparation. Probes can be inserted into laboratory reactors for in-line assays, or spectral data can be gathered through transparent sections of reactors [21]. By using in-line Fourier transform infrared (FTIR), which is used for monitoring solutions but not suspensions, it is possible to safely monitor O_2-free processes, such as carbonylations with CO and hydrogenations, and operations run at low temperature or high temperature. Under these circumstances personnel are not exposed to toxic reaction components; for instance, FTIR was used to monitor the formation and stability of dibenzyl chlorophosphate [22]. Probes can be inserted into pilot plant or manufacturing equipment reactors or into lines through which process streams circulate. Fiber optic probes can be used to position sensors deep in a reactor, even down to the minimum stirrable volume during concentration [23]. A high degree of sensitivity was found using Raman spectroscopy to follow the crystallization and disappearance of a metastable progesterone polymorph [24]. Data acquisition through fiber optics is also rapid, permitting rapid process control in manufacturing settings. An overview of some fiber optic capabilities is shown in Table 7.3.

Test strips may be useful as handy in-process tools, and are portable for troubleshooting on a plant floor. Test strips can be considered as convenient spot tests [25]. E. Merck Darmstadt has offered qualitative test strips for food and environmental analyses, including assays for Al, NH_4^+, ascorbate, Ca(II), carbonate/hardness, HCHO, H_2SO_3, glucose, Fe, lactate, Mg, malate, Mn, Mo, Ni, nitrate, nitrite, peracetic acid,

TABLE 7.3 Spectroscopy and Applications of Fiber Optics (1)

Type	Wavelength	Excitation	Applications
UV	<400 nm	Electronic vibrations (aromatic and conjugated)	Cleaning vessels (2)
Visible	400–800 nm	Electronic vibrations (aromatic and conjugated)	Colored materials, transition metals
Near-IR	800–2500 nm	Molecular vibrations	H_2O in solids (similar to LOD); H_2O in gases (scrubbing)
Mid-IR	2500–25,000 nm ($400–4000$ cm^{-1})	Functional group vibrations and bending	Reaction mechanisms (3) Trace amounts of impurities Temperature control (Milestone microwave reactor)
Raman	<400–25,000 nm	Functional group vibrations	Crystallization of a polymorph; less sensitive to H_2O than IR (4–6)

(1) Liauw, M. A.; Oderkerk, B.; Treu, D. Chem. Eng. **2006,** 113(7), 56.
(2) http://www.oceanoptics.com.
(3) Minnich, C. B.; Buskens, P.; Steffens, H. C.; Bauerlein, P. S.; Butvina, L. N.; Kupper, L.; Leitner, W.; Liauw, M. A.; Greiner, L. Org. Process Res. Dev. **2007,** 11, 94.
(4) Byrn, S. R.; Liang, J. K.; Bates, S.; Newman, A. W. J. PAT **2006,** 3, 14.
(5) Wang, F.; Wachter, J. A.; Antosz, F. J.; Berglund, K. A. Org. Process Res. Dev. **2000,** 4, 391.
(6) Raman spectroscopy uses light from a laser, which can be tuned to monitor the UV, visible, and IR spectra. Princeton Instruments, Raman Spectroscopy Basics: http://content.piacton.com/Uploads/Princeton/Documents/Library/UpdatedLibrary/Raman_Spectroscopy_Basics.pdf.

H_2O_2, phosphate, K, As, Cl, CrO_4, Co, CN, Cu, R_4N^+, Sn, Zn, and KI/starch [26]. Perhaps these strips can be used for in-process assays, if they are sensitive enough and other chemicals in the process samples (organic solvents, products, catalysts, and so on) do not interfere with the assay. It is tempting to consider trying test strips for uses other than the advertised ones; for instance, knowing that copper and palladium can be extracted by similar ligands [27], would a test strip for Cu assay for Pd?

> By paying attention to in-process data and IPCs it is possible to anticipate and avoid problems, and thus avoid troubleshooting. Better to generate excessive data during a run than to have to repeat an experiment to gather a few data points.

V. SPECIFICATIONS

Specifications range from guidelines for purchasing starting materials, reagents and solvents, to requirements for accepting intermediates and APIs. Some specifications are negotiable or "soft." For instance, if a CRO made a batch of material that was assayed at 92% AUC by HPLC instead of the 95% AUC specification proposed with the contract, a client may accept the batch if the deadline for making a drug candidate is looming. Under those circumstances the client is responsible for removing impurities associated with the batch; otherwise the CRO is responsible. Use-tests are often necessary to approve batches of reagents and intermediates early in the development of processes; as these processes mature, assays can be put in place to quantify key impurities that have been identified. If a chemical, such as MeLi or $LiAlH_4$, is deemed too reactive to assay in-house, a purchaser (or QA department) may just accept the supplier's specifications and certificate of analysis. Sometimes specifications are established for gathering information; for instance, "Record IR (diffuse reflectance mode)" or "Record melting point" may be used to assay products from initial pilot plant runs. "HPLC NLT 95% (AUC)" might also be used to characterize products from plant runs. "IR must conform to standard" and "HPLC purity 97.0–102.0% vs. standard" are more specific assays, and may be CQAs for products from pilot plants and manufacturing sites.

Specifications are also mandatory critical quality standards for regulatory starting materials, key filed intermediates, and APIs. Applicants propose and justify specifications, and these must be acceptable for regulatory authorities to approve a drug candidate [28]. The ICH Q6A guidance states that "A specification is defined as a list of tests, references to analytical procedures, and appropriate acceptance criteria which are numerical limits, ranges, or other criteria for the tests described." The generally applicable criteria include description, identification, purity, and impurities. The description of a compound should characterize the physical state (solid or liquid) and color; any batch not matching the specification description will be examined closely. Identification should be sufficient to distinguish similar compounds, using analyses that are accurate and suitable for that phase of development. One assay may be suitable to determine both the "strength" (purity) and related impurities in a batch. Other impurities, such as the contents of water, residual solvents, and residual metals, will be determined by separate and specific assays. For new drug substances physicochemical data must be generated, such as melting point, pH in solution, particle size, and polymorph. Chiral drug substances must also be tested for chirality [29]. A sample specification list is shown in Figure 7.5.

> Two or more options may be available to calculate limits for impurities. The values provided in tables in guidances sometimes are for 10 g daily dosages. Higher limits for impurities on a weight–weight basis may be permissible if the dosage is less than 10 g/day, thus easing concerns for purification.

The suitability of assays can change during the development of a compound. For instance, identifying an intermediate by ^1H NMR may be convenient and suitable early in the development of an NCE, but inappropriate for routine QC analysis of that same intermediate during manufacturing of an API. As another example, GSK workers used derivatization/mass spectrometry methods to analyze for dimethyl sulfate in undisclosed starting materials during process development of pazopanib (Figure 7.6). These methods proved sensitive down to very low levels, and aided development efforts to purge dimethyl sulfate from pazopanib process streams. Since mass spectrometry methods were not preferred for routine use in the QC laboratories, a derivatization/HPLC

CMO LETTERHEAD

CERTIFICATE OF ANALYSIS Page 1 of n

(Spectra and chromatograms follow)

Name:
Batch (Lot) Number:
Date of Manufacture:
Date of Release:
Analytical Reference Number(s):

TEST	SPECIFICATIONS (TEST METHOD NO.)	RESULT/REFERENCE
Appearance	Report result (TM 1)	Off-white solid
Identification:		
A. 300 MHz ^{1}H NMR Spectrum (DMSO-d_6/TFA-d_1)	Conforms to reference spectrum (TM 2)	Conforms to reference spectrum
B. IR Spectrum (ATR)	Conforms to reference spectrum (TM 3)	Conforms to reference spectrum
Purity: HPLC (area %)	≥98.0% (TM 4)	99.9% (area %)
Assay: HPLC (weight %)	Report result (TM 5)	100.0% (area %)
Impurities: HPLC (area %)	(TM 6)	
Total impurities	≤2.0%	0.06%
Individual impurities	≤0.2%	Impurity RRT area %
		Unknown 1 1.08 0.06%
		Unknown 2 1.11 <0.05%
Karl Fischer Analysis	Report result, wt/wt% (TM 7)	0.26%
Residue on Ignition	Report result, wt/wt% (USP<281>)	<0.1%
Differential Scanning Calorimetry	Report result (TM 8)	Endotherm 143.2 °C
Thermogravimetric Analysis	Report result (TM 9)	No weight loss before degradation onset 229.5 °C
X-Ray Powder Diffraction	Report result (compare to the reference standard form, TM 10)	Crystalline; XRD pattern consistent with reference standard pattern
Elemental Analysis	Report result (% composition)	Calcd: C, 54.41; H, 6.09; N, 3.52; S, 0.0; P, 7.79
		Found: C, 54.10; H, 6.43; N, 3.24; S, 0.21; P, 7.50 (CHOP&NS Labs, Inc.)
Phosphate Content by IC	Report result	7.8% (CHOP&NS Labs, Inc.)
Residual Solvents	ICH Q3C Option 1 (TM 11)	
Isopropyl Alcohol	<5000 ppm	<5000 ppm (670 ppm)*
Isopropyl Acetate	<5000 ppm	<5000 ppm (ND, 500 ppm limit of detection)
Melting Point	Report result (USP<741>)	202–203 °C
Heavy Metals	(USP <232>, USP <233>)	
Ni	<2.5 ppm	ND (0.05 ppm limit of detection)
Cu	<25 ppm	ND (0.05 ppm limit of detection)
Pd	<1.0 ppm	ND (0.05 ppm limit of detection)
As	<0.15 ppm	ND (0.05 ppm limit of detection)
Cd	<0.05 ppm	0.04 ppm
Pb	<0.1 ppm	0.05 ppm
Hg	<0.15 ppm	ND (0.05 ppm limit of detection)
Particle Size Analysis	Report result (TM 12)	X_{10} = 1.5 μm; X_{50} = 6.2 μm; X_{90} = 31 μm
Clarity of Solution (1% in water)	Report result (TM 13)	Slightly yellow clear solution

The method is a limit-test method. Values in parentheses are approximate.

Storage/Special Handling: Store at room temperature.

(Signed, dated)
Head, Quality Control

FIGURE 7.5 Sample specifications.

approach was used in the QC laboratories to assay for quantifying residual dimethyl sulfate in pazopanib starting materials [30].

With more data available, specifications can be set with greater confidence [29]. A typical specification for future batches of an API or intermediate could be calculated from the mean value of data for all routine batches ± three times the relative standard deviation [31]; such calculations could be applied to assays, such as particle size distribution, where a range of values is expected. For intermediates, minimum purity levels/maximum impurity levels should be based on the limits that allow routine conversion to the next step without requiring either the intermediates or the products to be reworked. Reasonable purity for an intermediate is at least 94% [32], and a diastereomeric ratio of 97:3 was said to be excellent quality [33]. For each batch of NCEs and APIs destined for human usage, the quality must be not lower than the quality of the tox batch, i.e., impurities at levels not greater than those qualified in the tox batch, and any new impurities must conform to the reporting, identifying, and qualifying thresholds mandated by Q3A (Table 13.4) [34]. Often a minimum purity of 98% is posed for a drug substance. Specifications may also be posed to comply with limits required by regulatory authorities, as for a residual solvent or a residual transition metal.

> Color and appearance specifications are required for APIs and key intermediates, but are subjective. What is "white" or "off-white"? If color is a CQA, establish an assay to measure absorbance of a solution at a given wavelength at a given concentration in a specified solvent. Grinding a solid with a mortar and pestle can turn a solid whiter.

The EMA recently clarified expectations for limits on impurities, in line with the thrust of QbD. Impurities in the categories of genotoxins, heavy metals, and residual solvents are treated similarly. A theoretical impurity which has not been found in studies during the development of a compound does not need to be included in the specifications. If an impurity is formed in a step prior to the final step it does not need to be included in an API specification, as long as the level of this PGI is controlled by a suitable limit in the intermediate and studies, such as process tolerance studies, have shown that the level of

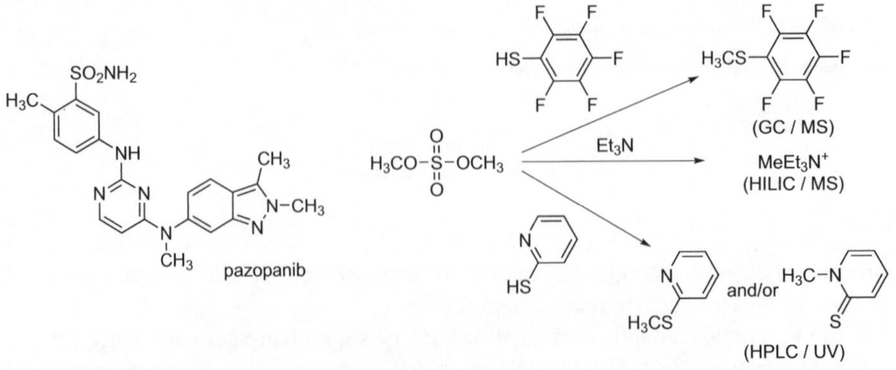

> Specifications should be based on data, not on goals that haven't yet been reached. "Interim" or "provisional" specifications may be acceptable if limited data are available, and tighter specifications can be established later. According to the ICH Q6A guideline, specifications may be tightened or loosened as more process understanding is demonstrated in manufacturing.

FIGURE 7.6 Derivatizing residual dimethyl sulfate for quantitation.

TABLE 7.4 PAT and Continuous Operations

Process Development Phase	Batch Operations	Off-line IPC	Continuous Operations	In-line IPC
Early—laboratory	X	X		
Large-scale laboratory	X	X	X	X
Pilot plant	X	X	X	X
Phases 2 and 3; manufacturing	X	X	X	X (PAT)

this impurity in the API is no higher than 30% of the TTC or other suitable limit. Otherwise this impurity must be included in the API specifications [35,36].

VI. PROCESS ANALYTICAL TECHNOLOGY

PAT operates through in-line or on-line analysis, without the need for samples to be withdrawn, manipulated or diluted. As continuous analysis, PAT marries well with continuous operations, QbD, and manufacturing of both APIs and drug products [37]; possible applications are shown in Table 7.4.

PAT has been strongly encouraged by the FDA and the ICH as a means to produce consistently high-quality products through continuous monitoring and control of critical operations. In this fashion quality is built into the processing (QbD), not tested into the products [2,5,38,39]. PAT and QbD also promote increased efficiency of operations. PAT has been reviewed [40]. PAT (on-line NIR and FTIR) has been used to monitor the SAFE additions of an alkyl chloride and an aryl bromide to magnesium metal during Grignard formation [41,42]. PAT seems to be used more to prepare the drug product than to prepare the drug substance [43], but the popularity of PAT is rising for manufacturing APIs.

Recently PAT has been described for monitoring and controlling crystallizations on scale. One of the key techniques is focused beam reflectance monitoring (FBRM^{TM}), which monitors the number, size, and shape of particles through chord length distribution. FBRM data can be gathered by probes inserted into crystallizers [44] or into external lines through which the crystal slurry circulates. FBRM can be used to monitor the loss of fines prior to filtration, which can markedly decrease the time needed for filtration. Other techniques employed include particle vision monitoring (PVM), which can be used to monitor the transition between oiling and crystallization, and attenuated total reflectance ultraviolet/visible (ATR/UV–vis), used to monitor API concentration, supersaturation, and hence crystallization. These techniques were used to reliably crystallize an API from EtOH/H_2O; previous attempts had been thwarted because the mixture tended to oil out [45]. In 2005 Pfizer researchers considered the power of PAT to understand processes as self-evident, and described usage of PAT for crystallizations on scale and redox reactions [23].

VII. SUMMARY AND PERSPECTIVE

Benefits of good in-process assays and IPCs include insights into kinetics of the formation of product and side products and being able to better develop processes to avoid scale-up problems. In-process assays and IPCs provide a basis for true comparison of processes, and are effective tools for troubleshooting. As windows into our operations, in-process assays and IPCs are very useful for process development and implementation on scale.

REFERENCES

1. Levi, P. *The Periodic Table;* Schocken Books: New York, 1984; pp 148–149.
2. ICH Q11, Development and Manufacture of Drug Substances (Chemical Entities and Biotechnological/ Biological Entities), May 2011. http://www.ich.org/fileadmin/Public_Web_Site/ICH_Products/Guidelines/ Quality/Q11/Step_2/Q11_Step_2.pdf
3. http://www.fda.gov/Drugs/DevelopmentApprovalProcess/HowDrugsareDevelopedandApproved/ ApprovalApplications/NewDrugApplicationNDA/default.htm
4. Guidance for Industry, Investigators, and Reviewers: Exploratory IND Studies: http://www.fda.gov/ downloads/Drugs/GuidanceComplianceRegulatoryInformation/Guidances/UCM078933.pdf
5. Guidance for Industry: PAT – A Framework for Innovative Pharmaceutical Development, Manufacturing, and Quality Assurance, http://www.fda.gov/downloads/Drugs/GuidanceComplianceRegulatoryInformation/ Guidances/ucm070305.pdf.
6. Mathew, J. S.; Klussmann, M.; Iwamura, H.; Valera, F.; Futran, A.; Emanuelsson, E. M. A. C.; Blackmond, D. G. *J. Org. Chem.* **2006,** *71*, 4711.
7. Massari, L.; Panelli, L.; Hughes, M.; Stazi, F.; Maton, W.; Westerduin, P.; Scaravelli, F.; Bacchi, S. *Org. Process Res. Dev.* **2010,** *14*, 1364.
8. Laird, T. *Org. Process Res. Dev.* **2011,** *15*, 305.
9. Wernerova, M.; Hudlicky, T. *Synlett* **2010,** *18*, 2701.
10. Gawley, R. E. *J. Org. Chem.* **2006,** *71*, 2411.
11. Anderson, G. W.; Zimmerman, J. E.; Callahan, F. M. *J. Am. Chem. Soc.* **1967,** *89*, 5012.
12. Anderson, G. W.; Callahan, F. M. *J. Am. Chem. Soc.* **1958,** *80*, 2902.
13. US Peroxide: http://www.h2o2.com/technical-library/analytical-methods/default.aspx?pid=70&name= Iodometric-Titration
14. Suffert, J. *J. Org. Chem.* **1989,** *54*, 509.
15. Furniss, B. S.; Hannaford, A. J.; Smith, P. W. G.; Tatchell, A. R. *Vogel's Textbook of Practical Organic Chemistry,* 5th ed.; Longman: Essex, 1989; pp 443–445.
16. Li, W.; Wayne, G. S.; Lallaman, J. E.; Chang, S.-J.; Wittenberger, S. J. *J. Org. Chem.* **2006,** *71*, 1725.
17. Anderson, N. G.; Ary, T. D.; Berg, J. L.; Bernot, P. J.; Chan, Y. Y.; Chen, C.-K.; Davies, M. L.; DiMarco, J. D.; Dennis, R. D.; Deshpande, R. P.; Do, H. D.; Droghini, R.; Early, W. A.; Gougoutas, J. Z.; Grosso, J. A.; Harris, J. C.; Haas, O. W.; Jass, P. A.; Kim, D. H.; Kodersha, G. A.; Kotnis, A. S.; LaJeunesse, J.; Lust, D. A.; Madding, G. D.; Modi, S. P.; Moniot, J. L.; Nguyen, A.; Palaniswamy, V.; Phillipson, D. W.; Simpson, J. H.; Thoraval, D.; Thurston, D. A.; Tse, K.; Polomski, R. E.; Wedding, D. L.; Winter, W. J. *Org. Process Res. Dev.* **1997,** *1*, 300.
18. Atkins, R. J.; Banks, A.; Bellingham, R. K.; Breen, G. F.; Carey, J. S.; Etridge, S. K.; Hayes, J. F.; Hussain, N.; Morgan, D. O.; Oxley, P.; Passey, S. C.; Walsgrove, T. C.; Wells, A. S. *Org. Process Res. Dev.* **2003,** *7*, 663.
19. Slade, J.; Parker, D.; Girgis, M.; Mueller, M.; Vivelo, J.; Liu, H.; Bajwa, J.; Chen, G.-P.; Carosi, J.; Lee, P.; Chaudhary, A.; Wambser, D.; Prasad, K.; Bracken, K.; Dean, K.; Boehnke, H.; Repic, O.; Blacklock, T. J. *Org. Process Res. Dev.* **2006,** *10*, 78.
20. Hileman, B. *Chem. Eng. News* **2006,** *84*(17), 10.

21. For detailed information on Fourier Transform Infrared Analyses and Equipment, see the Mettler-Toledo web pages: http://www.us.mt.com

22. Lee, G. M.; Eckert, J.; Gala, D.; Schwartz, M.; Renton, P.; Pergamen, E.; Whittington, M.; Schumacher, D.; Heimark, L.; Shipkova, P. *Org. Process Res. Dev.* **2001**, *5*, 622.

23. Sistare, F.; St. Pierre Berry, L.; Mojica, C. A. *Org. Process Res. Dev.* **2005**, *9*, 332.

24. Wang, F.; Wachter, J. A.; Antosz, F. J.; Berglund, K. A. *Org. Process Res. Dev.* **2000**, *4*, 391.

25. Feigl, F.; Anger, V. *Spot Tests in Organic Analysis*, 7th ed.; Elsevier, 1989.

26. http://www.emdchemicals.com

27. This was the basis for investigating TMT resins for removing Pd. Rosso, V. W.; Lust, D. A.; Bernot, P. J.; Grosso, J. A.; Modi, S. P.; Rusowicz, A.; Sedergran, T. C.; Simpson, J. H.; Srivastava, S. K.; Humora, M. J.; Anderson, N. G. *Org. Process Res. Dev.* **1997**, *1*, 311.

28. Argentine, M. D.; Owens, P. K.; Olsen, B. A. *Adv. Drug Del. Rev.* **2007**, *59*, 12.

29. ICH Q6A Specifications: Test Procedures and Acceptance Criteria for New Drug Substances and New Drug Products: Chemical Substances (issued May 2000), http://www.ema.europa.eu/docs/en_GB/document_library/Scientific_guideline/2009/09/WC500002823.pdf

30. Sun, M.; Liu, D. Q.; Kord, A. S. *Org. Process Res. Dev.* **2010**, *14*, 977.

31. This calculation acknowledges the reproducibility of the analytical method in use. For an example on the quality of a mock drug product, see http://www.fda.gov/downloads/Drugs/DevelopmentApprovalProcess/HowDrugsareDevelopedandApproved/ApprovalApplications/AbbreviatedNewDrugApplicationANDAGenerics/ucm120979.pdf

32. Belecki, K.; Berliner, M.; Bibart, R. T.; Meltz, C.; Ng, K.; Phillips, J.; Ripin, D. H. B.; Vetelino, M. *Org. Process Res. Dev.* **2007**, *11*, 754.

33. Federsel, H.-J. *Chirality* **2003**, *15*, S128.

34. http://www.fda.gov/RegulatoryInformation/Guidances/ucm127942.htm#i

35. http://www.ema.europa.eu/ema/index.jsp?curl=pages/regulation/q_and_a/q_and_a_detail_000071.jsp&murl=menus/regulations/regulations.jsp&mid=WC0b01ac058002c2af#section9

36. Robinson, D. *Org. Process Res. Dev.* **2011**, *15*, 325.

37. Mullin, R. *Chem. Eng. News* **2011**, *89*(30), 18.

38. Robinson, D. *Org. Process Res. Dev.* **2009**, *13*, 391.

39. Yu, L. X. *Pharm. Res.* **2008**, *25*, 781.

40. Byrn, S. R.; Liang, J. K.; Bates, S.; Newman, A. W. *J. PAT* **2006**, *3*, 14.

41. Wiss, J.; Länzlinger, M.; Wermuth, M. *Org. Process Res. Dev.* **2005**, *9*, 365.

42. Pesti, J.; Chen, C.-K.; Spangler, L.; DelMonte, A. J.; Benoit, S.; Berglund, D.; Bien, J.; Brodfuehrer, P.; Chan, Y.; Corbett, E.; Costello, C.; DeMena, P.; Discordia, R. P.; Doubleday, W.; Gao, Z.; Gingras, S.; Grosso, J.; Haas, O.; Kacsur, D.; Lai, C.; Leung, S.; Miller, M.; Muslehiddinoglu, J.; Nguyen, N.; Qiu, J.; Olzog, M.; Reiff, E.; Thoraval, D.; Totleben, M.; Vanyo, D.; Vemishetti, P.; Wasylak, J.; Wei, C. *Org. Process Res. Dev.* **2009**, *13*, 716.

43. Gade, S.; Jain, D. *Chem. Eng.* **2010**, *117*(9), 47.

44. Barrett, P.; Smith, B.; Worlitschek, J.; Bracken, V.; O'Sullivan, B.; O'Grady, D. *Org. Process Res. Dev.* **2005**, *9*, 348.

45. Deneau, E.; Steele, G. *Org. Process Res. Dev.* **2005**, *9*, 943.

Practical Considerations for Scale-Up

"A synthesis is not the same as a process…. Process chemistry…is not merely synthetic organic chemistry."

    – Trevor Laird, Editor, *Organic Process Research & Development* [1]

I. INTRODUCTION

The foundation for operations to make more material, whether in the kilo lab, on the pilot plant floor, or in manufacturing facilities, is thorough understanding of the process SAFETY issues (Chapter 2). Once the hazards of operating have been ascertained scale-up should be considered. For smooth operations we may take advantage of exotherms in reactions; for instance, we may appreciate that the H_2O byproduct from hydrogenating a nitroarene accelerates the reduction, and design conditions for faster reactions by adding H_2O to the hydrogenation [2]. If the SAFETY issues are too great for SAFE scale-up of a reaction such as a nitration through batch operations, we may opt for continuous operations.

Practical Process Research and Development. DOI: 10.1016/B978-0-12-386537-3.00008-3

The SAFETY considerations for laboratory investigations are extended for operations in the pilot plant and manufacturing facilities: whereas SAFETY glasses, gloves, laboratory coats and SAFETY shoes are routinely used in the laboratory, on scale-up these SAFETY considerations may be augmented with face shields, respirators or facemasks with supplied breathing air, sturdy gloves, heavy-duty protective aprons, and other protective personnel equipment (PPE). The necessary PPE often is cumbersome, limiting the field of vision and slowing or limiting routine operations such as physical transfers. For SAFETY the floor spaces in use may be roped off to discourage others from entering. If especially toxic materials must be used, access will be restricted to those involved in the operations. Charges for processes may be based using an entire container of a hazardous reagent such as KCN, thereby minimizing handling and storage issues [3]. All these limitations slow processing relative to what can be conveniently carried out on a gram-scale in the laboratory fume hood [4]. The ramifications of SAFE operations have been thoroughly discussed [5].

> For batch operations a 10-fold scale-up will require about twice as long to process.

Scale-up of batch processes takes longer than may be expected initially, due to the ability to remove heat from a reaction mass. Heat is removed from multipurpose cylindrical vessels by heat transfer fluids circulating through the external jacket of the vessel. The volume increases as a function of the radius squared ($V = \pi r^2 h$), while the surface area increases only as a function of the radius ($SA = 2\pi rh$, where r and h are the radius and height of the reactor). Thus as the radius of the tank increases, the volume increases faster than the surface area. As the volume increases 10-fold, the amount of time needed to remove the same amount of heat through the external jacket roughly doubles. (For an n-fold scale-up, the value is $n^{1/3}$. Thus for a 10-fold scale-up the value is 2.15.) This is essentially true for all steps requiring heat transfer, such as dose-controlled additions and removal of solvents by evaporation [6]. Some surprises may be found with extended processing on scale. For instance, Merck researchers found that upon extended evaporation of i-PrOAc and AcOH at an elevated temperature about 6% of the product ester was converted into the carboxylic acid (Figure 8.1). The ester may have been hydrolyzed to the carboxylic acid by H_2O partitioned into the organic extract from the aqueous washes [7], or by H_2O formed by the equilibrium between AcOH and acetic anhydride [8]. Trans-esterification of the ester with AcOH may also have produced the carboxylic acid. (Separating the carboxylic acid byproduct from the ester by extractions may not have been straightforward, since the ester is also moderately polar. For workup the mixture of acid and ester was treated with acid under conditions of Fischer esterification, with azeotropic removal of the byproduct H_2O [7]). Regardless of whether the carboxylic acid was generated by hydrolysis or transesterification, the extended operations on scale promoted the degradation of the product.

> Reaction conditions that synthetic organic chemists would consider unlikely to form useful amounts of material may generate significant amounts of impurities on scale. Extended operations provide time for byproducts to accumulate through slow equilibration. A change in solubility can drive the formation of a byproduct. Nature doesn't care about our opinions.

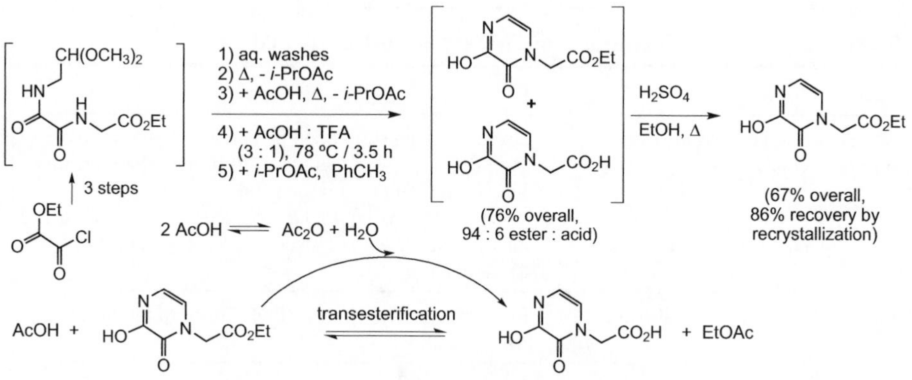

FIGURE 8.1 Extended evaporation of *i*-PrOAc and AcOH promoted degradation of an ester.

The basic considerations for practical scale-up in a pilot plant and manufacturing facility are shown in Table 8.1. Most of these considerations, which may seem obvious to a seasoned process chemist, are discussed in the following sections. A less experienced chemist may not become familiar with these concepts until a chemical engineer broaches these subjects just before a scale-up run. The excellent volumes by McConville [9], Sharratt [10], and Bonem [11] provide the perspective of chemical engineers. Near the end of this chapter is a discussion of tips for kilo lab operations.

TABLE 8.1 Basic Considerations for Efficient Scale-Up

Consideration	Comments
SAFETY is mandatory	Failure modes and effects analysis (FMEA) will be conducted before a process is approved for pilot plant introduction. N_2 atmosphere used to reduce risk of fires
Know how to control exothermic conditions	Dose-controlled additions are effective unless reaction is auto-catalytic Use exothermic additions to warm mixture to desired temperature
-40 to $+120\ °C$ generally preferred	Cryogenic temperatures can be achieved with Hastelloy reactors or continuous reactors; high-temperature processes can conducted safely with continuous reactors
Develop processes with tolerances readily attainable on scale	Suggestions: $\pm 5\%$ for solvents, $\pm 2\%$ for starting materials and reagents, $\pm 2\ °C$ for temperature ranges, or specify as required
Develop process conditions where all streams are mobile (non-viscous)	Physical aids, e.g., a long spatula, cannot be used to mobilize viscous streams in a stationary reactor with the manway closed
Minimize solvent charges	Excessive solvent charges can extend processing, decrease productivity, and increase cost of materials and waste disposal

(*Continued*)

TABLE 8.1 Basic Considerations for Efficient Scale-Up—cont'd

Consideration	Comments
Add liquids when possible over slurries or solids	Slurries may be suspended by agitation to ensure more even delivery to the reactor. With slurries charging all solids can be difficult if the solids are not readily suspended in the liquid. Devices are available for charging solids without exposing reactor contents to ambient air
Know stability of process streams	Extended operations on scale may lead to thermal decomposition
Use addition times of at least 20 minutes	Simulate addition times in the laboratory for best scale-up
Optimize V_{min} and V_{max}	Optimize productivity
Know solubilities of compounds	A difference of 20 °C can change the solubility of solids and liquids, or create emulsions. Small changes in pH can affect the partitioning of compounds during workup
Develop simple procedures	Repetitive steps, e.g., multiple extractions or treatments to absorb impurities and frequent pH adjustments, can extend processing and promote errors
Minimize number of IPCs needed	Optimize productivity
Consider physiochemical properties of reaction components	Operate under pressure or under reduced pressure as necessary. Anticipate fate of volatile reagents and byproducts
Develop efficient workups and direct isolation if possible	Transferring process streams risks physical losses and contamination

II. SAFETY: INERT ATMOSPHERES REDUCE RISKS

Operations are usually conducted under nitrogen for SAFETY, to eliminate the incursion of both O_2 and H_2O, and to dilute the concentrations of O_2 and reactive components such as hydrogen, butane and phosphine (PH_3). For instance, oxidations using or generating O_2 can lead to combustion of the solvent vapors above the reaction mixture, and the headspace of reactors may be swept with N_2 to dilute the O_2 content to less than 6–10% [12,13]. Virtually any release of H_2 has potential for ignition, as the flammability of H_2 in air is 4–75% [14]. A very low flow of N_2 over the surface of a reaction or a small over-pressure of N_2 in a sealed reactor may be referred to as a "nitrogen blanket."

SAFETY — A nitrogen cylinder became jet-propelled and broke concrete block walls when the valve separated from the tank [Reese, K. M. Chem. Eng. News 1984, 64 (47), 84]. Respect gas cylinders and secure them for SAFETY reasons.

There are several approaches to removed dissolved O_2. For hydrogenations reactors are subjected to a partial vacuum, and then filled with N_2; this process is usually repeated over 4–5 cycles until the calculated amount of O_2 is sufficiently low. Sparging with N_2 (or argon for laboratory use) while agitating may be effective enough and appropriate for some operations, especially if components decompose with heating. If removing O_2 is critical, the most effective process involves heating to reflux and cooling under N_2 [15,16]. Higher temperatures decrease the solubility of O_2 in aqueous alcohols [17,18] and other solvents. For critical operations O_2 may be monitored using a dissolved O_2 probe for reactions in water [19], or by headspace O_2 assay for reactions in organic solvents [20].

Off-gases can range from being annoying to being highly toxic, and so are scrubbed. Chemical engineers in the petroleum industry often use solutions that absorb gases for recovery and reuse, such as PEG dimethyl ether [21] or ethanolamine or methyldiethanolamine for the absorption of CO_2 and H_2S [22–24]. Chemical engineers in the pharmaceutical industry tend to use compounds that neutralize off-gases, and non-volatile components are generally preferred for scrubbing solutions. For instance, HCN and other toxic, volatile acids may be directed into aqueous basic solutions so that cyanide is contained in solution as non-volatile salts. (Treatment with NaOCl at pH 13–14 decomposes cyanide [25].) Volatile, neutral compounds, such as butane formed from reaction of n-BuLi, may be swept by N_2 flow to a thermal oxidizer and destroyed. Table 8.2 shows some scrubbing solutions that have been effective.

An atmosphere of N_2 over a reaction on scale is usually preferred over argon, as the latter is more expensive. Argon may be used in laboratory operations because it is denser than air, hence may keep air out of a reaction. (The densities of N_2 and Ar are about 0.87 and 1.38 times the density of air at room temperature [26].) Argon may also be preferred because N_2 can react with equipment and chemicals [27]; for instance, N_2 can be bound and split by complexes of vanadium [28] and molybdenum [29].

III. TEMPERATURE CONTROL

From laboratory investigations and hazard evaluations exothermic conditions can be identified, and the heat evolved can be used to reach the desired temperatures for processing. For instance, one might begin adding a reagent at room temperature so that the heat evolved will raise the reactor temperature to the desired range; this approach suffices as long as dose-controlled addition can be throttled and sufficient cooling is available to maintain desired temperature range for reaction. In the reaction shown in Figure 12.11 the solid thioester was added to 9.5 M NaOH at room temperature, raising the temperature of the resulting solution to 40–45 °C for the manufacture of captopril. In general the preferred temperature range for scale-up in multi-purpose reactors is −40 to +120 °C. For lower temperatures glass-lined reactors have been avoided, from fear that the glass lining will crack at cryogenic temperatures. Hastelloy reactors, no longer uncommon in pilot plant and manufacturing settings, are usually employed for reactions at cryogenic temperatures, sometimes with a pump to circulate dry ice/acetone in an additional vessel through the jacket of the Hastelloy reactor. Cryogenic temperatures can be reached by adding liquid N_2 to a reactor. Both cryogenic and high-temperature reactions may be pumped through

TABLE 8.2 Some Effective Scrubbing Treatments

Off-gas to be Scrubbed	Scrubbing Solution	Reference
Volatile bases, general	aq. H_2SO_4	
Volatile acids, general	aq. NaOH or aq. Na_2CO_3	
Isobutylene	MsOH in toluene	(1)
EtBr	aq. ethanolamine	(2)
MeBr	Na_2S with methyl tricaprylylammonium chloride; methanolic NH_3, then HCl	(3,4)
MeSH	$3NaOCl + NaOH$	(5)
BH_3 (diborane)	MeOH	(6,7)
Me_2S	Charcoal, NaOCl, Oxone®	(7,8)
HBr	AcOH	(9)
HCN	aq. K_2CO_3	(10)
H_2S	NaOCl	(11)
$C(S)Cl_2$, Cl_3CSH	NaOH	(12)
O_3	aq. $Na_2S_2O_5$	(13)

(1) Dias, E. L.; Hettenbach, K. M.; am Ende, D. J. Org. Process Res. Dev. **2005,** 9, 39. This approach was preferred over trapping isobutylene with an alcohol.

(2) Hettenbach, K.; am Ende, D. J.; Leeman, K.; Dias, E.; Kasthurikrishnan, N.; Brenek, S. J.; Ahlijanian, P. Org. Process Res. Dev. **2002,** 6, 407.

(3) Bielski, R.; Joyce, P. J. Org. Process Res. Dev. **2008,** 12, 781.

(4) Giles, M. E.; Thomson, C.; Eyley, S. C.; Cole, A. J.; Goodwin, C. J.; Hurved, P. A.; Morlin, A. J. G.; Tornos, J.; Atkinson, S.; Just, C.; Dean, J. C.; Singleton, J. T.; Longton, A. J.; Woodland, I.; Teasdale, A.; Gregertsen, B.; Else, H.; Athwal, M. S.; Tatterton, S.; Knott, J. M.; Thompson, N.; Smith, S. J. Org. Process Res. Dev. **2004,** 8, 628.

(5) Remy, B.; Brueggemeier, S.; Marchut, A.; Lyngberg, O.; Lin, D.; Hobson, L. Org. Process Res. Dev. **2008,** 12, pp 381.

(6) SAFETY: Quenching with MeOH and other protic compounds produces H_2.

(7) Atkins, W. J., Jr.; Burkhardt, E. R.; Matos, K. Org. Process Res. Dev. **2006,** 10, 1292. The authors provide detailed guidelines for quenching diborane and alkyl boranes.

(8) Parker, J. S.; Bowden, S. A.; Firkin, C. R.; Moseley, J. D.; Murray, P. M.; Welham, M. J.; Wisedale, R.; Young, M. J.; Moss, W. O. Org. Process Res. Dev. **2003,** 7, 67.

(9) Barton, B.; Hlohloza, N. S.; McInnes, S. M.; Zeelie, B. Org. Process Res. Dev. **2003,** 7, 571. Br_2 was regenerated by treating with air.

(10) Patel, I.; Smith, N. A.; Tyler, S. N. G. Org. Process Res. Dev. **2009,** 13, 49.

(11) Hull, J. W., Jr.; Romer, D. R.; Adaway, T. J.; Podhorez, D. E. Org. Process Res. Dev. **2009,** 13, 1125.

(12) Grayson, J. I. Org. Process Res. Dev. **1997,** 1, 240.

(13) Cabaj, J. E.; Kairys, D.; Benson, T. R. Org. Process Res. Dev. **2007,** 11, 378.

continuous reactors for efficient control of temperatures. Productive processes are designed for temperature ranges attainable on scale through minimal monitoring and adjusting. A convenient temperature range is ±2 °C. Operating at the reflux temperature of a mixture can be very convenient. A temperature range may be broader for some operations than for others. For instance, a temperature range of 5 °C may be specified for extractions, but the temperature range for seeding a crystallization may be optimal in a range of 2 °C.

> The reflux temperature of a mixture may change as components of the reaction are consumed or generated [30]. The temperature of a hot mixture in a reactor may rise when heating from a light reflux to a heavy reflux, as evidenced by condensate rapidly returning from the riser to the reactor. Hence "heating at reflux" may not be specific.

IV. HETEROGENEOUS PROCESSING AND CONSIDERATIONS FOR AGITATION

Viscous fluids or suspensions can entrap starting materials and invariably cause scale-up difficulties – rarely can a long spatula be used in large-scale equipment! In stationary equipment it is important to develop process conditions where all streams are mobile.

Efficient mixing can be critical in heterogeneous and (pseudo)homogeneous systems. Intuitively good mixing is necessary for reproducible, effective processing, and greater consideration is given to agitation of heterogeneous mixtures than to solutions. Gases should be introduced below the agitator, through a device with holes small enough to form small bubbles, thus increasing the surface area for more efficient reaction, but the holes must be large enough that they will not be plugged by any solids in the reaction mixture. Maintaining a positive pressure through the sparging device will help prevent plugging. The agitator should give good pumping action, i.e., mixing perpendicular to the plane of rotation of the impeller [31]. Solids with bulk densities much greater than that of the solvent may be difficult to suspend above the impeller, and substituting less dense reagents may be helpful (Table 8.3).

One difficulty with heterogeneous, solid–liquid processes is exemplified in Figure 8.2 with the Michael addition of cyanide, which was generated from acetone cyanohydrin under basic conditions. Although the Na_2CO_3-mediated reaction worked readily in the laboratory, on scale this reagent stayed in the vortex of the stirred reaction mass and the reaction was unacceptably slow. Using the acetone-soluble base tetramethylammonium hydroxide led to a faster reaction [32]. (Michael addition to this electron-rich enone may not be a rapid reaction. Under these PTC conditions the low solvation of the cyanide increases the reaction rate.) SAFETY: These basic conditions

TABLE 8.3 Densities of Some Inorganic Reagents

Reagent	Density (g/mL)	Reagent	Density (g/mL)	Reagent	Density (g/mL)
$NaHCO_3$	2.16	$KHCO_3$	2.17	$NaI \cdot 2H_2O$	2.45
Na_2CO_3	2.53	K_2CO_3	2.43	NaI	3.67

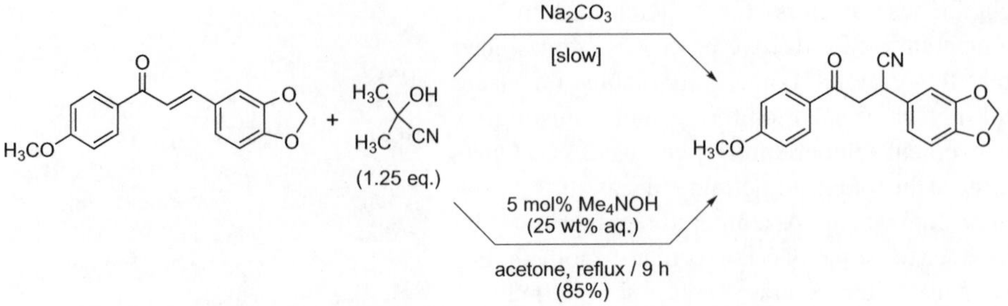

FIGURE 8.2 Effective scale-up using a homogeneous base.

discourage the formation of HCN, but a scrubber system should be in place to preclude any incidents from formation and escape of HCN from the reactor.

Reproducible mixing of liquid–liquid heterogeneous reactions in the laboratory, necessary to develop optimal procedures, can be a challenge. In Figure 8.3, agitation, addition rate, and heating rates were important in this sequence that comprised five one-pot reactions and one telescoped reaction. The authors noted that 1.2 equivalents of disodium sulfide (~ 0.5 M) was optimal, and in less than 2 min the solution of Na_2S was added to the active ester in CH_2Cl_2 at -18 °C, with vigorous agitation. Then the biphasic mixture was warmed to 25 °C over 5 min and stirred at that temperature for 2 h to effect formation of the thiolactone. After aqueous washes the thiolactone, an active ester, was opened with m-aminobenzoic acid in AcOH to give the amide. The diisopropylphosphoryl protecting group was readily hydrolyzed by aqueous HCl, affording the hydrochloride salt in an excellent isolated yield of 73%. This process was used on multi-kilogram scale to make the ertapenem sodium precursor [33]. Rapidly optimizing a process such as this in the laboratory would probably entail using the same glassware and stirrer for each run, to reproducibly mix the solutions with very different densities (water and CH_2Cl_2). Using a tachometer to gauge the speed of the agitator shaft might be helpful. Although controlling the agitation and temperature in the kilo lab might be easy enough, on scale having a 2-min addition time while controlling the temperature might prove challenging. For scale-up continuous processing could control the reaction temperature during rapid addition while ensuring efficient mixing of the two very different solutions.

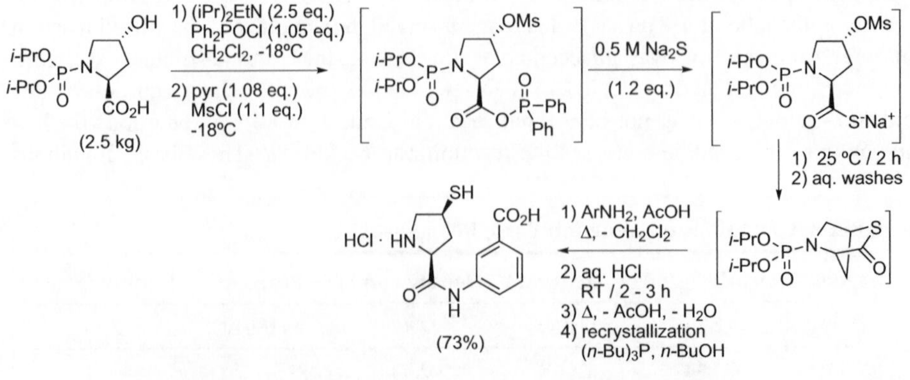

FIGURE 8.3 Efficient mixing needed for liquid–liquid processing.

V. ADDITIONS AND MIXING CONSIDERATIONS

On scale solids are charged by weight, and liquids may be charged by weight or volume. In pilot plants and manufacturing facilities rarely are solids and liquids directly weighed in vessels, although occasionally a reactor may be mounted on a load cell. A quantity of solids may be measured by weighing a container, such as a plastic bag, filling the container to the desired weight, and transferring the solids into a vessel. A small volume of solvent may be used to rinse out the container to complete the transfer. Usually on a large scale solids (and often liquids) are measured by difference, i.e., by taring a drum containing the solids and then transferring some solids into a vessel until the negative balance registered on the scale is the desired charge. The amount of liquids added may also be measured by a flow meter, a mass flow meter, by charging a liquid up to a desired volume in a calibrated vessel, or by difference from a calibrated container. In general processes should be designed to tolerate measured volumes of $\pm 5\%$ and measured weights of $\pm 2\%$. Greater accuracy of measuring small weights may be attainable by using a smaller scale.

> If solvents will be charged by weight on scale, charge solvents by weight in the laboratory. This approach may facilitate your thinking for process implementation. Note that the density of liquids will change slightly with temperature.

Slow distribution of components can produce large amounts of troublesome impurities on scale. The small-scale equipment used for screening process conditions, such as mixtures in 5 mL vials agitated by 8 mm × 1.5 mm diameter magnetic stirbars, may routinely provide reproducible conversions; however, any special benefits of these conditions may not be uncovered until larger runs are undertaken. For instance, the action of a magnetically driven stirbar may grind solids underneath it to smaller particles that encourage reaction through increased surface area. Spinning stirbars at high rpm may increase turbulence through a tumbling motion [34]. (The latter manuscript details a mixing test developed by Bourne, monitoring the amount of MeOH generated from mixing a stream of 2,2-dimethoxypropane and aqueous NaOH with HCl.) Iron leaching from magnetic stirbars has catalyzed reactions [35]. The benefits or problems of agitation in small-scale reactors are often uncovered by conducting reactions in larger laboratory equipment using mechanical stirrers.

Problems with micromixing (see below) can sometimes be anticipated by rapid additions and inverse additions. When the carboxylic acid in Figure 8.4 was added rapidly to Br_2 in aqueous NaOH, 22% of the bis-bromo impurity was formed, and the isolated yield of the desired monobrominated carboxylic acid was only 45%. To avoid handling Br_2 a mixture of NaBr and NaOCl was substituted. These brominating reagents were added to the starting material to decrease the level of the dibrominated impurity, but when the mixture of oxidants was added in less than 1 min the level of the dibrominated

> In the process development laboratory an overhead mechanical stirrer can be used to mimic agitation on scale. A crescent-shaped Teflon blade with length about half the diameter of the reaction flask should be raised slightly above the bottom of the flask. Agitation rates (N) of 60–200 rpm are usually sufficient. Magnetic stirbars are suitable for screening or recrystallizing a reference standard.

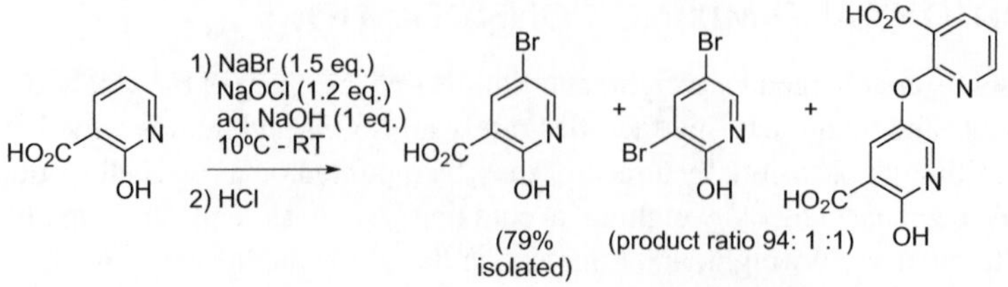

FIGURE 8.4 Optimal conditions for monobromination.

impurity rose. Under optimized conditions on 22 L scale NaBr and NaOCl were added over 14 min to a solution of the carboxylate [36]. The extended addition avoided side products formed from micromixing.

Short addition times should be examined to identify any areas of concern for scale-up or possible optimization. Longer addition times, e.g., 2 h vs. 20 min, can also help anticipate any byproducts that would form on scale; examining the impurity profile throughout the addition can be informative. In the laboratory if yields or impurity levels are dramatically different with high agitation or low agitation, or with short additions vs. "normal" additions, micromixing may be at play. At that point it may be useful to consider an inverse addition, or change to less reactive reagents. Continuous processing with a static mixer may be the optimal approach to control impurities from micromixing.

Even reactions that seem homogeneous may not be homogeneous on a micro-scopic scale. For instance, laminar mixing, which occurs at the shell of a reactor or tube, is not turbulent. Similarly the dropwise addition of a reagent to the surface of a reaction mass produces a localized high concentration, and such micromixing can lead to impurity formation if rapid reactions occur. With fast reactions the transient pH may differ from the measured pH [37]. Three types of mixing are shown in Table 8.4 [38].

Mesomixing of reagents is often a concern when adjusting the pH of a mixture during workup on scale. Rapid basification of a mixture of ethyl chloroacetate and HCl in water led to increased hydrolysis of the ester (Figure 8.5) [39]. When neutralizing a pH 2

TABLE 8.4 Characteristics of Micromixing, Mesomixing, and Macromixing

Mixing Type	Time Span	Comments
Micromixing	milliseconds	Independent of equipment size Depends on size of smallest eddies, mm Impurities may form due to transient localized high concentrations
Mesomixing	seconds	Depends on addition speed and pipe diameter May produce impurities during rapid pH adjustment
Macromixing	minutes	Depends on equipment size Similar to the calculated concentration overall

stream with 7 M NaOH excessive hydrolysis of the formate ester was seen with slow agitation. To minimize hydrolysis of the formate ester the aqueous NaOH was added to the interface between the two phases of the mixture (Figure 8.6) [40]. The position within a reactor for optimal turbulence and distribution of an addition stream is usually subsurface and between the tip of the impeller and the wall of the vessel, not in the vortex of a stirred mixture; such "sweet spots" can be determined for each reactor. (In passing the Merck authors noted that neutralization with aqueous $NaHCO_3$ produced less hydrolysis.)

Another example of what was probably mesomixing is illustrated in Figure 8.7. In the initial pilot plant batch the solution of n-BuLi was added above the reaction mixture, producing the product in 38% yield. By introducing the solution of n-BuLi subsurface a yield of 70% was obtained. The improved yield was probably due to better dispersion and mixing, avoiding the generation of localized "hot spots" that lead to side reactions [41].

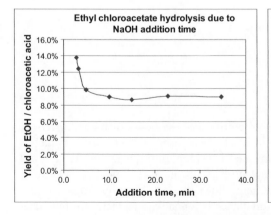

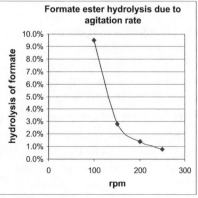

FIGURE 8.5   Hydrolysis due to mesomixing.

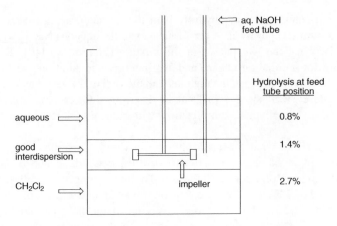

FIGURE 8.6 Optimal addition point for neutralizing acid.

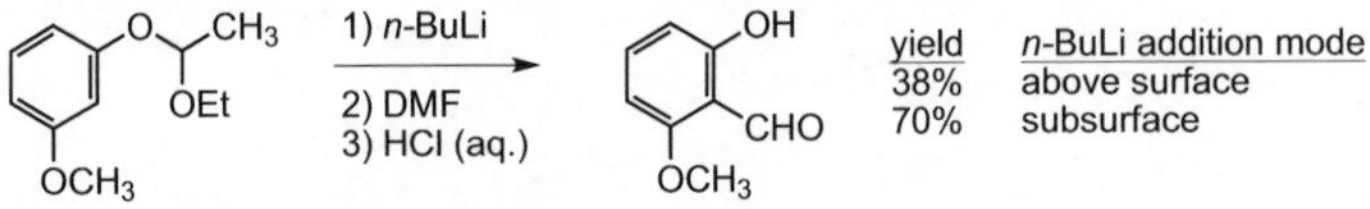

FIGURE 8.7 Improved yield due to subsurface addition of *n*-BuLi.

Turbulent mixing is usually needed for rapid, reliable processes. In homogeneous systems turbulence is often measured by the Reynolds number, *Re*. Turbulent mixing occurs at *Re* = 1000 for a stirred tank (Equation 8.1) or *Re* = 3500 for a plug flow reactor (Equation 8.2). Atherton and Carpenter have written an excellent book on such physicochemical concepts [42].

$$Re = [(ND)(D)](\rho/\mu) = (ND^2)(\rho/\mu) \tag{8.1}$$

$$Re = (du)(\rho/\mu) \tag{8.2}$$

where

Re = Reynolds number;
N = agitation rate (rotational speed);
D = impeller diameter;
d = tube diameter;
u = mean velocity;
ρ = bulk density; and
μ = dynamic viscosity.

In both Equations 8.1 and 8.2 *Re* is proportional to the bulk density and inversely proportional to the viscosity. For agitated vessels *Re* is based on the tip speed (which is proportional to *ND*) and the impeller diameter, hence proportional to D^2; in a large vessel a slow agitation speed can produce high turbulence. In a plug flow reactor *Re* is proportional to the mean velocity of the liquid flow, which is inversely proportional to the residence time, τ; hence increased turbulence is found with an increased flow rate, which translates to a decreased τ.

Micromixing issues should be identified early in laboratory development, in the initial stages where learning characteristics of a reaction are more important than obtaining a high yield. The system should be stressed by adding a reagent rapidly, slowly, in portions, or all at once; with high agitation or very slow agitation. These experiments can be run on a small scale and discarded after monitoring.

If side reactions due to micromixing are noted, many approaches are feasible (see below). Continuous operations, which may require some time initially to set up equipment and find optimal concentrations, temperatures, and flow rates, may prove the most rugged over time.

1. Increase agitation.
2. Extend addition time of reagent.
3. Dilute reagent solutions.
4. Moderate reactivity by cooling, or pH control.
5. Employ an inverse addition.
6. Resort to a less reactive reagent, e.g., Ac₂O instead of AcCl.
7. Use continuous operations.

Anticipate mixing issues early in laboratory development by adding a reagent slowly, rapidly, in portions, with high agitation or with very poor agitation. Yield is not important in early stages of investigation.

For batch processes typically addition time on scale range from 20 min to several hours, and longer addition times pose no problems – engineers can select from a wide assortment of pumps. The extended addition times are due to slower removal of heat from large vessels. If an addition is not especially exothermic then rapid additions may be possible, especially if the reagent is added to a cooled mixture and the reaction temperature is allowed to rise to the desired range. As shown in Figure 8.8, 30 kg of $POCl_3$ was added over 4 min, with the temperature rising to only 4 °C, and the aryl chloride was generated. This subsequent addition of the amine to form the drug substance was too

When confronted with mixing difficulties consult your engineer to determine what options are available for scale-up. Chemical engineers have additional tools to ensure good turbulence and mixing [43,44], e.g., specialized turbine agitators, static mixers, reciprocating Carr columns, BUSS hydrogenators, and centrifugal (CINC®) liquid–liquid separators. Timing is always an issue, and changing out an agitator in a vessel may not be feasible to meet a deadline.

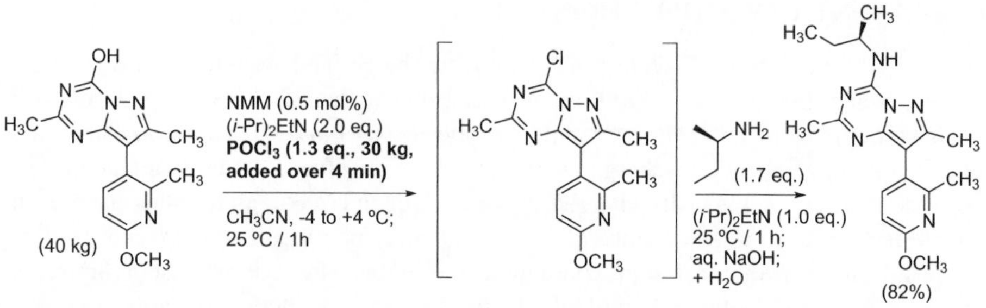

FIGURE 8.8 Rapid addition of phosphorus oxychloride on scale.

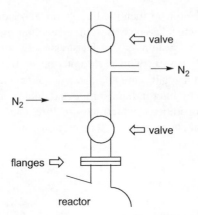

FIGURE 8.9   Schematic for charging solids to a reactor.

exothermic for rapid addition; certainly rapid additions on scale must be carried out with knowledge of how much heat is generated and whether the reaction conditions can control the temperatures at the desired range [45].

In general charging liquids and solutions is easier and preferred over charging solids. One concern with charging solids is that not all of the solids may be transferred to the reactor. Charging of liquids is usually completed by chasing residual liquids with a small charge of solvent; for instance, a solution may be prepared using 95% of the volume specified in the laboratory process description, with 5% of the latter solvent volume being used to complete the transfer.

More specialized equipment is used for charging solids on scale; such equipment may be known as alpha-beta valves, butterfly valves, shot valves, star valves, or other names. The concept for an apparatus used to charge lithium shot to a reactor is sketched in Figure 8.9 [46]. Glove boxes have also been erected on top of reactors for charging moisture-sensitive or oxygen-sensitive materials.

Specialized equipment is used to charge gases on scale, with special considerations to minimize any risk of releasing gases. To protect personnel and equipment from leaks of hazardous material such as H_2 and F_2, the lines used to charge such gases may be encased inside sealed lines or pipes [47]. Charging 37% aqueous HCl is more convenient using than charging gaseous HCl from cylinders.

VI. SOLVENT CONSIDERATIONS

Preventing the buildup and discharge of electrostatic charges underlies operations on scale. Causes of static electricity must be eliminated according to the ATEX 137 regulations [48], which were applied to the EU in 2006. Adding solids from a plastic bag or through a plastic chute can promote a static discharge, which could ignite solvent vapors in a reactor. For this reason solvents are usually charged under a nitrogen blanket after solids have been charged to a reactor. In the absence of O_2 reactors may be pitted and damaged. Hydrocarbon solvents, especially straight-chain molecules with an even number of carbons, are especially prone to electrostatic buildup. To increase the conductivity of solvents, anti-static agents such as Statsafe™ 5000 may be added to hydrocarbons, to increase the conductivity of the solvent [49] to at least 2000 pS/m [50]. Adding as little as 1–5 ppm of an anti-static agent may be as effective as adding 20 vol% of an oxygenated solvent such as MEK [51], and the benefits of introducing such an agent may outweigh the disadvantages. The potential for damage from electrostatic discharge must be respected, and cannot be thoroughly discussed here.

Prior to a scale-up run conduct a use-test with inputs and sequence of operations as planned, to antici-pate and avoid unsatisfactory scale-ups. "Obvious" conditions may be unfounded assumptions; for instance, it is useful to confirm that a suspension can be readily stirred if solvent is added on top of the solids in a vessel.

In every operation the amount of solvent charged should be minimized, as excessive use of solvents increases the costs of raw materials, waste disposal, and processing time (hence reducing productivity). Reaction concentrations of 0.3–0.4 M are good for initial laboratory studies, and investigations will show whether concentrations may be effectively increased. More dilute concentrations may be needed to minimize dimerization and relatively slow intermolecular reactions, e.g., β-lactam cyclization. (Characteristics of ring closure reactions have been reviewed [52].) More concentrated conditions will be helpful for one-pot reactions [53]. Mixtures of low viscosity are required for good mass transfer and heat transfer.

If more than one solvent is required, using as few solvents as possible simplifies solvent recovery operations. When changing from one solvent to another, solvent swaps will be faster if the lower-boiling solvent can be chased by a higher-boiling solvent, and if an azeotrope (e.g., iPrOH–iPrOAc or iPrOH–iPrOAc–H_2O) is involved. If two solvents are to be used, select them to minimize potential impurities. For example, to minimize side products from transesterification, extract a reaction mixture into EtOAc and crystallize from EtOH/H_2O, or extract into iPrOAc and crystallize from iPrOH/H_2O.

The concepts V_{min} and V_{max} can be used in several fashions: either the minimum agitation volume of a vessel, and maximum volume a vessel can contain; or, the minimum and maximum charges of solvent:substrate (L/kg) within a process, or within a reactor. In the context of optimizing solvent usage V_{max} usually occurs after quenching or during extractions. If the process V_{max} is much greater than V_{min}, sizing kettles may be difficult. To allow scale-up within an equipment train, a ratio of $V_{max}:V_{min} \leq 5:1$ within a given vessel may be helpful. For example, suppose a process with a $V_{max}:V_{min}$ ratio of 5:1 is to be implemented on scale to make clinical supplies. In reactor R2D2 the process could be conducted at the V_{min} of 10 vol%, the minimum agitation volume of the vessel, and subsequent operations would raise the volume to 50% of the vessel capacity. In principle the reaction could be run on twice the initial scale and the reactor could accommodate both V_{max} and V_{min}.

In general all solvent charges should be minimized, in order to minimize the V_{min}. Some key considerations are the minimum volumes needed to ensure complete dissolution of solids, good heat and mass transfer, minimal side reactions, ready phase splits, efficient extractions, and reliable crystallizations. Not always will these minimum volumes be sufficient; for instance, a product may crystallize more readily if impurities that interfered with crystallization in previous batches were removed. It is always a good idea to hedge by charging a small excess of solvent as insurance.

Software is available to predict optimal solvent composition and to potentially shorten development efforts. BMS researchers found that a mixture of *i*-BuOH, toluene, and water was desired to carry out a hydrogenolysis, extractive workup, and crystallization. Toluene was desired for the crystallization, but the solubility of the starting material was low in $PhCH_3$ and the reaction rate was slow. A solution of aqueous Na_2CO_3 was employed during the reaction to remove the HBr formed by hydrogenolysis. By adding a more-polar co-solvent the solubility of the starting material was increased, but the volume of the aqueous phase and the proportion of *i*-BuOH needed to be minimized to minimize loss of product to the aqueous phase. The ratio of *i*-BuOH/$PhCH_3$ needed to be low to ensure that azeotropic distillation would leave predominantly $PhCH_3$ in the crystallizer. Conditions and solvent loadings of this multicomponent system as determined by the model were consistent with optimal operating conditions [54].

> When assessing a route to an NCE, to ensure timely preparation of materials identify
> (1) any known SAFETY issues with reagents and starting materials;
> (2) the steps with the lowest yield and hence processing problems; and
> (3) the steps with the highest process V_{max} and lowest productivity.

VII. SIMPLE PROCEDURES

In general simple procedures are preferred, with the fewest repeated steps and fewest in-process controls (IPCs). For instance, gradually delivering aqueous HCl by pump over 8 h will be likely to give more reproducible results in a campaign of batches than a process in which the pH is monitored and adjusted by hand every 15 min over that same interval. This gradual acidification is useful to crystallize a product through pH adjustment [55]. Simple procedures are also desired in workups, an aspect of processing overlooked by many inexperienced people; for instance, if more than one treatment with TMT is needed to remove palladium to acceptable levels, perhaps another treatment to remove Pd should be examined. Of course the risk with simple processes is that they may be over-simplified or simplistic, and not prove to be rugged enough for routine operations.

The differences in boiling points of solvents and azeotropes are exploited in developing practical workups. When solvent displacements are required, as in changing from the solvent used for extraction into another solvent for crystallization, it is easier to chase a lower-boiling solvent with a higher-boiling solvent than to do the reverse. For instance, if a product were to be crystallized from EtOH (bp 78 °C) less time (and energy) would be required in removing MTBE (bp 55 °C) than in removing toluene (bp 111 °C); for this reason extractions using low-boiling solvents are generally preferred. Usually the rich extract is concentrated to a minimal stirrable volume, the lower-boiling solvent is added, and concentration is continued to the target volume. It may be required to concentrate again to the minimal stirrable volume and add another portion of the lower-boiling solvent. For example, in the solvent chasing shown in Figure 8.10a the rich organic extract of the ketone contained CH_3CN (bp 81 °C), and MTBE (bp 55 °C). The rich extract was concentrated at 50 °C from ~200 L to ~25 L, and 50 L of heptane (bp 98 °C) was added.

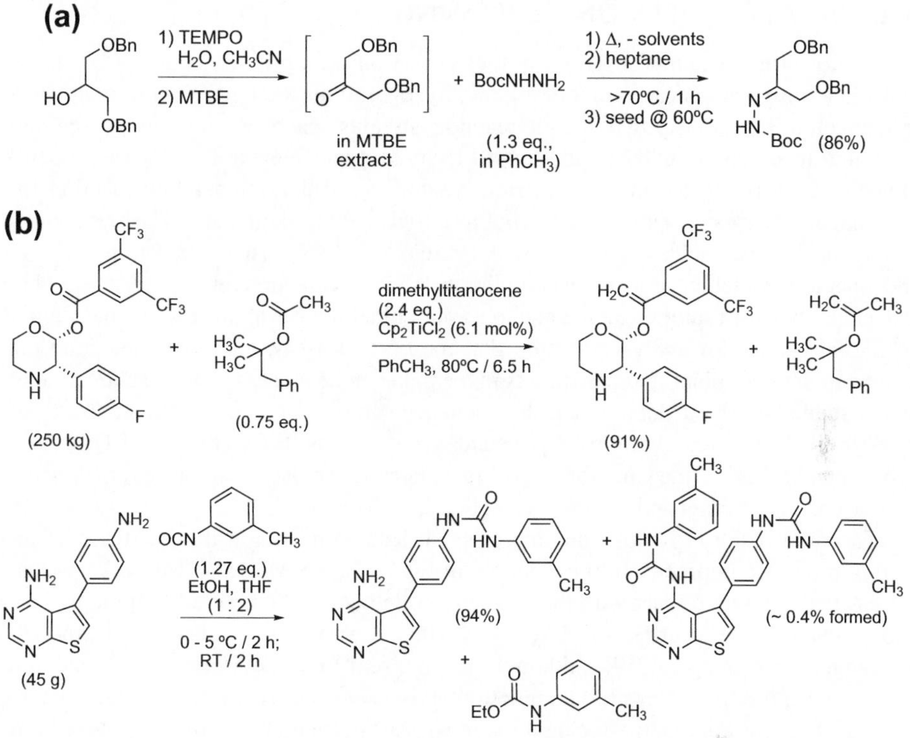

(a)

(b)

FIGURE 8.10 (a) Solvent chasing. (b) Using secondary substrates to minimize product decomposition.

Subsequent concentration probably removed first MTBE and then CH$_3$CN. After concentrating to a minimum stirrable volume heptanes was added, and the latter two steps were repeated twice before preparing the Boc hydrazone [56]. Chasing solvents by distilling at a constant volume has been shown to require less solvent, thus potentially reducing both solvent waste and efforts for recovery and recycling of solvents. After the initial concentration to remove solvent, the chasing solvent is added simultaneously as solvents are removed by distillation. In this fashion Abbott researchers reduced the volume of iPrOH needed to manufacture erythromycin A oxime by 30% [57]. If the amount of residual solvents was critical, a GC method would probably be established. When a higher-boiling solvent is chased with a lower-boiling solvent, the higher-boiling solvent is removed either by entrainment, or as an azeotrope with the lower-boiling solvent (see Section II of Chapter 5). More than one azeotropic distillation could be used. For instance, chasing AcOH with EtOH might take some time because AcOH and EtOH do not form an azeotrope. One could chase AcOH with toluene (azeotrope bp 101 °C, 32% AcOH) and then chase toluene with EtOH (azeotrope bp 77 °C, 32% PhCH$_3$); however, if one could run the desired reaction in EtOH processing could be quicker.

> TIME IS MONEY. Know goals of scale-up run. Is a pilot plant run a large-scale experiment, or is material being prepared for formulation studies? Minimize the number of shifts/batch to maximize productivity in manufacturing.

VIII. IMPACT OF IPCS ON PROCESSING

Since protecting personnel from hazardous conditions is a primary concern for operations in a pilot plant and manufacturing facility, operations are closely monitored when personnel might be exposed to hot reaction streams, such as splitting phases and sampling reactions. Usually a facility will have an upper temperature for such operations, often between 50 and 70 °C. Hence sampling a high-temperature reaction for external analysis can require time to cool to a SAFE temperature, and if the reaction is not complete then additional time will be required to return the reaction mass to a temperature suitable for reaction completion. A significant amount of time, e.g., 2–4 h, can elapse between sampling a reaction on a pilot plant floor, walking the sample aliquot to the laboratory for analysis, sample preparation and assay, and communicating the results to the pilot plant floor. (Artifacts in the data can be created if the reaction aliquot is not stable before quenching or if the quenched sample decomposes.) In contrast, in-line (in-line) analyses give virtually instantaneous results. IPCs are part of QC costs, which can be 30% of the CoG for an API [58]. For these reasons sampling for IPCs and use of IPCs are minimized.

The time required for off-line analysis can lead to decomposition of the product within process equipment. In the example in Figure 8.10b Merck scientists knew that excess reagent would react with the product enol ether, and to preclude degradation of the product while waiting for IPC assays they charged a secondary substrate to scavenge excess reagent [59]. Although on-line, real-time monitoring would shorten the time required to determine complete reaction, quenching on scale is not instantaneous, and this approach of using a secondary substrate allowed for better control of processing on scale. Abbott researchers similarly searched for a secondary substrate to scavenge excess isocyanate and thus minimize bis-acylation of the anilinopyrimidine. EtOH worked well as a competing nucleophile, with the ethyl urethane byproduct being solubilized in the mother liquor [60]. Scavenging substrates can be convenient approaches to minimizing impurities by exploiting different levels of reactivity.

IX. CONSIDER THE VOLATILITY OF REACTION COMPONENTS, AND USE OF PRESSURE

To effectively process operations that use volatile solvents, volatile reagents or generate volatile byproducts, controlling where these compounds reside is important. As needed, vessels can be sealed to contain volatile components. To remove volatile components reaction contents may be sparged or refluxed [61], or a vessel may be operated under reduced pressure. For instance, atorvastatin calcium trihydrate has been crystallized at 47–57 °C in a crystallizer sealed to contain MTBE (bp 55 °C, MTBE–H_2O azeotrope bp 53 °C). MTBE was charged to the aqueous hydrolysate to ensure saturation and routinely generate larger crystals, and by operating near the boiling point of MTBE in a sealed vessel more of the solvent resided in the liquid [62]. Gaseous reagents such as H_2 are contained by operating under pressure, in specialized equipment due to SAFETY concerns. The ammonolysis of an epoxide with aqueous NH_3 in MeOH at 60–65 °C was conducted in a pilot plant in a sealed reactor in order to contain the ammonia; the pressure resulting was less than 15 psig, well below the release pressure for the rupture disc, and the reaction was accelerated due to negligible loss of NH_3 [63]. BMS researchers found that for effective silylation of an aminoalcohol with less than one equivalent of hexamethyldisilazane (HMDS)

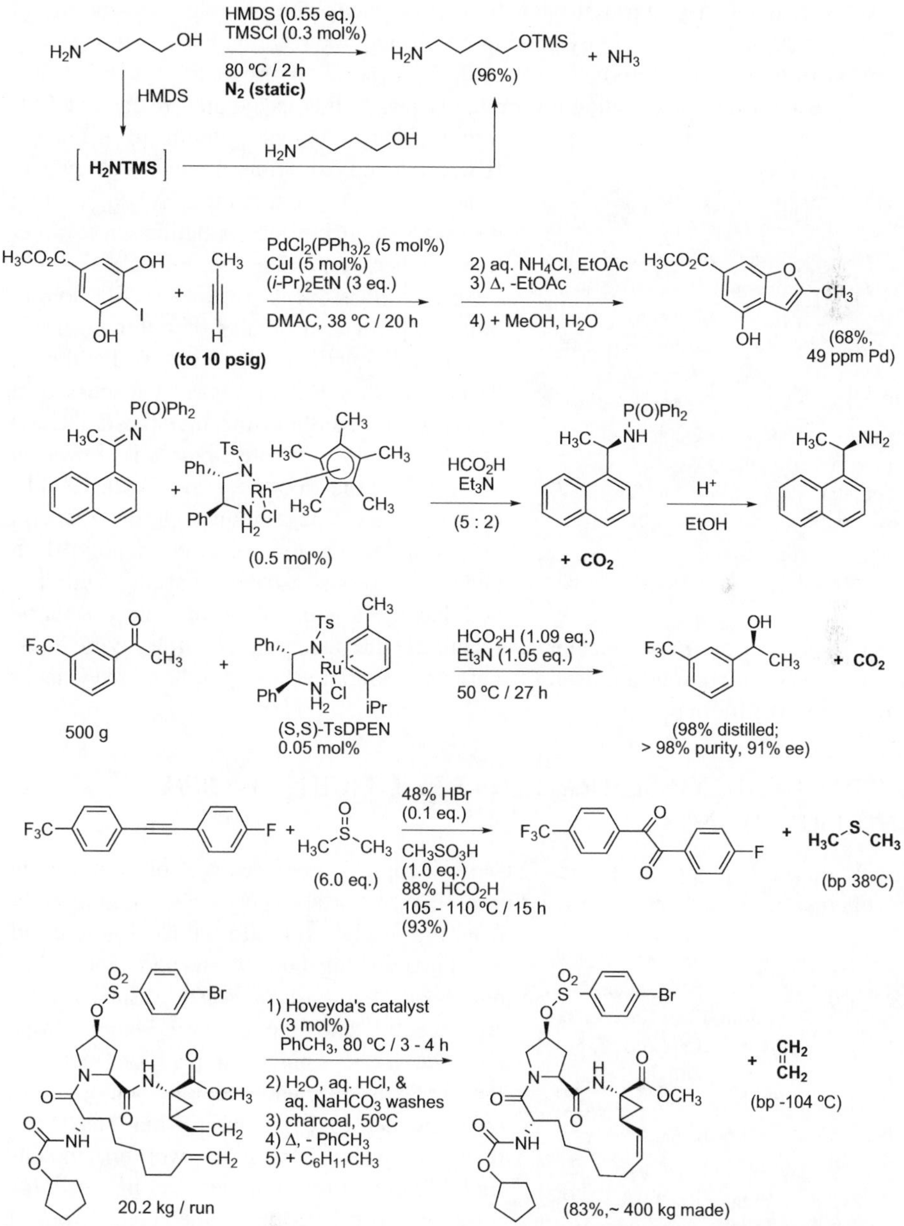

FIGURE 8.11 Additional processing considerations are needed for reactions with volatile reagents or byproducts.

processing without a nitrogen sweep was necessary for complete reaction (Figure 8.11); otherwise the trimethylsilylamine generated in situ was swept out of the reactor [64]. Pfizer researchers carried out Sonogashira reactions with propyne in sealed reactors at 38 °C; only 10 psig was required [65]. Processes such as the latter four would be accelerated by minimizing the volume in a reactor above the reaction mass. Alternatively, transfer hydrogenation of a diphenylphosphinylimine was carried out efficiently on a large scale by passing N_2 through the reaction mass with vigorous agitation to purge the byproduct CO_2 [66]. The transfer hydrogenation of an aceto-phenone was faster when scaled up, probably due to the accelerated release of CO_2 through more efficient agitation [67]. For the HBr-catalyzed oxidation of a diaryl acetylene, removing the byproduct dimethyl sulfide was necessary to drive the oxidation to comple-tion. Dimethyl sulfide (bp 38 °C) was co-distilled with solvents well above its boiling point [68]. (The difficulty of removing trace amounts of volatile components by distillation is described by Raoult's law, as discussed in Chapter 11). Ring-closing metathesis (RCM) of a BILN 2061 ZW precursor was slower on scale than in the laboratory, even with 1000 L/h of N_2 being passed through the reaction mass to purge the byproduct ethylene [69]. In general vigorous stirring, sparging with N_2, and having a great deal of empty volume above the reaction mass accelerate reactions that generate volatile byproducts. Headspace considerations are useful for efficient operations using volatile reagents or generating volatile byproducts.

> Using suction to charge volatile reagents, such as aqueous HCl and $SOCl_2$, will remove some of the reagent from the reactor. The loss of the reagent may adversely affect operations and contaminate equip-ment. The compatibility of the various chemicals and mixtures should be examined before chem-icals are committed to the reactors. For instance, THF is incompatible with gaskets made of NBR elasto-mers [9].

X. PRACTICAL CONSIDERATIONS FOR EFFICIENT WORKUPS AND ISOLATIONS

Simplified workups and isolations can afford rugged processes suitable for routine campaigns. By crystallizing the reaction product in suitable purity and yield directly from the reaction mixture the CoG is reduced through reducing both waste and labor [70]; until both the cost of waste disposal rises significantly and the cost of labor drops significantly the bigger impact on reducing the CoG is through reducing labor costs (see Chapter 3) [6]. Direct isolations avoid the time spent in extractive workups, poten-tially freeing the equipment to be used for other purposes. Even if the yield from a direct isolation is slightly lower than the yield from an extraction–crystallization process,

> Every transfer of a process stream presents opportunities for leaks (through gravity or pressure trans-fers) and contamination through contact with extraneous compo-nents (e.g., incorporation of air through transfers using suction). Minimize physical loss and contamination by isolating the product directly from the reactor.

over a campaign the CoG for producing a product through direct isolation may be less, due to decreased labor [6].

Extractive workups can afford valuable ways to remove impurities and provide a more effective isolation of the product. In general at most three extractions are preferred to extract a product from the aqueous phase (see Chapter 11). If more than three extractions are needed, a continuous counter-current extractive device may be employed; Karr reciprocating plate towers provide intimate mixing. Podbielniak centrifuges, which have been used in the dairy industry for decades to separate cream from whole milk, have been used to purify penicillins by rapidly separating streams that are prone to decompose under the extraction conditions [71]. The CINC liquid–liquid centrifugal separator has a small footprint and can be used in the laboratory and plant [72].

Extractions are generally preferred over filtrations to remove impurities. Extractions can often be performed in the vessel where the reaction occurred, while filtrations require additional equipment (filters, receivers, and transfer lines) and more operations. Filter elements can also plug, causing delays in processing. For instance, it is more convenient to remove $Et_3N \cdot HCl$ from a reaction run in toluene by extraction into H_2O than by filtration; however, if large volumes are needed for extractions, filtration may be preferred. Solvents that dissolve small amounts of water may be preferred if the product is water-sensitive. For instance, i-PrOAc may be preferred over EtOAc.

Case Study on an Optimized Workup: The process for the modified Arbuzov reaction in Figure 8.12 was optimized for manufacturing, and the volumes were minimized to increase the productivity (Table 8.5) [30]. The V_{max} of approximately 7.5 L/kg of starting material is low due to the relatively low charge of solvent (1.0 L/kg starting material, ~ 0.25 M considering volumes of reagents and the other starting material) and the small volume of water used to quench the reaction. (The V_{max} was calculated by assuming the densities of the phosphinic acid and chloroacetic acid were each 1 g/mL, and then summing the volumes. The V_{max} would be lower if the reaction components dissolved in each other to afford a smaller volume and if some component was volatilized and not condensed during the reflux period.) After removing the acetonitrile–hexamethyldisiloxane azeotrope the product was extracted into three portions of MIBK. The high [NH_4Cl] in the aqueous phase (1.4 equivalents of NH_4Cl is generated) probably assisted in partitioning the product into the organic extracts. Two extractions might have required the same total volume of MIBK, but

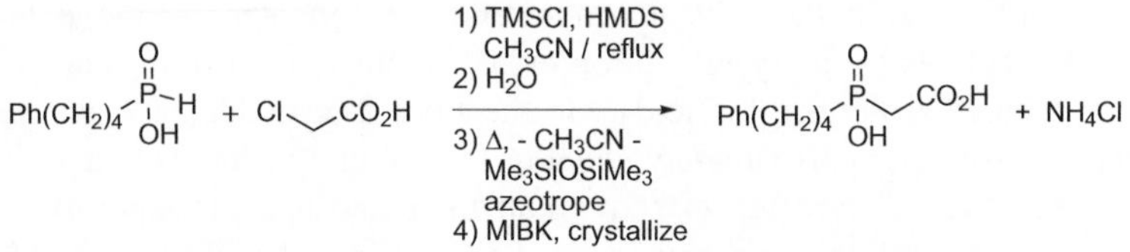

FIGURE 8.12 Processing for a modified Arbuzov reaction.

TABLE 8.5 Quantities for Modified Arbuzov Reaction

	Normalized Amounts
Starting material (phosphinic acid)	1.0 kg (~1 L)
Approximate combined volume of chloroacetic acid, TMSCl, and HMDS	3 L
Acetonitrile (reaction solvent)	1.0 L
H_2O quench	2.5 L
V_{max} at quench	~7.5 L
Aqueous residue after concentration	~3.5 L
MIBK extractions @ 35 °C	3 × 4 L
V_{max} during extractions	~6 L
Volume for crystallization (after concentration of combined extracts)	6.4 L
Chilled MIBK wash of wet cake	2 × 1 L
Product	1.06 kg (83%) 99.9% purity

by using three small extractions the V_{max} during the extractions was kept low, permitting more product to be produced within the equipment train (the vessel used for reaction, quenching, and azeotroping). By conducting the extractions slightly above room temperature a smaller amount of MIBK may have been needed; the space–time productivity of processing was slightly higher with warmer extractions because less time was required to cool to a temperature suitable for extractions. When the second MIBK extract was combined with the first extract a small aqueous phase separated at the bottom of the vessel, and since there was no appreciable amount of product in this phase it was discarded. By extracting at a slightly higher temperature than room temperature, more water may have partitioned into the organic phase, and the first extract, containing the highest concentration of the polar product, would have solubilized the highest concentration of water. Increased charges of the various solvents could have been employed to dissolve any solids during the reaction and workup. The crystallization was carried out at a productive charge of about 6 L/kg; higher solvent volumes decrease the space–time yield, but in general lower volumes for a crystallization require a closer control of temperature at filtration to achieve reproducible yields. On the filter the product was washed with two volumes of solvent approximately equal to the volume of the wet cake; washing with two cake volumes is often effective and uses minimal amounts of solvents. Another optimized workup and isolation has been described for eniluracil [73], and for crystallizations on scale [74].

For ease of operations during workups and isolations, biphasic mixtures, e.g., liquid–liquid or solid–liquid, are generally preferred over triphasic mixtures. (Sometimes triphasic process streams are developed, as in the crystallization of a product using a mixture of a water-insoluble solvent and an aqueous phase in the deCODE example, Figure 11.10.)

> Early in process development balance the equations to recognize how many equivalents of excess reagent must be decomposed during workup, and how byproducts may affect operations.

Generally a mixture is heated or more solvent is added to dissolve unwanted solids. By adding more solvent the V_{max} increases, but convenient processing, including efficient heat transfer and mass transfer, is usually more important than minimizing V_{max}.

XI. ADDITIONAL CONSIDERATIONS FOR KILO LAB OPERATIONS

For scale-up in the kilo lab any technique that works may be used, such as adding a solid reagent through an addition funnel. Work-up is often similar to that carried out in the process research laboratory, such as using Na_2SO_4 or $MgSO_4$ for drying extracts, rotary evaporation to dryness, and triturating oils and solids. (Note that polar compounds may absorb to desiccants.) Some useful equipment is shown in Table 8.6. As mentioned before, while running operations in the kilo lab the wise scientist will be observing operations and thinking about optimizations to shorten development time in the future.

TABLE 8.6 Some Equipment to Ease Kilo Lab Operations

Equipment	Use
Small-scale centrifuges	Spinning down a reaction aliquot can easily remove unwanted solids from the supernatant and ease sample preparation for HPLC analysis
Centrifugal separators	Allow intimate mixing of two liquid phases for efficient extraction, reducing the number of multiple extractions; continuous operations are amenable to scale-up (for CINC liquid–liquid separators, see http://www.coleparmer.com)
Custom centrifuges	Can be used to mimic centrifugation on scale (see http://www.westernstates.com/lab/labcent.htm)
Tissue homogenizers	Provide energy to suspensions or dispersions analogous to triturating (for Polytron benchtop homogenizer, see http://www.brinkmann.com)
Sonic horns	Provide energy to generate emulsions, accelerate reactions, promote mixing, and promote crystallization (see http://www.heilscher.com)

TIPS for kilo lab operations

- Make sure all equipment operable and fits in fume hood (including the length of the agitator shaft).
- Position catch basins in case vessels break.
- Measure key volumes in glassware before (hot) solutions are charged.
- Use polish filtration to remove insoluble impurities (good housekeeping).
- Recognize that operations will take longer.

XII. SUMMARY AND PERSPECTIVE

An underlying focus of efficient process development is to anticipate and avoid problems. Some problems can be readily anticipated and addressed: if a heterogeneous reaction slows, increase the agitation rate. Or, rapid addition of a concentrated NaOH stream will hydrolyze an ester when neutralizing a process stream during workup, so slow additions on scale are wise. Assessing the effects of other problems will require some experiments. For instance, how much of the carboxylate formed by too rapid addition of aqueous NaOH can be present for routine processing to continue? By knowing the impact on downstream processing based on laboratory experiments one can avoid troubleshooting on scale, which can be a stressful situation. Process tolerance studies can define the proven acceptable range and the desired or normal operating range.

Contingency plans can redirect processing as needed. For instance, if a reaction hasn't reached the desired IPC endpoint perhaps it can simply be stirred longer; if additional reagents need to be added, the optimal amounts, sequence and temperature ranges for charging need to be defined. Other considerations may be raised in discussions prior to introduction to a pilot plant or manufacturing facility, such as, whether the addition of a key reagent can be halted and resumed if necessary. Having developed a procedure to rework the isolated product may save time when a batch fails specifications.

Successful scale-up results from paying attention to details. The more details that can be collected and compiled the faster useful processes can be produced, speeding the development of an API. Process development should be fit for the purpose at hand. Rarely are processes completely optimized.

Further details on process implementation can be found in Chapter 16, and in a recent publication [75].

REFERENCES

1. Laird, T. *Org. Process Res. Dev.* **2007**, *11*, 783.
2. Girgis, M. J.; Kiss, K.; Ziltener, C. A.; Prashad, D. H.; Yoskowitz, R. S.; Basso, B.; Repic, O.; Blacklock, T. J.; Landau, R. N. *Org. Process Res. Dev.* **1997**, *1*, 339.
3. Wuts, P. G. M.; Ashford, S. W.; Conway, B.; Havens, J. L.; Taylor, B.; Hritzko, B.; Xiang, Y.; Zakarias, P. S. *Org. Process Res. Dev.* **2009**, *13*, 331.
4. *Prudent Practices in the Laboratory*; National Research Council of the National Academies. National Academies Press: Washington, DC, 2011.

5. Rose, V. E.; Cohrsson, B., Eds.; *Patty's Industrial Hygiene*, 6th ed.; Wiley, 2010.

6. Anderson, N. G. *Org. Process Res. Dev.* **2004,** *8*, 260.

7. Ashwood, M. S.; Alabaster, R. J.; Cottrell, I. F.; Cowden, C. J.; Davies, A. J.; Dolling, U. H.; Emerson, K. M.; Gibb, A. D.; Hands, D.; Wallace, D. J.; Wilson, R. D. *Org. Process Res. Dev.* **2004,** *8*, 192.

8. Fischer, R. W.; Misun, M. *Org. Process Res. Dev.* **2001,** *5*, 581.

9. McConville, F. X. *The Pilot Plant Real Book*, 2nd ed.; FXM Engineering and Design: Worcester, Massachusetts, 2007.

10. Sharratt, P. N., Ed.; *Handbook of Batch Process Design*; Blackie: London, 1997

11. Bonem, J. M. *Process Engineering Problem Solving.* Wiley: New York, 2008.

12. Slade, J.; Parker, D.; Girgis, M.; Mueller, M.; Vivelo, J.; Liu, H.; Bajwa, J.; Chen, G.-P.; Carosi, J.; Lee, P.; Chaudhary, A.; Wambser, D.; Prasad, K.; Bracken, K.; Dean, K.; Boehnke, H.; Repic, O.; Blacklock, T. J. *Org. Process Res. Dev.* **2006,** *10*, 78.

13. Caron, S.; Dugger, R. W.; Gut Ruggeri, S.; Ragan, J. A.; Brown Ripin, D. H. *Chem. Rev.* **2006,** *106*, 2943.

14. Hachoose *Chem. Eng.* **2006,** *113*(4), 54.

15. Königsberger, K.; Chen, G.-P.; Wu, R. R.; Girgis, M. J.; Prasad, K.; Repic, O.; Blacklock, T. J. *Org. Process Res. Dev.* **2003,** *7*, 733.

16. Rassias, G.; Hermitage, S. A.; Sanganee, M. J.; Kincey, P. M.; Smith, N. M.; Andrews, I. P.; Borrett, G. T.; Slater, G. R. *Org. Process Res. Dev.* **2009,** *13*, 774.

17. Cargill, R. W. *J. Chem. Soc. Faraday Trans. I.* **1976,** *72*, 2296.

18. Tokunaga, J. *J. Chem. Eng. Data* **1975,** *20*, 41.

19. For example, see http://www.hamiltoncompany.com/Sensors/.

20. For example, see http://www.oceanopticssensors.com/application/organicsolvent.htm.

21. Advertisement for Bryan Research & Engineering, Inc. *Chem. Eng.* **2010,** *117*(12),7. http://www.bre.com

22. Huttenhuis, P. J. G.; Agrawal, N. J.; Versteeg, G. F. *Ind. Eng. Chem. Res.* **2009,** *48*(8), 4051.

23. http://www.pressreleasepoint.com/black-amp-veatchpatented-technology-featured-asia%E2%80%99s-largest-natural-gas-plant.

24. Ondrey, G. *Chem. Eng.* **2010,** *117*(9), 11. Codexis is developing carbonic anhydrase enzymes to better capture CO_2 by methyl diethanolamine.

25. Atherton, J. H.; Carpenter, K. J. *Process Development: Physicochemical Concepts.* Oxford University Press, 1999; p 15.

26. Perry, R. H., Green, D. W.; Maloney, J. O., Eds.; *Perry's Chemical Engineers' Handbook*, 7th ed.; McGraw-Hill: New York, 1997; pp 2–9, 2–15, and 2–20

27. Argon has been used to preserve foods: http://encyclopedia.airliquide.com/Encyclopedia.asp?GasID=3

28. Vidyaratne, I.; Crewdson, P.; Lefebvre, E.; Gambarotta, S. *Inorg. Chem.* **2007,** *46*, 8836.

29. Schrock, R. R. *Acc. Chem. Res.* **2005,** *38*, 955.

30. Anderson, N. G.; Ciaramella, B. M.; Feldman, A. F.; Lust, D. A.; Moniot, J. L.; Moran, L.; Polomski, R. E.; Wang, S. S. Y. *Org. Process Res. Dev.* **1997,** *1*, 211.

31. Oldshue, J. Y. *Fluid Mixing Technology.* McGraw-Hill: New York, 1983; pp 4–6.

32. Ellis, J. E.; Davis, E. M.; Dozeman, G. J.; Lenoir, E. A.; Belmont, D. T.; Brower, P. L. *Org. Process Res. Dev.* **2001,** *5*, 226.

33. Brands, K. M. J.; Jobson, R. B.; Conrad, K. M.; Williams, J. M.; Pipik, B.; Cameron, M.; Davies, A. J.; Houghton, P. G.; Ashwood, M. S.; Cottrell, I. F.; Reamer, R. A.; Kennedy, D. J.; Dolling, U.-H.; Reider, P. J. *J. Org. Chem.* **2002,** *67*, 4771.

34. Dolman, S. J.; Nyrop, J. L.; Kuethe, J. T. *J. Org. Chem.* **2011,** *76*, 993.

35. Eisch, J. J.; Gitua, H. N. *Organometallics* **2007,** *26*, 778.

36. Haché, B.; Duceppe, J.-S.; Beaulieu, P. L. *Synthesis* **2002,** 528.

37. Bourne, J. R. *Org. Process Res. Dev.* **2003,** *7*, 471.

38. Bałdyga, J.; Pohorecki, R. *Chem. Eng. J.* **1995,** *58*, 183.

39. Baldyga, J.; Bourne, J. R.; Yang, Y. *Chem. Eng. Sci.* **1993,** *48*, 3383.

40. Paul, E. L.; Mahadevan, H.; Foster, J.; Kennedy, M.; Midler, M. *Chem. Eng. Sci.* **1992**, *47*, 2837. King, M. L.; Forman, A. L.; Orella, C.; Pines, S. H. *Chem. Eng. Progress* **1985**, *81*(5), 36.

41. Federsel, H.-J. *Org. Process Res. Dev.* **2000**, *4*, 362.

42. Atherton, J. H.; Carpenter, K. J. *Process Development: Physicochemical Concepts*. Oxford University Press: Oxford, UK, 1999; pp 36–45.

43. Atherton, J. H.; Carpenter, K. J. *Process Development: Physicochemical Concepts*. Oxford University Press: Oxford, UK, 1999.

44. Carpenter, K. J. Agitation. In *Handbook of Batch Process Design*; Sharratt, P. N., Ed.; Blackie: London, 1997, Chapter 4; p 107.

45. Broxer, S.; Fitzgerald, M. A.; Sfouggatakis, C.; Defreese, J. L.; Barlow, E.; Powers, G. L.; Peddicord, M.; Su, B.-N.; Tai-Yuen, Y.; Pathirana, C.; Sherbine, J. P. *Org. Process Res. Dev.* **2011**, *15*, 343.

46. Joshi, D. K.; Sutton, J. W.; Carver, S.; Blanchard, J. P. *Org. Process Res. Dev.* **2005**, *9*, 997.

47. Soundararajan, N. Personal communication.

48. http://www.hse.gov.uk/electricity/atex/general.htm

49. Innospec Active Chemicals: http://www.innospecinc.com/assets/_files/documents/may_08/cm__1210589634_Statsafe.pdf

50. Dixon, N. http://www.theengineer.co.uk/channels/process-engineering/electrostatics-danger-in-process-liquids/298123.article

51. Britton, L. G. *Avoiding Static Ignition Hazards in Chemical Operations*. AIChE: New York, 1999; pp 104–105.

52. Illuminati, G.; Mandolini, L. *Acc. Chem. Res.* **1981**, *14*, 95.

53. For example, see Tanaka, K.; Katsurada, M.; Ohno, F.; Shiga, Y.; Oda, M.. *J. Org. Chem.* **2000**, *65*, 432.

54. Hsieh, D.; Marchut, A. J.; Wei, C.; Zheng, B.; Wang, S. S. Y.; Kiang, S. *Org. Process Res. Dev.* **2009**, *13*, 690.

55. For details on pH measurement and control, see McMillan, G.; Baril, R. *Chem. Eng.* **2010**, *117*(8), 32.

56. Akao, A.; Hiraga, S.; Iida, T.; Kamatani, A.; Kawasaki, M.; Mase, T.; Nemoto, T.; Satake, N.; Weissman, S. A.; Tschaen, D. M.; Rossen, K.; Petrillo, D.; Reamer, R. A.; Volante, R. P. *Tetrahedron* **2001**, *57*, 8917.

57. Li, Y.-E.; Yang, Y.; Kalthod, V.; Tyler, S. M. *Org. Process Res. Dev.* **2009**, *13*, 73.

58. Mullin, R. *Chem. Eng. News* **2009**, *87*(39), 38.

59. Payack, J. F.; Huffman, M. A.; Cai, D.; Hughes, D. L.; Collins, P. C.; Johnson, B. K.; Cottrell, I. F.; Tuma, L. D. *Org. Process Res. Dev.* **2004**, *8*, 256.

60. Barnes, D. M.; Haight, A. R.; Hameury, T.; McLaughlin, M. A.; Mei, J.; Tedrow, J. S.; Toma, J. D. R. *Tetrahedron* **2006**, *62*, 11311.

61. See discussion in Chapter 10 on degassing reaction mixtures for oxygen-sensitive organometallic reactions.

62. Tully, W. U.S. Patent 6,600,051 B2, 2003 (to Warner-Lambert). Atorvastatin calcium can be recrystallized from EtOAc/hexane: Roth, B. D. U.S. Patent 5,273,995, 1993 (to Warner-Lambert Company).

63. Haas, O. W.; Nguyen, A.; Polomski, R. E.; Anderson, N. G. Unpublished. Oscar Haas, chemical engineer, raised this suggestion.

64. Davulcu, A. H.; McLeod, D. D.; Li, J.; Katipally, K.; Littke, A.; Doubleday, W.; Xu, Z.; McConlogue, C. W.; Lai, C. J.; Gleeson, M.; Schwinden, M.; Parsons, R. L., Jr. *J. Org. Chem.* **2009**, *74*, 4068.

65. A safe sparging process was also described in detail Berliner, M. A.; Cordi, E. M.; Dunetz, J. R.; Price, K. E. *Org. Process Res. Dev.* **2010**, *14*, 180.

66. Blacker, J.; Martin, J. In *Asymmetric Catalysis on Industrial Scale*; Blaser, H. U.; Schimdt, E., Eds.; Wiley-VCH: Weinheim, 2004; pp 201–216.

67. Miyagi, M.; Takehara, J.; Collet, S.; Okano, K. *Org. Process Res. Dev.* **2000**, *4*, 346.

68. Wan, Z.; Jones, C. D.; Mitchell, D.; Pu, J. Y.; Zhang, T. Y. *J. Org. Chem.* **2006**, *71*, 826.

69. Nicola, T.; Brenner, M.; Donsbach, K.; Kreye, P. *Org. Process Res. Dev.* **2005**, *9*(4), 513.

70. Chen, C.-K.; Singh, A. K. *Org. Process Res. Dev.* **2001,** *5*, 508.
71. Robbins, L. A.; Cusack, R. W. In *Perry's Chemical Engineers' Handbook*; Perry, R. H., Green, D. W.; Maloney, J. O., Eds.; 7th ed.; McGraw-Hill, 1997; p 15–1.
72. http://www.coleparmer.com/
73. Cooke, J. W. B.; Bright, R.; Coleman, M. J.; Jenkins, K. P. *Org. Process Res. Dev.* **2001,** *5*, 383.
74. Wieckhusen, D. In *Process Chemistry in the Pharmaceutical Industry*; Gadamasetti, K.; Braish, T., Eds.; 2nd ed.; CRC Press: Boca Raton, Florida, 2008; pp 295–312.
75. Anderson, N. G.; Burdick, D. C.; Reeve, M. M. *Org. Process Res. Dev.* **2011,** *15*, 162.

Chapter 9

Optimizing Processes by Minimizing Impurities

"Sounds like you found yourself in some pretty tall weeds there."

– Gary D. Madding, talking about difficulties in minimizing impurities in process optimization studies.

I. INTRODUCTION

The first efforts of process optimization are to become familiar with a new reaction, such as the SAFETY aspects of the starting materials, the physical states of the components, how the components mix, whether any exotherms result, relative solubilities of starting materials and products, and judging the quality of the product by a preliminary method such as NMR. Thereafter the order of optimization in general would first be to improve the yield, followed successively by efforts to improve product quality, improve productivity on scale, reduce the cost of the product, and to reduce

Practical Process Research and Development. DOI: 10.1016/B978-0-12-386537-3.00009-5

waste. This sequence may be changed by extenuating circumstances, such as conditions that produce a large amount of highly toxic waste. If the yield cannot be readily raised by improving the isolation conditions, then efforts may be undertaken to reduce impurities. As always goals are to minimize the expenditures of resources, time and chemicals.

The drive to minimize impurities is central to most optimization efforts. For instance, unreacted starting materials may be considered impurities, and consuming starting materials may raise the yield of a reaction. Troublesome impurities can decrease yields, slow down operations, force additional purification steps, and limit processing options. High-boiling solvents may be difficult to remove from crystalline products; a lower-boiling solvent may be applied to the wet cake to displace the higher-boiling solvent, or the wet cake may be used in the next step. The concentration of residual solvent impurities should be no higher than the limits filed with regulatory authorities. Heavy metal impurities must be reduced to low levels in drug substances. Consistent impurity profiles are needed for APIs destined for human use, and impurities should not exceed the levels seen in the qualifying tox batches. (Batches for tox studies are generally targeted at 97–98% purity, ideally purified by techniques anticipated for routine manufacturing. See Chapter 1.) This chapter focuses on minimizing impurities formed from starting materials and reagents.

To facilitate the development of an NCE one must consider how process changes can impact the efforts of others. Some steps to minimizing impurities have relatively little impact outside of ones own fume hood. For instance, one can readily control the impact from changing proportions of reaction components (different mole ratios, extended additions, removing products and byproducts as they form by crystallization or azeo-troping). Changing reagents and catalysts can affect downstream chemistry if these materials or side products are passed along to the next stage. Changing functional groups or routes can markedly change impurity profiles and impact chemistry and processing downstream.

Early in your process development efforts your goal is not to isolate product, but rather to find the best conditions to produce product. Vary processing conditions to discover best directions for process development, as determined by rapid and suitable in-process assays. Optimizing the workup and isolation too early may be non-productive if improved reactions are later developed.

Steps to optimizing a process are shown in Table 9.1.

A key aspect of process optimization can be identifying process impurities, through comparison to known compounds or unequivocal structural elucidation. The analytical tools at hand are often the fastest and most convenient. Useful analytical tools include HPLC, GC, LC/MS, GC/MS, LC/NMR, IR; preparative chromatography, followed by spectroscopy and/or MS. Samples may be conveniently isolated from mother liquors, washes, or extracts. Once a compound has been identified one can pose a mechanism for its formation, and poten-tially minimize its presence through process optimization. A statistical design of experiments (DoE) may be especially useful if a clear course for optimization is not intuitive.

TABLE 9.1 Steps to Optimizing a Process

Step	Comment
1. Establish baseline reaction conditions	Reproduce established procedure, IF NO SAFETY ISSUES ARE ANTICIPATED
2. Vary 2.1. Temperature	Baseline ±20 °C
2.2. Equivalents of reagents	1.05 equivalents is a good goal, unless reagent is inexpensive and easy to remove (such as MeOH or methanesulfonic acid) or reactions are slow
2.3. Addition of reagents	Rapid, extended, and portion-wise additions
2.4. Solvents/co-solvents	Consider for direct isolation, telescoping, or workup
2.5. Reaction concentration	50–200% of baseline
2.6. Change reagents	Consider availability, cost, ease of workup
2.7. Catalyst, [catalyst]; ligand, [ligand]	Reactions may be second-order in [catalyst]
2.8. Other operating parameters	pH, pressure, residence time, etc.
2.9. Stirring	Especially important for heterogeneous reactions
2.10. Extend operating conditions	Monitor for product decomposition, co-crystallization, attrition of particle size
3. Select conditions for complete reaction and minimal impurities	

Early in process development it is prudent to explore experimental conditions for parameters that can adversely affect scale-up if not understood and controlled. Warning signs include a change in product yield or quality with laboratory mixing speed, addition rate (portions or continuous), the position of a feed stream, or scale-up to vessel with different geometry. Micromixing is likely to be the cause of such difficulties. If impurities arise and yields drop with extended additions, some components may not be stable enough to make more material using batch operations developed to that time. Changing the addition protocol may solve the difficulties. Not all reactions can be readily scaled up using batch processing, and continuous operations may be necessary to provide more material (see Section XI and Chapter 14).

II. BENEFITS AND LIMITATIONS OF HETEROGENEOUS PROCESSING

Heterogeneous processes can be beneficial if byproducts precipitate as they form, removing them from further reaction, or if the product crystallizes as it forms. For instance, precipitation of NaCl drives the Finkelstein reaction (see example in Figure 9.4); a reactive crystallization can also drive a reaction to completion.

When mesylation of propargylic alcohol was conducted using a stoichiometric charge of Et$_3$N, 5–10% of propargyl chloride, an explosive, was formed (Figure 9.1).

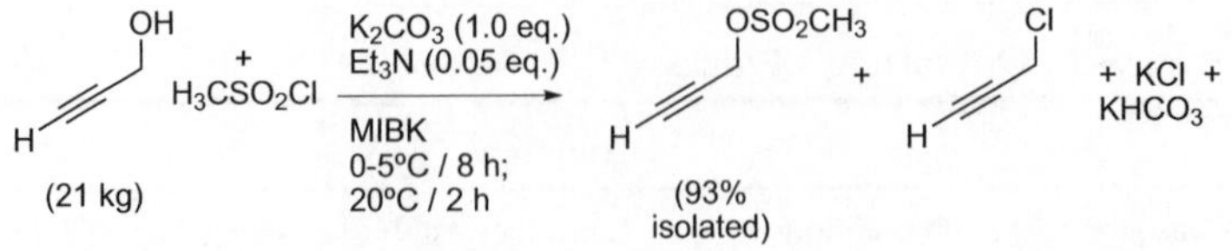

FIGURE 9.1 Heterogeneous conditions to minimize side reactions in a mesylation.

(SAFETY: AstraZeneca workers noted recently that the mesylate, benzenesulfonate, and tosylate of propargylic alcohol are also explosives [1].) When mesylation was conducted in MIBK using 5 mol% of Et_3N and one equivalent of K_2CO_3 the mesylate was isolated in high yield and the formation of propargyl chloride was significantly suppressed. Under these conditions the byproduct KCl was unavailable to form propargyl chloride because the inorganic salts precipitated from the reaction mixture. Lower yields were found when KOH or NaOH were substituted for K_2CO_3 [2]. The use of MTBE has also minimized reactions of $Et_3N \cdot HCl$ in mesylations with MsCl [3]. Another approach is to use methanesulfonic anhydride instead of MsCl to preclude formation of propargyl chloride.

An optimized heterogeneous process is the alkylation shown in Figure 9.2, as developed by researchers at Lilly's Indianapolis laboratories [4] and Tippecanoe laboratories [5]. When the chloride in 3-picolinyl chloride was substituted by bromide, iodide, or mesylate, some quaternization of the piperidine nitrogen occurred. When Et_3N or DBU was used, quaternization of the base by 3-picolinyl chloride occurred. More impurities were formed at room temperature than at higher temperatures; control experiments showed that piperidone·HCl is not split effectively below 50 °C. Due to concerns about rapid evolution of CO_2 the reaction mixture was heated in stages in early processing; on scale gradually heating to 70–72 °C proved satisfactory, even at higher concentrations.

With heterogeneous conditions as in Figures 9.1 and 9.2 small particle size and the presence of some H_2O in the reaction mixture may help provide reproducible operations. Other processes accelerated by insoluble reagents with smaller particle sizes include a Buchwald–Hartwig amination with Cs_2CO_3 [6], solid–liquid PTC with K_3PO_4 [7], and solid–liquid PTC with K_2CO_3 [8]. Small amounts of water may help increase the solubility of inorganic reagents and increase reaction rates, as was posed for

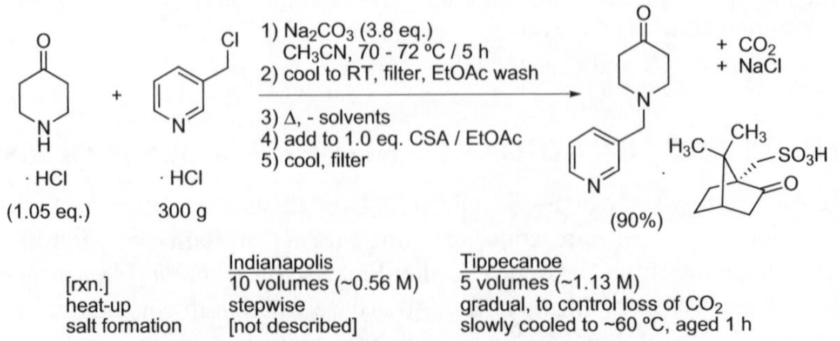

FIGURE 9.2 An optimized heterogeneous alkylation.

a KF-mediated Suzuki–Miyaura coupling [9]; excessive amounts of water can produce ineffective, viscous mud-like suspensions.

III. DECREASE SIDE PRODUCTS BY DECREASING DEGRADATION OF STARTING MATERIALS

Considering the relative reactivity of reagents and intermediates can be important when designing processes. Pfizer workers published some simple and effective optimization of a statin precursor by selecting solvents and a non-volatile acid (Figure 9.3). In polar solvents condensation of the benzyl ester with benzyl amine produced a significant amount of the benzyl ester arising from retro-Claisen fragmentation. Non-polar solvents such as heptanes and methylcyclohexane minimized formation of the latter impurity, but the starting material and desired product were poorly soluble. Triethylamine was found to increase the solubility of the starting material in non-polar heptanes. (Triethylamine is a non-polar solvent, with E_N^T of 0.043, and TWA of only 1 ppm.) When the reaction was conducted in heptanes–Et$_3$N the formation of the retro-Claisen byproduct was minimal, and the benzylamide was simply isolated by filtration. (Perhaps Et$_3$N formed the enolate of the starting material, inhibiting attack at the ketone carbonyl.) The imidazole was formed by condensing the benzylamide with the chiral primary amine often used in preparations of statins. Charging an acid minimized dimerization of the aminoester; acetic acid, which was probably the first acid chosen, was azeotropically removed during the reflux period and so was ineffective. Benzoic acid was selected because it was effective, decreased decomposition of the starting material, was not removed azeotropically, and was washed out of the reaction mixture [10].

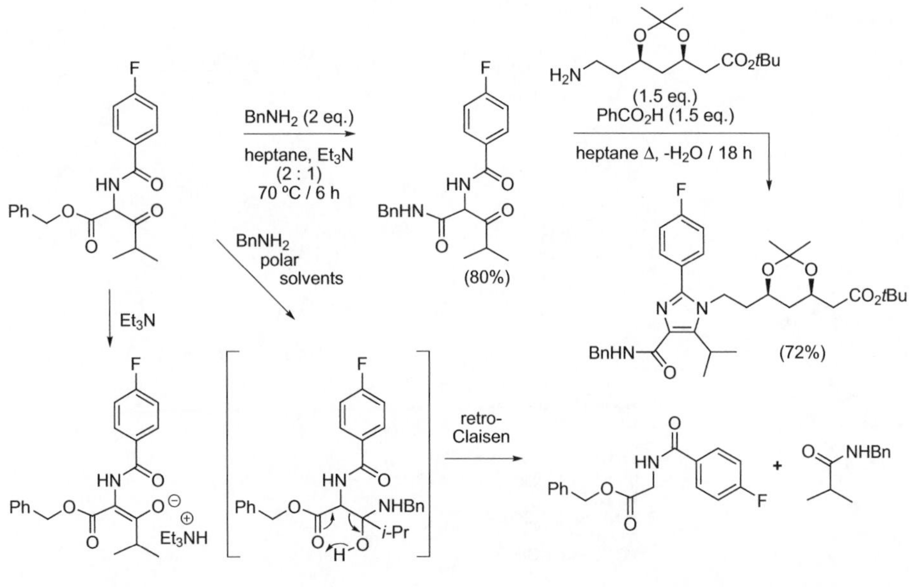

FIGURE 9.3 Process optimization by minimizing degradation of starting materials.

IV. CONTROLLING pH

Controlling pH can be a crucial part of optimization for processes that are aqueous or partially aqueous. Although knowledge of the pK_as of the molecules involved may help in predicting the optimal pH ranges, experimental data may show that optimal ranges for operations are considerably different. Through high-throughput screening the Sharpless group found that adding citric acid to an asymmetric osmium tetraoxide-catalyzed *cis*-dihydroxylation raised the yields considerably for electron-poor olefins (Figure 9.4). (For instance, dihydroxylation of diethyl maleate took place in less

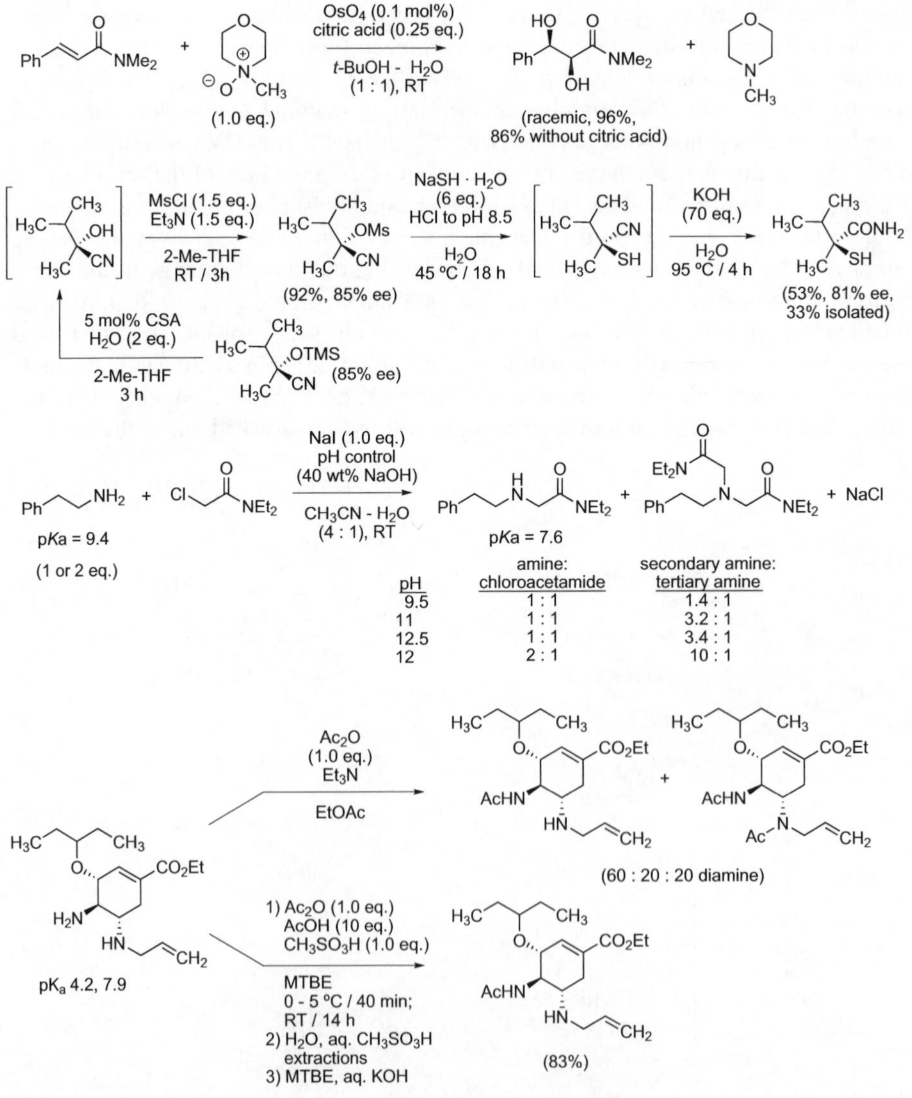

FIGURE 9.4   Influence of pH on process optimization.

than 10% yield without citric acid, and in 76% with citric acid.) In the presence of citric acid the pH rose during the reaction from pH 5.2 to pH 7.1. Without citric acid lower conversions resulted and dark precipitates formed, suggesting that the reaction had stalled as the osmium catalyst precipitated under the more basic conditions. The preferred acids were found to be α-hydroxy acids, which are more acidic than the un-hydroxylated acids, and the optimal pH at the beginning of the reaction was about pH 5 [11]. A practical processing option for this reaction is to maintain the pH in an optimal range, probably pH 5.2–6.8, by adding acid or by using a buffer. Amgen researchers found that pH and solvents were important in converting a trimethylsilyl cyanohydrin to a mercaptoamide. The cyanohydrin was completely racemized when the TMS-cyanohydrin was hydrolyzed with 1 M HCl in MeOH, but deprotection with catalytic camphorsulfonic acid (CSA) in the presence of two equivalents of water retained the desired chirality in the cyanohydrin. The mesylate was prepared and treated with excess sodium sulfide to generate the mercaptonitrile; the best conditions were found at about pH 8.5. At higher pH, which occurred in unbuffered reactions at the beginning of the reaction, the enantiomeric excess was very low, and the latter was seen to rise as the pH fell with time [12]. In another example of the importance of pH control, Novartis workers investigated the monoalkylation of phenethylamine with chloroacetamides. Using CsOH in DMF the ratio of secondary amine:tertiary amine was less than 1.3:1, and with K_2CO_3–NaI/DMF the ratio dropped to 0.8:1. Adding 5% H_2O to the latter system to partially solubilize the K_2CO_3 improved the ratio to 1.6:1, and by using the soluble base Et_3N the ratio improved to 3:1. The alkylation stalled when the solvent system was 1:1 acetonitrile:water, but when the solvents were adjusted to 20% water the byproduct NaCl from the Finkelstein exchange precipitated, driving the alkylation. The pH of the alkylation was maintained by adding 40 wt% NaOH, and poor selectivity was found when the conditions were held to pH 9.5. By maintaining pH 12, almost three pH units above the phenethylamine pK_a of 9.4, the ratio of secondary amine:tertiary amine rose to 3.4:1. By charging two equivalents of phene-thylamine a ratio of 10:1 was achieved. Under more acidic conditions, such as pH 9.5, most of the starting material was protonated, suppressing the desired monoalkylation [13]. Protonation in the absence of water can also exploit different degrees of reactivity in compounds. In the preparation of an oseltamivir intermediate Roche researchers found that acetylation of a diamine with one equivalent of acetic anhydride under basic conditions produced a 60:20:20 mixture of the desired aminoamide, the bisamide and starting material. Acetylation under acidic conditions provided the desired acetamide in 83% yield. With protonation of the more basic secondary amine (pK_a 7.9) acetylation occurred at the less basic amine (pK_a 4.2), β- to both the ether oxygen and the other amine [14].

> When aqueous conditions are involved monitor the pH for change during processing. Controlling the amount of acid or base present may be key to effective processing.

Controlling pH can be very important for biocatalytic reactions. For instance, better selectivity in hydrolysis of a diester was noted at pH 8 than at pH 6 using a lipase from *Candida rugosa*. Unfortunately facile rearrangement of the ester from the secondary alcohol to the primary alcohol occurred, even at pH 7. Immobilized *Pseudomonas cepacia* lipase (Amano PS-C1) was found to be optimal, working at pH 5.2. Workup was simplified by using Amano PS-C1 to conduct a transesterification instead of a hydrolysis; the solvent mixture employed was *n*-BuOH–MTBE, with enough H_2O to sufficiently hydrate the enzyme [15].

V. ADDITION SEQUENCE, DURATION OF ADDITION, AND TIME BETWEEN ADDITIONS

Optimal addition sequence can be very important to the success of processes. Rao has devoted one chapter to addition sequences [16]. Gradual or portion-wise additions are easy operations and can dramatically raise the yield of a process. For the double Michael-addition of Meldrum's acid to divinyl sulfone, batch operations were not amenable to scale-up: when all components were charged and heated to reflux over 30 minutes the yield was over 90%, but when the mixture was heated to reflux over 120 min the yield was only 57% (Figure 9.5). Decomposition of Meldrum's acid occurred

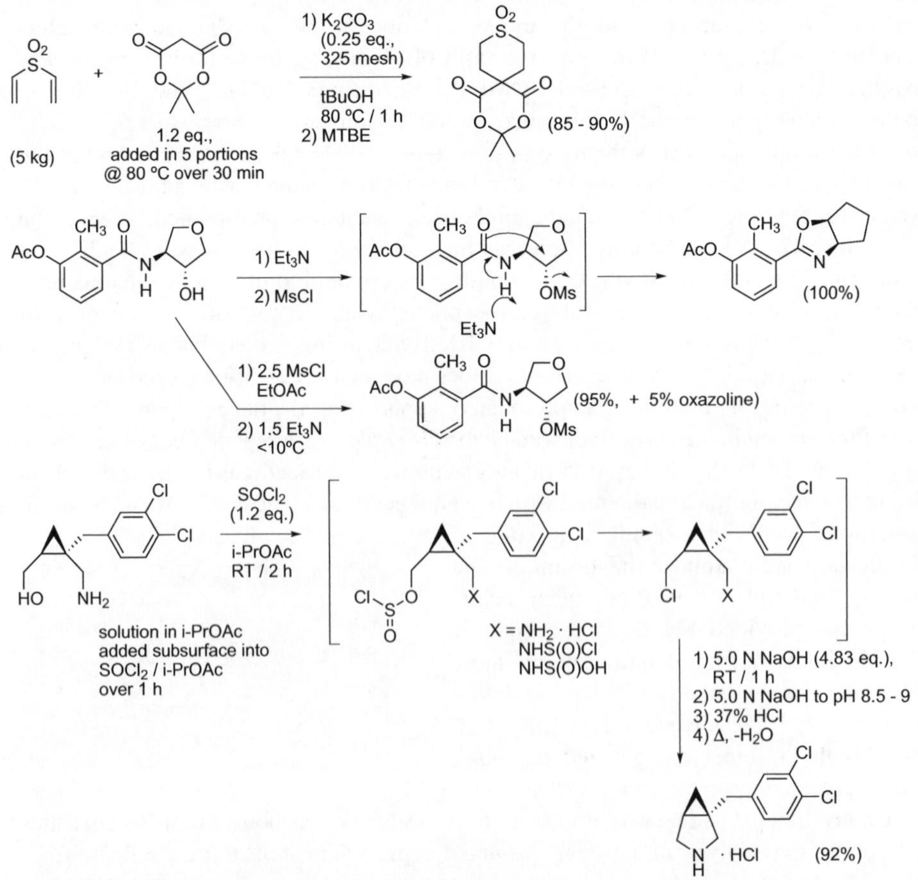

FIGURE 9.5   Successful processes using a portion-wise and inverse additions.

below the temperature needed for effective reaction. High yields were obtained by adding Meldrum's acid in portions to the hot mixture [17] (semi-batch operations). Inverse additions can also be powerful. Under routine conditions for the preparation of a mesylate (a nelfinavir mesylate precursor), the sole product was the oxazoline shown in Figure 9.5. To minimize the contact of the mesylate with base, methanesulfonyl chloride was added first, followed by the slow addition of triethylamine, and the desired mesylate was formed in excellent yield [18]. When thionyl chloride was added to a solution of the amino alcohol in Figure 9.5 chlorination was incomplete, because the starting material crystallized as the hydrochloride salt. The optimal sequence was to gradually add a solution of the amino alcohol subsurface into a solution of $SOCl_2$ in i-PrOAc over 1 h, then age for 2 h at RT. The reasoning behind the inverse addition was that if only a small amount of the amino alcohol were present at any moment then crystallization of the amino alcohol hydrochloride would be slowed, allowing formation of the chlorosulfite and ultimately chlorination. No sticky solids were formed, and probably crystallization was inhibited by the multiple species present (as detected by NMR). The fluid reaction mixture was directly quenched with aqueous NaOH and cyclized to the pyrrolidine at room temperature; the product was extracted into i-PrOAc at pH 8.5–9 [19].

> Five-membered rings are very readily formed (Illuminati, G.; Mandolini, L. *Acc. Chem. Res.* **1981,** *14*, 95). Expect impurities from five-membered ring formation to be present in processing.

When a catalyst is used a gradual addition of one or more components to the vessel containing the catalyst may minimize impurities and afford the best yield. AstraZeneca researchers mentioned that when the amine was added to a slurry of the carboxylic acid and N,N-dimethylaminopropyl-N'-ethyl-carbodiimide (ECDI) in CH_2Cl_2 the product was formed in only 56% yield, with about 34% of the yield being the byproducts from O- to N-acyl shift (Figure 9.6). In subsequent reactions the catalyst N-hydroxybenzotriazole (HOBt) was charged to intercept the O-acyl adduct of EDCI and the carboxylic acid and thus minimize rearrangement. (SAFETY NOTE: Anhydrous HOBt is explosive, and transportation of $HOBt \cdot H_2O$ is restricted. Alternatives to HOBt are discussed in Chapter 4, along with other peptide coupling regimens.) By gradually adding a solution of EDCI as the limiting reagent the formation of the unproductive side products was dramatically reduced. At this stage closer examination showed the presence of 10% of the late-eluting ester amide, which inhibited crystallization of the product. The ester amide was hydrolyzed by treating an MTBE solution of the crude product with a solution of NaOH, and the product amide was in 86% yield [20]. Bäckvall's group showed that rapid addition of a Grignard reagent to a cuprate-catalyzed reaction produced primarily the product from S_N2 reaction, but extended additions greatly increased the amount of S_N2' product [21]. In a ruthenium-catalyzed cyclopropanation ethyl diazoacetate was added over 16 h, to minimize the concentration of the corresponding carbene and the formation of diethyl fumarate. (Often the styrene component has been charged in excess for such cyclopropanations.) The *cis*-product from the cyclopropanation hydrolyzed more slowly affording some purification. The product was isolated in excellent chiral purity as the salt with dehydroabietylamine (DAA) [22]. Parallel addition of both starting material and reagent to

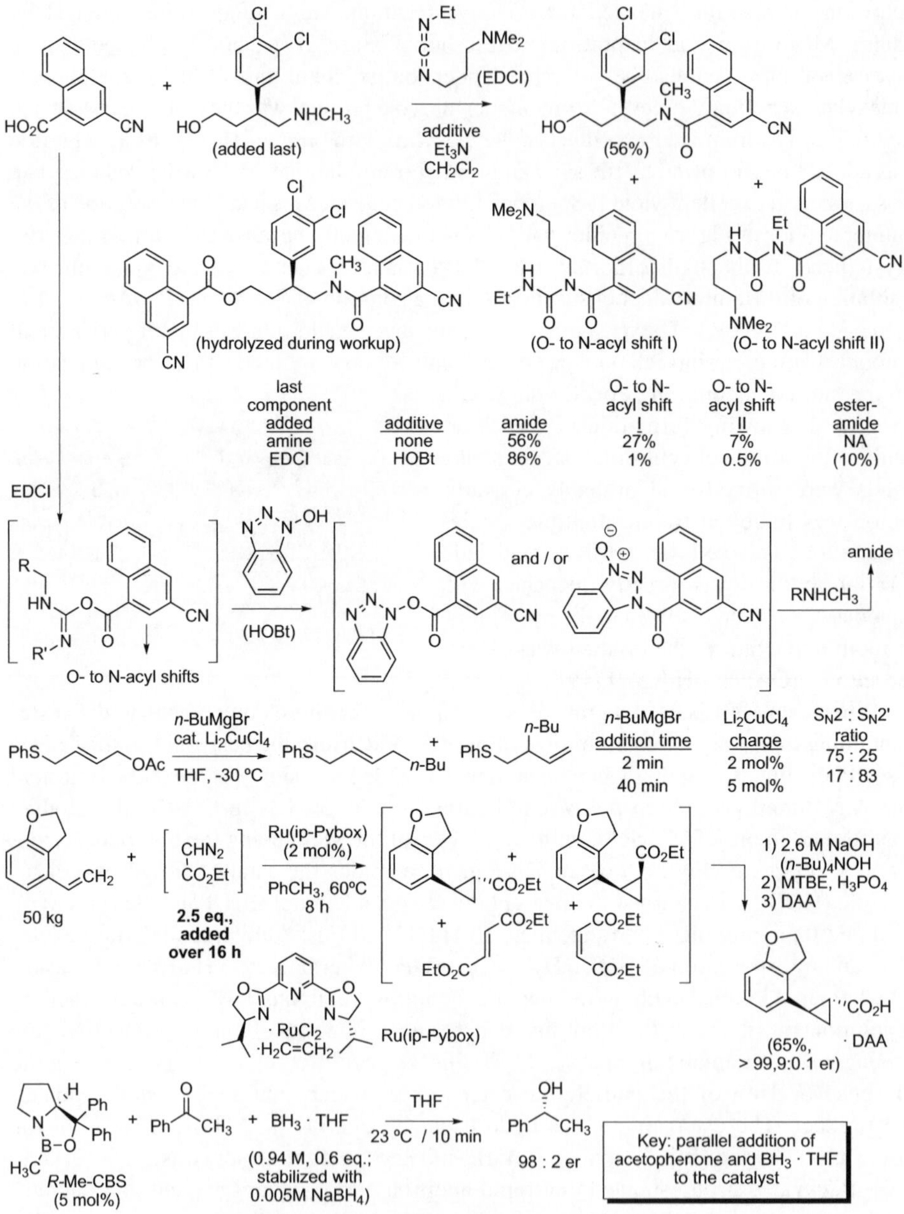

FIGURE 9.6 Benefits of adding reaction streams to a catalyst.

the vessel containing the catalyst can also optimize the yield. (See discussion with Figure 7.3 for a parallel addition in an uncatalyzed process.) With the *R*-Me-CBS-catalyzed reduction a parallel addition of ketone and 1 M $BH_3 \cdot THF$ produced the secondary alcohol with an enantiomeric ratio (er) of 98:2; the er eroded with other addition modes. The commercial lot of 1 M $BH_3 \cdot THF$, stabilized with 0.005 M $NaBH_4$, was examined and found to be 0.94 M. Only NaB_3H_8 was detected in this solution that had been stabilized with $NaBH_4$, and the addition of NaB_3H_8 to a reaction in non-stabilized 1 M $BH_3 \cdot THF$ reduced the er to 85.5:14.5. By using the parallel addition mode the concentration of R-Me-CBS was always high relative to the other components, and the catalytic, enantioselective reduction predominated. (The presence of $NaBH_4$-related species could also be negated by adding Lewis acids such as $BF_3 \cdot Et_2O$ [23]). Optimized addition modes as shown in Figure 9.6 can greatly improve yields and simplify purifications, with relatively little effort expended in operations.

Optimization of what seems at first glance to be a simple sulfide formation offered considerable challenges for GSK workers, and optimizing an addition sequence became key (Figure 9.7). In investigations for the first scale-up a small excess of Cs_2CO_3 and the tosylate were used, and the reaction sometimes would inexplicably stall. Despite investigations into the particle size and moisture content of the hygroscopic Cs_2CO_3 and other factors, no "quick fix" was apparent. In addition to the disulfide formed from the starting mercaptan, four impurities were identified: the acrylate and mercaptan arising from β-elimination, the disulfide from oxidation of the byproduct mercaptan, and the sulfide from reaction of the cysteine starting material with the acrylate byproduct. These impurities, especially the disulfides, inhibited the crystallization of the product. Excess base in the reaction was decomposing the product, leading to the formation of most of these impurities. Oxygen was removed to minimize disulfide formation by heating the solvent (acetonitrile) to reflux [24] or by distilling one or two volumes of solvent. By monitoring the formation and competing decomposition of the product the reaction could be stopped when the level of the product had been maximized, and the product was isolated in 78% yield in the kilo lab [25].

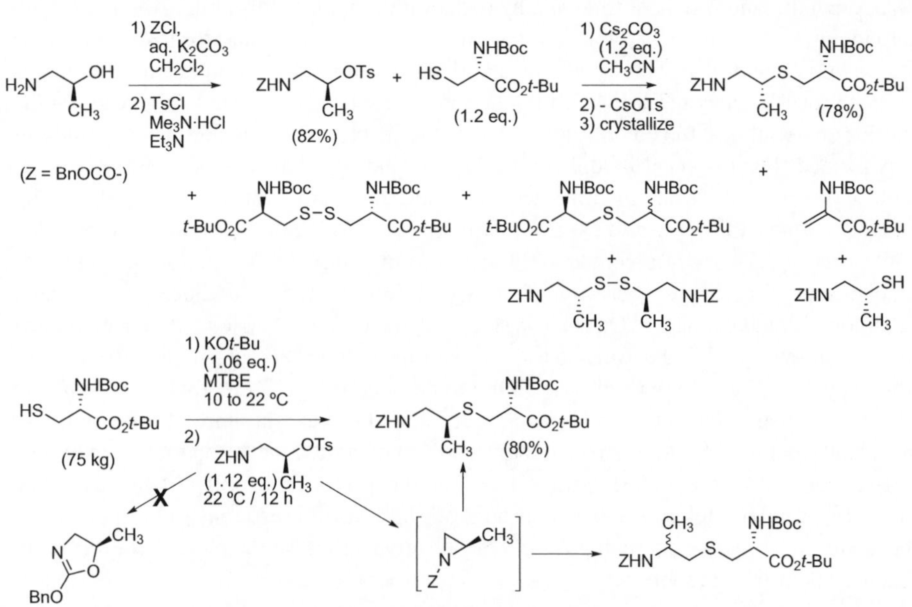

FIGURE 9.7 Optimizing formation of a mixed sulfide.

Further investigations to make the sulfide were undertaken before committing further resources to a scale-up. A wide range of bases and conditions were screened, including PTC. K_3PO_4 provided a rate similar to Cs_2CO_3, but reaction rates were inconsistent due to the particle size and moisture content. DBU provided fast reactions, but unfortunately led to some epimerization α- to the ester in the product. KOt-Bu, soluble in many organic solvents, was selected. Sulfide formations using KOt-Bu in MTBE were homogeneous, did not stall, were complete within 12 h, and the product did not decompose with extended stirring. Further investigations led to the identification of two diastereomeric impurities that were isomeric with the product, implicating the aziridine as an intermediate. There was no evidence for formation of the oxazoline. To minimize formation of the aziridine, a small excess of KOt-Bu was added to form the anion of the mercaptan and after 10 min the tosylate was added. (Longer times before adding the tosylate led to the formation of the acrylate and KHS.) Using these optimized conditions the sulfide was isolated on scale in 80% yield [25].

When yields drop upon extended addition of a key reagent, byproducts may be forming from reaction of unstable intermediates, or through the slow build up of helpful intermediates generated from addition of the reagent. Since extended additions are likely for batch operations on scale, extended additions should be considered in the laboratory for routine process development. For instance, extended additions of $POCl_3$ in the chlorination of a hydroxypyrimidine lead to the formation of a pseudo-dimer, and HPLC analysis showed that the pseudo-dimer formed during the initial addition of $POCl_3$ (see discussion with Figure 7.3).

If an unstable intermediate is involved, one option is to develop continuous operations so that the unstable intermediate is consumed as it is generated (Chapter 14). This situation may be encountered with one-pot processes such as the generation of a reactive organometallic intermediate followed by the addition of an electrophile. Another option is to identify an additive that will stabilize the reactive intermediate; for instance, Roche researchers found that $MgCl_2$ would stabilize 2-lithio-3,4-dichloroanisole, which is prone to form a benzyne [26]. A third option is to add the key reagent to a mixture of the starting material and the electrophile, so that the reactive intermediate is consumed as it is formed. For instance, iodophenyllithium displayed a half-life of about 40 min at $-60\,^{\circ}C$; on a 1000 L scale the time needed to add the n-BuLi was calculated at 90 min, and the ketone addition would have required at least 2 h (Figure 9.8). Good temperature control was necessary to preclude yield losses from competing formation of 1-butyl-4-iodobenzene, formed from reaction of the aryl lithium with butyl iodide, the byproduct of iodide–lithium exchange. On scale, a selective reaction was carried out under Barbier-like conditions by adding n-BuLi to a solution of the ketone and 1,4-diiodobenzene [27]. Similarly the anion of a pyridine was found to be too unstable for ready boronation, and Merck scientists showed that the base could be added last to a mixture of 3-Br–pyridine and tributylborate [28]. Merck researchers also demonstrated the boronation of an indole by treating an N-Boc indole with n-BuLi in the presence of $B(O\text{-}i\text{-}Pr)_3$ [29]. The Snieckus group has telescoped such boronations into Suzuki reactions [30]. In changing the addition sequence these highly chemoselective processes exploited the different reactivities of the components.

Controlling dilution can be extremely important for exploiting chemoselectivity and minimizing impurity formation, and extended additions can be considered a form of

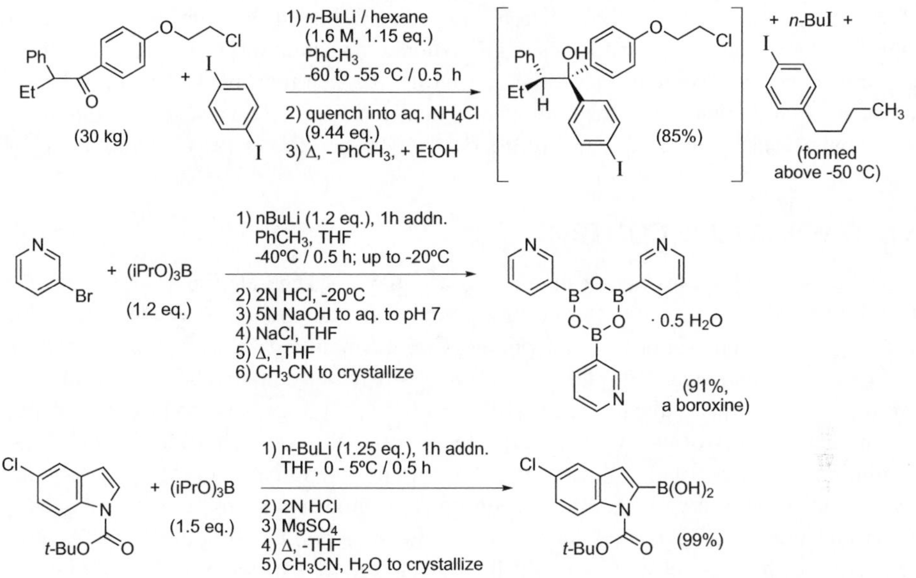

FIGURE 9.8 Benefits of adding a base to mixture of starting material and electrophile.

dilution. Mesomixing and micromixing can play large roles, especially on scale (Chapter 8). The selective hydrolysis of the diester in Figure 9.9 is probably due to the very dilute conditions (~ 0.04 M), low temperature, and dropwise addition of the reagent; formation of a microemulsion may also play a role to minimize over-hydrolysis [31]. AstraZeneca researchers found that an extended addition of EtMgBr was necessary to exchange a methoxy group for an ethyl group [32]. (In the discovery route deme-thylation of the methyl ether using Mg and I_2 was troublesome and produced methyl iodide. After ready conversion of the resulting naphthol to the triflate, the reaction mixture from Pd-catalyzed coupling with triethylborane required chromatographic purification, and the catalyst was deemed expensive for long-term use.) The researchers noted that *ipso*-displacement of aryl methoxy groups was facilitated in the presence of *o*-oxazoline groups, and wondered if the *o*-ester group would provide similar activation.

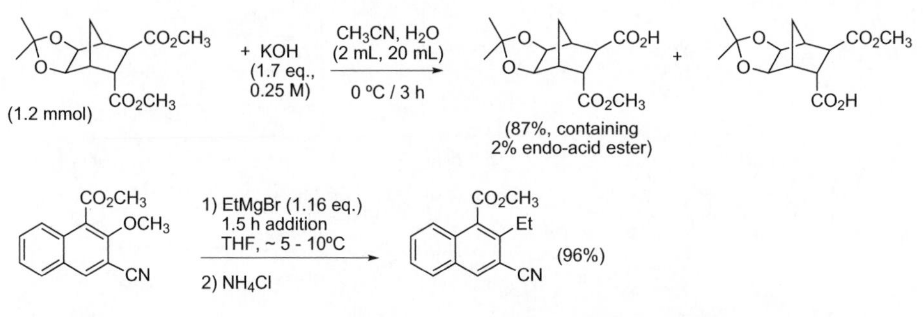

FIGURE 9.9 Dilution and extended addition as important factors in process development.

With an extended addition the reaction with ethyl Grignard was successful, with no significant reaction at the nitrile or ester group under the conditions shown. The resulting process is very creative and practical, substituting one step for three steps, and is based on an association that others may not have considered. Optimistic researchers ran these reactions because generating data are the best way to confirm or deny a hypothesis.

VI. TEMPERATURE CONTROL

Reaction at temperatures higher than necessary may not generate the desired product. Conducting a reaction above the boiling point of one component may encourage the loss of that material. For instance, in process C of the modified Arbuzov reaction shown in Figure 12.2 the reaction mixture containing TMSCl was held at 60–70 °C during the addition of HMDS; gradual loss of TMSCl (bp 57 °C) may have contributed to the overall unreliability of this process [33]. The addition of thionyl chloride (bp 79 °C) as a solution in toluene to a mixture of pyridine and phenoxyethanol in refluxing toluene was effective on a large scale; these conditions ensured rapid, dose-controlled generation and release of SO_2, and there was no pressure buildup (Figure 9.10). At this temperature the reagent $SOCl_2$ could have been vaporized. Perhaps the solution of $SOCl_2$ in $PhCH_3$ had sufficient heat capacity so that the $SOCl_2$ was not volatilized in the

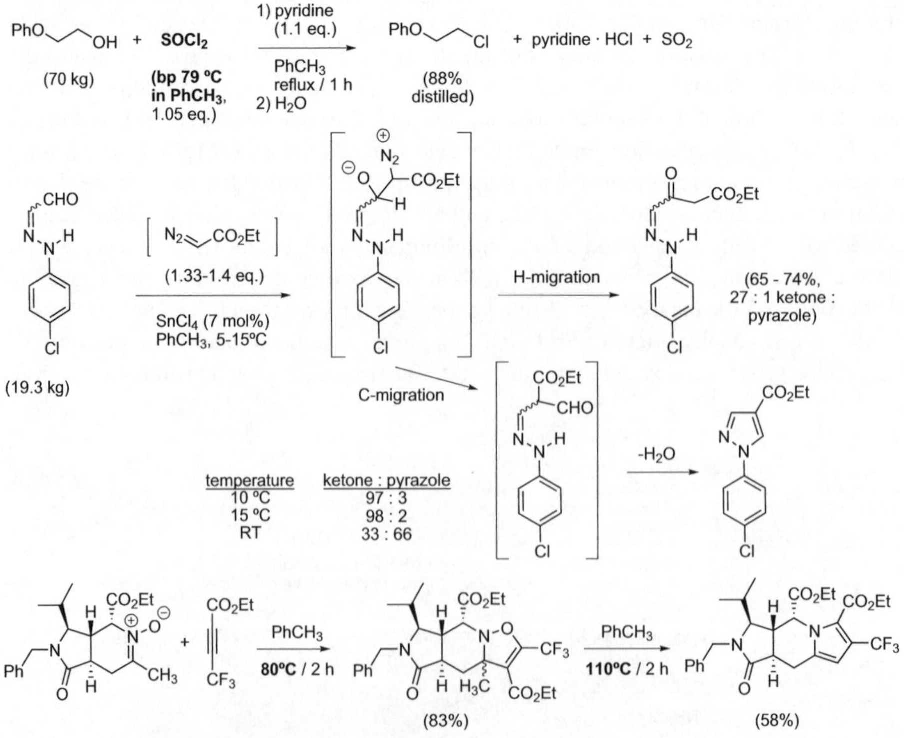

FIGURE 9.10 Temperature considerations in optimizing processes.

headspace of the reactor before reaching the surface of the reaction [34]. Other helpful factors may have been that the SOCl$_2$ reacted virtually instantaneously in solution, due to a subsurface addition, or a pyridine·SOCl$_2$ complex formed that helped keep the SOCl$_2$ in solution. Small temperature changes can greatly affect byproduct formation, by increasing the rate of a side reaction more than the rate of the desired reaction. In the SnCl$_4$-catalyzed reaction of an imine with ethyl diazoacetate the product ratios changed markedly between 15 °C and room temperature [35]. Excessively high temperatures can also lead researchers to overlook interesting reaction products. Reaction of the nitrone with the acetylene at about 110 °C produced the pyrrole shown in Figure 9.10. Subsequently the J&J researchers showed that reaction at a lower temperature produced the isoxazoline, which was converted to the pyrrole at higher temperature [36]. Minimizing the consumption of energy is one of the principles of green chemistry, and such results confirm that excess heat can be deleterious.

VII. MINIMIZING IMPURITIES FORMED DURING WORKUP

At the end of the reaction, generating a stable reaction mixture is necessary to prevent formation of significant amounts of impurities during subsequent processing, especially on scale. In the manufacturing of paclitaxel (Taxol®) by the semi-synthetic route a number of impurities were identified in converting the penultimate to the API (Figure 9.11). Acid-mediated deprotection removed the methoxypropyl (MOP) group first, followed by the triethylsilyl (TES) group. Reactions were significantly shorter in AcOH than in EtOH or acetone, and higher levels of 10-des-acetyl-paclitaxel (the major impurity) were found with H$_2$SO$_4$ than with trifluoroacetic acid, trichloroacetic acid, or

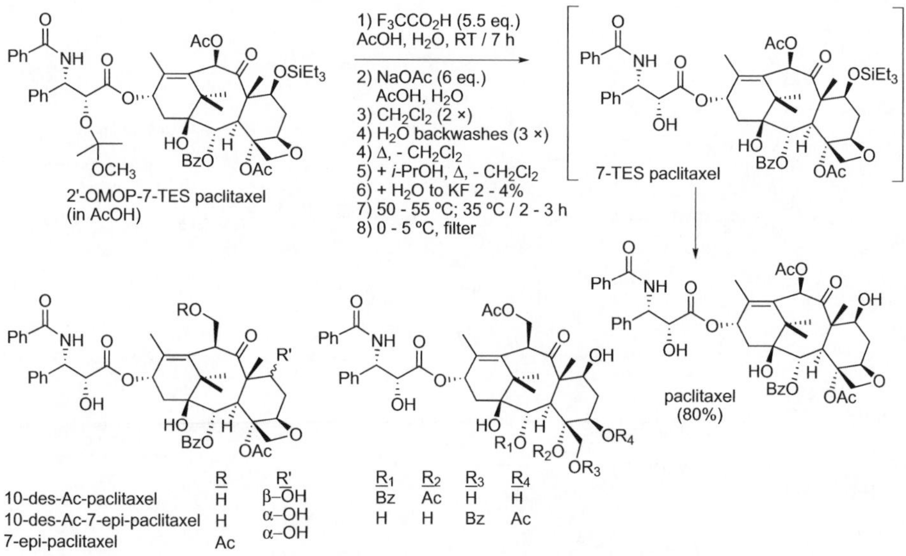

FIGURE 9.11 Quenching a TFA-mediated reaction reduced impurities in the manufacture of semi-synthetic paclitaxel.

HCl. TFA was chosen as the reagent for deprotection, as it is non-corrosive to the stainless steel equipment that was used in manufacturing, and by adding water to the reaction medium (3:1 AcOH:H$_2$O) the reactions were faster. Reaction mixtures degraded to various levels of the five impurities shown, along with a ring-contracted impurity, so the reaction was quenched with aqueous NaOAc. (Quenching with NaHCO$_3$ or Na$_2$CO$_3$ formed the epimeric impurities at the 7-position from retro-aldol–aldol reactions.) The product was extracted into CH$_2$Cl$_2$, concentrated, and the CH$_2$Cl$_2$ was chased with i-PrOH. After adjusting the moisture content, the concentrate was heated to provide a solution, cooled, and stirred at 30–35 °C for several hours for transformation to the desired polymorph [37,38]. Quenching the acid-mediated reaction made the process suitable for manufacturing semi-synthetic paclitaxel.

VIII. FACTORS AFFECTING HYDROGENATIONS

Physicochemical factors can affect the success of hydrogenations. In the chiral hydrogenation of the imine precursor to metolachlor (Figure 9.12) the temperature influenced mainly the enantioselectivity, while the hydrogen pressure influenced primarily the reaction rate. The catalyst-to-substrate loading was 1:2 × 10^6, requiring

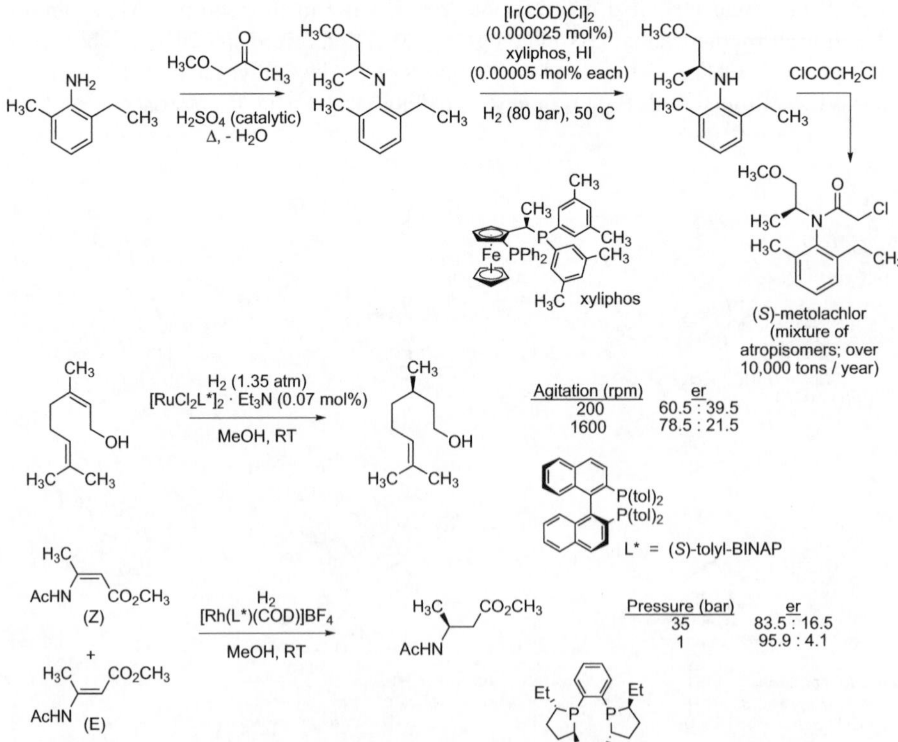

FIGURE 9.12 Factors involved in successful hydrogenations.

high quality for the input imine to preclude catalyst poisoning [39]. In hydrogenation of an alkylbenzene using Rh/C, small increases in temperature increased the reaction rate more than did small increases in pressure [40]. Blackmond and co-workers found that agitation can affect the enantiomeric ratio (er) in the hydrogenation of an allylic alcohol. The key factor was the amount of H_2 dissolved in the liquid phase, not the pressure of H_2 in the gaseous phase [41]. Catalytic hydrogenation of a mixture of enamines using the (S,S)-ethyl DuPhos catalyst efficiently produced the chiral β-amino ester; hydrogenation at an increased pressure lowered the er [42]. These examples illustrate the importance of screening pressure, temperature, and agitation in hydrogenations.

IX. STATISTICAL DESIGN OF EXPERIMENTS

Statistical design of experiments (DoE) provides an organized approach to generate data for process optimization, for any process with multiple parameters. In a DoE approach experiments may be run in random order while changing several variables at once, in contrast to optimization by the one-variable-at-a-time approach (OVAT, aka one-factor-at-a-time, or OFAT). DoE is generally used in three areas: relatively early in process optimization when the best approach is not obvious; in parallel screening efforts to find optima, as for QbD filings; and to optimize operations in routine manufacturing where an improved yield of a few percentages can afford large savings. DoE studies have been carried out using geometry, as in the simplex method and the Nelder–Mead algorithm, to find local optima or minima [43]. Software packages [44] are routinely used for chemical process development. The classic text for DoE is that from Box et al. [45]. The quality of DoE data will be no better than the reproducibility of processing, and the accuracy and reproducibility of assays.

The first example shown in Figure 9.13 serves to discuss the value of DoE in some detail. Researchers from the University of Bristol and GSK used DoE to raise the yield of a Heck reaction from 57% to 89%, while reducing the loading of the palladium catalyst from 3 mol% to 0.5 mol% [46]. Initially high and low levels of five variables were examined: catalyst loading, amount of base, time, concentration and temperature. With a full factorial study 2^5 experiments would be needed; a half-factorial study was undertaken, with additional "center points," for a total of 20 experiments. Based on the initial results a second set of 30 experiments was carried out to define non-linear interactions using a response surface model. Reactions were conducted in parallel reactor mode, which allowed for good reproducibility and rapid generation of data. The authors discussed that randomization is necessary to eliminate bias, and detailed approaches to minimize the number of experiments needed. Concentration and temperature were found to be interdependent variables. The optimal conditions were found to be relatively dilute, which probably prolonged catalyst lifetime by decreasing the formation of palladium mirror; [47] under dilute conditions higher temperatures were necessary for good conversion. These high-temperature, relatively dilute conditions were similar to those developed for some other ligand-free Heck reactions [47,48]. Optimization by the researchers from the University of Bristol and GSK was completed in only three weeks [46].

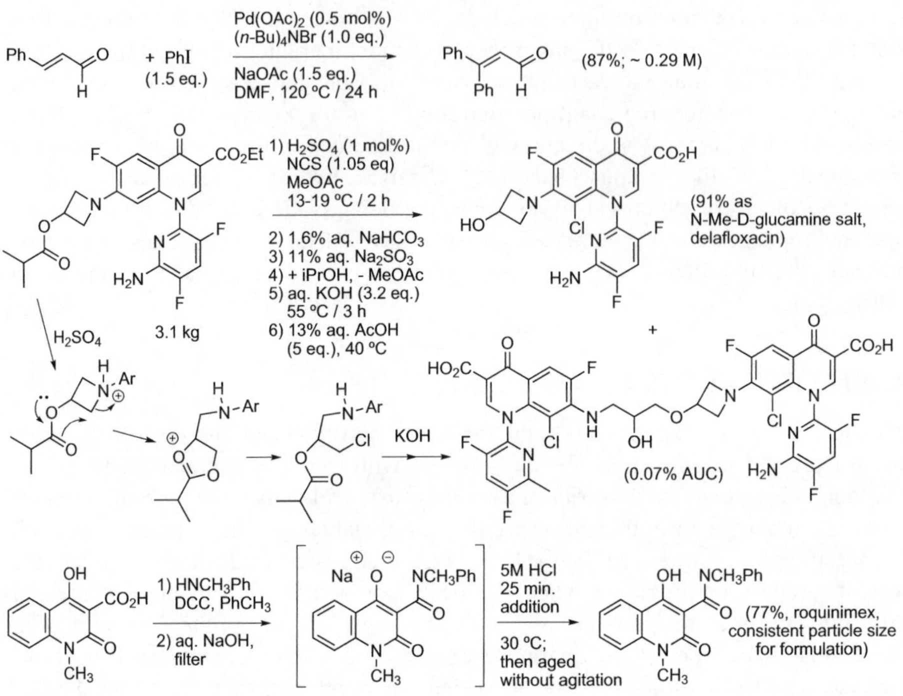

FIGURE 9.13 Some reactions optimized by DoE.

DoE has been used to optimize a wide diversity of processes. For instance, DoE was used to optimize a ring-closing metathesis [49], an oxidation using Na_2WO_4 and H_2O_2 [50], and a palladium-catalyzed conversion of aryl bromides to styrenes [51]. AstraZeneca researchers exploited DoE studies for a Swern oxidation and a reductive amination, rapidly preparing kilogram quantities of an NCE within the development timeframes [20]. For a sulfuric acid-catalyzed chlorination [52] of an aminoquinolone, DoE was used to optimize the yield and minimize the acid-mediated decomposition of the acyloxyazetidine (Figure 9.13); DoE predicted that with 1 mol% H_2SO_4 not more than 0.1% of the dimer would be produced, and runs in a kilo lab confirmed that [53]. In the sequential removal of THP and t-Bu protecting groups the yield was raised from 50–65% to 85% using DoE [54]. DoE has been used to increase the isolated yield of an alkylated product that initially decomposed under the workup conditions [55]. DoE has been used to optimize crystallization of an anhydrate from conditions containing water [56]. Using DoE robust conditions were developed to crystallize Roquinimex by adding aqueous HCl to a solution of the sodium salt [57] (Figure 9.13), and to define proven acceptable ranges (PARs) in QbD investigations [58]. DoE has been used to define optimal conditions for operating continuous counter-current distillation columns in preparing ketals [59]. Dozens of papers in *Organic Process Research & Development* detail DoE investigations.

X. REACTION KINETICS

"Kinetics" is a term used loosely within process development activities, often referring to how a reaction is progressing toward the accepted endpoint. Reaction kinetics is a powerful tool for understanding reaction mechanisms, and optimizing processes [60]. McQuade's group investigated the mechanism of the Baylis–Hillman reaction using reaction rate and isotopic data, and found that the reaction is second-order in aldehyde. The mechanism they proposed (Figure 9.14) explains why Baylis–Hillman reactions slowly proceed to completion when only one equivalent of aldehyde is used, and is consistent with the formation of dioxanone side products [61].

Blackmond has advanced reaction progress kinetic analysis to determine reaction mechanisms in complex catalytic systems, and a reaction mechanism can be deduced through only two well-chosen experiments [62,63]. Data from several orthogonal analyses has given further credence to the interpretations. For instance, studies by the Blackmond, Buchwald and Hartwig groups on a Pd(BINAP)-mediated amination of amines with 3-bromoanisole generated data from spectroscopy and kinetic rate studies (Figure 10.1). These studies indicated that the active catalyst was formed from the precatalyst components, and helped explain the induction period occasionally seen with this reaction [64].

XI. ALTERNATIVE ADDITION PROTOCOLS, OR CONTINUOUS PROCESSING

Continuous processing may be necessary to make more material safely or in high yield due to generation of a reactive intermediate or product. Nitration of the hydroxyaminopyrimidine in Figure 9.15 was found to be very exothermic [65], and for suitable temperature control the maximum scale for batch operations was a 22 L reactor. For SAFETY reasons continuous processing was developed to make more material, with continuous quenching into H_2O at 5 °C. The researchers optimized molar ratios,

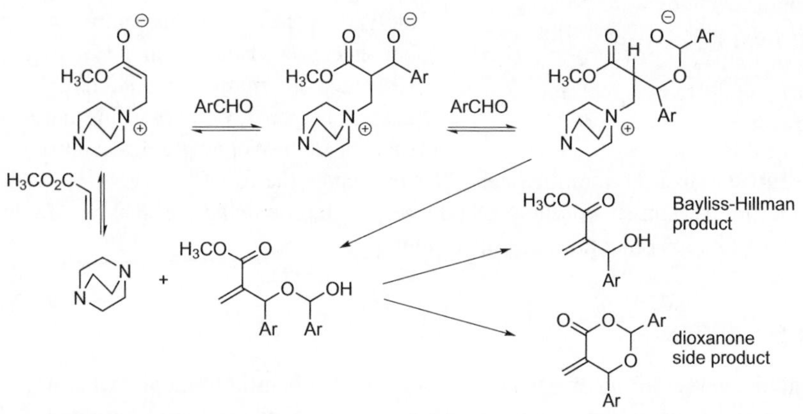

FIGURE 9.14 Mechanism of the Baylis–Hillman reaction, second-order in aldehyde.

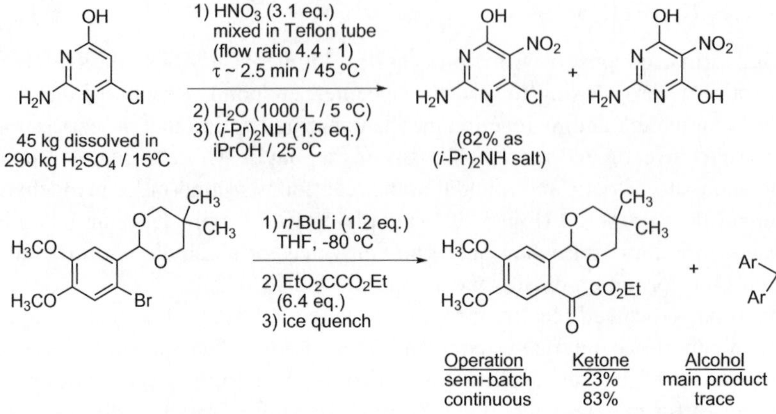

FIGURE 9.15   Practical applications of continuous processes.

temperature, flow rate, and [HNO$_3$]. The dihydroxy impurity was produced if the starting material was dissolved in warm H$_2$SO$_4$. If the residence time (τ) was much less than 2.5 min, a significant amount of unreacted starting material remained. Bis-nitration or decomposition occurred if τ was much longer than 2.5 min or if 98% HNO$_3$, were used. A stable salt was prepared by slurrying the wet cake with (iPr)$_2$NH [66]. (These details were described to illustrate the similar issues in optimizing continuous and batch operations.) For the acylation of the intermediate aryl lithium, the product ketoester was more reactive than diethyl oxalate, and low yields resulted from batch operations. A greatly improved yield was possible by mixing the aryl lithium intermediate with an excess of diethyl oxalate and physically separating the ketoester from the aryl lithium. The researchers adapted a 1-mL syringe barrel with a T-junction as a continuous reactor; by suction streams were drawn through the reactor and quenched immediately into ice. From this simple set-up the ketoester was isolated in 83% yield [67]. Chemical alternatives would have been to change reagents, perhaps using an arylzinc species or a Weinreb amide, introduce an additive to stabilize the intermediates [26], or change the route.

> Some chemists consider continuous operations as engineering solutions to problems. Dumping our problems on other people is irresponsible. Continuous operations may provide the best approaches to making more material, and should be considered.

Some comparisons of continuous and semi-batch processing are shown in Table 9.2. Chapter 14 focuses on continuous operations.

XII. PERSPECTIVE

To minimize levels of impurities in process R&D it is helpful to predict what impurities will form and how to detect them, in addition to minimizing the impurities that have been identified by routine analyses. Predicting impurity formation is also useful in

TABLE 9.2 Comparison of Parameters for Batch and Continuous Operations

Parameter	Batch or Semi-batch Operations	Continuous Operations
Endpoint	Reaction time	τ (Residence time in reactor, flow rate for a plugged flow reactor)
Comparative reaction temperatures	Cryogenic	-20 to $+40$ °C possible, due to shorter contact time
To increase turbulence	Increase agitation rate, use baffles	Increase flow rate (decrease τ)
[Reaction]	Routine	[Component] in each stream may be higher before combining
In-process assays	Usually external, but can be on-line (PAT)	Readily adaptable for PAT

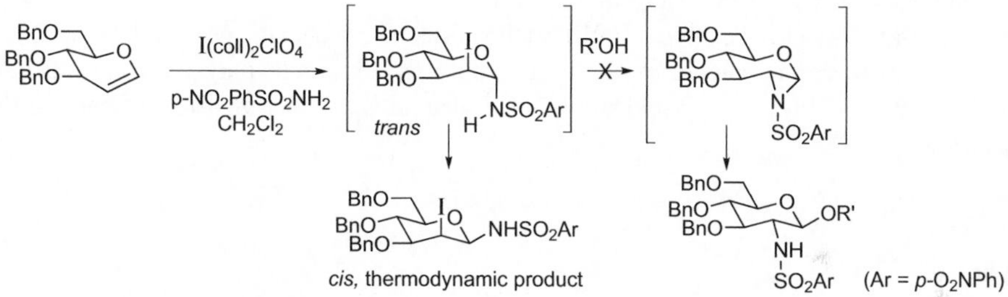

FIGURE 9.16 Expected formation of a β-glycoside.

troubleshooting (Chapter 17). Continuing to read the literature on analogous chemistry and processing is helpful to anticipate impurities. For instance, when using carbodiimides, epimerization and byproducts from *O*- to *N*-acyl migration can be minimized by employing additives such as HOBt (Figure 9.6). Efficient conditions for cross-couplings of aryl boronic acids minimize protodeboronation, as discussed in Chapter 10. Alkyl chloride byproducts can be minimized in mesylations, as discussed with Figure 9.1. Byproducts from Swern oxidation, Pummerer rearrangement, and other name reactions [68] have been discussed. By broadly reading the literature one may come across information that others might not consider in identifying and minimizing impurities; making associations between bits of information that at first seem unrelated can provide creative and valuable approaches to problem-solving. Unfortunately researchers often feel overwhelmed by the great deal of information available today. Ones experience is always extremely valuable.

In optimizing processes one can be fooled by false negatives and false positives, and

Don't discard negative results too quickly, and don't accept positive results too quickly. Corroborate results with independent analyses. Even scientists can be subjective in viewing data.

interpret data in a way that confirms ones hypotheses. For example, after iodinations the glycal repetitive couplings "failed" to afford the expected β-glycoside (Figure 9.16). Based on initial TLC and NMR analyses the product of these reactions was the expected *trans*-iodosulfonamide intermediate. Closer NMR analysis (nuclear Overhauser effects, with slight differences seen for H1 & C2 signals) revealed that the product had rearranged to the *cis*-iodosulfonamide. TLC showed very similar R_fs for the more stable *cis*-product and the *trans*-intermediate. (Further investigations into the scope of this reaction showed that the rearrangement also occurred upon chromatography and when benzenesulfonamide was reacted with the 3,4,6-tri-(*O*-pivalate)-glycal or with the 3-TES-4,6-*O*-acetonide-glycal [69].) Such subtle differences in spectral data, such as an NMR or IR signal outside of the range routinely expected, may indicate unusual shielding by an aryl ring or unusual strain in a molecule, and cause an astute researcher to reconsider data.

> To speed process development, optimize work-up and isolation after optimizing a reaction. Split experiments, varying processing on each stream, and compare results. Minimize duplication, unless reproducibility is the goal. If extended processing produces byproducts, more attention will be required for scale-up. Robotics and automation are the present and the future means of generating data, but the results can be no better than the reproducibility and accuracy of the reaction conditions and assays.

REFERENCES

1. Federsel, H.-J.; Sveno, A. In *Process Chemistry in the Pharmaceutical Industry*; Gadamasetti, K.; Braish, T., Eds., 2nd ed.; CRC Press: Boca Raton, Florida, 2008; p 111.
2. Tanabe, Y.; Yamamoto, H.; Yoshida, Y.; Miyawaki, T.; Utsumi, N. *Bull. Chem. Soc. Jpn.* **1995**, *68*, 297.
3. Wuts, P. G. M.; Ashford, S. W.; Conway, B.; Havens, J. L.; Taylor, B.; Hritzko, B.; Xiang, Y.; Zakarias, P. S. *Org. Process Res. Dev.* **2009**, *13*, 331.
4. Faul, M. M.; Kobierski, M. E.; Kopach, M. E. *J. Org. Chem.* **2003**, *68*, 5739.
5. Boini, S.; Moder, K. P.; Vaid, R. K.; Kopach, M.; Kobierski, M. *Org. Process Res. Dev.* **2006**, *10*, 1205.
6. Meyers, C.; Maes, B. U. W.; Loones, K. T. J.; Bal, G.; Lemiere, G. L. F.; Dommisse, R. A. *J. Org. Chem.* **2004**, *69*, 6010.
7. Qafisheh, N.; Mukhopadhyay, S.; Joshi, A. V.; Sasson, Y.; Chuah, G.-K.; Jaenicke, S. *Ind. Eng. Chem. Res.* **2007**, *46*, 3016.
8. Wilk, B. K.; Mwisiya, N.; Helom, J. L. *Org. Process Res. Dev.* **2008**, *12*, 785.
9. Fray, M. J.; Gillmore, A. T.; Glossop, M. S.; McManus, D. J.; Moses, I. B.; Praquin, C. F. B.; Reeves, K. A.; Thompson, L. R. *Org. Process Res. Dev.* **2010**, *14*, 263.
10. Bowles, D. M.; Bolton, G. L.; Boyles, D. C.; Curran, T. T.; Hutchings, R. H.; Larsen, S. D.; Miller, J. M.; Park, W. K. C.; Schineman, D. C. *Org. Process Res. Dev.* **2008**, *12*, 1183.
11. Dupau, P.; Epple, R.; Thomas, A. A.; Fokin, V. V.; Sharpless, K. B. *Adv. Synth. Catal.* **2002**, *344*(3+4), 421.
12. Caille, S.; Cui, S.; Hwang, T.-L.; Wang, X.; Faul, M. M. *J. Org. Chem.* **2009**, *74*, 3833.
13. Loeser, E.; Prasad, K.; Repic, O. *Synth. Commun.* **2002**, *32*, 403.
14. Karpf, M.; Trussardi, R. *J. Org. Chem.* **2001**, *66*, 2044.
15. Barnwell, N.; Cheema, L.; Cherryman, J.; Dubiez, J.; Howells, G.; Wells, A. *Org. Process Res. Dev.* **2006**, *10*, 644.

16. Rao, C. S. *The Chemistry of Process Development in Fine Chemical & Pharmaceutical Industry*, 2nd ed.; Wiley: New York, 2007.

17. Bio, M. M.; Hansen, K. B.; Gipson, J. *Org. Process Res. Dev.* **2008**, *12*, 892.

18. Zook, S. E.; Busse, J. K.; Borer, B. C. *Tetrahedron Lett.* **2000**, *41*, 7017.

19. Xu, F.; Simmons, B.; Reamer, R. A.; Corley, E.; Murry, J.; Tschaen, D. *J. Org. Chem.* **2008**, *73*, 312.

20. Parker, J. S.; Bowden, S. A.; Firkin, C. R.; Moseley, J. D.; Murray, P. M.; Welham, M. W.; Wisedale, R.; Young, M. J.; Moss, W. O. *Org. Process Res. Dev.* **2003**, *7*, 67.

21. Bäckvall, J.-E.; Sellén, M. *JCS Chem. Commun.* **1987,** 827.

22. Simpson, J. H.; Godfrey, J.; Fox, R.; Kotnis, A.; Kacsur, D.; Hamm, J.; Totelben, M.; Rosso, V.; Mueller, R.; Delaney, E.; Deshpande, R. P. *Tetrahedron: Asymmetry* **2003**, *14*, 3569.

23. Nettles, S. M.; Matos, K.; Burkhardt, E. R.; Rouda, D. R.; Corella, J. A. *J. Org. Chem.* **2002**, *67*, 2970; see also Hilborn, J. W.; Lu, Z.-H.; Jurgens, A. R.; Fang, Q. K.; Byers, P.; Wald, S. A.; Senanayake, C. H. *Tetrahedron Lett.* **2001**, *42*, 8919.

24. Königsberger, K.; Chen, G.-P.; Wu, R. R.; Girgis, M. J.; Prasad, K.; Repic, O.; Blacklock, T. J. *Org. Process Res. Dev.* **2003**, *7*, 733.

25. Rassias, G.; Hermitage, S. A.; Sanganee, M. J.; Kincey, P. M.; Smith, N. M.; Andrews, I. P.; Borrett, G. T.; Slater, G. R. *Org. Process Res. Dev.* **2009**, *13*, 774.

26. Rawalpally, T.; Ji, Y.; Shankar, A.; Edwards, W.; Allen, J.; Jiang, Y.; Cleary, T. P.; Pierce, M. E. *Org. Process Res. Dev.* **2008**, *12*, 1293.

27. Ennis, D. S.; Lathbury, D. C.; Wanders, A.; Watts, D. *Org. Process Res. Dev.* **1998**, *2*, 287; D. Ennis. Personal communication.

28. Li, W.; Nelson, D. P.; Jensen, M. S.; Hoerrner, R. S.; Cai, D.; Larsen, R. D.; Reider, P. J. *J. Org. Chem.* **2002**, *67*, 5394.

29. Vazquez, E.; Davies, I. W.; Payack, J. F. *J. Org. Chem.* **2002**, *67*, 7551.

30. Alessi, M.; Larkin, A. L.; Ogilvie, K. A.; Green, L. A.; Lai, S.; Lopez, S.; Snieckus, V. *J. Org. Chem.* **2007**, *72*, 1588.

31. Niwayama, S.; Cho, H.; Zabet-Moghaddam, M.; Whittlesey, B. R. *J. Org. Chem.* **2010**, *75*, 3775.

32. Parker, J. S.; Smith, N. A.; Welham, M. J.; Moss, W. O. *Org. Process Res. Dev.* **2004**, *8*, 45.

33. Anderson, N. G.; Ciaramella, B. M.; Feldman, A. F.; Lust, D. A.; Moniot, J. L.; Moran, L.; Polomski, R. E.; Wang, S. S. Y. *Org. Process Res. Dev.* **1997**, *1*, 211.

34. Ace, K. W.; Armitage, M. A.; Bellingham, R. K.; Blackler, P. D.; Ennnis, D. S.; Hussain, N.; Lathbury, D. C.; Morgan, D. O.; O'Connor, N.; Oakes, G. H.; Passey, S. C.; Powling, L. C. *Org. Process Res. Dev.* **2001**, *5*, 479.

35. Clark, J. D.; Heise, J. D.; Shah, A. S.; Peterson, J. C.; Chou, S. K.; Levine, J.; Karakas, A. M.; Ma, Y.; Ng, K.-Y.; Patelis, L.; Springer, J. R.; Stano, D. R.; Wettach, R. H.; Dutra, G. A. *Org. Process Res. Dev.* **2004**, *8*, 176.

36. Murray, W. V.; Francois, D.; Maden, A.; Turchi, I. *J. Org. Chem.* **2007**, *72*, 3097.

37. Singh, A. K.; Weaver, R. E.; Powers, G. L.; Rosso, V. W.; Wei, C.; Lust, D. A.; Kotnis, A. S.; Comezoglu, F. T.; Liu, M.; Bembenek, K. S.; Phan, B. D.; Vanyo, D. J.; Davies, M. L.; Mathew, R.; Palaniswamy, V. A.; Li, W.-S.; Gadamsetti, K.; Spagnuolo, C. J.; Winter, W. J. *Org. Process Res. Dev.* **2003**, *7*, 25.

38. Powers, G. L.; Rosso, V. W.; Singh, A. K.; Weaver, R. E. U.S. 6,184,395, 2001 (to Bristol-Myers Squibb).

39. Blaser, H.-U.; Hanriech, R.; Schneider, H.-D.; Spindler, F.; Steinacher, B. *Asymmetric Catalysis on Industrial Scale*; Wiley-VCH, 2004; p 55. This chapter also describes the equipment used on large scale to make metolachlor: water separators, heat exchangers, loop hydrogenator, thin film evaporator, and hydrogen generator.

40. Anderson, N. G.; Burke, L. Unpublished.

41. Sun, Y.; Landau, R. N.; Wang, J.; LeBlond, C.; Blackmond, D. G. *J. Am Chem. Soc.* **1996**, *118*, 1348.

42. Heller, D.; Holz, J.; Drexler, H.-J.; Lang, J.; Drauz, K.; Krimmer, H.-P.; Börner, A. *J. Org. Chem.* **2001**, *66*, 6816; see also Jin, F.; Wang, D.; Confalone, P. N.; Pierce, M. E.; Wang, Z.; Xu, G.; Choudhury, A.; Nguyen, D. *Tetrahedron Lett.* **2001**, *42*, 4787.

43. http://www.scholarpedia.org/article/Nelder-Mead_algorithm. Prospectors may have used this approach in mining ores.

44. For example, see http://www.jmp.comfor JMP software, and http://www.statease.com for DesignExpert software

45. Box, G. E. P.; Hunter, W. G.; Hunter, J. S. *Statistics for Experimenters*; Wiley, 1978.

46. Aggarwal, V. K.; Staubitz, A. C.; Owen, M. *Org. Process Res. Dev.* **2006**, *10*, 64.

47. de Vries, A. H. M.; Mulders, J. M. C. A.; Mommers, J. H. M.; Henderickx, H. J. W.; de Vries, J. G. *Org. Lett.* **2003**, *5*, 3285.

48. Arvela, R. K.; Leadbeater, N. E.; Sangi, M. S.; Williams, V. A.; Granados, P.; Singer, R. D. *J. Org. Chem.* **2005**, *70*, 161.

49. Wang, H.; Goodman, S. N.; Dai, Q.; Stockdalem, G. W.; Clark, W. M., Jr. *Org. Process Res. Dev.* **2008**, *12*, 226.

50. Hida, T.; Fukui, Y.; Kawata, K.; Kabaki, M.; Masui, T.; Fumoto, M.; Nogusa, H. *Org. Process Res. Dev.* **2010**, *14*, 289.

51. Denmark, S. E.; Butler, C. R. *J. Am. Chem. Soc.* **2008**, *130*, 3690.

52. For AcOH catalysis of a chlorination using NCS, see Pesti, J. A.; Hunn, G. F.; Yin, J.; Xing, Y.; Fortunak, J. M.; Earl, R. A. *J. Org. Chem.* **2000**, *65*, 7718.

53. Hanselmann, R.; Johnson, G.; Reeve, M. M.; Huang, S.-T. *Org. Process Res. Dev.* **2009**, *13*, 54.

54. Kuethe, J. T.; Tellers, D. M.; Weissman, S. M.; Yasuda, N. *Org. Process Res. Dev.* **2009**, *13*, 471.

55. de Koning, P. D.; McManus, D. J.; Bandurek, G. R. *Org. Process Res. Dev.* **2011**, *15*, 1081.

56. Black, S. N.; Phillips, A.; Scott, C. I. *Org. Process Res. Dev.* **2009**, *13*, 78.

57. Sjövall, S.; Hansen, L.; Granquist, B. *Org. Process Res. Dev.* **2004**, *8*, 802.

58. Cimarosti, Z.; Bravo, F.; Castoldi, D.; Tinazzi, F.; Provera, S.; Perboni, A.; Papini, D.; Westerduin, P. *Org. Process Res. Dev.* **2010**, *14*, 805.

59. Clarkson, J. S.; Walker, A. J.; Wood, M. A. *Org. Process Res. Dev.* **2001**, *5*, 630.

60. Singh, U. K.; Orella, C. J. In *Chemical Engineering in the Pharmaceutical Industry: R&D to Manufacturing*; am Ende, D. J., Ed.; Wiley, 2010; p 79.

61. Price, K. E.; Broadwater, S. J.; Walker, B. J.; McQuade, D. T. *J. Org. Chem.* **2005**, *70*, 3980.

62. Mathew, J. S.; Klussmann, M.; Iwamura, H.; Valera, F.; Futran, A.; Emanuelsson, E. A. C.; Blackmond, D. G. *J. Org. Chem.* **2006**, *71*, 4711.

63. Blackmond, D. G. *Angew. Chem. Int. Ed.* **2005**, *44*, 4302.

64. Shekhar, S.; Ryberg, P.; Hartwig, J. F.; Mathew, J. S.; Blackmond, D. G.; Strieter, E. R.; Buchwald, S. L. *J. Am. Chem. Soc.* **2006**, *128*, 3584.

65. For another example of an exothermic reaction and safety considerations, see Marterer, W.; Prikoszovich, W.; Wiss, J.; Prashad, M. *Org. Process Res. Dev.* **2003**, *7*, 318.

66. De Jong, R. L.; Davidson, J. G.; Dozeman, G. J.; Fiore, P. J.; Kelly, M. E.; Puls, T. P.; Seamans, R. E. *Org. Process Res. Dev.* **2001**, *5*, 216.

67. Hollwedel, F.; Koßmehl, G. *A. Synthesis* **1998**, 1241.

68. Li, J. J. *Name Reactions: A Collection of Detailed Reaction Mechanisms*; 3rd ed.; Springer, 2006.

69. Owens, J. M.; Yeung, B. K. S.; Hill, D. C.; Petillo, P. A. *J. Org. Chem.* **2001**, *66*, 1484.

Optimizing Organometallic Reactions

"The identification of side-products, observed regioselectivity, reactivity or lack of reactivity, should be viewed merely as a symptom of the underlying complicated mechanistic pathways involving Pd…On one hand, it is of significant interest to understand how to minimise unwanted reactions and their side-products, but on the other, one can seek to exploit such reaction pathways and develop new synthetic methodologies."

– Gerard P. McGlacken and Ian J. S. Fairlamb [1]

I. INTRODUCTION

Many books have been published recently on optimizing catalytic, organometallic reactions [2–14]. Recently excellent reviews have been published on patented cross-couplings [15], cross-coupling reactions used on scale to make APIs [16] and products in the pharmaceutical, agrochemical and fine chemicals industries [17], cross-couplings of heteroarenes [18], side products identified in Pd-catalyzed cross-couplings [1], and an extensive chapter reviewing cross-coupling reactions that could be adapted for further scale-up [19]. The breadth and depth of these reviews will lend insight into other process R&D investigations. Other reviews on dehydrative cross-couplings [20], oxidative cross-couplings [21], cobalt-catalyzed cross-couplings [22], and cross-couplings of alkyl organometallics [23] may be applied more to future process R&D in the pharmaceutical industry. The field of organometallic chemistry is burgeoning. This chapter will mention some approaches for process optimization, with minimal repetition of information from those excellent reviews.

Practical Process Research and Development. DOI: 10.1016/B978-0-12-386537-3.00010-1

To apply organocatalytic methods for cost-effective scale-up there are additional practical considerations. First of all, expensive ligands and metal precatalysts will increase the CoGs, especially if high loadings are necessary. Secondly, sensitivity of the precatalysts and the catalysts to O_2 and H_2O can complicate handling on scale. Thorough degassing is often required for metal-catalyzed processes to minimize formation of side products. Techniques are available to reduce O_2 and H_2O in reactors to very low levels, and to monitor these levels. Dry boxes can even be placed on top of reactors in manufacturing facilities if necessary to charge components under controlled conditions. Finally, difficulties in removing the metals, ligands and side products will extend workups, reducing productivity and possibly reducing isolated yields. Charging extremely low levels of catalysts presents a lower burden for purification, but at lower catalyst loadings processes are more sensitive to low levels of inhibitors. Recovering and recycling the metals and ligands can decrease the CoGs, but some effort may be required to develop such methods.

For scale-up insoluble (heterogeneous) catalysts classically have been preferred over homogeneous catalysts, because heterogeneous catalysts are more readily recovered and recycled [24]. Insoluble supports can include activated carbon, alumina, silica, polymers such as PdEnCat®, and carbon nanotubes [25]. The use of heterogeneous palladium catalysts for C–C bond formation has been reviewed [26]. Since ligands are rarely added to a process using heterogeneous catalysts, it is not necessary to monitor the decomposition of the ligands and the presence of ligands and side products in the product, and some heterogeneous catalysts are less expensive. Recovery of the catalyst and removal of the metals from the product may be facilitated if the metal is readsorbed to the solid support of the heterogeneous catalyst at the end of the reaction. However, heterogeneous catalysts may not provide the reactivity or selectivity necessary for a successful process, and homogeneous catalysts are widely used today.

The mechanisms involved in organocatalytic reactions are complex. The Blackmond, Buchwald, and Hartwig groups investigated in detail a palladium-catalyzed amination, using NMR spectroscopy, rate data, and kinetic rate studies (Figure 10.1). Under the basic conditions the amine added to dibenzylidene acetone (dba) in a Michael fashion, and other side products were also formed. The most rapid coupling reaction with 3-methoxyphenyl bromide occurred when $Pd_2(dba)_3$ and BINAP were incubated with NaOt-Am and the amine before adding the aryl bromide. From the precatalysts $Pd_2(dba)_3$ and BINAP three more precatalyst species were identified. Of the latter three precatalysts, [Pd(BINAP)]$_2$(dba) most readily produced the active, catalytic species [27]. The scheme in Figure 10.1 may serve to illustrate the complexity of metal-catalyzed coupling reactions [28,29].

To enter the catalytic cycle above Pd(II) must be reduced to Pd(0). Phosphine ligands and amines may be the reductants, as shown in Figure 10.2 [30]. Phosphine oxide byproducts may readily be detected by HPLC. If these components are not charged in relatively small amounts the byproducts may be significant impurities. For instance the side product acetone from oxidation of diisopropylamine produced 9% of a 2-methyl indole impurity, and analogous impurities were formed when triethylamine or dicyclohexylamine was used as a base [31]. Treatment of the precatalyst PdCl$_2$[PPh$_3$]$_2$ with 4-fluorophenylboronic acid produced triphenylphosphine oxide and the biaryl shown in Figure 10.3 [32]. In Pd-catalyzed cyanations Pd(II) has been reduced by zinc metal [33],

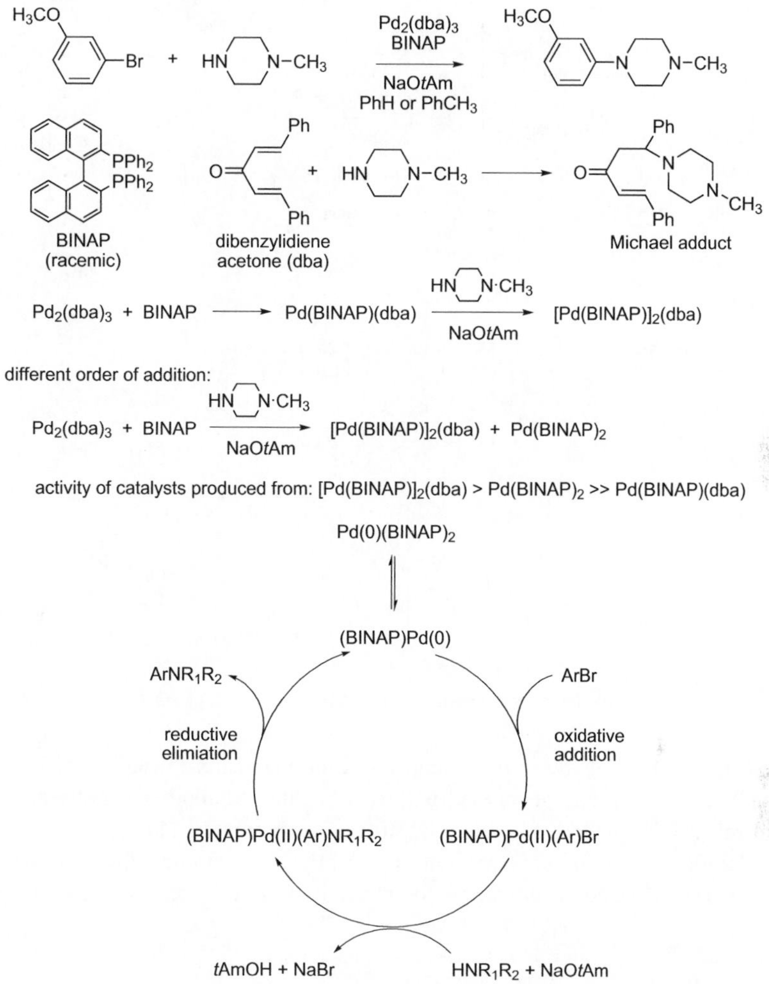

FIGURE 10.1 Proposed reaction mechanism and side reactions for a Pd-catalyzed amination.

ZnEt$_2$ [34], and electrolysis [35]. Catalysts for iron-mediated cross-coupling have been prepared by reducing Fe(acac)$_3$ with Grignard reagents [36]. Catalysts for nickel-mediated cross-coupling have been prepared by using Vitride (sodium bis-2-methoxyethoxy aluminum hydride) [37], triphenylphospine [38] or n-BuLi [39] as reductants.

The mechanism of a Suzuki–Miyaura coupling of a trifluoroborate [40] was probed by NMR studies (Figure 10.3). Controlling the amount of water present was essential to achieving an excellent yield. Even without degassing essentially none of the phenol and homocoupled side products were obtained from cross-coupling of the fluoroborate with the aryl bromide, whereas those two impurities were formed with inadequate degassing when the coupling was conducted with the boronic acid. Protodeboronation was seen in the absence of Pd, during titration studies. The trifluoroborate functioned as a sustained release form of the boronic acid, the active species for transmetalation. Slow hydrolysis

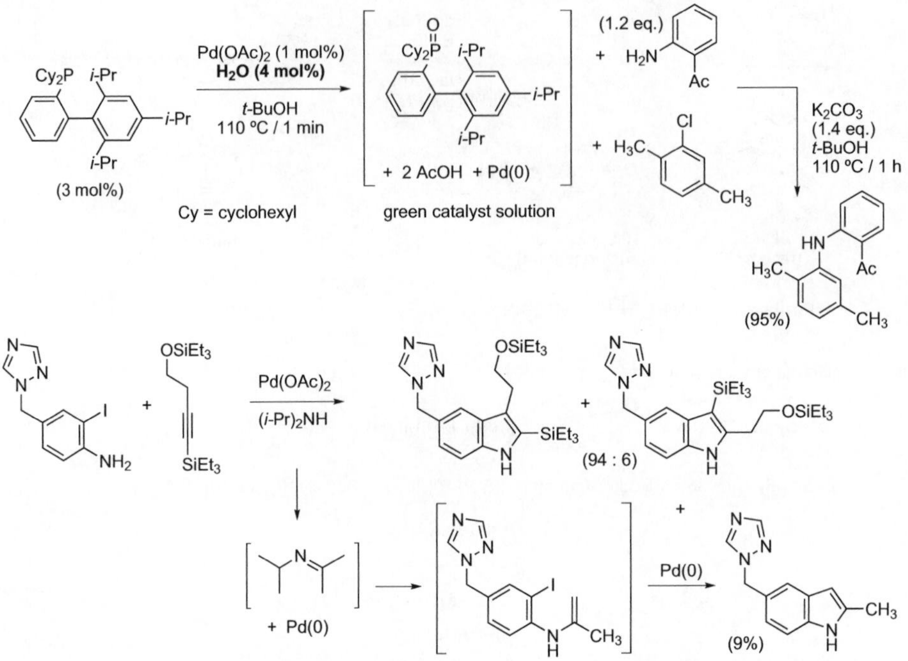

FIGURE 10.2 Side products formed from reduction of Pd(II).

prevented the buildup of the boronic acid and minimized the formation of impurities [32]. The Burke group has developed MIDA boronates as another beneficial form of sustained release for Suzuki–Miyaura couplings (see Figure 10.11) [41].

Considerations about ligands are central to catalytic organometallic chemistry. New ligands may be developed to address synthetic needs, as highlighted by research from the

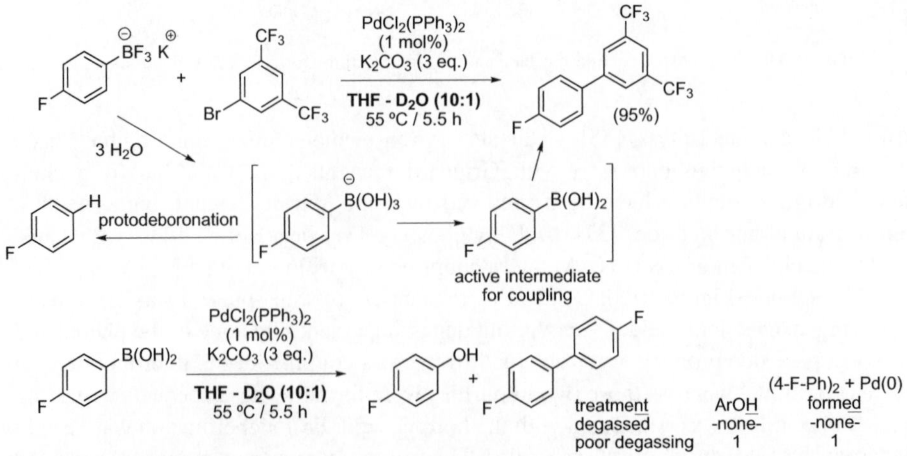

FIGURE 10.3 Mechanism of Suzuki–Miyaura coupling of a trifluoroborate.

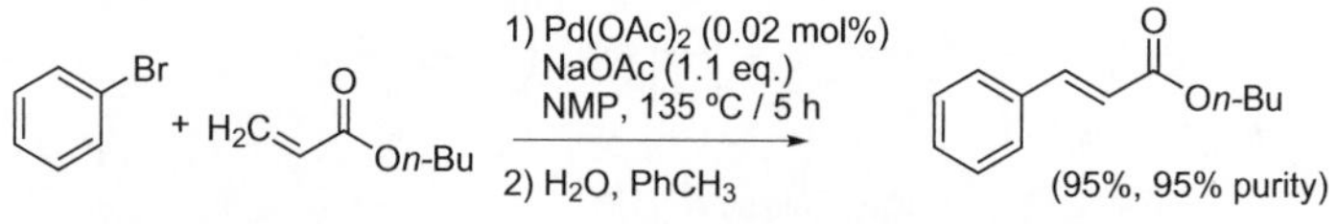

FIGURE 10.4 Homeopathic, ligand-free Heck reaction.

groups of Buchwald, Hartwig, and Nolan and others. The excellent review on patented processes from Corbet and Mignani discusses some of these efforts [15]. Some transition metal-catalyzed reactions have been conducted without ligands. Reasoning that ligands help to solubilize metals, and keep them from forming inactive species, such as Pd mirrors, a Heck coupling was studied with only 0.02 mol% of Pd(OAc)$_2$ (Figure 10.4). Under these "homeopathic Heck" conditions, with a very high substrate to catalyst ratio, the Heck reaction is faster than aggregation [42–44]. Although not using ligands can lower the CoGs and simplify product purification, at such low catalyst loadings, small amounts of impurities may have a big effect on inhibiting reactions, and agitation may affect such heterogeneous reactions.

Changing the metals employed can significantly change reaction products. Perhaps the most familiar area where this is true is in hydrogenations over catalysts, as in Figure 10.5 [45]. Substituting a metal while keeping the ligand constant can also change the outcome [46], as with the catalyzed cyanation of acetophenone using a ligand derived from D-glucose [47]; such reactions may be second-order in catalysts [48,49].

Developing optimal conditions for catalytic reactions can require a great deal of screening to identify optimal combinations of metal, ligands, and solvents. Pfizer researchers used high-throughput experimentation to screen 200 reaction conditions on a 10 mg scale to find suitable conditions for Suzuki couplings of a family of three aryl boronic acids. This screening was complicated by the low solubility of the starting materials and products, and the electronic differences in the aryl boronic acids. Reaction conditions found to give over 98% conversion for these substrates are shown in Figure 10.6 [50]. Merck [51] and GSK [52] researchers have also discussed examples of extensive and successful screening.

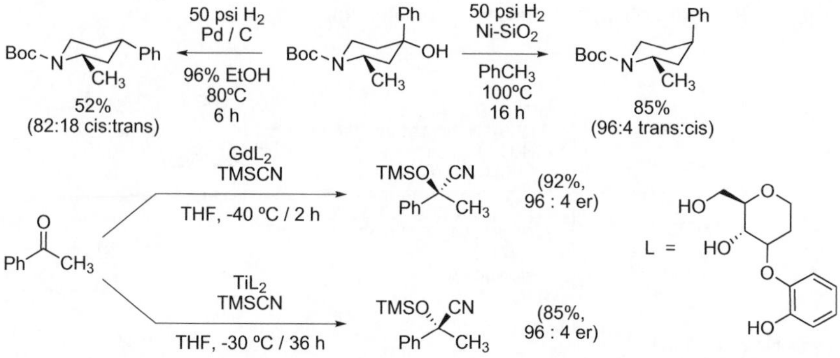

FIGURE 10.5 Influence of changing metals in a reduction and cyanosilylation.

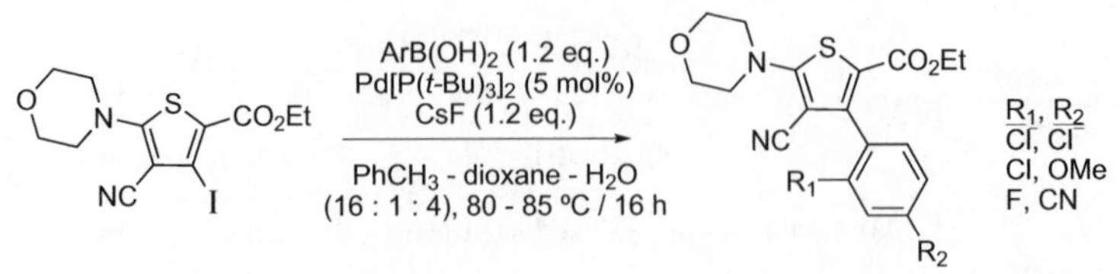

FIGURE 10.6   Optimal conditions for a Suzuki–Miyaura coupling found through high-throughput screening.

Optimal addition sequences and temperatures may lead to unusual or relatively narrow sets of optimal conditions. Extensive optimization may be undertaken if significant economic benefits are possible. Persistent process R&D led to a reliable oxidation process to prepare esomeprazole, in an example of chiral switching [53] (Figure 10.7). Water and base played key roles in the titanium-catalyzed oxidation. A wide range of chiral diols effectively provided the necessary chirality, and S,S-diethyl tartrate was the most available chiral diol. The best composition of a complex with titanium tetra-isopropoxide was 1:2:0.5 Ti:S,S-diethyl tartrate:H_2O. Introducing H_2O raised the enantiomeric ratio (er), but higher levels produced more sulfone side product. Equilibration of the catalyst complex in the presence of starting material at 50 °C for 1 h was found to be optimal, and the reaction was cooled to 30 °C and charged with diisopropylethylamine. Organic bases also increased the er, and of the 10 examined (i-Pr)$_2$NEt gave the process that was easiest to operate, with the best results when (i-Pr)$_2$NEt was added with the oxidant. Then 0.97 equivalents of cumene hydroperoxide in cumene was added, and oxidation was completed in 30–45 min. A wide range of non-polar solvents was effective, and toluene was preferred, in order to telescope the oxidation with the formation of the sulfide starting material. Workup involved extracting into basic aqueous solutions, perhaps aqueous ammonia, removing titanium salts, and upgrading the chirality by dissolving the crude product in acetone and removing the precipitated sulfoxide racemate. The magnesium salt was formed by adding a solution of $MgSO_4$ to a basic solution of esomeprazole. The resources of many people over many years led to this rather complex, optimized process for esomeprazole [54].

This chapter will be broken down into four main areas to improve reaction conditions: chemical activation, operations to minimize side reactions, the impact of impurities, and some considerations on heterogeneous catalysis.

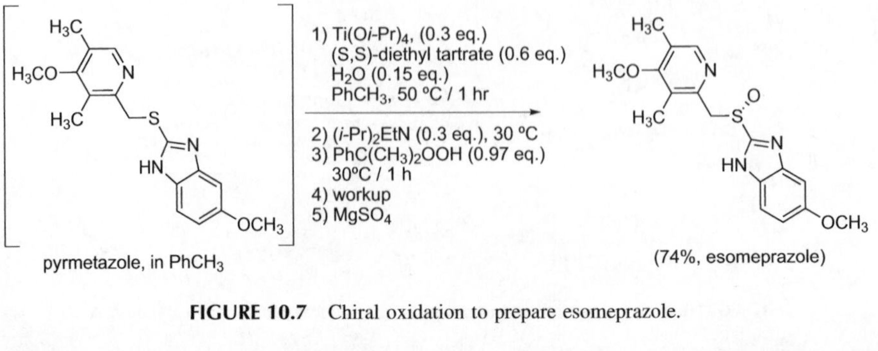

FIGURE 10.7   Chiral oxidation to prepare esomeprazole.

Transition metal catalysts, precatalysts, and ligands can be hazardous. Anhydrous catalysts can be highly pyrophoric, which is why commercial catalysts such as Pd/C, Rh/C, and Pt/C are supplied as moist solids containing about 50% water. Filtering off a Pd/C catalyst and suctioning air though the cake can cause a fire. Ligands such as tri-t-butylphosphine are pyrophoric. Organometallic reactions can be very exothermic, as has been shown for a Suzuki cross-coupling [55] and Heck reactions [43,56]. Metal-catalyzed processes should be evaluated for SAFE operating conditions before scale-up.

II. CHEMICAL ACTIVATION TO IMPROVE REACTION CONDITIONS

For stoichiometric organometallic reactions Grignard, Barbier, Reformatsky, and organolithium reactions are foremost. Approaches to optimize organolithium reactions have been reviewed [57]. Magnesium and zinc metals, as chips, shavings or powders, are activated to ensure that formation of the Grignard and Reformatsky reagents proceeds controllably, without an induction period. Some activating reagents are shown in Table 10.1. Rao has devoted one chapter in his book to additives [58].

III. OPERATIONS TO MINIMIZE SIDE REACTIONS

Attention to details is critical for most organometallic processes. Degassing, solvent choice, temperature, addition sequence, and extended additions may be key operations.

Degassing has been critical with many Pd-catalyzed [15,16] and Ru-catalyzed processes [59], and the need to degas may be dependent on the catalyst and the process conditions [60]. Operations in the laboratory using Schlenk tubes may degas solutions using freeze–thaw cycles; this may be why some academic researchers have chosen benzene as the reaction solvent. Oxygen is thoroughly removed by heating the solvent to reflux and cooling under N_2. A color change may be seen, depending on the ligands; for instance, Novartis researchers noted that a Pd(II) catalyst was reddish, while the Pd(0) form was yellow [61]. Distilling one or two volumes of solvent was also found to be effective in removing O_2 [62]. For systems sensitive to high temperatures, oxygen in a reactor may be removed by subsurface sparging with N_2 [63]. Oxygen sensors can be installed in the headspace of vessels for use with organic solvents [64], or a dissolved O_2 probe can be employed for reactions in water [65].

Excluding O_2 can also be critical for reactions using stoichiometric amounts of metals. Modi *et al.* found that O_2 reacted with Grignard reagents faster than H_2O [66]. Researchers at Johnson & Johnson studied the conversion of an Mtr-protected arginine derivative to the corresponding ketones using Grignard reagents [67]. (The 4-methoxy-2,3,6-trimethylbenzenesulfonyl (Mtr) protecting group is removed by TFA and other strong acids. Other protecting groups have superseded the use of the Mtr group for arginine residues [68].) The J&J researchers found that when flasks were opened for sampling O_2 was inadvertently introduced and led to the formation of the benzo[d]thiazole trimer shown in Figure 10.8; in this case one molecule of O_2

TABLE 10.1 Some Activating Reagents for Stoichiometric and Catalytic Organometallic Reagents

Reaction Type	Activating Reagent	Function of Reagent	Reference
Grignard (Mg)	TMSCl DIBAL-H I_2 + DIBAL-H $BrCH_2CH_2Cl$ Vitride (Red-Al) Grignard from prior batch	Removes H_2O bound to surface of metal, "cleans" metal surface	(1—6)
Reformatsky (Zn)	$(i\text{-Bu})_2AlH$ (DIBAL-H) CuCl, H_2SO_4	Activates Zn(0) in Barbier-like fashion	(7,8)
Adams catalyst $(PtO_2 \cdot H_2O)$	AcOH wash	Activates catalyst for hydrogenation	(9)
Heck coupling	$(n\text{-Bu})_4NBr$ (TBAB)	Catalyst	(10)
Pd-catalyzed cyanation	$ZnBr_2$, H_2O i-PrOH $Zn(O_2CH)_2 \cdot 2H_2O$	Increased solubility May reduce Pd(II) to Pd(0) May reduce Pd(II) to Pd(0)	(11—13)
Suzuki coupling	KF	Catalyst	(14)

(1) Ace, K. W.; Armitage, M. A.; Bellingham, R. K.; Blackler, P. D.; Ennis, D. S.; Hussain, N.; Lathbury, D. C.; Morgan, D. O.; O'Connor, N.; Oakes, G. H.; Passey, S. C.; Powling, L. C. Org. Process Res. Dev. **2001,** 5, 479.
(2) Tilstam, U.; Weinmann, H. Org. Process Res. Dev. **2002,** 6, 906.
(3) Kopach, M. E.; Singh, U. K.; Kobierski, M. E.; Trankle, W. G.; Murray, M. M.; Pietz, M. A.; Forst, M. B.; Stephenson, G. A.; Mancuso, V.; Giard, T.; Vanmarsenille, M.; DeFrance, T. Org. Process Res. Dev. **2009,** 13, 209.
(4) Wiss, J.; Länzlinger, M.; Wermuth, M. Org. Process Res. Dev. **2005,** 9, 365.
(5) Vitride is sodium (bis-methoxyethoxy)aluminum hydride: Farina, V.; Groziner, K.; Müller-Bötticher, H.; Roth, G. in Process Chemistry in the Pharmaceutical Industry; Gadamasetti, K. G., Ed.; Marcel-Dekker; 1999; p 114. Refer to Simanek, V.; Klasek, K. Coll. Czech. Chem. Commun. **1973,** 38, 1614.
(6) Silverman, G. Rakita, P. E. Handbook of Grignard Reagents; CRC Press: Boca Raton, FL, 1996; pp 79—88.
(7) Girgis, M. J.; Liang, J. K.; Du, Z.; Slade, J.; Prasad, K. Org. Process Res. Dev. **2009,** 13, 1094.
(8) Scherkenbeck, T.; Siegel, K. Process Res. Dev. **2005,** 9, 216.
(9) Ikemoto, N.; Tellers, D. M.; Dreher, S. D.; Liu, J.; Huang, A.; Rivera, N. R.; Njolito, E.; Hsiao, Y.; McWilliams, J. C.; Williams, J. M.; Armstrong, J. D., III; Sun, Y.; Mathre, D. J.; Grabowski, E. J. J.; Tillyer, R. D. J. Am. Chem. Soc. **2004,** 126, 3048.
(10) TBAB is referred to as the Jeffrey catalyst: Jeffrey, T. Tetrahedron **1996,** 52, 10113. For the use of TBAB in coupling and its removal by extraction, see: Lipton, M. F.; Mauragis, M. A.; Maloney, M. T.; Veley, M. F.; VanderBor, D. W.; Newby, J. J.; Appell, R. B.; Daugs, E. D. Org. Process Res. Dev. **2003,** 7, 385.
(11) Buono, F. G.; Chidambaram, R.; Mueller, R. H.; Waltermire, R. E. Org. Lett. **2008,** 10, 5325.
(12) Ren, Y.; Liu, Z.; He, S.; Zhao, S.; Wang, J.; Niu, R.; Yin, W. Org. Process Res. Dev. **2009,** 13, 764.
(13) Yu, H.; Richey, R. N.; Miller, W. D.; Xu, J.; May, S. A. J. Org. Chem. **2011,** 76, 665.
(14) Fray, M. J.; Gillmore, A. T.; Glossop, M. S.; McManus, D. J.; Moses, I. B.; Praquin, C. F. B.; Reeves, K. A.; Thompson, L. R. Org. Process Res. Dev. **2010,** 14, 263.

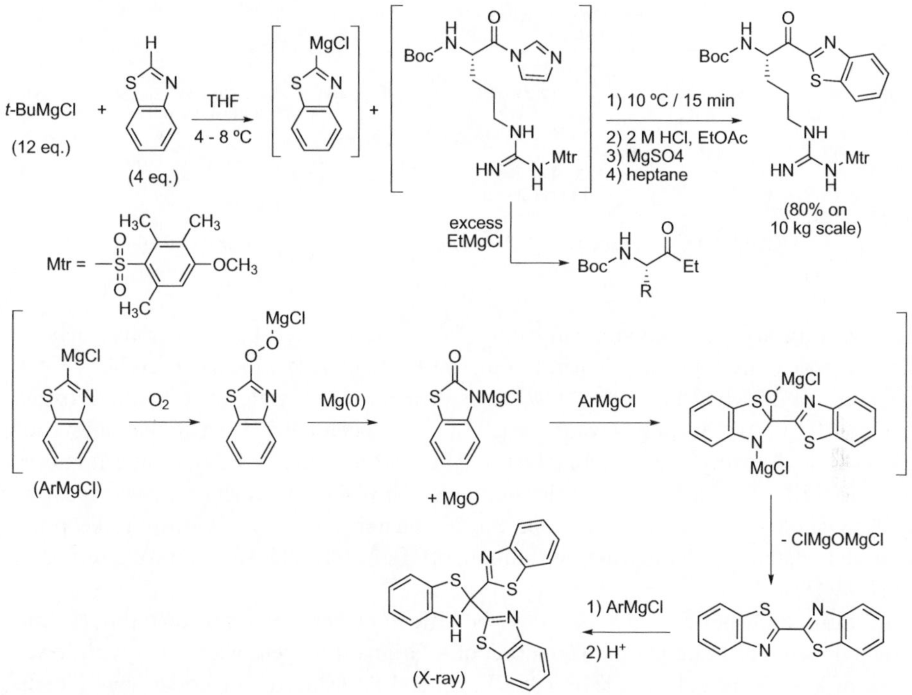

FIGURE 10.8 Impurities formed by reaction of a Grignard reagent with O_2.

consumed three molecules of benzothiazolyl magnesium chloride. The researchers posed that reaction of the benzothiazolyl Grignard with O_2 generated the peroxide derivative, which was then reduced by Mg(0) present in reaction mixtures to the benzothiazolone. Subsequent reaction of the latter with the benzothiazolyl Grignard generated the dimer, and reaction with a third molecule of the benzothiazolyl Grignard produced the trimer. This reaction pathway was confirmed when commercially available benzothiazolone was subjected to the reaction conditions and produced the trimer. By adding the benzo-

SAFETY — Grignard reagents generate hydrocarbons when they are used as bases or when they contact protic impurities; for instance, i-PrMgBr generates propane. Hence precautions such as N_2 atmospheres should be used to reduce the risk of fires. Positive pressures of N_2 should be used in the laboratory, pilot plant and manufacturing equipment to prevent intrusion of O_2 into reactions.

thiazole to a solution of excess t-BuMgCl the generation of the trimer was minimized. (EtMgCl reacted with the imidazolide to produce the ethyl ketone, but the bulkier t-BuMgCl did not react with the imidazolide.)

As with all process R&D activities solvent choice can be a crucial aspect, as mentioned many times in the excellent review by Magano and Dunetz [16]. Often polar aprotic solvents such as DMAc have been preferred. The Buchwald group has developed many catalysts that function well in toluene. Dioxane, a suspected carcinogen [69], has

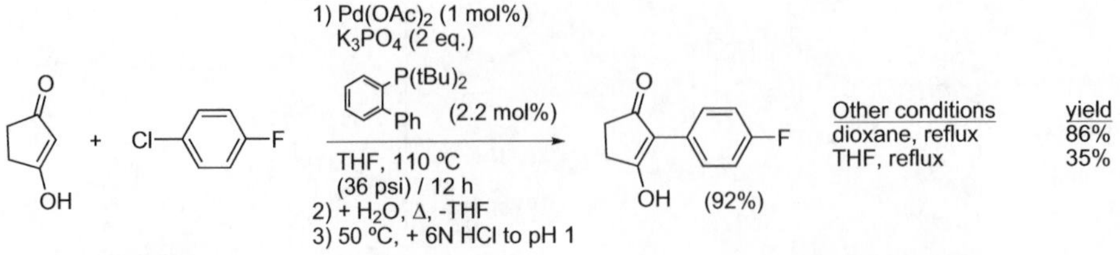

FIGURE 10.9 Dioxane and THF gave equivalent yields for a Heck reaction.

been used for many transition metal-catalyzed cross-couplings, but temperature may be more important than solvent choice. For instance, Merck researchers carried out a Heck reaction in 92% yield in THF at 110 °C, equivalent to the yield in refluxing dioxane (Figure 10.9) [70]. Suzuki–Miyaura reactions have been carried out in alcohol–water mixtures, in polyethylene glycol at 110 °C [71] and in water or EtOH–water mixtures at up to 150 °C with microwave irradiation [72,73]. For a ruthenium-catalyzed *cis*-hydroxylation of a maleimide on scale Pfizer researchers used acetonitrile to keep the toxic and relatively volatile RuO_2 in solution; off-gases from this reaction were scrubbed by $Na_2S_2O_3$ [74].

Controlling water contents has made the difference between a practical process and a stalled reaction. Initiating the formation of a Grignard reagent was easier with lower levels of H_2O in the solvent (THF) [75]. In general Pd-catalyzed processes can tolerate water, and some water may be beneficial; however, reactions catalyzed by nickel, a more reactive metal, may require anhydrous conditions or very low levels of water. Some examples are shown for the cross-coupling reactions in Figures 10.10 and 10.11. The importance of water in a Pd-catalyzed coupling was noted in one route to losartan, by activating the K_2CO_3 and the boronic acid for a Suzuki coupling [76]. Adding one equivalent of H_2O to a KF-catalyzed Suzuki–Miyaura coupling eliminated an induction period and produced a complete reaction in only 2 hours [55]. The role of H_2O has been described in detail for the Suzuki–Miyaura coupling of a potassium trifluoroborate, with the gradual generation of the corresponding boronic acid identified as the key to minimizing side products (Figure 10.3) [32]. Using DoE studies GSK researchers found that a Suzuki–Miyaura coupling could be best carried out in primarily aqueous conditions, with short reaction times at a moderate temperature [52,77]. The Suzuki–Miyaura coupling of a pyrazolyl boronate that was prone to deboronation was best carried out by not adding any H_2O [78]. In the case of the Pd-catalyzed amidation of *o*-bromotoluene 2.5 equivalents of H_2O was optimal when the reaction was run with Cs_2CO_3 in toluene; in dioxane or with PhONa or NaO*t*-Bu as base the amidation was less sensitive to H_2O. The authors posed that H_2O increased the solubility of the base in the reaction solvent [79]. Snieckus and co-workers found that control of the overall water content was essential in the Suzuki–Miyaura couplings of carbamates and other phenolic derivatives using air-stable $NiCl_2(PCy_3)_2$. Anhydrous K_3PO_4 was needed, and kinetic studies showed that the optimal ratio of boroxine:boronic acid:H_2O was about 1.0:0.1:0.1; with higher contents of water present reactions slowed and inactive $NiO/Ni(OH)_2$ formed [80]. Similarly the Ellman group found that rhodium-catalyzed couplings of a boroxine proceeded optimally when the water content of the boroxine was less than 30% [81].

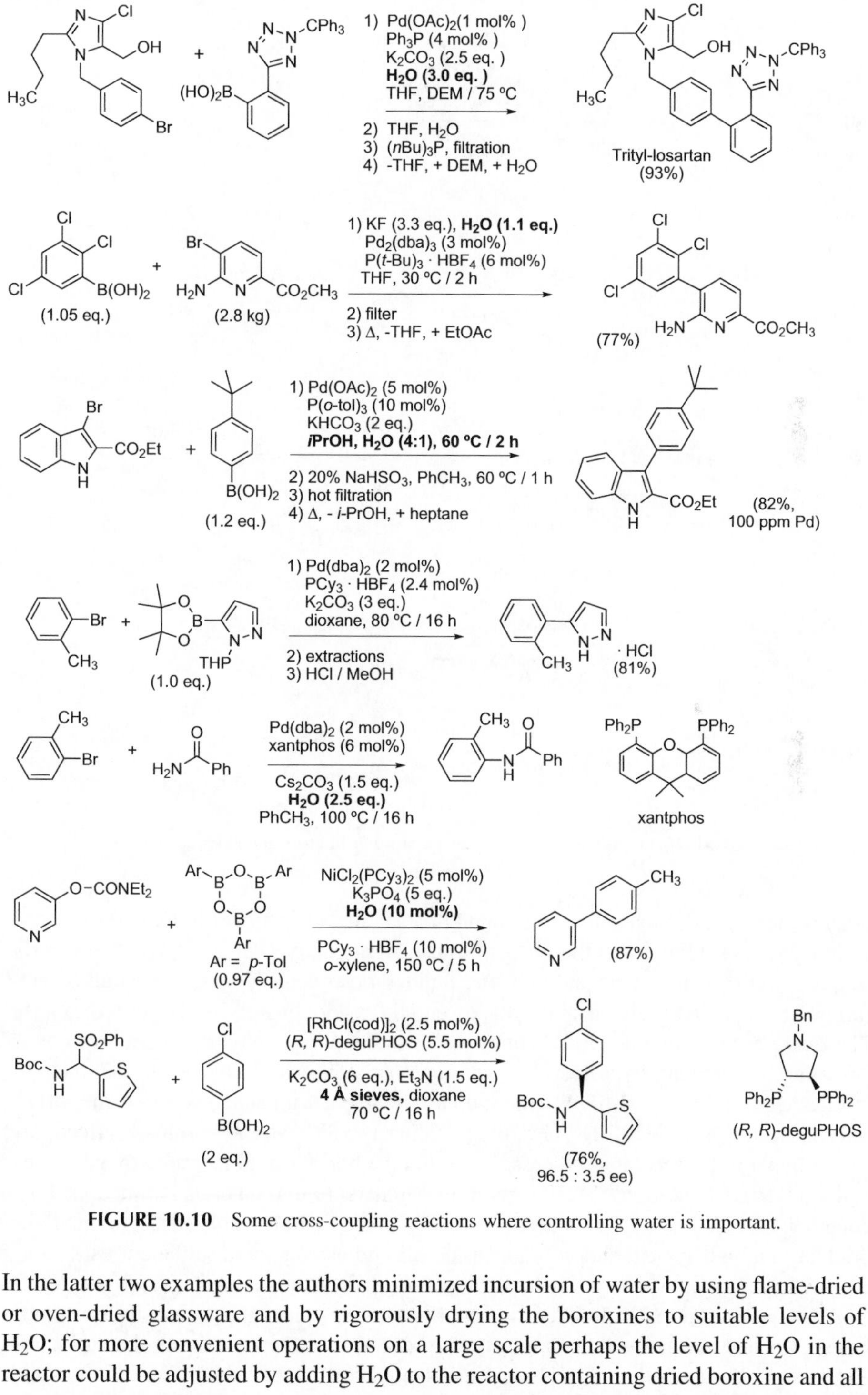

FIGURE 10.10 Some cross-coupling reactions where controlling water is important.

In the latter two examples the authors minimized incursion of water by using flame-dried or oven-dried glassware and by rigorously drying the boroxines to suitable levels of H_2O; for more convenient operations on a large scale perhaps the level of H_2O in the reactor could be adjusted by adding H_2O to the reactor containing dried boroxine and all

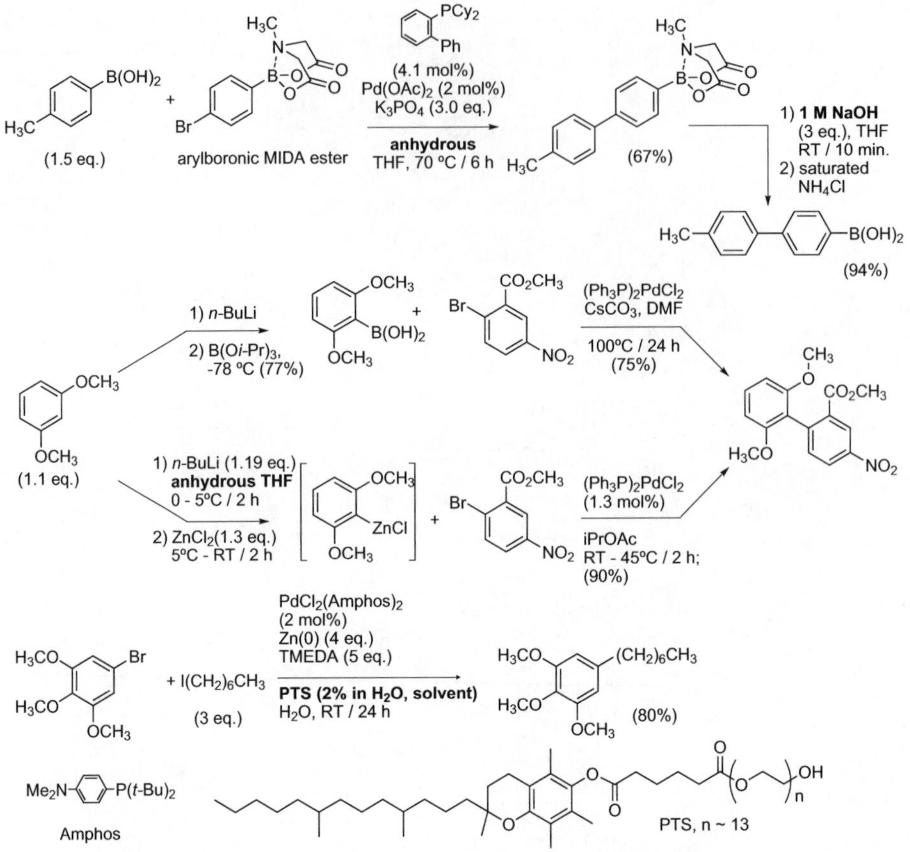

FIGURE 10.11 Some considerations about water in cross-coupling reactions.

other contents. The Buchwald group found that heating a mixture of H_2O, Pd(OAc)$_2$ and a hindered phosphine led to the formation of a highly active Pd(0) catalyst that readily coupled aryl chlorides with anilines and amides or aryl mesylates with amides; preforming the catalyst in this fashion was essential for good reactivity (Figure 10.2) [30]. The Burke group found that in Pd-catalyzed cross-couplings anhydrous conditions were necessary to prevent hydrolysis of aryl boronic acids protected with the methyl-iminodiacetic acid ester (MIDA) group; under treatment with aqueous NaOH the MIDA ester is readily hydrolyzed for iterative coupling [82,83]. MIDA boronates effectively cross-couple in Pd-mediated processes wherein the boronic acid is gradually generated under alkaline conditions [41]. Abbott researchers found that a zinc-mediated Pd coupling (Negishi coupling) required no cryogenic temperatures and provided the product in a higher yield and with an easier workup as compared to the corresponding Suzuki reaction; in the experimental procedure anhydrous solvent (THF) was used, which is typically required for Negishi couplings [84,85]. In contrast, the Lipshutz group has shown that Zn-mediated couplings can proceed in water containing the surfactant PTS using their work on aqueous microemulsions [86]. Such conditions hold considerable promise for developing practical processes on scale [87,88]. For his work with surfactants Lipshutz received a US Presidential Green Chemistry Award in 2011 [89]. In summary, considerations about water have played a big role in process development of organometallic reactions.

Addition sequences and extended additions can be key operations. Examples of preferred addition sequences are shown with the amination investigated by Blackmond, Buchwald, and Hartwig (Figure 10.1) and with the oxidation to make esomeprazole (Figure 10.7). For a Pd-catalyzed cyanation in DMF Ryberg found that adding $Zn(CN)_2$ last at 50 °C was key to successful manufacturing; at 50 °C catalytic cyanation was faster than cyanide poisoning of the catalyst, and the process was driven by the gradual dissolution of $Zn(CN)_2$ [90]. The extended addition of a slurry of $Zn(CN)_2$ and starting material to the reactor containing the Pd catalyst was effective in a cyanation, probably by minimizing the amount of cyanide present at any moment [34]. Extended additions also create a high concentration of catalyst relative to the starting materials, and rapid reactions to give products may decrease the formation of side products. Extended additions were helpful for a ruthenium-catalyzed cyclopropanation (Figure 9.6) and for a Suzuki coupling of a 3-bromoindole [52] (Figure 10.10). An extended addition minimized protodeboronation in another Suzuki coupling [91].

IV. IMPACT OF IMPURITIES

With any catalytic process small amounts of impurities can be helpful or harmful. The key component in a catalyst could be an impurity, as may be the case for iron-catalyzed processes [92,93]. Ensuring that ammonium chloride was present at a concentration equimolar to that of the catalyst was optimal for the chiral hydrogenation to manufacture sitagliptin (Figure 10.16) [94]. Chloride was found to be helpful in a Pd-catalyzed $NaBH_4$ reduction of a benzyl phosphate; the benefit of residual chloride was discovered when the process was de-telescoped (Figure 10.12) [95]. As shown by extensive analytical studies, palladium impurities at the ppb level in commercial Na_2CO_3 were responsible for a Suzuki coupling that was initially carried out without charging Pd [96]. (This situation is analogous to the "homeopathic Heck" reaction conditions, as in Figure 10.4.) Morpholine in the solvent toluene at concentrations less than 20 ppm inhibited the ring-closing metathesis to make BILN 2061 ZW (a hepatitis C protease inhibitor); at the dilute concentration of the reaction about one equivalent of morpholine was present relative to the ruthenium catalyst [59]. A succinamic acid isolated by acidification with HCl contained about 1 wt% chloride, producing a slow Rh-catalyzed hydrogenation; after workup of the succinamic acid with H_2SO_4, the hydrogenation was 26 times faster. Notably, at the Rh charge of 0.001 mol% the product as a residue contained only 9 ppm Rh [97]. The buildup of soluble cyanide has stalled many Pd-catalyzed cyanations, probably by precipitation of Pd as $Pd(CN)_2$ [34]. This has led to cyanations using poorly soluble reagents such as $Zn(CN)_2$ [34], potassium ferrocyanide, $K_4Fe(CN)_6$ [98] or additives such as i-PrOH [99] or zinc formate [100] that may provide Pd(0) in situ. Merck researchers reported that the speed of a Pd-mediated cyanation with $Zn(CN)_2$ was dependent on the quality of the $Zn(CN)_2$ [63]. In the enantioselective hydrogenation to manufacture sitagliptin the product amine was found to inhibit the reduction

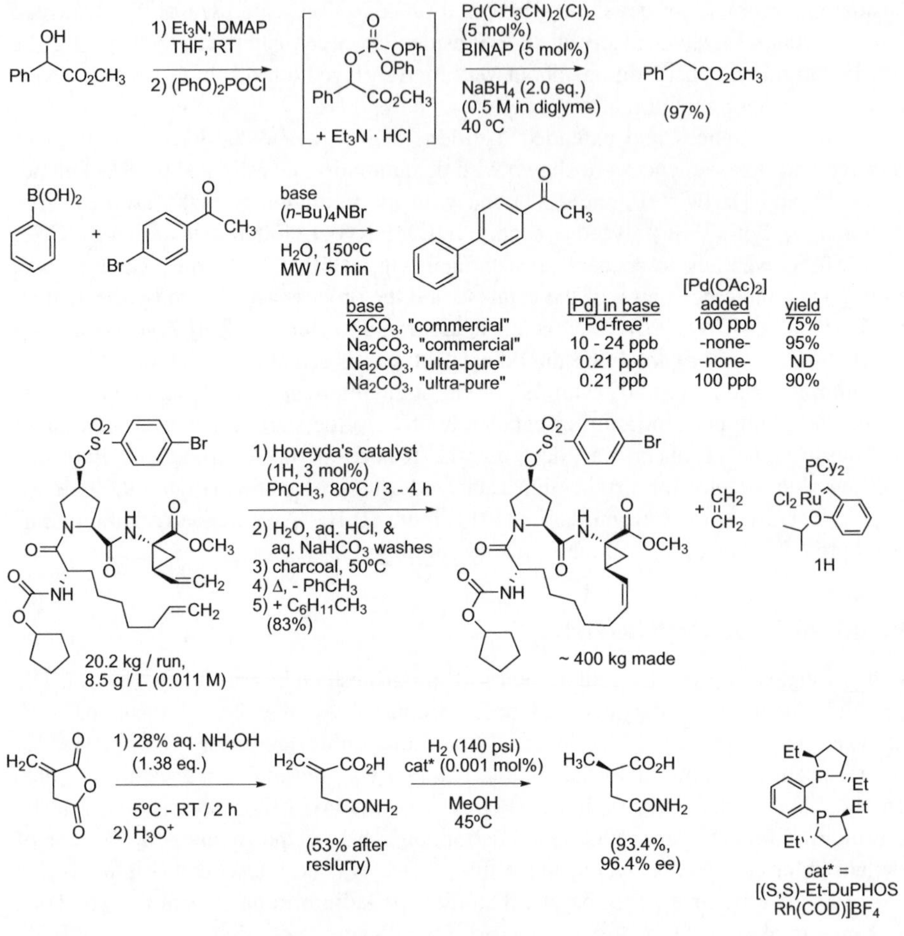

FIGURE 10.12 Beneficial and inhibitory impurities.

(Figure 10.16) [101]; hydrogenation of model enamines in the presence of Boc$_2$O was rapid, as the Boc-amine generated in situ did not inhibit the reaction [102]. Pretreatment of reaction streams with Chelex® resins may remove troublesome metal impurities [103]. Trace amounts of impurities containing sulfur may be problematic; hydrogenolysis required large amounts of Perlman's catalyst after Corey–Kim oxidation using Me$_2$S, despite extensive workup, including activated carbon treatment, chromatography, and crystallization [104]. Understanding the role of impurities can be essential for designing reliable processes.

Transition metals and ligands can pose challenges in purifying products. Researchers from Lilly have described a method to recover a key ligand [105]. Treatments to remove transition metals during workups are discussed in Chapter 11. Rhodia researchers have found that when catalyst loadings are restricted to about 1000 ppm recovering catalysts is no longer economically feasible, and the metals can be purged by routine workup [15].

Loss of a halide or decomposition of a sulfonate from an aromatic ring can produce problematic impurities and reduce yields. Two examples in which loss of a sulfonate and a fluoride were minimized are shown in Figure 10.13. In a Hiyama coupling adding 0.25 equivalents of AcOH increased yields, perhaps by neutralizing the nucleophilic byproduct ethoxide [106]. In the manufacturing route to aprepitant Merck researchers found that scale-up led to significant amounts of the des-fluoro intermediate, which was not purged by crystallization. By reducing the charge of Pd/C, increasing the hydrogen pressure, and optimizing agitation the rates of imine reduction and hydrogenolysis were increased, decreasing the formation of the des-fluoro impurity [107].

Deboronation can be a significant side reaction in transition metal-catalyzed reactions, because boronic acids are a component of the popular Suzuki–Miyaura coupling [108]. Boronic acids and boronate esters are generally more stable than other organoborane derivatives, and thus are preferred for cross-coupling reactions. Suzuki noted that deborylation is more significant with aryl dialkyl boranes than with the corresponding arylboronic acids [109]; nonetheless suitable stability of 3-(diethylboryl) pyridine was present for Pharmacia researchers to carry out a Suzuki coupling on a 175 kg scale [110]. Deboronation (aka deborylation) is a common side reaction, and deboronated side products not only decrease the isolated yields but may also complicate product isolation. Deboronation of low molecular weight boronic acids may produce

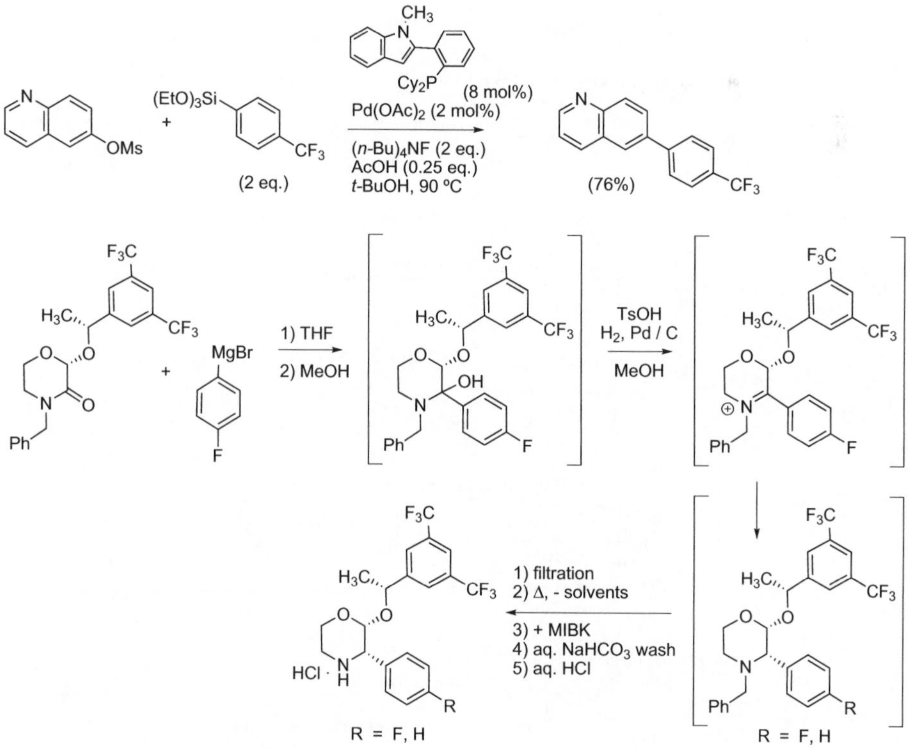

FIGURE 10.13 Minimizing side products from loss of a sulfonate and a fluoride.

volatile and flammable side products, as with the ready deboronation of cyclopropyl boronic acid to afford cyclopropane [41]. Figure 10.14 gives some examples of preparations of boronic acids and esters, and attendant side products. In the preparation of a boronic acid used to make losartan, tri-isopropyl borate was preferred over trimethyl borate, as the latter generated more borinic acid. The reaction was worked up through facile hydrolysis of the intermediate boronate with saturated aqueous ammonium

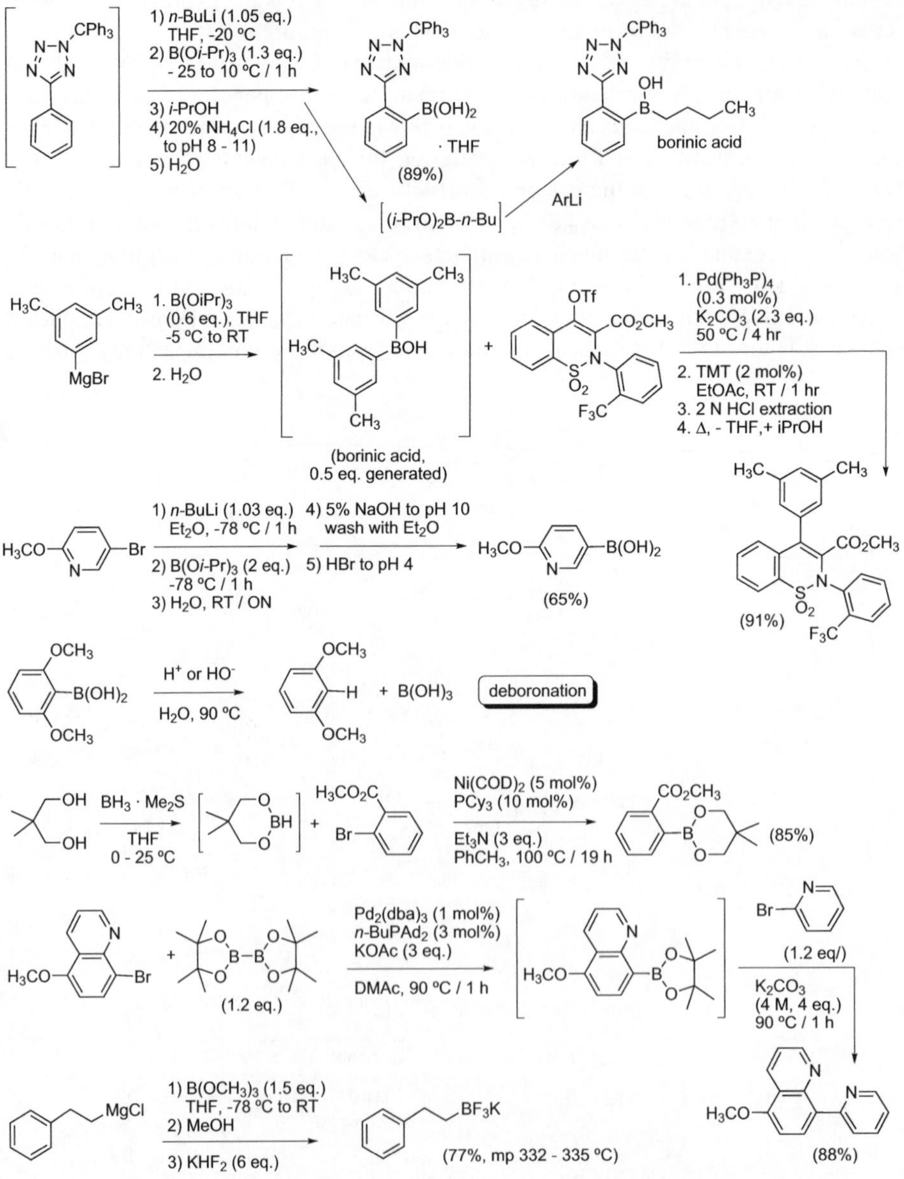

FIGURE 10.14 Preparation of some boronic acids and attendant side products.

chloride, and filtering off the crystalline boronic acid. Under these conditions the mixture was at about pH 8–11, to minimize loss of the trityl group. Formation of the borinic acid side product was minimized by titrating the solution of the trityl tetrazole to a red color endpoint, then adding 1.05 equivalents more of *n*-BuLi [111]. Kuivila and co-workers found that protodeboronation of 2,6-dimethoxybenzeneboronic acid was slowest at about pH 5, and rapid under more acidic or basic conditions [112]. Hence transferring an intermediate boronate into a quench buffered at about pH 5 at room temperature or below may minimize deboronation in sensitive boronic acids and derivatives. A pyridineboronic acid was precipitated at pH 4 by acidification with HBr [113]. The presence of residual HCl in boronic acids has been found to inhibit cross-coupling of hindered boronic acids [1]. Boroxines (see Figure 9.8) have sometimes crystallized instead of the boronic acids [114]; the presence of some water generally converts the boroxines to active coupling partners. To circumvent these issues a number of alternative boronic acid derivatives have been investigated. The Percec group extensively investigated the Ni-catalyzed formation of the neopentylglycol boronates; they found that deboronation was catalyzed by nickel species, especially for the *ortho* ester shown. Since these reactions were run under rigorously dry conditions proto-deboronation is unlikely [115]. Coupling an 8-bromoquinoline with bis(pinacolato) diboron [116] afforded the pinacol boronate, which was telescoped into a Suzuki–Miyaura coupling in high yield [117]. The Burke group noted that MIDA boronates (see Figure 10.11) are significantly more stable than the boronic acid analogs [41]; the Ellman group found that MIDA boronates were advantageous in the rhodium-catalyzed alkylation of sulfinyl imines [118]. Molander's group has developed potassium tri-fluoroborate salts [119] as air-stable reagents for cross-couplings [40]. The increased availability of boronic acids and boronate esters on a large commercial scale has expanded the utility of reactions with boronic acids.

Borinic acids (Figure 10.14) are sometimes troublesome impurities. Researchers at Pfizer found that avoiding formation of a borinic acid was not feasible under their reaction conditions: even at −70 °C the ratio of boronic acid:borinic acid was no better than 2.5:1. When the mixture of boronic acid and borinic acid was subjected to coupling with the triflate, both aryl groups were transferred from the borinic acid. Non-cryogenic conditions were used to optimize preparation of the borinic acid and exploit this serendipitous coupling [120]. Such couplings with borinic acids have been mentioned in the patent literature [121] and may be more prevalent than is appreciated.

Deboronated side products often form in cross-coupling reactions and decrease yields. Sterically hindered, electron-rich aryl boronates and heteroaromatic boronates are especially prone to deboronation [122,123]. Protodeboronation was found for a boronate anion, but not for the corresponding boronic acid [32]. Merck researchers found that a side product from protodeboronation began to form in appreciable amounts after about 6 hours of a Pd-mediated Suzuki coupling, and did not require Pd [51]; perhaps deactivation of the catalyst, through gradual accumulation of inhibitors or through decomposition of the ligands by C–P bond cleavage [124], was responsible for the increased rate of protodeboronation. Using Raman spectroscopy Leadbeater's group found that under microwave-promoted conditions a Suzuki reaction was competing with deboronation [73]. Buchwald's group found that by developing an efficient catalyst rapid Suzuki coupling of unstable aryl boronic acids proceeded faster than deboronation [125]. The Lautens group found that in aqueous media sodium dode-cylsulfate (SDS) suppressed deboronation [126]. Fluoride-promoted deboronation in the presence of O_2 gave rise to a colored radical [127]. Nickel-catalyzed couplings produced significant deboronation, and in couplings with Pd catalysts adding H_2O to the reactions decreased deboronation, perhaps because Suzuki couplings were faster [128]. Steric congestion near the boronic acid residue was found to accelerate proto-deboronation for some boronic acid derivatives of polycyclic aromatic hydrocarbons [129]. The Snieckus group found that deboronation occurred readily when pinacol was added to 4-pyridyl boronic acids [130]. One equivalent of H_2O was added to a KF-catalyzed Suzuki reaction, minimizing deboronation and reducing the reaction time to only 2 hours. When the catalyst was deactivated by running the reaction in the presence of air, the amount of deboronated side product increased to 40% [55]. Suzuki–Miyaura coupling of a pyrazolyl boronate prone to deboronation was optimally performed in the absence of H_2O to minimize deboronation [78]. Deboronation of a hindered difluor-obenzene boronic acid was accelerated in the presence of Pd, and to minimize deboronation anhydrous conditions were needed for a Suzuki coupling [131]. Amgen researchers found that deboronation of an *ortho*-substituted aryl boronic acid was a significant competing side reaction in a Suzuki coupling, and deboronation proceeded in the absence of catalyst. Deboronation was highest in the presence of K_2CO_3, and was progressively less with K_3PO_4, KOAc, and alkylamines. The lowest level of deboro-nation was found with dicyclohexylamine, and four equivalents were charged for a rapid reaction [132].

In summary deboronation often competes with the desired metal-catalyzed coupling. Deboronation of *ortho*-substituted aryl boronic acids is facile, especially if there are electron-withdrawing groups present. In some cases deboronation is metal-catalyzed. As was suggested in 1997, the best approach may be to develop conditions that favor coupling in order to out-compete deboronation [122]. Creating homogeneous reaction conditions may significantly increase the rate of cross-coupling reactions.

A recent concern is the mutagenicity of aryl boronic acids. Aryl boronic acids, but not the corresponding deboronated arenes, have been found to be mutagenic in microbial assays [133]. Diethanolaminomethyl polystyrene (DEAM-PS) [134,135] and immobi-lized catechol [136] have been used to scavenge boronic acids. Complex formation with diethanolamine may solubilize residual boronic acids in mother liquors. Since aryl-boronic acids ionize similarly to phenols, basic washes of an API solution may remove arylboronic acids. A selective crystallization can purge an arylboronic acid from the API. The best means to control residual aryl boronic acids at the ppm level in APIs may be to decompose them through deboronation. Perhaps simply extending the reaction conditions or heating with aqueous hydroxide may suffice to decompose an arylboronic acid. By knowing the kinetics of the decomposition of the aryl boronic acid it may be possible to show by QbD that analyses for residual aryl boronic acid in an API are not necessary.

V. SOME CONSIDERATIONS FOR HETEROGENEOUS REACTIONS

Heterogeneous catalysts have classically been preferred in principle for metal-catalyzed processes, as they afford more efficient recovery and reuse of valuable materials, and

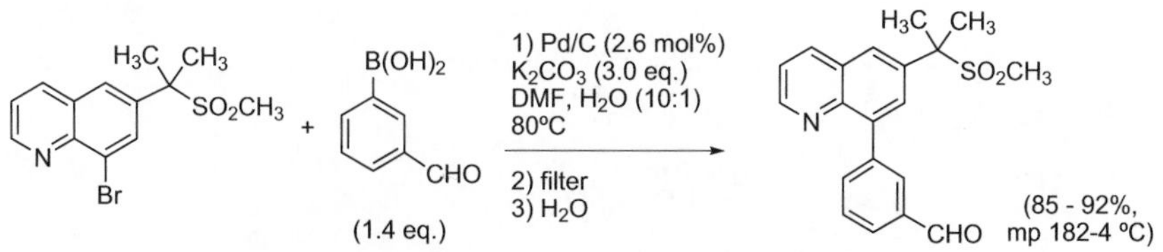

FIGURE 10.15 Suzuki coupling using Pd/C.

may ease removal of transition metals from the products. A key aspect for the successful recovery and reuse of heterogeneous catalysts is low attrition of particle size: if the solids become very small they may plug the pores of filter elements and slow operations. This consideration applies to hydrogenations, hydrogenolyses, and cross-couplings using heterogeneous catalysts. Unfortunately heterogeneous conditions for cross-couplings have sometimes been found to be unacceptably slow, and homogeneous catalysts are widely used [137].

Cross-coupling reactions have been mediated by Pd/C [138–142] and by PdEnCat® catalysis [143]. The kinetics of catalytic reactions can be very revealing. In the Suzuki coupling shown in Figure 10.15 reaction rates were found to depend on the type of carbon support, and the reactor was degassed after each addition. Based on ICP-

> Extending the contact time of a reaction with Pd/C catalyst can allow the Pd more opportunity to be readsorbed to the carbon, and hence decrease the burden of removing Pd from the API.

MS analysis, the concentration of soluble Pd species peaked before the maximum formation of the Suzuki product, and the concentration of soluble Pd species was minimized about 0.5 h after the conversion to product had stabilized [144]. Similarly Lipshutz found that nickel-mediated couplings were effected in solution, and the nickel species were readsorbed to the carbon support [145].

Merck researchers detailed some investigations into optimizing the enantioselective hydrogenation to sitagliptin using a homogeneous catalyst (Figure 10.16), and some considerations from that process optimization can apply to heterogeneous catalytic processes. Initial screening showed promise with rhodium catalysts, and poor results with iridium and ruthenium catalysts. Ferrocenyl ligands gave the best reaction rates, yields, and enantioselectivities. A key factor for consistent conversion rates and enantiomeric ratios (er) was adding one equivalent of ammonium chloride relative to the Rh charge. Small amounts of NH_4Cl were generated during the preparation of the crystalline enamine. The importance of this side product in the hydrogenation was determined during production of 18 batches on scale [94]. Acids other than NH_4Cl increased the rate of hydrogenation but also increased the formation of dimer-like side products. Running the hydrogenation at 70 °C instead of 50 °C decreased the er and formed some of the elimination side product. For efficient throughput the hydrogenation began as a slurry in MeOH (8 L/kg of enamine), and the mixture became homogeneous after about 7 h of hydrogenation. At that point the reaction was about 80% complete, and the remaining 10–11 h of reaction was needed for complete conversion due to weak inhibition of the catalyst by the product. By increasing the pressure of H_2 to 250 psig the catalyst loading

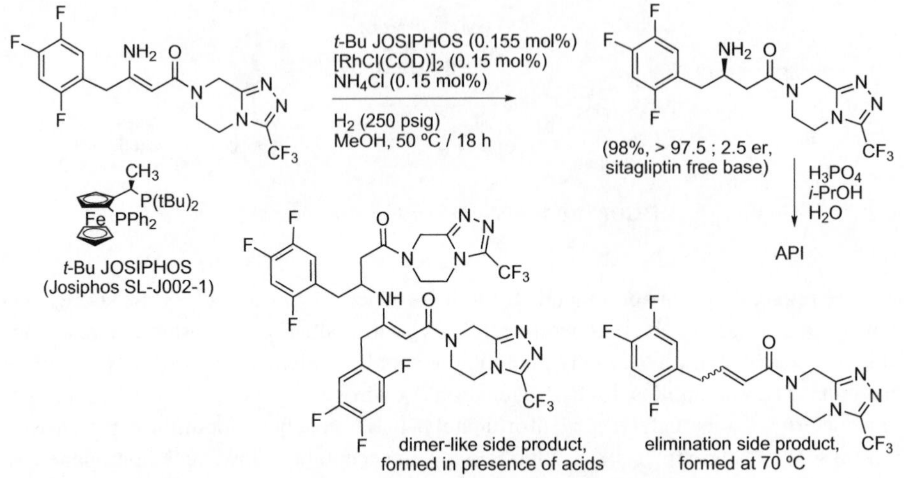

FIGURE 10.16 Details on the enantioselective hydrogenation to sitagliptin.

could be reduced by half to 0.15 mol%. The rhodium could be efficiently recovered by adsorption to the activated carbon product Ecosorb C-941 when the latter was charged in the presence of H_2 [101]. Substantial efforts were invested into developing this catalytic, organometallic reaction into an economical process run on scale.

Rylander's text is still the classic for catalytic hydrogenations [146]. Merck researchers have reviewed their asymmetric hydrogenation efforts, using homogeneous catalysts [147]. In another form of heterogeneous reaction conditions hydrogenations have been described for organic–aqueous conditions, in which water-soluble ligands such as tris-sulfonylated triphenylphosphine constrain the metal catalysts to the aqueous phase for recycling [61,148,149].

VI. PERSPECTIVE AND SUMMARY

The Suzuki–Miyaura coupling is generally preferred over other cross-couplings. There are four main benefits of this cross-coupling: first, the levels of homocoupled side products are lower; secondly, the intermediates (generally boronic acids and boronate esters) may be stable enough to be isolated; third, anhydrous conditions are often not required; and lastly, the byproducts boric acid and related esters are readily removed from the products [108]. The main disadvantage of the Kosugi–Migita–Stille coupling is the stoichiometric production of toxic organotin byproducts. The Negishi reaction can tolerate mild conditions and be effective with electron-rich boronic acid derivatives, but generates large amounts of zinc waste. The Sonogashira reaction can accommodate many functional groups, but the use of alkynes may raise SAFETY concerns. The Kumada–Corriu–Tamao reaction can be carried out under mild conditions, but generates a great deal of magnesium waste; this reaction is sensitive to moisture, as are many other nickel-catalyzed reactions and the Negishi reaction. The Hiyama reaction can be limited by the availability of appropriate silanes. The Mizoroki–Heck reaction is quite flexible, but generally is limited to vinyl halides and sulfonates. If intermediates for cross-coupling reactions are prepared from Grignard reagents the reactivity of these reagents toward other functional groups in the molecules must be considered. Corbet and Mignani have reviewed the advantages and limitations of cross-couplings [15]. As with byproducts of all reagents, boric acid and borates pose some environmental concerns as reproductive toxicants [150] and being harmful to plants [151].

More cross-coupling reactions without transition metals may be seen in the future. Aryl iodides have been coupled with benzene using KOt-Bu and 1,2-dimethylethylene diamine without any transition metal [152]. Verkade's group has used tetrabuty-lammonium acetate as base for ligand-, copper- and amine-free Sonogashira couplings [153]. Transition metal-free Suzuki-type couplings have been conducted in water with polyethylene glycol (170 °C, microwave) [154]. Dual catalysts have also been found useful [155]. As catalyst systems change, the presence of beneficial impurities may be critical, even at the ppb level [96]; such thinking may encourage the purchase of reagents that are not highly purified. Unfortunately contaminants in reagents may prove difficult to remove from products, as Wennerberg found for residual lead entering process streams from Cu_2O [156].

Academic researchers have often patented the catalysts they have discovered, and these may be aggressively marketed by chemical supply houses [157]. Contracts for exclusive sales of compounds can drive up the prices of reagents. Licenses may be required to buy patented catalysts used to prepare marketed drugs, driving up the CoG of a drug substance, or complicating its manufacture by circumventing the patented catalyst. Some patented cross-coupling technologies have been reviewed [15]. Such restrictions have led researchers to develop their own catalysts and ligands in-house [158], as Pfizer did for the "trichickenfootphos" ligand to avoid using the (R,R)-Rh-Me-DuPhos ligand for the highly productive hydrogenation of a pregbalin intermediate [159,160] (Figure 10.17). Researchers at Boehringer-Ingelheim have also described ligands developed in-house [161].

Catalysts using less expensive or less toxic transition metals may be developed more extensively. For instance, copper-catalyzed reactions may be preferred over palladium-catalyzed reactions due to the higher limits for residual copper in APIs and the reduced cost of copper metal. Couplings with electron-rich aryl halides and sulfonates can be problematic, especially with piperidines and morpholines [162], encouraging other catalysts to be developed. Copper-catalyzed coupling reactions have been reviewed [163]. Iron catalysis promises to be less expensive than Pd-catalyzed reactions.

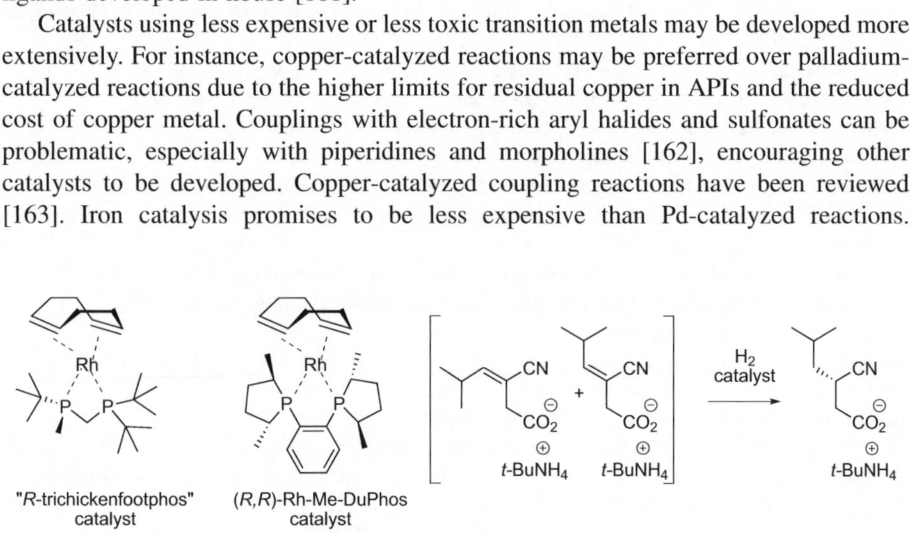

"R-trichickenfootphos"   (R,R)-Rh-Me-DuPhos
       catalyst   catalyst

FIGURE 10.17 Catalysts for hydrogenation of a pregbalin intermediate.

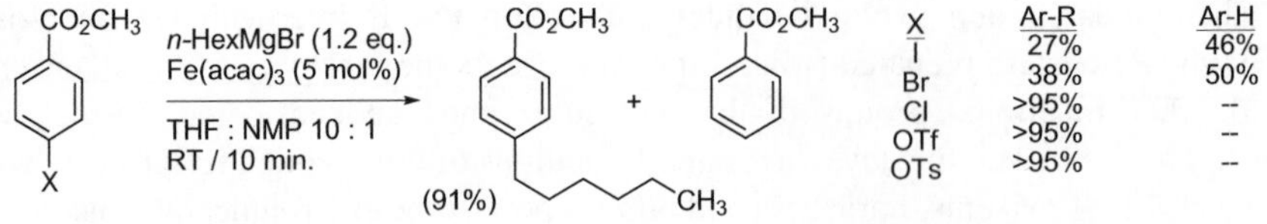

X	Ar-R	Ar-H
I	27%	46%
Br	38%	50%
Cl	>95%	--
OTf	>95%	--
OTs	>95%	--

FIGURE 10.18   Iron-mediated couplings.

Fürstner's group found that aryl tosylates and chlorides were better coupling partners than the more expensive aryl bromides and iodides [36] (Figure 10.18). Trace amounts of other transition metals may be necessary in iron-catalyzed reactions [92,164]. Iron-catalyzed reactions [93] and iron-catalyzed C–H insertions [165] have been reviewed. Iron-catalyzed nitrene insertions have been used to form sulfonylamides [166]. Nickel catalysts may be preferred for cost reasons. A catalytic nickel on carbon system has coupled morpholine to an aryl chloride in good yield [39]. Ni-catalyzed cross-couplings involving carbon-oxygen bonds have been reviewed [167].

Congratulations to Professors Heck, Negishi, and Suzuki on receiving the Nobel Prize for their pioneering research into metal-catalyzed cross-coupling reactions. Their influential work has led to many processes for APIs, intermediates, drug candidates, and other products.

REFERENCES

1. McGlacken, G. P.; Fairlamb, I. J. S. *Eur. J. Org. Chem.*, **2009**, 4011.
2. Oestreich, M., Ed.; *The Mizoroki–Heck Reaction*; Wiley, 2009.
3. *Palladium in Heterocyclic Chemistry. A Guide for the Synthetic Chemist*, 2nd ed.; Li, J. J.; Gribble, G. W., Eds; Elsevier, 2006.
4. Miyaura, N., Ed.; *Cross-Coupling Reactions: A Practical Guide*; Springer, 2010.
5. Müller, T. J. J., Ed.; *Metal Catalyzed Cascade Reactions (Topics in Organometallic Chemistry)*; Springer, 2010.
6. Rapoport, Z. Z.; Marek, I., Eds.; *The Chemistry of Organozinc Compounds*; Wiley, 2007.
7. Knochel, P.; Jones, P., Eds.; *Organozinc Reagents: A Practical Approach*; Oxford, 1999.
8. Harmsen, A. *Fundamentals of Organometallic Catalysis*; Wiley-VCH, 2011.
9. Sheldon, R. A.; Arends, I.; Hanefeld, U. *Green Chemistry and Catalysis*; Wiley-VCH, 2007.
10. Crabtree, R. H. *The Organometallic Chemistry of the Transition Metals*, 5th ed.; Wiley, 2009.
11. Hartwig, J. *Organotransition Metal Chemistry: From Bonding to Catalysis*; University Science Press, 2009.
12. Chatani, N., Ed.; *Directed Metallation (Topics in Organometallic Chemistry)*; Springer, 2010.
13. Hegedus, L.; Soderberg, B. *Transition Metals in the Synthesis of Complex Organic Molecules*, 3rd ed.; University Science Books, 2009.
14. Hall, D. G. *Boronic Acids: Preparation and Applications in Organic Synthesis and Medicine*; Wiley, 2006.
15. Corbet, J.-P.; Mignani, G. *Chem. Rev.* **2006,** *106,* 2651.
16. Magano, J.; Dunetz, J. R. *Chem. Rev.* **2011,** *111,* 2177.
17. Torborg, C.; Beller, M. *Adv. Synth. Catal* **2009**, *351,* 3027.
18. Slagt, V. F.; de Vries, A. H. M.; de Vries, J. G.; Kellogg, R. M. *Org. Process Res. Dev.* **2010, ** *14,* 30.
19. Caron, S.; Ghosh, A.; Gut Ruggeri, S.; Ide, N. D.; Neslon, J. D.; Ragan, J. A. In *Practical Synthetic Organic Chemistry: Reactions, Principles, and Techniques*; Caron, S., Ed.; Wiley, 2011; pp 279–340.

20. Yeung, C. S.; Dong, V. M. *Chem. Rev.* **2011,** *111,* 1215.

21. Liu, C.; Zhang, H.; Shi, W.; Lei, A. *Chem. Rev.* **2011,** *111,* 1780.

22. Cahiez, G.; Moyeux, A. *Chem. Rev.* **2010,** *110,* 1435.

23. Jana, R.; Pathak, T. P.; Sigman, M. S. *Chem. Rev.* **2011,** *111,* 1417.

24. Benaglia, M., Ed.; *Recoverable and Recyclable Catalysts*; Wiley, 2009.

25. Chen, Z.; Guan, Z.; Li, M.; Yang, Q.; Li, C. *Angew. Chem. Int. Ed.* **2011,** *50,* 4913.

26. Yin, L.; Liebscher, J. *Chem. Rev.* **2007,** *107,* 133.

27. Shekhar, S.; Ryberg, P.; Hartwig, J. F.; Mathew, J. S.; Blackmond, D. G.; Strieter, E. R.; Buchwald, S. L. *J. Am. Chem. Soc.* **2006,** *128,* 3584.

28. For discussion of the mechanisms of other cross-couplings, see Corbet, J.-P.; Mignani, G. *Chem. Rev.* **2006,** *109,* 2651.

29. Technically speaking the catalyst is the species that facilitates the synthetic transformation, which could comprise a metal, ligands, and solvent molecules. Catalysts are usually quite different from the materials, precatalysts, charged to a reactor. The distinction between a catalyst and a precatalyst is blurred in the literature. In this chapter "catalyst" may refer to the precatalysts.

30. Fors, B. P.; Krattiger, P.; Strieter, E.; Buchwald, S. L. *Org. Lett.* **2008,** *10,* 3505; Dooleweert, K.; Fors, B. P.; Buchwald, S. L. *Org. Lett.* **2010,** *12,* 2350.

31. Chen, C. In *The Art of Process Chemistry*; Yasuda, N., Ed.; Wiley-VCH, 2011; p 117.

32. Butters, M.; Harvey, J. N.; Jover, J.; Lennox, A. J. J.; Lloyd-Jones, G. C.; Murray, P. M. *Angew. Chem. Int. Ed.* **2010,** *49,* 5156.

33. Eckert, M. U.S. 6,784,308 B2, 2004 (to Bayer).

34. Marcantonio, K. M.; Frey, L. F.; Liu, Y.; Chen, Y.; Strine, J.; Phenix, B.; Wallace, D. J.; Chen, C. *Org. Lett.* **2004,** *6,* 3723.

35. Davidson, J. B.; Jasinski, R. J.; Peerce-Landers, P. J. U.S. 4,499,025, 1985 (to Occidental Chemical).

36. Fürstner, A.; Leitner, A. *Angew. Chem. Int. Ed.* **2002,** *41,* 609; Fürstner, A.; Leitner, A.; Méndez, M.; Krause, H. J. *Am. Chem. Soc.* **2002,** *124,* 13856.

37. Miller, J. A.; Farrell, R. P. *Tetrahedron Lett.* **1998,** *39,* 6441.

38. Tasler, S.; Lipshutz, B. H. *J. Org. Chem.* **2003,** *68,* 1190.

39. Lipshutz, B. H.; Ueda, H. *Angew. Chem. Int. Ed.* **2000,** *39,* 4492.

40. Molander, G. A.; Canturk, B. *Angew. Chem. Int. Ed. Engl.* **2009,** *48,* 9240; Molander, G. A.; Ellis, N. *Acc. Chem. Res.* **2007,** *40,* 275; Molander, G. A.; Biolatto, B. *J. Org. Chem.* **2003,** *68,* 4302.

41. Knapp, D. N.; Gillis, E. P.; Burke, M. D. *J. Am. Chem. Soc.* **2009,** *131,* 6961.

42. deVries, A. H. M.; Mulders, J. M. C. A.; Mommers, J. H. M.; Henderickx, H. J. W.; deVries, J. G. *Org. Lett.* **2003,** *5,* 3285.

43. Camp, D.; Matthews, C. F.; Neville, S. T.; Rouns, M.; Scott, R. W.; Truong, Y. *Org. Process Res. Dev.* **2006,** *10,* 814; These Pfizer researchers described the SAFE scale-up of a Heck reaction, as detailed in Figure 2.12.

44. For a ligand-less Heck reaction run on a 2400 mol scale, see Schils, D.; Stappers, F.; Solberghe, G.; van Heck, R.; Coppens, M.; Van den Heuvel, D.; Van der Donck, P.; Callewaert, T.; Meeussen, F.; De Bie, E.; Eersels, K.; Schouteden, E. *Org. Process Res. Dev.* **2008,** *12,* 530.

45. Merschaert, A.; Delhaye, L.; Kestemont, J.-P.; Brione, W.; Delbeke, P.; Mancuso, V.; Napora, F.; Diker, K.; Giraud, D.; Vanmarsenille, M. *Tetrahedron Lett.* **2003,** *44,* 4531.

46. Bartók, M. *Chem. Rev.* **2010,** *110,* 1663.

47. Yabu, K.; Masumoto, S.; Yamasaki, S.; Hamashima, Y.; Kanai, M.; Du, W.; Curran, D. P.; Shibasaki, M. *J. Am. Chem. Soc.* **2001,** *123,* 9908.

48. Nugent, W. A. *J. Am. Chem. Soc.* **1998,** *120,* 7139.

49. Mouri, S.; Chen, Z.; Mitsunuma, H.; Furutachi, M.; Matsunaga, S.; Shibasaki, M. *J. Am. Chem. Soc.* **2010,** *132,* 1255.

50. Huang, Q.; Richardson, P. F.; Sach, N. W.; Zhu, J.; Liu, K. K.-C.; Smith, G. L.; Bowles, D. M. *Org. Process Res. Dev.* **2011,** *15,* 556.

51. Cai, C.; Chung, J. Y. L.; McWilliams, J. C.; Sun, Y.; Shultz, C. S.; Palucki, M. *Org. Process Res. Dev.* **2007,** *11,* 328.

52. Bullock, K. M.; Mitchell, M. B.; Toczko, J. F. *Org. Process Res. Dev.* **2008,** *12,* 896.

53. Rouhi, A. M. *Chem. Eng. News* **2003,** *81*(18), 56.

54. Federsel, H.-J.; Larsson, M., In *Asymmetric Catalysis on Industrial Scale: Challenges, Approaches and Solutions*; Blaser, H.-U.; Schmidt, E., Eds.; Wiley-VCH: Weinheim; 2004; pp 413–436; Cotton, H.; Elebring, T.; Larsson, M.; Li, L.; Sörensen, H.; von Unge, S. *Tetrahedron: Asymmetry* **2000,** 11, 3819.

55. Fray, M. J.; Gillmore, A. T.; Glossop, M. S.; McManus, D. J.; Moses, I. B.; Praquin, C. F. B.; Reeves, K. A.; Thompson, L. R. *Org. Process Res. Dev.* **2010,** *14,* 263.

56. Yasuda, N.; Hsiao, Y.; Jensen, M. S.; Rivera, N. R.; Yang, C.; Wells, K. M.; Yau, J.; Palucki, M.; Tan, L.; Dormer, P. G.; Volante, R. P.; Hughes, D. L.; Reider, P. J. *J. Org. Chem.* **2004,** *69,* 1959.

57. Rathman, T. L.; Bailey, W. F. *Org. Process Res. Dev.* **2009,** *13,* 144.

58. Rao, C. S. *The Chemistry of Process Development in Fine Chemical & Pharmaceutical Industry,* 2nd ed.; Wiley: New York, 2007.

59. Nicola, T.; Brenner, M.; Donsbach, K.; Kreye, P. *Org. Process Res. Dev.* **2005,** *9,* 513.

60. In the first scale-up of an RCM using a Hoveyda catalyst aggressive sparging was necessary to remove the byproduct ethylene. In the optimized RCM, using a Grela catalyst, aggressive sparging was not necessary. Farina, V.; Shu, C.; Zeng, X.; Wei, X.; Han, Z.; Yee, N. K.; Senanayake, C. H. *Org. Process Res. Dev.* **2009,** *13,* 250.

61. Königsberger, K.; Chen, G.-P.; Wu, R. R.; Girgis, M. J.; Prasad, K.; Repic, O.; Blacklock, T. J. *Org. Process Res. Dev.* **2003,** *7,* 733.

62. Rassias, G.; Hermitage, S. A.; Sanganee, M. J.; Kincey, P. M.; Smith, N. M.; Andrews, I. P.; Borrett, G. T.; Slater, G. R. *Org. Process Res. Dev.* **2009,** *13,* 774.

63. Chen, C.; Frey, L. F.; Shultz, S.; Wallace, D. J.; Marcantonio, K.; Payack, J. F.; Vazquez, E.; Springfield, S. A.; Zhou, G.; Liu, P.; Kieczykowski, G. R.; Chen, A. M.; Phenix, B. D.; Singh, U.; Strine, J.; Izzo, B.; Krska, S. W. *Org. Process Res. Dev.* **2007,** *11,* 616.

64. For example, see http://www.oceanopticssensors.com/application/organicsolvent.htm

65. For example, see http://www.hamiltoncompany.com/Sensors/

66. Modi, S. P.; Gardner, J. O.; Milowsky, A.; Wierzba, M.; Forgione, L.; Mazur, P.; Solo, A. J.; Duax, W. L.; Galdecki, Z. *J. Org. Chem.* **1989,** *54,* 2317.

67. Kenney, B. D.; Breslav, M.; Chang, R.; Glaser, R.; Harris, B. D.; Maryanoff, C. A.; Mills, J.; Roessler, A.; Segmuller, B.; Villani, F. J., Jr. *J. Org. Chem.* **2007,** *72,* 9798.

68. Lloyd-Williams, P.; Albericio, F.; Giralt, E. *Chemical Approaches to the Synthesis of Peptides and Proteins*; CRC Press: Boca Raton, 1997; p 27.

69. Report on Carcinogens, 11th ed.; http://www.pall.com/pdf/11th_ROC_1_4_Dioxane.pdf

70. Kuethe, J. T.; Wong, A.; Wu, J.; Davies, I. W.; Dormer, P. G.; Welch, C. J.; Hillier, M. C.; Hughes, D. L.; Reider, P. J. *J. Org. Chem.* **2002,** *67,* 5993.

71. Li, J.-H.; Liu, W.-J.; Xie, Y.-X. *J. Org. Chem.* **2005,** *70,* 5409.

72. Arvela, R.; Leadbeater, N. E. *Org. Lett.* **2005,** *7,* 2101.

73. Leadbeater, N. E.; Smith, R. J. *Org. Lett.* **2006,** *8,* 4589.

74. Couturier, M.; Andresen, B. M.; Jorgensen, J. B.; Tucker, J. L.; Busch, F. R.; Brenek, S. J.; Dubé, P.; am Ende, D. J.; Negri, J. T. *Org. Process Res. Dev.* **2002,** *6,* 42.

75. Ace, K. W.; Armitage, M. A.; Bellingham, R. K.; Blackler, P. D.; Ennis, D. S.; Hussain, N.; Lathbury, D. C.; Morgan, D. O.; O'Connor, N.; Oakes, G. H.; Passey, S. C.; Powling, L. C. *Org. Process Res. Dev.* **2001,** *5,* 479.

76. Smith, G. B.; Dezeny, G. C.; Hughes, D. L.; King, A. O.; Verhoeven, T. R. *J. Org. Chem.* **1994,** *59,* 8151; Two impurities were identified in this process, the *o*-bromo isomer of the one starting material and the deboronated side product. By understanding how much deboronation took place under the reaction conditions and setting specifications on the *o*-bromo impurity they could have controlled downstream processing.

77. Bullock, K. M.; Burton, D.; Corona, J.; Diederich, A.; Glover, B.; Harvey, K.; Mitchell, M. B.; Trone, M. D.; Yule, R.; Zhang, Y.; Toczko, J. F. *Org. Process Res. Dev.* **2009**, *13*, 303.

78. McLaughlin, M.; Marcantonio, K.; Chen, C.; Davies, I. W. *J. Org. Chem.* **2008**, *73*, 4309.

79. Dallas, A. S.; Gothelf, K. V. *J. Org. Chem.* **2005**, *70*, 3321.

80. Antoft-Finch, A.; Blackburn, T.; Snieckus, V. *J. Am. Chem. Soc.* **2009**, *131*, 17750.

81. Storgaard, M.; Ellman, J. A. *Org. Synth.* **2009**, *86*, 360; Note 21 details recrystallization of boroxines from water and subsequent drying.

82. Ballmer, S. G.; Gillis, E. P.; Burke, M. D. *Org. Synth.* **2009**, *86*, 344.

83. Burke has also described the *N*-pinane iminodiacetic acid (PIDA) boronates as chiral reagents for iterative coupling Li, J.; Burke, M. D. *J. Am. Chem. Soc.* **2011**, *133*, 13774.

84. Kin, Y.-Y.; Grieme, T.; Raje, P.; Sharma, P.; Morton, H. E.; Rozema, M.; King, S. A. *J. Org. Chem.* **2003**, *68*, 3238.

85. Miller has described nickel-catalyzed cross-couplings to make *o*-tolyl benzonitrile (OTBN), which has been elaborated into intermediates used for the manufacture of sartans. Ref. Miller, J. A.; Farrell, R. P. *Tetrahedron Lett* **1998**, *39*, 6441.

86. Krasovskiy, A.; Duplais, C.; Lipshutz, B. H. *J. Am. Chem. Soc.* **2009**, *131*, 15592.

87. Nishikata, T.; Lipshutz, B. H. Fujiwara–Moritani reactions can be conducted at room temperature in water. *Org. Lett.* **2010**, *12*, 1972.

88. Lipshutz, B. H.; Ghorai, S.; Abela, A. R.; Moser, R.; Nishikata, T.; Duplais, C.; Krasovskiy, A.; Gaston, R. D.; Gadwood, R. C. *J. Org. Chem.* **2011**, *76*, 4379.

89. Ritter, S. K. *Chem. Eng. News* **2011**, *89*(26), 11. http://pubs.acs.org/cen/news/89/i26/8926notw1b.html

90. Ryberg, P. *Org. Process Res. Dev.* **2008**, *12*, 540.

91. Payack, J. F.; Vazquez, E.; Matty, L.; Kress, M. H.; McNamara, J. *J. Org. Chem.* **2005**, *70*, 175.

92. Laird, T. *Org. Process Res. Dev.* **2009**, *13*, 823.

93. Bolm, C.; Legros, J.; Le Paih, J.; Zani, L. *Chem. Rev.* **2004**, *104*, 6217.

94. Clausen, A. M.; Dziadul, B.; Cappuccio, K. L.; Kaba, M.; Starbuck, C.; Hsiao, Y.; Dowling, T. M. *Org. Process Res. Dev.* **2006**, *10*, 723.

95. Milne, J. E.; Murrry, J. A.; King, A.; Larsen, R. D. *J. Org. Chem.* **2009**, *74*, 445.

96. Arvela, R. K.; Leadbeater, N. E.; Sangi, M. S.; Williams, V. A.; Granados, P.; Singer, R. D. *J. Org. Chem.* **2005**, *70*, 161. A trace of Pd was also found to be necessary for a AuI-catalyzed Sonogashira coupling: Lauterbach, T.; Livendahl, M.; Rosellón, A.; Espinet, P.; Echavarren, A. E. *Org. Lett.* **2010**, *12*, 3006.

97. Cobley, C. J.; Lennon, I. C.; Praquin, C.; Zanotti-Gerosa, A.; Appell, R. B.; Goralski, C. T.; Sutterer, A. C. *Org. Process Res. Dev.* **2003**, *7*, 405.

98. Weissman, S. A.; Zewge, D.; Chen, C. *J. Org. Chem.* **2005**, *70*, 1508.

99. Ren, Y.; Liu, Z.; He, S.; Zhao, S.; Wang, J.; Niu, R.; Yin, W. *Org. Process Res. Dev.* **2009**, *13*, 764.

100. Yu, H.; Richey, R. N.; Miller, W. D.; Xu, J.; May, S. A. *J. Org. Chem.* **2011**, *76*, 665.

101. Hansen, K. B.; Hsiao, Y.; Xu, F.; Rivera, N.; Clausen, A.; Kubryk, M.; Krska, S.; Rosner, T.; Simmons, B.; Balsells, J.; Ikemoto, N.; Sun, Y.; Spindler, F.; Malan, C.; Grabowski, E. J. J.; Armstrong, J. D., III. *J. Am. Chem. Soc.* **2009**, *131*, 8798.

102. Hanson, K. B.; Rosner, T.; Kubryk, M.; Dormer, P. G.; Armstrong, J. D., III. *Org. Lett.* **2005**, *7*, 4935.

103. The capricious nature of a V_2O_5-catalyzed oxidation of an α-hydroxy acid to the α-keto acid has been attributed to the presence of trace metals. Branchaud, B. P.; Meier, M. S. *J. Org. Chem.* **1989**, *54*, 1320

104. Li, B.; Berliner, M.; Buzon, R.; Chiu, C. K.-F.; Colgan, S. T.; Kaneko, T.; Keene, N.; Kissel, W.; Le, T.; Leeman, K. R.; Marquez, B.; Morris, R.; Newell, L.; Wunderwald, S.; Witt, M.; Weaver, J.; Zhang, Z.; Zhang, Z. *J. Org. Chem.* **2006**, *71*, 9045.

105. Magnus, N. A.; Anzeveno, P. B.; Coffey, D. S.; Hay, D. A.; Laurila, M. E.; Schkeryantz, J. M.; Shaw, B. W.; Staszak, M. A. *Org. Process Res. Dev.* **2007**, *11*, 560.

106. So, C. M.; Lee, H. W.; Lau, C. P.; Kwong, F. Y. *Org. Lett.* **2009**, *11*, 317.

107. Brands, K. M. J.; Krska, S. W.; Rosner, T.; Conrad, K. M.; Corley, E. G.; Kaba, M.; Larsen, R. D.; Reamer, R. A.; Sun, Y.; Tsay, F.-R. *Org. Process Res. Dev.* **2006,** *10,* 109.

108. Pérez-Balado, C.; Willemsens, A.; Ormerod, D.; Aelterman, W.; Mertens, N. *Org. Process Res. Dev.* **2007,** *11,* 237.

109. Miyaura, N.; Suzuki, A. *Chem. Rev.* **1995,** *95,* 2457.

110. Lipton, M. F.; Mauragis, M. A.; Maloney, M. T.; Veley, M. F.; VanderBor, D. W.; Newby, J. J.; Appell, R. B.; Daugs, E. D. *Org. Process Res. Dev.* **2003,** *7,* 385.

111. Larsen, R. D.; King, A. O.; Chen, C. Y.; Corley, E. G.; Foster, B. S.; Roberts, F. E.; Yang, C.; Lieberman, D. R.; Reamer, R. A.; Tschaen, D. M.; Verhoeven, T. R.; Reider, P. J.; Lo, Y. S.; Rossano, L. T.; Brookes, A. S.; Meloni, D.; Moore, J. R.; Arnett, J. F. *J. Org. Chem.* **1994,** *59,* 6391. See also Reference 85.

112. Kuivila, H. G.; Reuwer, J. F., Jr.; Mangranite, J. A. *Can. J. Chem.* **1963,** *41,* 3081.

113. Parry, P. R.; Wang, C.; Batsanoy, A. S.; Bryce, M. R.; Tarbit, B. *J. Org. Chem.* **2002,** *67,* 7541.

114. Li, W.; Nelson, D. P.; Jensen, M. S.; Hoerrner, R. S.; Cai, D.; Larsen, R. D.; Reider, P. J. *J. Org. Chem.* **2002,** *67,* 5394.

115. Moldoveanu, C.; Wilson, D. A.; Wilson, C. J.; Leowanawat, P.; Resmerita, A.-M.; Liu, C.; Rosen, B. M.; Percec, V. *J. Org. Chem.* **2010,** *75,* 5438.

116. Available in bulk from AllyChem Co. Limited.

117. Zhang, Y.; Gao, J.; Li, W.; Lee, H.; Lu, B. Z.; Senanayake, C. H. *J. Org. Chem.* **2011,** *76,* 639.

118. Brak, K.; Ellman, J. A. *J. Org. Chem.* **2010,** *75,* 3147.

119. Molander, G. A.; Petrillo, D. E. In *Organic Syntheses,* Vol. 84; Miller, M. J., Ed.; Organic Syntheses, Inc., 2007; p 317.

120. Winkle, D. D.; Schaab, K. M. *Org. Process Res. Dev.* **2001,** *5,* 450.

121. Chekroun, I.; Ruiz-Montes, J.; Bedoya-Zurita, M.; Rossey, G. U.S. 5,278,312, 1994 (to Synthelabo).

122. Goodson, F. E.; Wallow, T. I.; Novak, B. M. In *Organic Syntheses,* Vol. 74; Smith, A. B., III, Ed.; Organic Syntheses, Inc., 1997; p 64.

123. Baudoin, O.; Cesario, M.; Guénard, D.; Guéritte, F. *J. Org. Chem.* **2002,** *67,* 1199.

124. Alcazar-Roman, L. M.; Hartwig, J. F.; Rheingold, A. L.; Liable-Sands, L. M.; Guzei, I. A. *J. Am. Chem. Soc.* **2000,** *122,* 4618.

125. Kinzel, T.; Zhang, Y.; Buchwald, S. L. *J. Am. Chem. Soc.* **2010,** *132,* 14073.

126. Lautens, M.; Roy, A.; Fukuoka, K.; Fagnou, K.; Martín-Matute, B. *J. Am. Chem. Soc.* **2001,** *123,* 5358.

127. Galbraith, E.; Fyles, T. M.; Marken, F.; Davidson, M. G.; James, T. D. *Inorganic Chem.* **2008,** *47,* 6236.

128. Gøgsig, T. M.; Søbjerg, L. S.; Lindhardt (neé Hansen), A. T.; Jensen, K. L.; Skrydstrup, T. *J. Org. Chem.* **2008,** *73,* 3404.

129. Dai, Q.; Xu, D.; Lim, K.; Harvey, R. G. *J. Org. Chem.* **2007,** *72,* 4856.

130. Alessi, M.; Larkin, A. L.; Ogilvie, K. A.; Green, L. A.; Lai, S.; Lopez, S.; Snieckus, V. *J. Org. Chem.* **2007,** *72,* 1588.

131. Cammidge, A. N.; Crépy, K. V. L. *J. Org. Chem.* **2003,** *68,* 6832.

132. Achmatowicz, M.; Thiel, O. R.; Wheeler, P.; Bernard, C.; Huang, J.; Larsen, R. D.; Faul, M. M. *J. Org. Chem.* **2009,** *74,* 795.

133. O'Donovan, M. R.; Mee, C. D.; Fenner, S.; Teasdale, A.; Phillips, D. H. *Mutat. Res.: Genet. Toxicol. Environ. Mutagen* **2011,** *724*(1–2), 1.

134. Antonow, D.; Cooper, N.; Howard, P. W.; Thurston, D. E. *J. Comb. Chem.* **2007,** *9,* 437.

135. Hall, D. G.; Tailor, J.; Gravel, M. *Angew. Chem. Int. Ed.* **1999,** *38,* 3064.

136. Yang, W.; Gao, X.; Springsteen, G.; Wang, B. *Tetrahedron Lett.* **2002,** *43,* 6339.

137. Chen, C.; Dagneau, P.; Grabowski, E. J. J.; Oballa, R.; O'Shea, P.; Prasit, P.; Robichaud, J.; Tillyer, R.; Wang, X. *J. Org. Chem.* **2003,** *68,* 2633.

138. Jiang, Z.; She, J.; Lin, X. *Adv. Synth. Catal.* **2009,** *351,* 2558.

139. Tagata, T.; Nishida, M. *J. Org. Chem.* **2003,** *68,* 9412.

140. Novák, Z.; Szabó, A.; Répási, J.; Kotschy, A. *J. Org. Chem.* **2003,** *68,* 3327.

141. Sakurai, H.; Tsukuda, T.; Hirao, T. *J. Org. Chem.* **2001,** *67,* 2721.

142. For a Fukuyama coupling of a thioester with an organozinc compound using Pd(OH)$_2$/C, see Mori, Y.; Seki, M. *J. Org. Chem.* **2003,** *68,* 1571.

143. Broadwater, S. J.; McQuade, D. T. *J. Org. Chem.* **2006,** *71,* 2131.

144. Conlon, D. A.; Pipik, B.; Ferdinand, S.; LeBlond, C. R.; Sowa, J. R., Jr.; Izzo, B.; Collins, P.; Ho, G.-J.; Williams, J. M.; Shi, Y.-J.; Sun, Y. *Adv. Synth. Catal.* **2003,** *345,* 931.

145. Lipshutz, B. H.; Tasler, S.; Chrisman, W.; Spliethoff, B.; Tesche, B. *J. Org. Chem.* **2003,** *68,* 1177.

146. Rylander, P. *Catalytic Hydrogenation in Organic Syntheses*; Academic, 1979.

147. Shultz, C. S.; Krska, S. W. *Acc. Chem. Res.* **2007,** *40,* 1320.

148. Joó, F. *Acc. Chem. Res.* **2002,** *35,* 738.

149. For Rh chemistry with a TPPTS analog where one aryl group was substituted by 2-pyridyl, see Lautens, M.; Yoshida, M. *J. Org. Chem.* **2003,** *68,* 762.

150. Questions and Answers on 30th Adaptation to Technical Progress (ATP) of Council Directive 67/548/ EEC on the classification, labelling of dangerous substances. http://ec.europa.eu/environment/chemicals/ dansub/pdfs/qa_30_atp.pdf

151. http://www.boraxfree.com/borax.html

152. Liu, W.; Cao, H.; Zhang, H.; Zhang, H.; Chung, K. H.; He, C.; Wang, H.; Kwong, F. Y.; Lei, A. *J. Am. Chem. Soc.* **2011,** *132,* 16737.

153. Urgaonkar, S.; Verkade, J. A. *J. Org. Chem.* **2004,** *69,* 5752.

154. Leadbeater, N. E.; Marco, M. *J. Org. Chem.* **2003,** *68,* 5660.

155. Trost, B. M.; Luan, X. *J. Am. Chem. Soc.* **2011,** *133,* 1706.

156. Malmgren, H.; Bäckström, B.; Sølver, E.; Wennerberg, J. *Org. Process Res. Dev.* **2008,** *12,* 1195.

157. For instance, Sigma–Aldrich had full-page advertisements in Chemical & Engineering News for 12 ligands prepared by Buchwald's group (May 30, 2011) and for 14 MIDA boronates, which originated from the Burke laboratories (June 13, 2011).

158. Hawkins, J. M.; Watson, T. J. N. *Angew. Chem. Int. Ed.* **2004,** *43,* 3224.

159. Hoge, G.; Wu, H.-P.; Kissel, W. S.; Pflum, D. A.; Greene, D. J.; Bao, J. *J. Am. Chem. Soc.* **2004,** *126,* 5966.

160. For a detailed case study on the preparation of the olefin precursor and chiral hydrogenation using the (*R,R*)-Rh-Me-DuPhos ligand, see Jennings, R.; Kissel, W. S.; Le, T. V.; Lenoir, E.; Mulhern, T.; Wade, R. In *Fundamentals of Early Clinical Drug Development*; Abdel-Magid, A. F., Caron, S., Eds.; Wiley: Hoboken, NJ, 2006.

161. Lorenz, J. Z.; Busacca, C. A.; Feng, X.; Grinberg, N.; Haddad, N.; Johnson, J.; Kapadia, S.; Lee, H.; Saha, A.; Sarvestani, M.; Spinelli, E. M.; Varsolona, R.; Wei, X.; Zeng, X.; Senanayake, C. H. *J. Org. Chem.* **2010,** *75,* 1155.

162. For coupling under Ullmann-like conditions, see Jiao, J.; Zhang, X.-R.; Chang, N.-H.; Wang, J.; Wei, J.-F.; Shi, X.-Y.; Chen, Z.-G. *J. Org. Chem.* **2011,** *76,* 1180.

163. Rao, H.; Fu, H. *Synlett.* **2011,** 745.

164. Buchwald, S. L.; Bolm, C. *Angew. Chem. Int. Ed.* **2009,** *48,* 5586.

165. Sun, C.-L.; Li, B.-J.; Shi, Z.-J. *Chem. Rev.* **2011,** *111,* 1293.

166. Chen, G.-Q.; Xu, Z.-J.; Liu, Y.; Zhou, C.-Y.; Che, C.-M. *Synlett.* **2011,** 1174.

167. Rosen, B. M.; Quasdorf, K. W.; Wilson, D. A.; Zhang, N.; Resmerita, A.-M.; Garg, N. K.; Percec, V. *Chem. Rev.* **2011,** *111,* 1346.

Workup

Chapter Outline

"In a typical chemical operation, 60 to 80% of both capital expenditures and operating costs go to separations."

– Charles A. Eckert, Professor of chemical engineering, Georgia Institute of Technology [1]

I. INTRODUCTION

Separations of solids, liquids and gases are operations essential to work up reactions and isolate products. In many published reaction schemes only the reagents and solvents for a transformation are shown, with no indications of the workups, but the details of workups are extremely important for rugged, effective and efficient scale-up operations. Since workup contributes greatly to operating costs, capital expenditures, productivity, and cost of goods, an inefficient workup often leads to a route or step being de-selected for scale-up [2]. Operating costs are reduced by developing simple but effective workups that require relatively few hours of labor and lead to isolated products that do not need to be reworked to raise product quality. Capital costs are reduced by using multipurpose equipment that can be used to process more than one product or intermediate. Optimizing separations can be key to minimizing the cost of goods [3].

Workup comprises the operations after the reaction has been declared complete. Such operations include quenching the reaction both to remove impurities and facilitate product isolation and to allow SAFE handling of process streams, even after product isolation. Some typical workup operations are shown in Table 11.1.

Practical Process Research and Development. DOI: 10.1016/B978-0-12-386537-3.00011-3

TABLE 11.1 Some Typical Workup Operations

Quenching reactive species	Removing solvents by distillation
pH adjustment	Removing solvents by membranes
Filtration (recovery of reagents and removal of impurities)	Deionization
Precipitation of byproducts and filtration	Treatment with activated carbon or other adsorbents
Extractions	Chromatography
Azeotroping to remove volatile components, including solvents	Polish filtration

The selection of the best workup depends on the use of the reaction product. One option is to telescope one step into the next, by transferring process streams as extracts or concentrates and not isolating the reaction product. Telescoping, also termed concatenation or through-processing, is an attractive option for intermediates, as long as any impurities passed along with the intermediate do not cause difficulties in downstream purifications and such operations are consistent with the processes filed with regulatory authorities. With telescoping there are fewer operations, hence less chance for physical losses and contamination of process streams. Usually fewer pieces of equipment are used, thus freeing up equipment for other uses. If changing solvents is necessary, chasing a lower-boiling solvent with a higher-boiling solvent requires less time and energy (see discussion around Figure 8.10a). Telescoping also decreases the exposure of operators to potent, toxic materials [4]. By azeotroping out H_2O process streams and equipment can be dried for the next step; process streams may be treated with Na_2SO_4, $MgSO_4$, or molecular sieves if temperatures required for azeotropic removal of H_2O cannot be tolerated (see Chapter 5). Volatile byproducts and reagents may be removed without extractive workup, as in the formation of the unsymmetrical urea in Figure 11.1 [5]. (This approach could streamline the preparation of libraries of compounds.) If a product or byproduct is to be isolated as a solid, conditions for crystallization should be developed in order to control the physical and chemical characteristics of the solids, as precipitates may filter slowly. Precipitations have been optimized to isolate peptides on a large scale [6]. Direct isolations are more productive than isolation after extractive

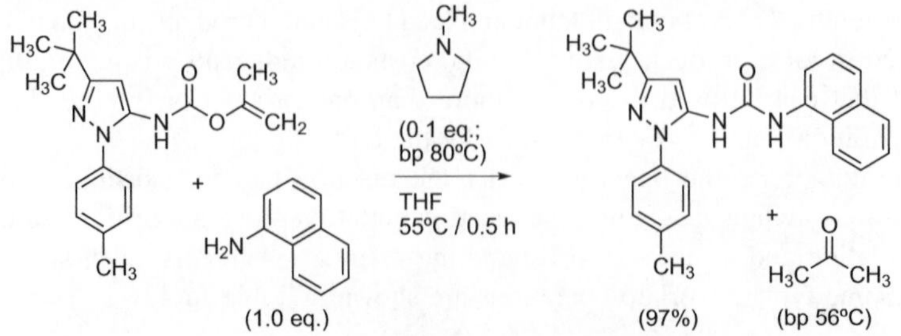

FIGURE 11.1 A reaction with volatile reagents and byproducts suitable for telescoping.

workups (see discussion with Figure 3.21) [7]. With direct isolation the product is crystallized from the reaction mixture, often by the addition of an antisolvent [8]. Extractive workups are often used to remove troublesome impurities, or to raise the overall yield by removing impurities that impede crystallization. Preparative chromatography may also be used to ease supply of materials for initial studies, but may be expensive for routine manufacturing on scale. Chromatography has been routinely used to purify peptides manufactured on a large scale [6].

II. QUENCHING REACTIONS

SAFETY is the foremost consideration in quenching a reaction. Simply diluting a reaction will not stop reactions, unless a product or reactive component is physically removed by crystallization, precipitation, or partitioning into another liquid phase. Quenching neutralizes reactive components, often quenching reagents are charged in excess, to provide SAFE conditions for subsequent operations and to minimize degradation of the product stream. In this regard SAFE operations take into account chemical hazards and toxicological hazards. For instance, in a manufacturing facility solid oxalic acid was charged to quench excess dicyclohexylcarbodiimide (DCC), which is a sensitizer and extremely irritating (Figure 11.2) [9,10]. After byproducts such as dicyclohexylurea (DCU) are removed the product may be crystallized and isolated.

Carbodiimides are a category of reagents that should be viewed as reactive and hence quenched. Popular carbodiimides include DCC, diisopropylcarbodiimide, and ethyl dimethylaminopropylcarbodiimide (EDC or EDCI). Acids, such as oxalic, acetic, and phosphoric, will react with carbodiimides, and hydrolysis of the resulting iminoesters yields the corresponding ureas. The intermediate iminoesters of carboxylic acids can rearrange to the corresponding N-acyl ureas (Figure 9.6). DCU is poorly soluble in DMF, leading to the selection of DMF as a solvent for reactions using DCC. Perhaps difficulties removing DCU from reaction mixtures are due to incomplete quenching of the more-soluble DCC.

A detonation in the laboratory illustrates the danger that can be posed by the difficulty in removing residual solvents. During the workup of the crude azide in Figure 11.3 the condensate in a rotary evaporator detonated, and the explosive agent was subsequently identified as diazidomethane, $CH_2(N_3)_2$. Excess dichloromethane (bp 42 °C) from the mesylation had been removed under reduced pressure by rotary evaporation, DMF

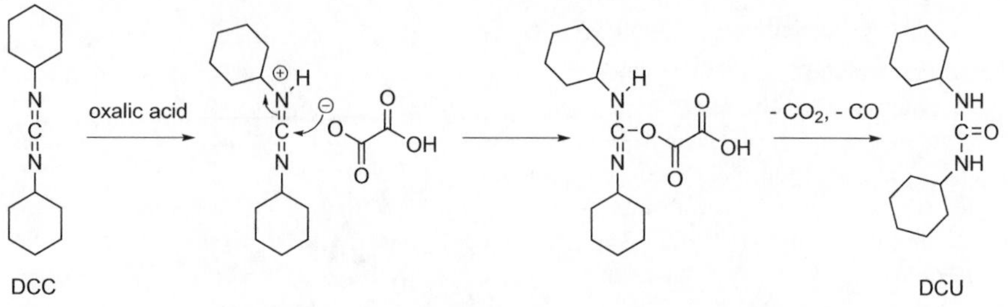

FIGURE 11.2 Quenching a DCC reaction with oxalic acid.

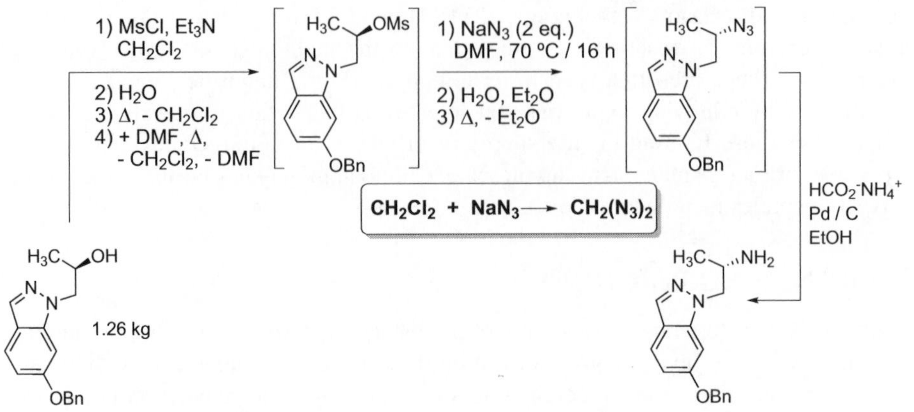

FIGURE 11.3 Generation of diazidomethane.

(bp 152 °C) was added, and solvent stripping was resumed ($\sim$20 Torr, bath temperature about 35 °C). When no more condensate was noted, the absence of any CH_2Cl_2 in the flask was inferred; evaporation was ceased and more DMF and NaN_3 were charged. The subsequent reaction was heated at 70 °C for 16 h, cooled, and partitioned between Et_2O and H_2O. The rich organic phase was concentrated on a rotary evaporator, and the receivers were emptied. The next morning the rotary evaporator detonated. Further analysis showed that 8–10 mol% of diazidomethane was present in the crude azide, which was then subjected to reducing conditions to destroy the diazidomethane present [11].

In the example in Figure 11.3 the approach of ending evaporation when no more condensate formed seemed very reasonable, but in the final analysis not all of the CH_2Cl_2 was removed, despite the 110 °C difference in boiling point. GC analysis or NMR analysis of the residue may have confirmed the presence of a significant amount of CH_2Cl_2 [12]. The difficulty in removing final mol fractions of a volatile component is consistent with Raoult's law [13], which extends processing. Another approach to developing inherently SAFE reaction conditions is to consume reactive species such as NaN_3, by undercharging NaN_3 relative to the mesylate in this case. For SAFE operations on scale, destroying the excess NaN_3 in the "spent" aqueous extract would be essential.

Significance of Raoult's law for removing volatile components

Raoult's law: P_A = partial pressure of A from solution
$P_A = P_A^0 \times N_A$ P_A^0 = vapor pressure of pure A
                       N_A = mol fraction of A

Vapor pressure is essentially the tendency for molecules of A to escape from solution into vapor. When $\sum(P_i)$ = atmospheric pressure, a mixture boils. P_A decreases as the mol fraction of A decreases, and hence near the end of a distillation or concentration it becomes more difficult to remove the small amount of A remaining. When drying a crystallized product in a dryer the rate of removing a residual solvent can slow dramatically as the solvent content approaches the limit. Other low-boiling components that may be more difficult to remove than expected include CO_2, Me_2S, and Me_2NH; in some cases actively purging the system may be essential (Chapter 8).

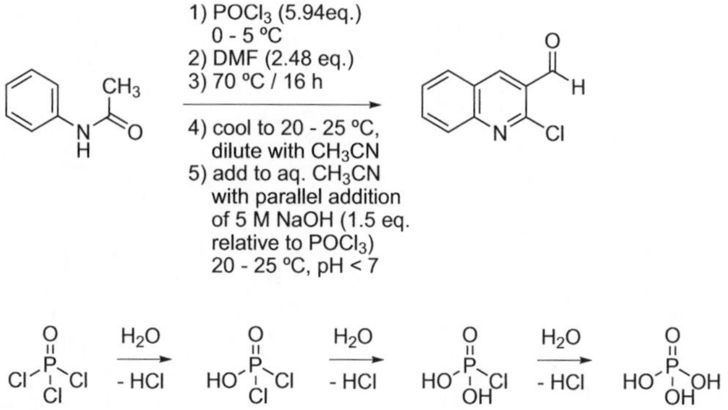

FIGURE 11.4 Neutralizing excess POCl₃ by inverse quench with aqueous NaOH.

In another incident two galvanized steel drums containing a condensate of thionyl chloride and EtOAc unexpectedly detonated after 45 minutes of storage. Subsequent analysis by accelerated rate calorimetry (ARC) showed that although a mixture of SOCl₂ and EtOAc produced no exotherm, a mixture of SOCl₂, EtOAc, and Zn produced a runaway reaction, exothermically releasing SO₂, EtCl, acetyl chloride, and solid sulfur. Thus the zinc coating in the carbon-steel drums led to an exothermic reaction of SOCl₂ and EtOAc [14]. The latter paper mentioned that SOCl₂ is also known to react with THF and DMF, and advised caution when generating mixtures containing reactive components such as SOCl₂, tosyl chloride, and phosphorus oxychloride. Thionyl chloride was also shown to react with MTBE [15]. Amgen workers raised concerns about quenching POCl₃ from a preparation of a quinoline by the Meth-Cohn reaction (Figure 11.4): after filtering off the product they observed that the temperature of the aqueous filtrates slowly rose to 50 °C. Investigations showed that the second stage of hydrolysis of excess POCl₃ was much slower than the first stage, producing a latent exotherm [16]. Neutralizing process streams conscientiously and promptly is prudent.

Simply concentrating a reaction may not be an effective workup. In early studies to prepare an aprepitant precursor Merck scientists activated a lactol as the trichloroacetimidate (Figure 11.5). Upon scale-up removal of CH₂Cl₂ and excess trichloroacetonitrile by distillation took significantly longer than in the laboratory, and with slow distillation the equilibrium between the imidate and the lactol shifted back toward the starting materials as excess trichloroacetonitrile was removed by concentration. To avoid difficulties in using trichloroacetonitrile the Merck scientists activated the lactol as the trifluoroacetate [17].

> "Rich" and "spent" phases usually refer to phases that contain virtually all or essentially no product, respectively. For SAFE operations neutralize any reactive components of "spent" reaction streams soon after they have been generated.

Organic chemists rely on many reactions that are equilibria, driven to completion by two fundamental approaches. The first is to add a large excess of a reagent, as in preparing a methyl ester by Fischer esterification in a large volume of MeOH. The second approach is to physically remove a product (or byproduct) from the

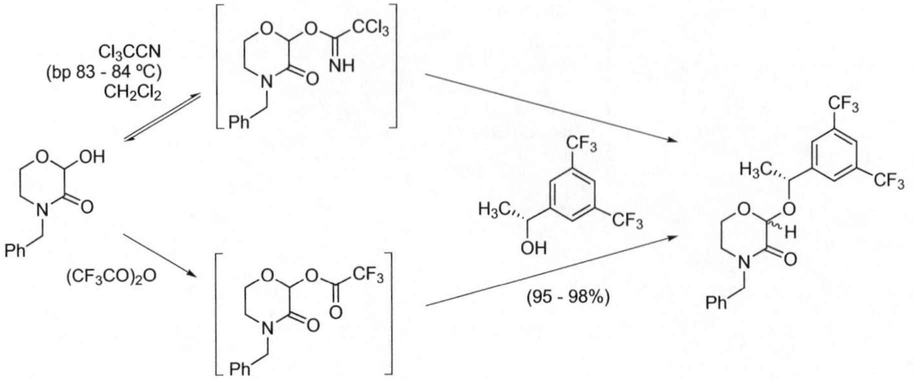

FIGURE 11.5 Avoiding difficulties of working up an equilibrium reaction by changing a reagent.

equilibrium, as in removing H_2O using a Dean–Stark trap, crystallizing the product from a CIAT or CIDR process (Chapter 12), or deprotecting a Boc-amine by volatilizing CO_2 and isobutylene. Equilibrium processes, including metathesis and many enzymatic reactions, must be actively quenched before workup in order to achieve desired yields. In the laboratory pouring a reaction mixture on top of a column of silica gel can be an effective quench: the catalyst is absorbed from the reaction mixture. But on a large scale transfers and such "plug chromatography" may require considerable time, even hours, allowing the equilibria of reactions to degrade product. An example is shown in the ruthenium-catalyzed ring-closing metathesis (RCM) in Figure 11.6 [18]; the yield of the first batch on scale was much lower than anticipated because the reactive complex was not quenched. The reaction mixture was actively quenched with 2-mercaptonicotinic acid and/or imidazole [19,20]. Workers at Array found that diethylene glycol monovinyl ether very effectively quenched ruthenium reagents used in metathesis [21].

Exothermic quenches may be conducted in stages or by transferring the reaction mixture into a quench, an inverse quench, which may be chilled. (A familiar example of an inverse quench is the adage that adding acid to water is safer than adding water to acid.)

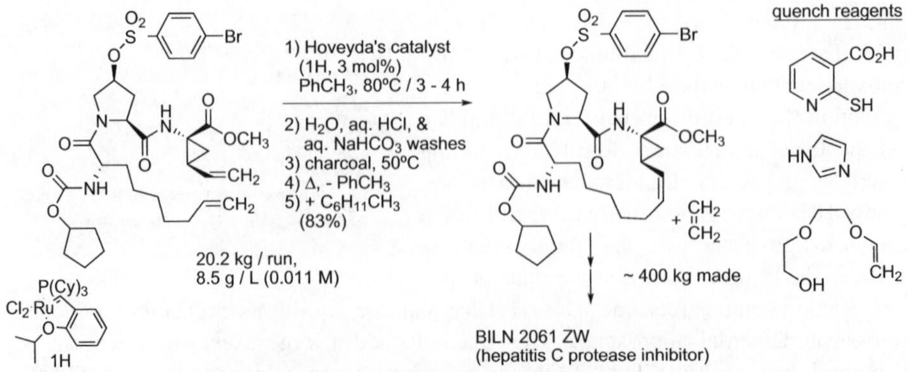

FIGURE 11.6 Quenching a Ru-catalyzed RCM on scale.

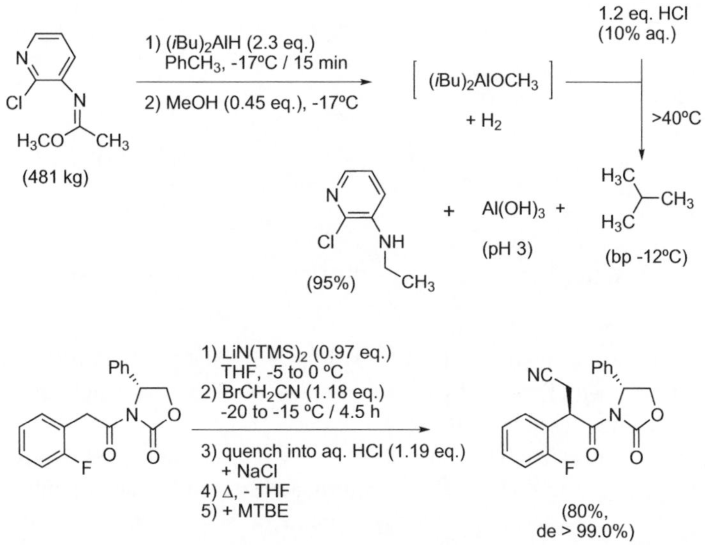

FIGURE 11.7 Controlling quench conditions by quenching in stages and using inverse quenches.

These features are demonstrated in Figure 11.7. The slow addition of MeOH to the diisobutylaluminum hydride reaction controlled the evolution of H_2, and the subsequent, controlled addition of the quenched reaction mass into aqueous HCl at 40 °C lead to rapid hydrolysis with controlled evolution of isobutane. Under these acidic conditions the aluminum salts were dissolved in the pH 3 quench, and the non-basic product (the substituents withdrew electron density from the pyridine) was extracted into the reaction solvent [22]. In the second example in Figure 11.7 the reaction was quenched by inverse addition into aqueous HCl, because when aqueous HCl was added to the reaction mass some epimerization α- to the carbonyl was noted [23]. Inverse addition minimized epimerization because in that case the reaction mixture was always acidic during the quenching operation.

SAFETY — Quench reactions well below the boiling point of a reaction mixture or solvent. Consider a dissolving metal reaction in liquid ammonia (bp −33 °C): if the reaction temperature is too high, quenching with H_2O or NH_4Cl can cause the ammonia to boil out of the reactor. When quenching at temperatures near −33 °C solid NH_4Cl can act as boiling stones, with the reaction mixture "bumping" out of the reactor. Water should not be added to a refluxing mixture if a lower-boiling azeotrope is formed: the condenser could be flooded, with physical loss of the reactor contents. Consider the physicochemical nature of reaction components and mixtures for SAFETY.

Excluding even traces of oxygen from entering a reaction can facilitate workups. Rigorous exclusion of O_2 in quenching a copper-catalyzed addition of Grignard reagents to steroidal dienones improved isolated yields, and the authors proposed that the

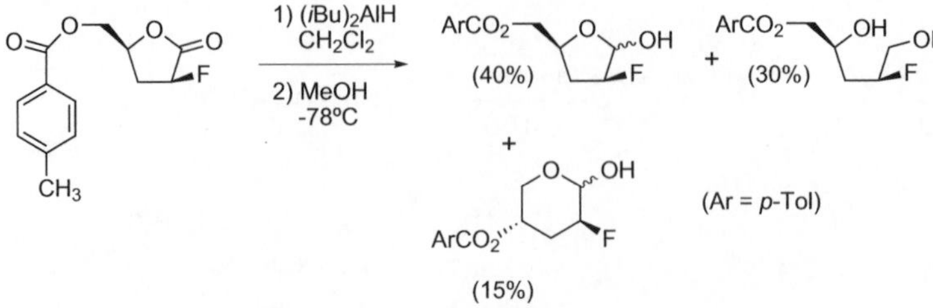

FIGURE 11.8 Quenching a DIBAH reduction with AcOH.

intermediate enolates reacted faster with O_2 than with H_2O [24]. Merck researchers noted that treatment with Ecosorb® C-941 to remove rhodium from a hydrogenation was more effective in the presence of H_2 [25].

The products from quenching can subsequently lead to undesired byproducts. For instance, quenching $LiAlH_4$ with EtOAc can form acetate, which can be neutralized with an acidic workup and partition into an organic extract; subsequent processing, such as concentrating an organic extract with azeotropic removal of H_2O, can form acetamides of primary or secondary amines. (EtOAc can also react with amines, as discussed in Chapter 5.) Quenching a $LiAlH_4$ reduction with EtOH produced alkoxides, which reacted with the product; quenching with AcOH did not produce such decomposition [26]. Similarly, quenching a $(iBu)_2AlH$ reduction with MeOH produced methoxide, which promoted acyl migration (Figure 11.8); quenching with AcOH produced the less-basic LiOAc with no formation of the pyranose. (In this case the preferred reagent was $LiAlH_4$/pyridone, with reduction in THF at $-30\,°C$, producing the desired lactol in 93% yield [27]).

> SAFETY — Reactions should not be scaled up until the potential for exothermic reactions has been assessed. Reaction products can also be destroyed during uncontrolled quenching.
>
> Quenching a reaction under conditions that produce a mass that cannot be stirred can cause SAFETY concerns on scale-up. For instance, adding an aqueous solution of a quench to a cryogenic reaction can produce a shell of ice inside a reactor; such fouling slows down heat transfer from a fluid in an external jacket. If too much heat is applied to the reactor the shell of ice may suddenly break free of the reactor wall, subjecting the reaction mass to excessively high temperatures. If quenching with a reagent produces solids, consider alternatives. For instance, substituting propionic acid (mp $-24\,°C$) for AcOH (mp $17\,°C$) may be helpful. Dissolving the quench in an organic solvent may also lower the freezing point.

Table 11.2 lists some reagents that have been used to quench reactions and neutralize reactive components. This list is not exhaustive, and treatments to quench a reagent may be described more fully in the *Encyclopedia of Reagents for Organic Synthesis* (e-EROS) [28] or in a recent book on scale-up procedures [29]. Reagents to remove metals from reaction products, as discussed in Section VI, may also serve as useful quenches.

TABLE 11.2 Some Reagents to Quench Reactions. This list is a general guideline, and all quench operations should be conducted cautiously for SAFE operations. Suitable hazard evaluations should be conducted before scale-up

Reactant	Neutralizing Reagent	Reference
HO−, RO−, organic bases	Organic or inorganic acids	
H+	Organic or inorganic bases	
RCOCl, SOCl$_2$	H$_2$O, then base	
POCl$_3$	aq. NaOH	(1)
H$_2$O$_2$, NaOCl, Cl$_2$, Br$_2$, I$_2$, NCS, NBS, other oxidants	Activated carbon (Darco G-60)	Quenching performic acid (2)
	Ascorbic acid	(3)
	H$_3$PO$_2$ (hypophosphorus acid)	(4)
	NaHSO$_3$, Na$_2$S$_2$O$_3$, Na$_2$S$_2$O$_5$	
	MnO$_2$	Decomposing H$_2$O$_2$ (5)
Os(VI), Os(VIII)	Na$_2$SO$_3$	(6)
O$_3$, ozonides	H$_3$SCH$_3$, Ph$_3$P, NaBH$_4$	(7,8)
H$_2$NNH$_2$, CN−, NaN$_3$,	NaOCl	
BH$_4^-$	Acetone	
LiAlH$_4$	Acetone, EtOAc, Rochelle salt	(9)
	Na$_2$SO$_4$·10H$_2$O	(10)
	H$_2$O, NaOH, H$_2$O	(11)
	AcOH, EtOH	(12)
(iBu)$_2$AlH	MeOH, then aq. HCl	(13)
	AcOH	(14)
	EtOAc, then aq. NaHCO$_3$	(15)
	Oxalic acid	
CH$_3$I, H$_3$COSO$_2$OCH$_3$	NH$_3$	(16)
CH$_3$OTs	DABCO	(17)
H$_3$SCH$_3$, RSH	NaOCl	
DCC, carbodiimides	Oxalic acid, AcOH, H$_3$PO$_4$	(18)
Na(0), Li(0), etc.	NH$_4$Cl, then H$_2$O; or H$_2$O	(19)
RLi, R$_2$NLi	AcOH, NH$_4$Cl, tartaric acid	(20)

(Continued)

TABLE 11.2 Some Reagents to Quench Reactions. This list is a general guideline, and all quench operations should be conducted cautiously for SAFE operations. Suitable hazard evaluations should be conducted before scale-up—cont'd

Reactant	Neutralizing Reagent	Reference
RMgX	AcOH, NH$_4$Cl, tartaric acid	(21,22)
RZnX	Brine	(23)
BR$_3$	H$_2$O$_2$, oxone, other oxidants, often with HCl or AcOH	(24,25)
Ru (metathesis)	Imidazole	(26)
	2-Mercaptopicolinic acid	(27)
	Diethylene glycol monovinyl ether	(28)

References

(1) Achmatowicz, M. M.; Thiel, O. R.; Colyer, J. T.; Hu, J.; Elipe, M. V. S.; Tomaskevitch, J.; Tedrow, J. S.; Larsen, R. D. Org. Process Res. Dev. **2010**, 14, 1490.

(2) Ripin, D. H. B.; Weisenburger, G. A.; am Ende, D. J.; Bill, D. R.; Clifford, P. J.; Meltz, C. N.; Phillips, J. E. Org. Process Res. Dev. **2007**, 11, 762.

(3) Xiang, Y.; Caron, P.-Y.; Lillie, B. M.; Vaidyanathan, R. Org. Process Res. Dev. **2008**, 12, 116. Quenching an iodination with Na$_2$S$_2$O$_3$, left residual sulfur in the rich extract, which inhibited the subsequent Pd-catalyzed coupling.

(4) Chen, C.-K.; Singh, A. K. Org. Process Res. Dev. **2001**, 5, 508.

(5) Vaino, A. R. J. Org. Chem. **2000**, 65, 4210.

(6) Prasad, J. S.; Vu, T.; Totleben, M. J.; Crispino, G. A.; Kacsur, D. J.; Swaminathan, S.; Thornton, J. E.; Fritz, A.; Singh, A. K. Org. Process Res. Dev. **2003**, 7, 821.

(7) Van Ornum, S. G.; Champeau, R. M.; RPariza, R. Chem. Rev. **2006**, 106, 2990.

(8) Allian, A. D.; Richter, S. M.; Kallemeyn, J. M.; Robbins, T. A.; Kishore, V. Org. Process Res. Dev. **2011**, 15, 91.

(9) Hirokawa, Y.; Hirokawa, T.; Noguchi, H.; Yamamoto, K.; Kato, S. Org. Process Res. Dev. **2002**, 6, 28.

(10) For an example, see Brückner, C.; Xie, L. Y.; Dolphin, D. Tetrahedron **1998**, 54, 2021.

(11) By the sequential addition of n mL of H$_2$O, n mL of 15% NaOH, and 3n mL of H$_2$O a granular precipitate of hydroxides may be obtained: Fieser, L. F.; Fieser, M. Reagents for Organic Synthesis; Vol. 1; Wiley: New York, 1967; p 584.

(12) Benoit, G.-E.; Carey, J. S.; Chapman, A. M.; Chima, R.; Hussain, N.; Popkin, M. E.; Roux, G.; Tavassouli, B.; Vaxelaire, C.; Webb, M. R.; Whatrup, D. Org. Process Res. Dev. **2008**, 12, 88.

(13) Perrault, W. R.; Shephard, K. P.; LaPean, L. A.; Krook, M. A.; Dobrowolski, P. J.; Lyster, M. A.; McMillan, M. W.; Knoechel, D. J.; Evenson, G. N.; Watt, W.; Pearlman, B. A. Org. Process Res. Dev. **1997**, 1, 106.

(14) Jin, F.; Wang, D.; Confalone, P. N.; Pierce, M. E.; Wang, Z.; Xu, G.; Choudhury, A.; Nguyen, D. Tetrahedron Lett. **2001**, 42, 4787; Stork, G.; Boeckrnann, R. K., Jr.; Taber, D. F.; Still, W. C.; Singh, J. J. Am. Chem. Soc. **1979**, 101, 7107; Selliah, R. D.; Hellberg, M. R.; Sharif, N. A.; McLaughlin, M. A.; Williams, G. W.; Scott, D. A.; Earnest, D.; Haggard, K. S.; Dean, W. D.; Delgado, P.; Gaines, M. S.; Conrow, R. E.; Klimko, P. G. Bioorg. Med. Chem. Lett. **2002**, 14, 4525; Delgado, P.; Conrow, R. E.; Dean, W. D.; Gaines, M. S. U.S. Patent 6,828,444, 2004 (to Alcon).

(15) Shieh, W.-C.; Chen, G.-P.; Xue, S.; McKenna, J.; Jiang, X.; Prasad, K.; Repic, O.; Straub, C.; Sharma, S. K. Org. Process Res. Dev. **2007**, 11, 711.

(16) Prashad, M.; Har, D.; Hu, B.; Kim. H.-Y.; Repic, O.; Blacklock, T.J. Org. Lett. **2003**, 5, 128; Prashad, M.; Har, D.; Hu, B.; Kim. H.-Y.; Girgis, M. J.; Chaudhary, A.; Repic, O.; Blacklock, T. J. Org. Process Res. Dev. **2004**, 8, 331.

(17) Camp, D.; Matthews, C. F.; Neville, S. T.; Rouns, M.; Scott, R. W.; Truong, Y. Org. Process Res. Dev. **2006**, 10, 814.

(18) Fieser, L. F.; Fieser, M. Reagents for Organic Synthesis; Vol. 1; Wiley: New York, 1967; p 231.

TABLE 11.2 Some Reagents to Quench Reactions. This list is a general guideline, and all quench operations should be conducted cautiously for SAFE operations. Suitable hazard evaluations should be conducted before scale-up—cont'd

Reactant	Neutralizing Reagent	Reference

(19) Joshi, D. K.; Sutton, J. W.; Carver, S.; Blanchard, J. P. Org. Process Res. Dev. **2005**, 9, 997. SAFETY NOTE: Reactions should be quenched at about −40°C, below the boiling point of NH₃ (−33°C). Solid NH₄Cl can act like boiling stones to cause the NH₃ to boil out of the reactor.

(20) Excess AcOH should be used to quench excess alkyl lithium reagents, as two equivalents of alkyl lithium can react with carboxylic acids to form the tertiary carbinol.

(21) Quenching a Grignard reaction into cold aqueous AcOH afforded good control of pilot plant operations: Kopach, M. E.; Singh, U. K.; Kobierski, M. E.; Trankle, W. G.; Murray, M. M.; Pietz, M. A.; Forst, M. B.; Stephenson, G. A. Org. Process Res. Dev. **2009**, 13, 209.

(22) The greater solubility of Mg(OAc)₂ over that of (NH₄)MgCl₃ has been exploited in working up a Weinreb amide: Davies, I. W.; Marcoux, J.-F.; Corley, E. G.; Journet, M.; Cai, D.-W.; Palucki, M.; Wu, J.; Larsen, R. D.; Rossen, K.; Pye, P. J.; DiMichele, L.; Dormer, P.; Reider, P. J. J. Org. Chem. **2000**, 65, 6415.

(23) Girgis, M. J.; Liang, J. K.; Du, Z.; Slade, J.; Prasad, K. Org. Process Res. Dev. **2009**, 13, 1094.

(24) Atkins, W. J., Jr.; Burkhardt, E. R.; Matos, K. Org. Process Res. Dev. **2006**, 10, 1292. Alkyl boranes can be highly flammable. The authors provide detailed guidelines for quenching diborane and alkyl boranes.

(25) Astbury, G. W. Org. Process Res. Dev. **2002**, 6, 893.

(26) Brenner, M.; Meineck, S.; Wirth, T. US Patent 7,183,374 B2, 2007 (to Boehringer Ingelheim).

(27) Yee, N. K.; Farina, V.; Houpis, I. N.; Haddad, N.; Frutos, R. P.; Gallou, F.; Wang, X.; Wei, X.; Simpson, R. D.; Feng, X.; Fuchs, V.; Xu, Y.; Tan, J.; Zhang, L.; Xu, J.; Smith-Keenan, L. L.; Vitous, J.; Ridges, M. D.; Spinelli, E. M.; Johnson, M. J. Org. Chem. **2006**, 71, 7133.

(28) Liu, W.; Nichols, P. J.; Smith, N. Tetrahedron Lett. **2009**, 50, 6103.

The process chemist or engineer may explore adapting reagents from Table 11.2 for other purposes. For instance, diazabicyclooctane (DABCO) was used to decompose methyl *p*-toluenesulfonate, and certainly would be expected to react readily with methyl iodide or dimethyl sulfate. Since the amine groups of DABCO are accessible for nucleophilic reaction one might try adding DABCO to decompose other alkylating reagents, such as ethyl chloroacetate or benzyl bromide. Such quaternary ammonium salts should be water-soluble and removed by aqueous extraction, especially under mildly acidic conditions in which there may be some tendency to protonate the other amine group of DABCO.

LiAlH₄ reductions pose handling concerns due not only to the reactivity of that moisture-sensitive, flammable reagent but also due to the exothermic quenching and emulsions formed on workups [30]. The recipe of quenching with n mL of H_2O, n mL of 15% NaOH, then $3n$ mL of H_2O can produce aluminum salts that can be removed by filtration [31]. Saturated aqueous NaHCO₃ has been added below 15 °C to precipitate aluminum salts [32]. Citric acid can be used to solubilize the salts under basic conditions [33] (Figure 11.9). Rochelle salt has produced solids that were filtered off [34] or dissolved in water [26]; aqueous tartaric acid may produce a biphasic organic–aqueous mixture upon stirring [35]. Other reagents may be effective reductants and provide easier workups, such as LiBH₄ [34], NaBH₄/ZnCl₂ [36], or BH₃·THF [37].

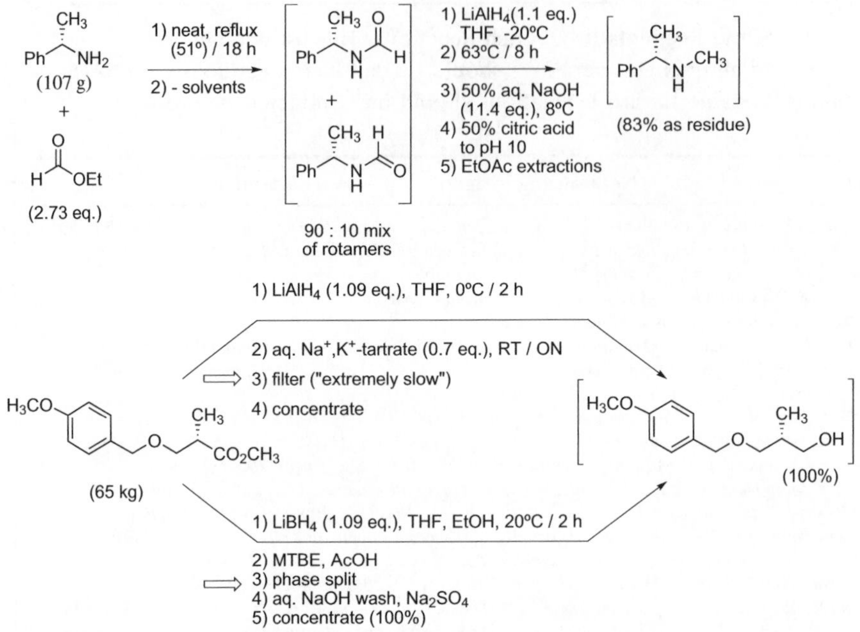

FIGURE 11.9 Workups for LiAlH₄ reactions.

SAFETY — Lithium aluminum hydride is extremely reactive toward H_2O. Handling, charging, and quenching this reagent should be undertaken with due caution. Quenching by adding H_2O to a reaction containing excess hydride may be extremely exothermic and may require a slow addition on a large scale. Transferring the reaction mixture into a quench may afford better temperature control, if such a reaction mixture is fluid enough to transfer. However, inverse quenches carry the risks of contamination and material loss, and adding a quench mixture to the vessel may minimize occupational exposure to LiAlH₄.

In removing inorganic salts different reagents may be substituted to take advantage of solubilities in water to ease workup. Crystallization (or precipitation) of byproducts from water separates those impurities from products that are highly water-soluble and that may be extracted with difficulty into an organic solvent; such filtrations may be used on scale if the formation of byproduct and filtrations are rapid and effective. For instance, quenching Grignard reactions with ammonium chloride can form poorly soluble species that produce stiff emulsions; Grignard reactions quenched with AcOH, citric acid or tartaric acid may lead to reduced emulsions [38,39].

Table 11.3 shows solubilities of some salts that may be encountered in process chemistry. In general salts of metals are less soluble under more basic conditions. Salts of oxalates, sulfates, and phosphates may precipitate or crystallize. Phosphate substitutes being developed by the detergent industry may be effective and inexpensive treatments to remove metal contaminants [40]. Citrates and tartrates are often highly soluble. Such solubility differences can be exploited in designing workups for other metal salts, such as salts of aluminum.

TABLE 11.3 Selected Solubilities of Inorganic Salts in H_2O. (1)

Family	Solubility in 100 g H_2O				
Fluorides	LiF 0.13	NaF 4.13	KF 101.65	MgF_2 0.01	CaF_2 0.002
Chlorides	LiCl 84.54	NaCl 35.96	KCl 34.54	$MgCl_2$ 56.01	$CaCl_2$ 81.26
Bromides	LiCl 180.90	NaBr 94.55	KBr 67.79	$MgBr_2$ 102.43	$CaBr_2$ 156.41
Iodides	LiI 165.25	NaI 184.09	KI 148.14	MgI_2 146.31	CaI_2 215.46
Bicarbonates	$LiHCO_3$ 5.74 (15 °C)	$NaHCO_3$ 9.32	$KHCO_3$ 36.24		$Ca(HCO_3)_2$ 16.6 (15 °C)
Carbonate	Li_2CO_3 1.30	Na_2CO_3 30.72	K_2CO_3 111.42	$MgCO_3$ 0.039	$CaCO_3$ 0.0007
Acetates	LiOAc 44.97	NaOAc 50.38	KOAc 269.28	$Mg(OAc)_2$ 65.59	$Ca(OAc)_2 \cdot$ $2H_2O$ 34.7
Oxalates	$Li_2C_2O_4$ 6.24	$Na_2C_2O_4$ 3.61	$K_2C_2O_4 \cdot H_2O$ 33	MgC_2O_4 0.004	CaC_2O_4 0.0007
Citrates	$Li_3C_6H_5O_7 \cdot 4H_2O$ 74.5	$Na_3C_6H_5O_7 \cdot 2H_2O$ 92.6	$K_3C_6H_5O_7 \cdot H_2O$ 167 (15 °C)		
Sulfates	$LiHSO_4 \cdot H_2O$ 34.41 Li_2SO_4 334.41	$NaHSO_4$ 28.53 Na_2SO_4 28.11	$KHSO_4$ 50.60 K_2SO_4 11.98	$Mg(HSO_4)_2$ $MgSO_4$ 35.69	$Ca(HSO_4)_2$ $CaSO_4$ 0.21
Phosphates	LiH_2PO_4 126.24 Li_3PO_4 0.03	NaH_2PO_4 94.86 $Na_2HPO_4 \cdot 12H_2O$ 4.15 Na_3PO_4 14.42	KH_2PO_4 24.95 K_2HPO_4 168.10 K_3PO_4 105.76	$Mg_3(PO_4)_2$ 0.0003 (15 °C)	$Ca(H_2PO_4)_2$ 1.8 (15 °C) $CaHPO_4$ 0.004 (15 °C) $Ca_3(PO_4)_2$ 0.002 (15 °C)
Hydroxides	LiOH 12.49	NaOH 100	KOH 120.75	$Mg(OH)_2$ 0.0009 (15 °C)	$Ca(OH)_2$ 0.173 (15 °C)

(1) Unless specified otherwise, solubilities were measured at 25° C (CRC Handbook of Chemistry and Physics; Lide, D. R., Ed.; 88th ed.; CRC Press: Boca Raton, 2008; pp 8-116e 8-121) and converted to units of g/100 g H_2O. Other data were abstracted from http://en.wikipedia.org/wiki/Solubility_table. When multiple forms (hydrates) are possible, the form with the lowest solubility is shown. In practice solubilities will vary with pH, temperature, and impurities that are present in solution.

Insoluble and highly crystalline solids contain components with high affinity for each other, which may be exploited for effective workups. For instance, the mineral struvite [(NH₄) MgPO₄·6H₂O] is poorly soluble in alkaline water, and is found in kidney stones; struvite is generated for recovering phosphates from waste water [41]. Researchers have noted that Grignard reagents that were quenched with aqueous KH_2PO_4 produced precipitates [42] or gums; [43] perhaps quenching a Grignard reaction with a salt of ammonium phosphate and pH adjustment would remove the magnesium byproducts as a crystalline byproduct. As a second example, $NiCl_2 \cdot 1,2$-dimethoxyethane (mp > 300 °C) is commercially available; perhaps nickel salts could be removed from reaction products by treatment with dimethoxyethane or other ethylene glycol derivatives to afford precipitates or complexes with high affinity for nickel.

III. EXTRACTIONS

Extractive workups can be used to halt a reaction and to remove impurities so that a reaction product can be crystallized efficiently. Extractions may be employed to recover even small amounts of a high value-added product. For manufacturing operations additional extractions are frequently avoided to save time and increase productivity in isolating a product.

Extractive workups are generally employed to separate neutral compounds from water-soluble compounds dissolved in an aqueous phase, including small molecules, salts and ionizable byproducts, as illustrated in Figure 11.10. The oxime of erythromycin A was formed from reaction of the ketone with aqueous hydroxylamine in i-PrOH, and further reaction was stopped by extracting the product into i-PrOAc [44]. The modified Mitsunobu reaction (sulfonation with inversion) saved three manufacturing steps and produced high-quality cis-mesylate carboxylic acid (less than <0.1% trans-isomer impurity) in high yield without chromatography. The carboxylate was separated at pH 6–7 by extraction from the toluene phase containing the neutral byproduct triphenylphosphine oxide [45]. In this process triphenylphosphine oxide crystallized as a complex with the hydrazide from the biphasic mother liquor, and the crystallization was promoted because most of the other products had been removed from the toluene phase [46]. In the third example excess 2,4-dichlorobenzyl chloride (a potential genotoxic impurity, or PGI) was partitioned into the organic phase as the product was extracted into the aqueous phase as the carboxylate [47]; crystallizing the product in the presence of a small amount of organic solvents further enhanced efforts to purge the PGI from the product.

Crystallizing a product from a mixture of two immiscible liquids may efficiently remove impurities.

Removing triphenylphosphine oxide from a product. The best approach may be to generate an ionizable intermediate that can be extracted into an aqueous phase, as in Figure 11.10. Deprotection of a Boc-amine may generate a salt that can be extracted into H_2O or slurried in a polar solvent to remove triphenylphosphine oxide [48]. Crystallizing a product from MeOH or MeOH–H₂O may leave the TPPO in the mother liquor [45]. TPPO has formed complexes with $MgCl_2$ that can be removed by filtration [49]. TPPO has also been extracted into AcOH, with hexanes as the organic phase [50]. Triturating a product with a hydrocarbon solvent such as heptanes is generally less effective to remove TPPO.

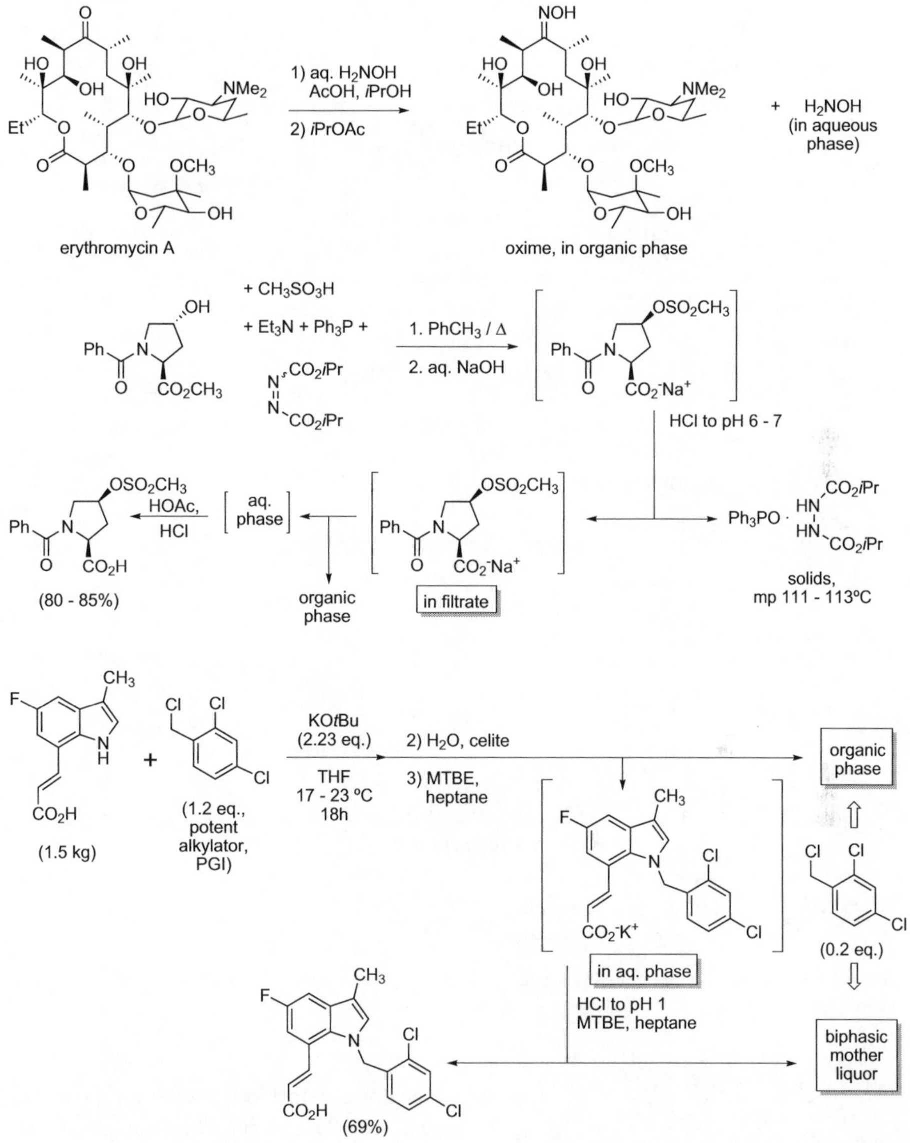

FIGURE 11.10 Extractive workups to stop a reaction and remove troublesome impurities.

Controlling pH may be essential for developing a robust extractive workup. In Figure 11.11 by basifying to pH 5.0 ± 0.2 the acidic phosphinic acid product was extracted into the aqueous phase, leaving also the acidic but more lipophilic phosphinic acid side product in the organic phase [51]. By adding water to the latter organic phase and extracting at pH 5.6 ± 0.15 a small amount of the phosphinic acid product was recovered. These pH ranges for extractions were determined experimentally by monitoring by HPLC the partitioning of product and byproduct into the organic and aqueous phases. Similarly, the byproduct dicarboxylic acid was left in the aqueous phase by extraction at pH 5.5 [52].

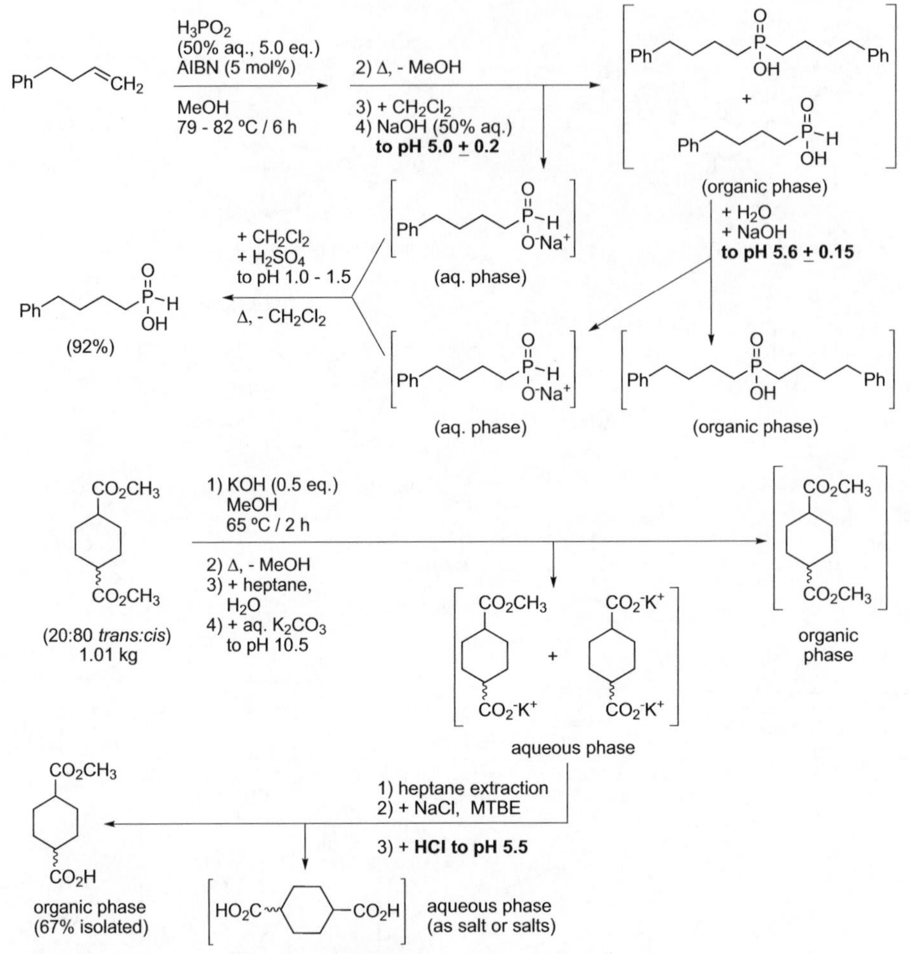

FIGURE 11.11 Examples of removing an acidic byproduct from an acidic product by pH adjustment and extractions.

Monitor and control the pH of aqueous phases during extractive workups to develop rugged processes; such data can also help with troubleshooting. Significantly more accurate than pH strips are pH meters, inexpensive but valuable additions to the process chemistry laboratory. The portable, non-flammable pH strips may be useful for troubleshooting on the floor of a plant.

In an analogous fashion, zwitterions can be extracted from water into organic phases at relatively narrow pH ranges. Amines can be extracted from an aqueous phase that is more acidic than the pK_a of the amine. As shown in Figure 9.5, a secondary amine (p$K_a \sim$ 10.5–11) was extracted into *i*-PrOAc at pH 8.5–9; extraction under more basic conditions would be expected to hydrolyze more *i*-PrOAc. With all such extractions, factors such as pH, ionic strength, choice of organic solvent, concentrations, and temperature must be optimized.

In general extractions are performed at room temperature, which is convenient for laboratory process development. Depending on the restrictions, both physical and those implemented for SAFETY, higher temperatures can be employed. Physical restrictions include factors such as the boiling point of a mixture and the solubility of solutes; an example may be seen with the filtration of the montelukast precursor in Figure 11.15. SAFETY restrictions are developed out of considerations for the outcomes from leaks; for instance, spray or fumes from a phase split of heptanes solutions at 50 °C could injure personnel or lead to combustion. SAFETY restrictions, which may vary from site to site, should be discussed well before processes are introduced.

The solvents selected for extracting products from an aqueous phase are generally those solvents that are immiscible in water. Sometimes an organic phase comprising a water-miscible solvent will separate from an aqueous phase if the ionic strength of the aqueous phase is high. Ionic strength is proportional to the concentration of an ion and the charge of that ion squared (Equation 11.1). Pfizer workers described extracting a polar sugar derivative into an organic phase comprising DME, and the partitioning was probably favored by the high ionic strength of the aqueous phase [53] (Figure 11.12). Sepracor workers showed that an aminoalcohol could be extracted into i-PrOH [54]; maintaining suitably basic conditions for extraction is necessary in this case also. i-PrOH (bp 82 °C), 2-butanol (aka sec-butanol, bp 98 °C) and isobutanol (bp 108 °C) are easier to remove than is n-BuOH (bp 118 °C), which has classically been used to extract polar molecules from aqueous phases.

$$I = \frac{1}{2} \sum_{i=1}^{n} c_i (z_i)^2 \tag{11.1}$$

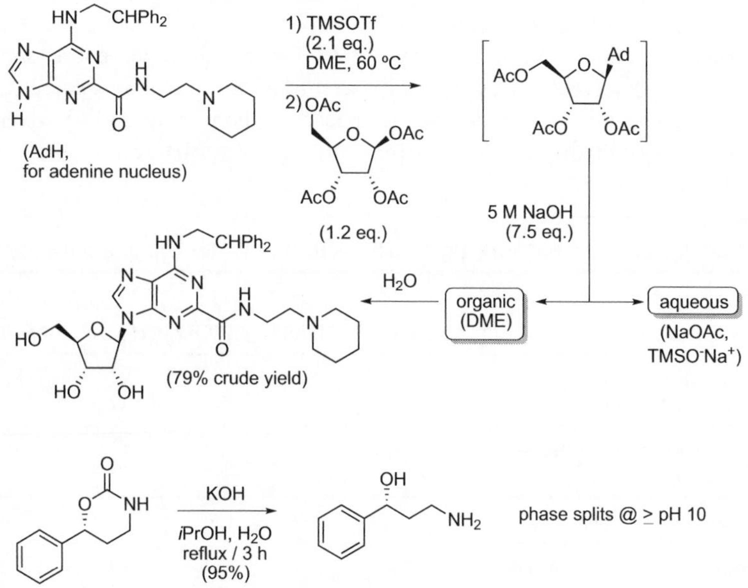

FIGURE 11.12 Separation of organic and aqueous phases promoted by high ionic strength.

where I = ionic strength, c = concentration of an ion (mol/L), and z = charge of that ion.

> Consider acidifying basic solutions with H_2SO_4 or H_3PO_4 instead of HCl. Because H_2SO_4 and H_3PO_4 are less corrosive than HCl they may be used in vessels that are not glass-lined, and may be favored by the owners of large equipment. Acidifying a 0.3 M solution of an ionized carboxylic acid ($pK_a \sim 4.75$) with 1.0 molar equivalent of either concentrated HCl or concentrated H_2SO_4 ($pK_a \sim -3$ and 1.92) produces a solution with identical ionic strength I of about 0.3 M; acidifying with 0.5 molar equivalents of concentrated H_2SO_4 produces a solution with I of about 0.375 M. The increased ionic strength in the latter case may promote extraction of the carboxylic acid into an organic phase. To reach the same pH endpoint more equivalents of H_3PO_4 (pK_a 2.12, 7.21, and 12.67) will be required, thus increasing both the ionic strength and the burden of waste disposal. Since H_3PO_4 is a relatively weak acid it may decrease decomposition of other reaction components under acidic conditions [55].

A typical approach in academia to work up a reaction conducted in a water-miscible solvent is to first remove the solvent by rotary evaporation and then partition the products between water and a water-immiscible solvent [56]; the examples above indicate that the extra step of removing a water-miscible solvent may not be necessary. Solvent removal also requires additional processing time and equipment. The key in simplifying such workups lies in understanding the distribution of product and impurities between the phases in order to optimize the extractions; for productive processing minimizing the charge of the water-miscible solvent may be considered. Possible approaches to streamlined extractions are summarized in Table 11.4.

When reactions are run in polar solvents and the products are extracted into another organic solvent, removing the polar reaction solvent can be tedious. Solvents that partition water-miscible solvents primarily into the aqueous phase are shown in Table 11.5 [57]. No solvent works especially well to partition etheric solvents such as THF, dioxane, and diglyme, into an aqueous phase; in these cases adding water as an antisolvent and crystallizing the product from the aqueous mixture may be preferred.

TABLE 11.4 Approaches to Work Up Reactions Run in Water-Soluble Solvents

Isolation Approach	MeOH	EtOH	*i*-PrOH	CH_3CN	THF or 2-Me-THF
Concentrate off solvent, add H_2O and immiscible solvent for phase split	X	X	X	X	X
Add H_2O + immiscible solvent for phase split	X	X	X	X	X
Add H_2O (or brine) for phase split			X	X	X
Direct isolation (e.g., cool, concentrate, or add antisolvent)	X	X	X	X	X

TABLE 11.5 Solvents that Poorly Extract Polar Aprotic Solvents From H_2O and May Be Useful for Extractive Workups (1)

	EtOAc	*i*-PrOAc	PhCH$_3$	Heptane
DMSO	X	X	X	X
DMF		X	X	X
NMP			X	X
Dimethyl acetamide (DMAc)		X	X	X
Tetramethylurea (TMU)				X
Dimethyl imidazolone (DMI)		X	X	X
THF				
Dioxane				
Diglyme				
Acetonitrile				X

(1) Adapted from Delhaye, L.; Ceccato, A.; Jacobs, P.; Köttgen, C.; Merschaert, A. Org. Process Res. Dev. **2007**, 11, 160.

The presence of solutes may modify how effectively the water-miscible solvents partition into the aqueous phase.

In the process of extracting a product into an organic phase often other components of the reaction mixture are partitioned into the organic phase, and these may affect operations in a positive manner. For instance, the carboxylic acid in Figure 11.13 was extracted into EtOAc, and subsequent concentration led to the crystallization of the product. The composition of the rich organic extract was examined to understand what prevented premature crystallization from the solvent used for extraction. ^{1}H NMR of the rich EtOAc extract showed H_2O and EtOH were present at 5 wt% and 2 wt%, respectively, and extended aging of this extract at 100 mg/mL gave no crystals of product: the solubility of the product was 10 times higher in the extract than in anhydrous EtOAc. The subsequent concentration at atmospheric pressure removed an azeotrope of EtOAc, H_2O and EtOH (bp 70 °C: 9% H_2O, 8% EtOH), providing conditions for the product to crystallize [58].

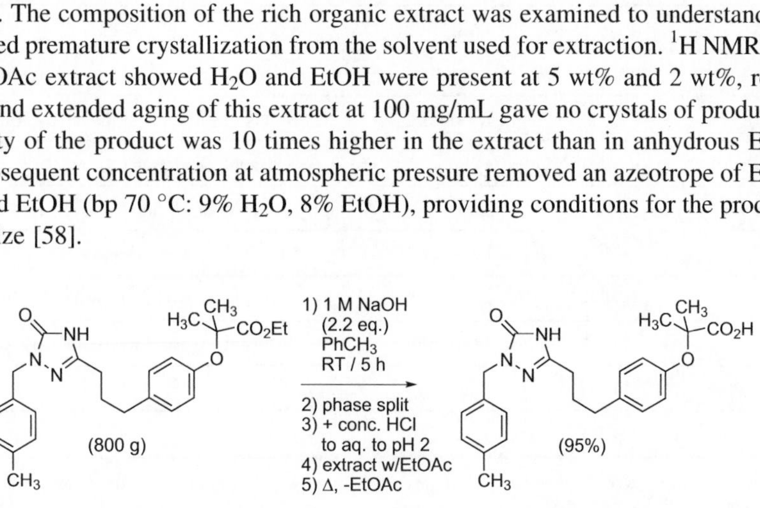

FIGURE 11.13   Extraction and crystallization of a product from EtOAc.

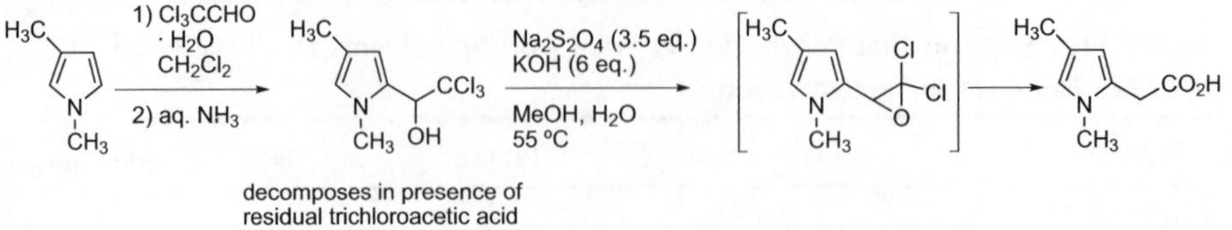

FIGURE 11.14 Extraction with dilute NH₃ to remove trichloroacetic acid.

Screening various bases for extractions may be necessary. Removing trichloroacetic acid (an impurity in the reagent chloral hydrate that was probably beneficial for this condensation) was necessary to prevent acid-catalyzed decomposition of the chloral adduct (Figure 11.14). Strongly basic conditions were avoided, as these conditions led to the premature formation of the epoxide, the intermediate which was reduced to the desired acetic acid by sodium dithionite [59]. Various bases were screened, such as NaHCO₃ and KHCO₃ as solids and aqueous solutions; dilute aqueous ammonia (~0.5%) was found to be optimal to remove residual trichloroacetic acid [46].

Often dilute aqueous solutions of reagents (3–10%) are employed for extractive workups. Some convenient aqueous solutions for extractions are shown in Table 11.6. These solutions are not saturated, for several reasons. First of all, high concentrations of reagents generally are not necessary, and pose additional burdens on purchasing, handling, and waste disposal. Secondly, a large excess of undissolved solids at the bottom of

TABLE 11.6 Convenient Aqueous Solutions for Extractions

	pH of 0.1 N Soln.	Relative Solubility in Organic Solvents	Comments
HCl	1.1	High	Corrosive, volatile
H₂SO₄	1.2	Low	
H₃PO₄	1.5	Low	Weak acid, hard to over-acidify
Citric acid	2.0	Moderate	Used for splitting DCHA salt (1)
AcOH	2.9	High	Weak acid, hard to over-acidify
NaH₂PO₄	~5	Low	Weak acid, hard to over-acidify
Na₂HPO₄	~8.5	Low	
NaHCO₃	8.4	Low	Weak base, hard to over-basify
NH₃	11.1	Moderate	Volatile, good for extracting RCO₂H
Na₂CO₃	11.6	Low	
Na₃PO₄	12.0	Low	

(1) Camp, D.; Matthews, C. F.; Neville, S. T.; Rouns, M.; Scott, R. W.; Truong, Y. *Org. Process Res. Dev.* **2006,** 10, 814.

a reactor may prevent an agitator from stirring. Finally, on a large scale preparing saturated solutions poses a logistics problem: solids at the bottom of the vessel would have to be removed by siphoning or filtration, steps which require additional operations and equipment.

On scale generally more than one extraction of an aqueous phase is employed, often to ensure complete recovery of the product. If the first extract removes 90% of the product from the aqueous phase, to decrease solvent usage the second extract may be only 25% of the volume of the first extract. Sometimes a third, small extract may be used as insurance on scale. A case study on an optimized extraction is described with Figure 8.12.

Extracting with brine may be unnecessary and even counter-productive. Check the moisture content of the organic extract before and after washing with brine to determine whether that extraction removed any water. Extracting with brine may partition some NaCl into the organic phase of a solvent with high water-solubility, such as 2-Me-THF; subsequent treatment with H_2SO_4 may lead to formation of both the desired sulfate salt and the hydrochloride salt.

Emulsions formed during extractions can delay processing and lead to product losses. To break emulsions convenient approaches are to raise the temperature slightly, to increase the ionic strength of the aqueous phase by adding brine (not solid NaCl), to adjust the pH slightly, or to filter off small particles such as those formed from polymerization of isobutylene generated from removal of a Boc group [60]. Increasing the charge of solvent present in the greatest proportion may help, or adding a small amount of a more polar solvent, e.g., 2−10 vol% of MeOH, i-PrOH, or EtOAc. To avoid forming an emulsion, gentler agitation may be required next time.

IV. TREATMENT WITH ACTIVATED CARBON AND OTHER ABSORBENTS

Trace amounts of impurities can impede crystallization, and absorbents are often used to remove these impurities and ease crystallization of the product. Impurities are removed from solutions of the product (absorbed) by the absorbent; these impurities are bound to (adsorbed to) the absorbent, which may also be termed an adsorbent. Absorbents include forms of activated carbon (commonly known as charcoal), other solids such as clays, solid chromatography phases such as SiO_2, Al_2O_3, Florisil®, diatomaceous earth, and ion exchange resins and other derivatized polymers [61].

Many types of activated carbon are available, in both powder and granular forms [62]. Activated carbon is often targeted toward industrial uses such as removing color from sucrose or removing mercury from the gas flues of coal-fired power plants [63]. Preparing charcoal is a "black" art, and procedures are kept proprietary. Centuries ago charcoal makers smothered fires of sticks and logs by piling dirt on top of them. Today carbon feedstocks are pyrolyzed at 400–600 °C essentially in the absence of oxygen, generating H_2O, CO, low-molecular weight alkenes (syn gas), and alkanes [64]. When these products exit the solids holes are formed, and the remaining solids become more brittle. Usually the solids are activated by steam, a byproduct of pyrolysis, but additives such as H_3PO_4 or $ZnCl_2$ may also be used [65]. The sites of manufacturing and sources of carbon (primarily wood, nut shells, or coal) will invariably change, possibly leading to variability in some products or inavailability of other products. (According to the consortium Sugar Knowledge International Limited, bone char was once used extensively to decolorize raw cane sugar, and has been essentially replaced with granular activated carbon [66].) Use-tests may be more important for activated carbon than for other components used by the pharmaceutical and fine chemicals industries.

Impurities are adsorbed to pores of activated carbon through van der Waals attractive forces, making pore size a key characteristic of activated carbon. Other important characteristics are surface area and the pH of an aqueous suspension [67]. Activated carbons are commonly classified as macroporous (1000–100,000 Å), mesoporous (100–1000 Å), and microporous (<100 Å) [68]. Because of the difficulty in cleaning stationary equipment coated with activated carbon, it is usually localized in filter cartridges, such as those manufactured by Cuno. Particles with small surface area may absorb more impurities, but also blind filter elements and produce high back pressure and low flow rates. The Ecosorb® line of products binds high-surface area carbons onto filaments, allowing high absorption of impurities with an enhanced flow rate [69]. The latter products have also been used to remove precious metals from APIs (Section VI).

Optimizing treatment with activated carbon. Impurity removal is often more effective using polar solvents [70], and solvents with and without H_2O should be examined [71]. Removal of Pd may be more effective under basic conditions, e.g., impurity removal may be more effective on a carboxylate than for the carboxylic acid [72]. Perhaps basic conditions lead to the formation of less-soluble metal oxides. Treating reaction streams with absorbents is more productive than treatments first requiring a solvent switch [71]. Adsorption of impurities to a solid phase is an equilibrium process, which may be improved by slower flow rates through a bed of activated carbon and increased contact time.

Merck scientists have examined absorbents as convenient and efficient treatments to remove impurities [73]. Absorbents may be added to a batch and removed by filtration after suitable contact with the solution, or a process stream may be passed through a filter containing the absorbent. Rapidly eluting a solution through a plug of absorbent may be called batch chromatography, plug chromatography, or plug-flow treatment. By agitating mini-centrifuge tubes containing solutions of analytes and absorbents and then centrifuging, Merck researchers readily prepared supernatant samples for HPLC analysis of the organic contents and inductively coupled plasma-mass spectroscopy (ICP-MS) analysis of the metals contents. In this paper they list 30 absorbents that they determined were widely useful and commercially available [71]. Colored impurities can be very elusive, with tiny amounts tinting batches of products. Welch and co-workers described high-throughput, microplate-based screening to identify adsorbents for removing colored impurities from API candidates, screening 31 commercially available absorbents [74]. The ideal combination of absorbent, solvent and perhaps pH would produce a mixture in which the supernatant contained none of the impurity and the absorbent contained none of the product. The selectivity of impurity removal can be calculated by the selectivity factor α (Equation 11.2), with higher α values indicating better selectivity [75]. Such absorbents can be useful to quickly remove impurities that are otherwise difficult to purge, thus allowing timely preparation of key supplies for early development decisions. Absorbents have also found use in manufacturing processes as reliable treatments to control the amounts of toxic materials at low levels, e.g., heavy metals limited to no more than 20 ppm. With increased volume of demand the price of such absorbents should decrease.

$$\alpha = \frac{(1 - X_{imp})/X_{imp}}{(1 - X_{prod})/X_{prod}} = \frac{1 - X_{imp}}{1 - X_{prod}} \times \frac{X_{prod}}{X_{imp}} \qquad (11.2)$$

where X_{imp} = fraction of impurity remaining in supernatant ($0 < X_{imp} \leq 1.0$) and X_{prod} = fraction of product remaining in supernatant ($0 \leq X_{prod} \leq 1.0$).

V. FILTRATION TO REMOVE IMPURITIES

Fine particles can plug filtration media, slowing or stopping filtration and producing a high back pressure that can cause leaks. Sources of fine particles can include batches of activated carbon, solids formed from rapid precipitation or crystallization, dust, dirt, and other impurities. To enhance filtration filter aids (often diatomaceous earth such as Celite®, Hyflo®, Kieselguhr®, or Sulkafloc®) may be charged; these filter aids effectively increase the surface area of the filtration unit. Often filter aids are slurried and added to a filter before a suspension is transferred into the filter, and sometimes filter aids are added directly to a vessel containing a slurry.

"Polish filtration" is an operation to remove trace amounts of insoluble impurities before other operations, such as product crystallization, occur. Such extraneous impurities can cause a cGMP batch to fail. On scale polish filtration is often carried out by passing a process stream through in-line filters, perhaps a 10 μm filter followed by a 1 μm filter to aid filtration. Additional benefits of removing trace amounts of such impurities include avoiding emulsions, controlling the crystallization of a desired polymorph through seeding, and improving the quality of the crystallized product. Polish filtration is simply "good housekeeping."

Filtrations often cool process streams, plugging transfer lines and causing premature crystallization in filtrate receivers. Hot (warm) polish filtrations are often employed, and different sites may set different upper temperature limits for these operations. Premature crystallization in transfer lines before and after the filter is always a concern; on scale these lines may be warmed to suitable temperatures by rapidly transferring the process streams at the beginning of the filtration. Other alternatives to warm the transfer lines are to circulate heat transfer fluid around the lines with or to wrap the lines with electrical heat tape. Hot polish filtrations in the kilo lab are often more difficult than in the pilot plant or manufacturing facility. To avoid such premature crystallization conduct a polish filtration at least 10 °C above the temperature at which crystallization will occur.

Not all filter media and gaskets may be compatible with a reaction stream; for instance, streams containing HNO_3 may dissolve a filter paper, and THF may make gaskets brittle. If in doubt about how a filtration will handle on scale, obtain a piece of the filter medium to be used and examine the filtration in the laboratory.

Other filtrations may be employed, chiefly on manufacturing scale. Clarifying beer involves many filtrations [76]. Ultrafiltration through membranes absorbs proteins and other large molecules. Microfiltration through ceramic or polymeric filters removes soluble proteins and bacteria, sterilizing solutions without heat; such treatment is critical for the formulation of drugs to be delivered by injection. The use of membranes for processing APIs will increase as "green" techniques to save energy and processing time [77].

> A polish filtration should be required prior to isolating an API. The potential benefits outweigh the additional work in setting up a filter, filtrate receiver, and transfer lines.

VI. REMOVING METAL SALTS AND METALS

Today's specifications for APIs are moving away from the classic "heavy metal" test, which involved measuring the amount of precipitate formed after mixing Na_2S with a solution of the API. The latter assay was imprecise and did not distinguish between the elements, and has largely been replaced by ICP-MS or atomic absorption spectroscopy [78]. Limits for "heavy metals" in APIs may be as low as 3 ppm, depending on the metal and the dosage, as discussed with Table 13.6. Biocatalysis using low loadings of enzymes, which may themselves contain trace amounts of first-row transition metals, may replace routes using transition metal catalysts and eliminate the burden of controlling trace amounts of transition metals. The biocatalytic route to sitagliptin developed by Codexis and Merck, for which they won a 2010 Presidential Green Chemistry Award [79], does not have the burden of controlling levels of Rh and Fe that made up the catalyst for chiral hydrogenation [80].

Transition metals are largely the metals of concern today, along with residual alkali metal salts such as NaCl that might contaminate APIs. Transition metals that may be encountered in organic synthesis include titanium, iron, cobalt, nickel, copper, zinc, zirconium, ruthenium, rhodium, palladium, tungsten, osmium, iridium, platinum, and gold; generally compounds of these metals are used in catalytic amounts. The prices of these compounds will drive research for alternative reagents, including metals and ligands [81]. Chromium and manganese oxides, which decades ago were used in excess as oxidants, have been largely replaced by catalytic, greener oxidants. Compounds of the rare earth elements, some of which are not so rare [82], have been largely used in material sciences applications, and may become strategic issues [83].

> The burden of removing a heavy metal from an API is reduced by positioning the metal-catalyzed step earlier in the route, and by charging less of the metal. The feasibility of the latter was demonstrated in a Heck reaction. The authors reasoned that ligands may be charged to increase the solubility of a metal such as Pd and reduce the tendency to precipitate the Pd as a "mirror"; by charging 0.02 mol% of Pd the reaction was successful [87]. A ligand-free cyanation with low loading of Pd has also been described [88].

A general table for removing metal salts is shown in Table 11.7. Many of these treatments seem like "common sense," and can be adapted for other applications. For instance, basic conditions promote the formation of insoluble metal oxides, such as iron oxides (rust, the primary constituent of the "red mud," pH 13 waste from alumina processing that escaped containment reservoirs in Hungary [84]). Many carbonates are poorly soluble in water. These treatments are guidelines, and more than one operation or treatment followed by crystallization of the product may be needed to purge metal contaminants to acceptable levels. A number of papers have reviewed the subject of removing metals [85,86].

> Mixtures of heptanes–acetonitrile or heptanes–anhydrous MeOH are immiscible. If the product readily dissolves in the more polar solvent, extracting with heptanes may remove non-polar impurities such as $(n\text{-}Bu)_3SnCl$ or mineral oil.

TABLE 11.7 Treatments for Removing Metals from Reaction Streams

Metal Ion	Treatment	Means of Removal	Comments & References
M(II), M(III), M(IV), …	Basify	Precipitation	
	Acidify	Extraction into aq. acid	
	Activated carbon	Absorption	
	Ion-exchange chromatography (Including chelating resins)	Absorption Absorption	(1)
	Polystyrene-supported ligands		Table 11.9
Cu(I) and Cu(II)	Oxalic acid, TMT, or KH_2PO_4	Precipitation	(2–4)
	Dilute aq. pyridine, NH_3 or $(NH_4)_2SO_4$	Extract into aq. phase	Wash until blue color disappears from aq. (5)
Al(III)	Rochelle salt, malic acid	Extraction into aqueous phase	(6,7)
Cu(II), Pd(II), Mn(II), Zn(II), …	Chelation, e.g., EDTA	Extraction	(8–12); May be pH-sensitive
	2,4,6-s-trimercaptotriazine (TMT)	Precipitation	
Fe	Ecosorb® C-941	Absorption	(13)
	Lactic acid	Extraction	(14)
Mg(II)	$(NH_4)_2SO_4$	Extraction	(15)
	Citric acid	Extraction	(16)
Ti(IV)	Citric acid	Precipitation	(17)
$(n\text{-}Bu)_3SnBr$, $(n\text{-}Bu)_3SnCl$	Heptane	Extraction vs. CH_3CN	(18)
	Aqueous KF	Precipitation	(19)
Zn(II)	Ethylenediamine	Extraction	(20,21)
	EDTA	Extraction	(22)
	H_2SO_4; aqueous NH_3	Precipitation; solubilization	(23)
$B(OH)_3$	MeOH, acid	Azeotrope $B(OMe)_3$	(24)
	IRA743 (methyl glucamine resin)	Absorption	(25)

(Continued)

TABLE 11.7 Treatments for Removing Metals from Reaction Streams—cont'd

Metal Ion	Treatment	Means of Removal	Comments & References
Ni	Versene™ ligands	Extraction	(26)
	Chelex™, Lewatit™ TP 207 resin	Absorption	(27,28)
Rh	Ecosorb® C-941	Absorption	(12: may be better under H_2 atmosphere)
Ru	Imidazole	Complexation; extraction	(29)
	Citric acid	Extraction	(30)
	Nuchar RGC	Absorption	(31) MeOH better than 1:1 MeOH:H_2O
	Activated carbon	Absorption	(32)
	Ecosorb® C-908	Absorption	(33)
	Amine-functionalized silica	Absorption	(34)
	Super-critical CO_2	Absorption	(35)
	Diethylene glycol monovinyl ether + chromatography	Fischer carbene Formation + absorption	(36)
	2-Mercaptonicotinic acid	Complexation; extraction	(37)
	Cysteine	Complexation; extraction	(38)
	H_2O_2	Oxidation to RuO_2, filtration	(39)
Pd	Various		See Table 11.8

References

(1) Branchaud, B. P.; Meier, M. S. J. Org. Chem. **1989,** 54, 1320.

(2) Sedergran, T. C.; Anderson, C. F. U.S. 4,675,398, 1987 (to E. R. Squibb & Sons).

(3) Malmgren, H.; Bäckström, B.; Sølver, E.; Wennerberg, J. Org. Process Res. Dev. **2008,** 12, 1195.

(4) Ashwood, M. S.; Alabaster, R. J.; Cottrell, I. F.; Cowden, C. J.; Davies, A. J.; Dolling, U. H.; Emerson, K. M.; Gibb, A. D.; Hands, D.; Wallace, D. J.; Wilson, R. D. Org. Process Res. Dev. **2004,** 8, 192.

(5) Schwartz, A.; Madan, P.; Whitesell, J. K.; Lawrence, R. M. Org. Synth. **1990,** 69, 1.

(6) Fieser, L. F.; Fieser, M. Reagents for Organic Synthesis; Vol. 1; Wiley: New York, 1967; p 963.

(7) Wu, G.; Wong, Y.; Steinman, M.; Tormos, W.; Schumacher, D. P.; Love, G. M.; Shutts, B. Org. Process Res. Dev. **1997,** 1, 359.

(8) Butters, M.; Ebbs, J.; Green, S. P.; MacRae, J.; Morland, M. C.; Murtiashaw, C. W.; Pettman, A. J. Org. Process Res. Dev. **2001,** 5, 28.

(9) Conway, M. Chem. Eng. **1999,** 106(3), 86.

(10) Rosso, V. W.; Lust, D. A.; Bernot, P. J.; Grosso, J. A.; Modi, S. P.; Rusowicz, A.; Sedergran, T. C.; Simpson, J. H.; Srivastava, S. K.; Humora, M. J.; Anderson, N. G. Org. Process Res. Dev. **1997,** 1, 311.

(11) To remove Fe(II) salts extraction with EDTA at pH 10 was preferred: McCamley, K.; Ripper, J. A.; Singer, R. D.; Scammells, P. J. J. Org. Chem. **2003,** 68, 9847.

(12) TMT was found to be more effective for removing Cu(II) than Cu(I), so a solution was sparged with air to remove copper species during TMT treatment: Malmgren, H.; Bäckström, B.; Sølver, E.; Wennerberg, J. Org. Process Res. Dev. **2008,** 12, 1195.

(13) Hansen, K. B.; Hsiao, Y.; Xu, F.; Rivera, N.; Clausen, A.; Kubryk, M.; Krska, S.; Rosner, T.; Simmons, B.; Balsells, J.; Ikemoto, N.; Sun, Y.; Spindler, F.; Malan, C.; Grabowski, E. J. J.; Armstrong, J. D., III J. Am. Chem. Soc. **2009,** 131, 8798.

(14) Chen, C.; Dagneau, P.; Grabowski, E. J. J.; Oballa, R.; O'Shea, P.; Prasit, P.; Robichaud, J.; Tillyer, R.; Wang, X. J. Org. Chem. **2003,** 68, 2633.

(15) Karpf, M.; Trussardi, R. J. Org. Chem. **2001,** 66, 2044.

(16) Harrington, P. J.; Brown, J. D.; Foderaro, T.; Hughes, R. C. Org. Process Res. Dev. **2004,** 8, 86.

(17) Gao, Y.; Hanson, R. M.; Klunder, J.M.; Ko, S. Y.; Masamune, H.; Sharpless, K. B. J. Am. Chem. Soc. **1987,** 109, 5765.

TABLE 11.7 Treatments for Removing Metals from Reaction Streams—cont'd

(18) Ragan, J. A. in Fundamentals of Early Clinical Development: From Synthesis Design to Formulation; Abdel-Magid, A.; Caron, S., Eds.; Wiley, 2006; pp 33–34.

(19) For recovery of (n-Bu)$_3$SnCl, see Wang, G.; Sun, B.; Peng, C. Org. Process Res. Dev. **2011**, 15, 986.

(20) Denni-Dischert, D.; Marterer, W.; Bänziger, M.; Yusuff, N.; Batt, D.; Ramsey, T.; Geng, P.; Michael, W.; Wang, R.-M. B.; Taplin, F., Jr.; Versace, R.; Cesarz, D.; Perez, L. B. Org. Process Res. Dev. **2006**, 10, 70.

(21) Aqueous ZnCl$_2$ was also used to precipitate an impurity related to ethylenediamine: Ace, K. W.; Armitage, M. A.; Bellingham, R. K.; Blackler, P. D.; Ennis, D. S.; Hussain, N.; Lathbury, D. C.; Morgan, D. O.; O'Connor, N.; Oakes, G. H.; Passey, S. C.; Powling, L. C. Org. Process Res. Dev. **2001**, 5, 479.

(22) The trisodium salt was found to be optimal for removing Zn: Pérez-Balado, C.; Willemsens, A.; Ormerod, D.; Aelterman, W.; Mertens, N. Org. Process Res. Dev. **2007**, 11, 237.

(23) Scherkenbeck, T.; Siegel, K. Org. Process Res. Dev. **2005**, 9, 216.

(24) Repic, O. Principles of Process Research and Chemical Development in the Pharmaceutical Industry; Wiley: New York, 1998; pp 90–92.

(25) Chem. Eng. **2006**, 113 (3), 13.

(26) Dow's Versene™ ligands are salts of EDTA, in liquid (tetrasodium salt) and dry forms (disodium salt and calcium disodium salt): http://www.dow.com/versene

(27) http://www.bio-rad.com/webmaster/pdfs/9184_Chelex.PDF

(28) Lewatit™ TP 207 is a weekly acidic, macroporous cation exchange resin with imino diacetate groups. It is used to remove Cu(II) and Ni(II) from municipal water: http://www.bakercorp.com/pdfs/TP_207.pdf; http://lanxess.com/en/industries-products-india/industries-india/water-treatment-india/lewatitR-tp-207-india/

(29) Brenner, M.; Meineck, S.; Wirth, T. U.S. Patent 7,183,374 B2, 2007 (to Boehringer Ingelheim).

(30) Citrate washes have been used to remove Ru from reactors: Couturier, M.; Andresen, B. M.; Jorgensen, J. B.; Tucker, J. L.; Busch, F. R.; Brenek, S. J.; Dubé, P.; am Ende, D. J.; Negri, J. T. Org. Process Res. Dev. **2002**, 6, 42, Ref. 7.

(31) Welch, C. J.; Albanese-Walker, J.; Leonard, W. R.; Biba, M.; DaSilva, J.; Henderson, D.; Laing, B.; Mathre, D. J.; Spencer, S.; Bu, X.; Wang, T. Org. Process Res. Dev. **2005**, 9, 198.

(32) Fleitz, F. J.; Lyle, T; J.; Zheng, N.; Armstrong, J. D., III; Volante, R. P. Synth. Commun. **2000**, 30, 3171.

(33) Welch, C. J.; Leonard, W. R.; Henderson, D. W.; Dorner, B.; Childers, K. G.; Chung, J. Y. L.; Hartner, F. W.; Albanese-Walker, J.; Sajonz, P. Org. Process Res. Dev. **2008**, 12, 81.

(34) McEleney, K.; Allen, D. A.; Alison E. Holliday, A. E.; Cathleen M. Crudden, C. M. Org. Lett. **2006**, 8, 2663.

(35) Gallou, F.; Saim, S.; Koenig, K. J.; Bochniak, D.; Horhota, S. T.; Yee, N. K.; Senanayake, C. H. Org. Process Res. Dev. **2006**, 10, 937. The Ru residue was adsorbed to the walls of the pressure reactor.

(36) Liu, W.; Nichols, P. J.; Smith, N. Tetrahedron Lett. **2009**, 50, 6103.

(37) Yee, N. K.; Farina, V.; Houpis, I. N.; Haddad, N.; Frutos, R. P.; Gallou, F.; Wang, X.; Wei, X.; Simpson, R. D.; Feng, X.; Fuchs, V.; Xu, Y.; Tan, J.; Zhang, L.; Xu, J.; Smith-Keenan, L. L.; Vitous, J.; Ridges, M. D.; Spinelli, E. M.; Johnson, M. J. Org. Chem. **2006**, 71, 7133.

(38) Wang, H.; Matsuhashi, H.; Doan, B. D.; Goodman, S. N.; Ouyang, X.; Clark, W. M., Jr. Tetrahedron **2009**, 65, 6291.

(39) Knight, D. W.; Morgan, I. R.; Proctor, A. J. Tetrahedron **2010**, 51, 638.

Many approaches have been developed to remove palladium from reaction streams, chiefly because no one approach works for every application [85,89,90]. Colloidal Pd may be removed by centrifugation, especially on a small scale, or by filtration through diatomaceous earth [91]. Unfortunately removing colloidal Pd by filtration has not always been effective, probably since Pd(0) is readily oxidized to Pd(II). Merck researchers found that the product from a Pd/C-mediated hydrogenolysis assisted in extracting Pd(II) from the carbon support, and by pre-reducing the Pd/C catalyst before adding the substrate the amount of Pd in the product was greatly reduced [92]. Heteroatoms in APIs may coordinate to Pd species and increase the level of palladium in a product; acidifying a process stream may protonate an amine serving as a ligand for Pd, hence making removal of Pd from the process stream easier [92]. Removing Pd from basic streams has been easier in other cases, e.g., treating a stream of a carboxylate salt with charcoal was more effective than treating a solution of the carboxylic acid [93]. The solubility of Pd in a Pd/C-mediated Suzuki–Miyaura cross-coupling was found to maximize at about 90% of the product formation, then fall [94]; by extending a hydrogenolysis for 30–60 min after complete reaction and then beginning workup the burden of removing Pd from the process streams was reduced [95]. (Similarly, adsorption of Rh to Ecosorb® C-941 activated carbon was facilitated by the presence of H_2 [25].) Sometimes a product crystallizes without the need for treatment to remove the Pd [96]. General approaches to separate the API from the metal impurities include precipitation, extracting the metal(s) into one phase when the API is dissolved in another phase, reslurrying the solid API to extract the metal into the supernatant, and recrystallizing the API, perhaps as a salt. The latter approach may be the most practical, as it does not require the purchase, qualification, handling, or analysis in the API for yet another substance. The options available are dependent on the solubilities of the API and salts; reslurrying the API to remove metals may prompt formation of another polymorph [97]. The reagents in Table 11.8 have been used to remove Pd from APIs. Merck researchers have shown that a combination of approaches may be more effective than one approach [98].

TABLE 11.8 Treatments to Remove Pd From Process Streams

Treatment	Means of Removing Pd	Comments & References
Charge less Pd		(1)
Crystallization of product	Solubilize in mother liquor	Acetonitrile may be especially effective
Activated charcoal Darco KBB, Darco G-60 Ecosorb® C-941	Absorption	(2–5); pH control may increase efficiency of removal (3,6–8)
$NaHSO_3$ + celite	Produces Pd(0) which gets absorbed	(9)
$(nBu)_3P$	Complexation and dissolution in mother liquor	(10,11); has been used in manufacturing
Et_3N	Complexation and dissolution in mother liquor	(12)
Lactic acid	Extraction	(11)
2,4,6-trimercaptotriazine (TMT)	Precipitation	(13)
N-Ac-cysteine	Solubilize in aq. extracts or as additive to crystallize product	(14)
1,2-diaminopropane + DPPE	Solubilize in mother liquor	(15)
$BH_3 \cdot Me_3N$	Precipitation as Pd(0), filtration	(16)

TABLE 11.8 Treatments to Remove Pd From Process Streams—cont'd

Treatment	Means of Removing Pd	Comments & References
Solid-phase reagents	Absorption	(17)
Harbolite clay (filtration aid)		(18)
mercaptopropyl silica		(19–21)
triamine bound to silica		(9); www.silicycle.com
diethanolaminomethyl polystyrene (PS-DEAM)		(22)
macroporous TMT (MP-TMT)		(23)
polymer-bound ethylene-diamine		(24)
Smopex® 110 fibers		(25)
QuadraPure™ resins		(26); Table 11.9
S(II)-derivatized SiO$_2$		(27)
Deloxan® THP (thiourea)		(28)
Membranes	Nanofiltration	(29)

References

(1) deVries, A. H. M.; Mulders, J. M. C. A.; Mommers, J. H. M.; Henderickx, H. J. W.; deVries, J. G. Org. Lett. **2003**, 5, 3285.

(2) Ripin, D. H. B.; Bourassa, D. E.; Brandt, T.; Castaldi, M. J.; Frost, H. N.; Hawkins, J.; Johnson, P. J.; Massett, S. S.; Neumann, K.; Phillips, J.; Raggon, J. W.; Rose, P. R.; Rutherford, J. L.; Sitter, B.; Stewart, A. M., III; Vetelino, M. G.; Wei, L. Org. Process Res. Dev. **2005**, 9, 440.

(3) Stewart, G. W.; Brands, K. J. M.; Brewer, S. E.; Cowden, C. J.; Davies, A. J.; Edwards, J. S.; Gibson, A. W.; Hamilton, S. E.; Katz, J. D.; Keen, S. P.; Mullens, P. R.; Scott, J. P.; Wallace, D. J.; Wise, C. S. Org. Process Res. Dev. **2010**, 14, 849. Basic conditions were found to be more effective to remove Pd from a sulfamide.

(4) Federsel, H.-J. Org. Process Res. Dev. **2000**, 4, 352. Weakly acidic aqueous conditions were found to be more effective to remove Pd from an amine.

(5) Hobson, L. A.; Nugent, W. A.; Anderson, S. R.; Deshmukh, S. S.; Haley, J. J., III; Liu, P.; Magnus, N. A.; Sheeran, P.; Sherbine, J. P.; Stone, B. R. P.; Zhu, J. Org. Process Res. Dev. **2007**, 11, 985.

(6) Williams, J. M.; Brands, K. M. J.; Skerlj, R. T.; Jobson, R. B.; Marchesini, G.; Conrad, K. M.; Pipik, B.; Savary, K. A.; Tsay, F.-R.; Houghton, P. G.; Sidler, D. R.; Dolling, U.-H.; DiMichele, L. M.; Novak, T. J. J. Org. Chem. **2005**, 70, 7479.

(7) Graver Technologies, http://www.gravertech.com

(8) Welch, C. J.; Albaneze-Walker, J.; Leonard, W. R.; Biba, M.; DaSilva, J.; Henderson, D.; Laing, B.; Mathre, D. J.; Spencer, S.; Bu, X.; Wang, T. Org. Process Res. Dev. **2005**, 9, 198.

(9) Bullock, K. M.; Mitchell, M. B.; Toczko, J. F. Org. Process Res. Dev. **2008**, 12, 896.

(10) Smith, G. B.; Dezeny, G. C.; Hughes, D. L.; King, A. O.; Verhoeven, T. R. J. Org. Chem. **1994**, 59, 8151.

(11) Chen, C.; Dagneau, P.; Grabowski, E. J. J.; Oballa, R.; O'Shea, P.; Prasit, P.; Robichaud, J.; Tillyer, R.; Wang, X. J. Org. Chem. **2003**, 68, 2633.

(12) Li, B.; Buzon, R. A.; Zhang, Z. Org. Process Res. Dev. **2007**, 11, 951.

(13) Rosso, V. W.; Lust, D. A.; Bernot, P. J.; Grosso, J. A.; Modi, S. P.; Rusowicz, A.; Sedergran, T. C.; Simpson, J. H.; Srivastava, S. K.; Humora, M. J.; Anderson, N. G. Org. Process Res. Dev. **1997**, 1, 311.

(14) Königsberger, K.; Chen, G.-P.; Wu, R. R.; Girgis, M. J.; Prasad, K.; Repic, O.; Blacklock, T. J. Org. Process Res. Dev. **2003**, 7, 733.

(15) Flahive, E. J.; Ewanicki, B. L.; Sach, N. W.; O'Neill-Slawecki, S. A.; Stankovic, N. S.; Yu, S.; Guinness, S. S.; Dunn, J. Org. Process Res. Dev. **2008**, 12, 637.

(16) Jensen, M. R.; Hoerrner, R. S.; Li, W.; Nelson, D. P.; Javadi, G. J.; Dormer, P. G.; Cai, D.; Larsen, R. D. J. Org. Chem. **2005**, 70, 6034.

(17) Barbaras, D.; Brozio, J.; Johannsen, I.; Allmendinger, T. Org. Process Res. Dev. **2009**, 13, 1068.

(18) Rassias, G.; Hermitage, S. A.; Sanganee, M. J.; Kincey, P. M.; Smith, N. M.; Andrews, I. P.; Borrett, G. T.; Slater, G. R. Org. Process Res. Dev. **2009**, 13, 774.

(Continued)

TABLE 11.8 Treatments to Remove Pd From Process Streams—cont'd

Treatment	Means of Removing Pd	Comments & References

(19) Crudden, C. M.; Sateesh, M.; Lewis, R. J. Am. Chem. Soc. **2005,** 127, 10045. This reagent has catalyzed a Mizoroki-Heck reaction.

(20) Ryberg, P. Org. Process Res. Dev. **2008,** 12, 540. http://www.silicycle.com

(21) Houpis, I. N.; Shilds, D.; Nettekoven, U.; Schnyder, A.; Bappert, E.; Weerts, K.; Canters, M.; Vermuelen, W. Org. Process Res. Dev. **2009,** 13, 598.

(22) Tsukamoto, H.; Suzuki, T.; Sato, M.; Kondo, Y. Tetrahedron Lett. **2007,** 48, 8438; http://www.biotage.com

(23) Huang, J.-P.; Chen, X.-X.; Gu, S.-X.; Zhao, L.; Chen, W.-X.; Chen, F.-E. Org. Process Res. Dev. **2010,** 14, 939, http://www.biotage.com

(24) Urawa, Y.; Miyazawa, M.; Ozeki, N.; Ogura, K. Org. Process Res. Dev. **2003,** 7, 191.

(25) Jiang, X.; Lee, G. T.; Villhauer, E. B.; Prasad, K.; Prashad, M. Org. Process Res. Dev. **2010,** 14, 883, http://www.chemicals.matthey.com/products

(26) Girgis, M. J.; Kuczynski, L. E.; Berberena, S. M.; Boyd, C. A.; Kubinski, P. L.; Scherholz, M. L.; Drinkwater, D. E.; Shen, X.; Babiak, S.; Lefebvre, B. G. Org. Process Res. Dev. **2008,** 12, 1209.

(27) Galaffu, N.; Man, S. P.; Wilkes, R. D.; Wilson, J. R. H. Org. Process Res. Dev. **2007,** 11, 406.

(28) Dorow, R. L.; Herrinton, P. M.; Hohler, R. A.; Maloney, M. T.; Mauragis, M. A.; McGhee, W. E.; Moeslein, J. A.; Strohbach, J. W.; Veley, M. F. Org. Process Res. Dev. **2006,** 10, 493.

(29) Pink, C. J.; Wong, H.; Ferreira, F. C.; Livingston, A. G. Org. Process Res. Dev. **2008,** 12, 589.

TABLE 11.9 Treatments With QuadraPure® Resins (Functionalized Macroporous Polystyrene) to Remove Metal Salts (1)

Reaction	Metal Reagent(s)	Solvent for Flow Treatment	[Metal] Initially	[Metal] Final	Scavenger Resin
Suzuki	$Pd(OAc)_2$	CH_2Cl_2	60 ppm	<1 ppm	A
Sonogashira	$Pd(Ph_3P)_2Cl_2$, CuI	THF	NA	<1 ppm Pd	B
Rosenmund—von Braun	CuCN	Pyridine—CH_2Cl_2	345 ppm	<1 ppm	C
Michael	$FeCl_3$	THF	61 ppm	<0.2 ppm	D
Hydrogenation w/ Wilkinson's catalyst	$RhCl(Ph_3P)_3$	$PhCH_3$	309 ppm	4 ppm	E

(1) Hinchcliffe, A.; Hughes, C.; Pears, D. A.; Pitts, M. R. Org. Process Res. Dev. **2007,** 11, 477.

VII. QUENCHING AND WORKING UP BIOCATALYTIC REACTIONS

Because many biocatalytic reactions are equilibrium processes, quenching may be crucial to ensure that the product can be isolated in the expected yield. Quenching operations should be designed based on the characteristics of the enzyme(s) used and the nature of the products. In some cases simply extracting the product into an organic solvent suffices. Often proteins are denatured by adjusting the pH, or adding salt, or organic solvents, and the proteins are removed by filtration. Heating a suspension of proteins with agitation can cause the proteins to denature and flocculate (the "hot break" in brewing beer, although if the mixture is too acidic such flocculation may not occur [98]). In the case of the KRED-catalyzed reduction of the poorly soluble ketone in Figure 11.15 the reaction was driven by crystallization of the enantiopure monohydrate product in a slurry-to-slurry conversion; EtOAc and brine were added, and the enzyme was removed by pressure filtration at 60–70 °C. This relatively high temperature was probably required to dissolve the product in the solvent mixture; workup continued in a straightforward fashion, and the alcohol, a precursor to montelukast, was isolated in excellent yield [99]. In another KRED process reduction of the racemic aldehyde was taken to 45% completion, and EDTA was charged to quench the reaction by sequestering the magnesium ions required by the enzyme. The boiling points of the unreacted aldehyde and the product are similar, so chemical conversion of the undesired aldehyde was necessary. Treatment of the mixture with aqueous NaOH converted the aldehyde to the dimer-like Knoevenagel side product for removal; unfortunately the product alcohol was isolated with lower chiral purity, and this treatment precluded recovery and recycle

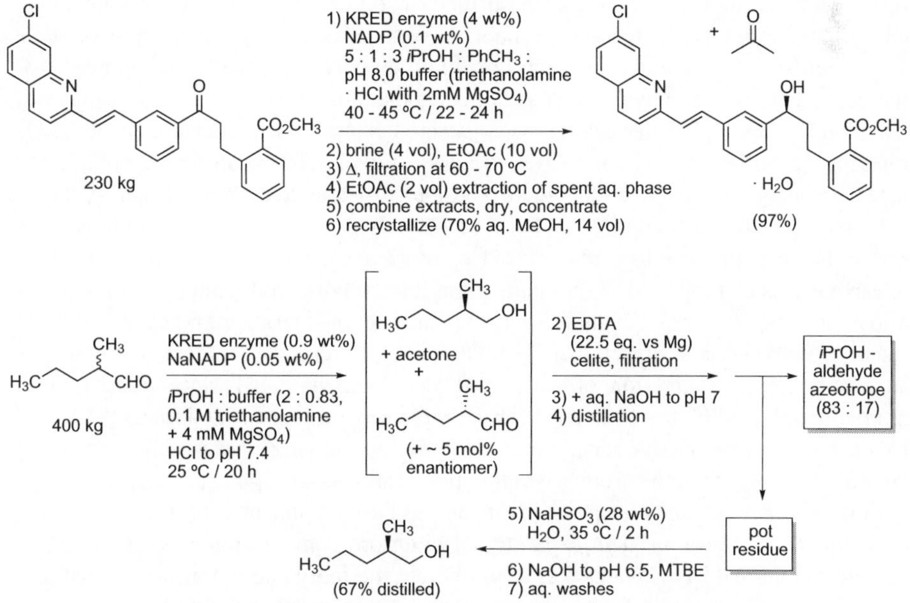

FIGURE 11.15 Quenching and working up biocatalytic reactions.

of the aldehyde. Laboratory investigations found that the aldehyde could be removed by atmospheric distillation as an azeotrope with i-PrOH. On scale most of the undesired aldehyde was removed as the azeotrope, and sodium bisulfite was added to convert the remaining unreacted aldehyde to a H_2O-soluble species. The product was extracted into MTBE and worked up in a straightforward fashion (Figure 11.15) [100]. The ketone in another KRED-mediated reduction was also quenched with sodium metabisulfite to facilitate workup [101].

VIII. CHROMATOGRAPHY

In some industries chromatographic purifications have been employed for decades, even for low value-added products. Epigallocatechin gallate (EGCG), a nutraceutical, is isolated by chromatographic purification [102]. Paclitaxel (Taxol®) is extracted from plant tissue culture, purified by chromatography, and crystallized; this approache manufactures paclitaxel with a CoG 80% of the CoG of semisynthetic paclitaxel [103]. Ion exchange resins and aqueous elutions have been used to purify annually over 100 MT of cephalosporins [104]. Simulated moving bed chromatography (SMB) has been used to manufacture 500,000 tons per year of p-xylene [105], and fructose and sucrose have been separated by SMB since the 1960s [106]. In a joint venture NovaSep and Rohm and Haas have developed an SMB process to purify glycerol generated from biodiesel manufacturing [107]. SMB has also been shown to be practical for resolutions of high value-added compounds [108], such as Lundbeck's escitalopram [105], Cephalon's R-modafinil [109], and Pfizer's sertraline [110]. SMB requires a significant capital investment, and routine processing would become problematic if impurity profiles change. Nucleotides have been purified by ion-exchange chromatography [111] and by membranes [112]. Natural products have been isolated on a preparative scale using high-speed countercurrent chromatography [113]. Usually equipment for preparative chromatography is found at dedicated manufacturing facilities, at companies that manufacture high value-added APIs isolated from natural sources, or at CROs preparing quantities of drug candidates for early studies. These chromatographic techniques are being refined by vendors to afford more cost-effective approaches to APIs.

Preparative chromatography has been used for high value-added compounds. This may be the best approach to manufacture tens of kilograms of an orphan drug [114]. In the early stages of process development, when relatively few kilograms are needed for various studies, the drug candidate may be purified by chromatography at multiple stages [115]. Welch has written that chromatography to remove troublesome impurities can enable rapid progression of early drug development [116]. Super-critical fluid chromatography (SFC) has been used for the rapid separation of enantiomers [117] and provided a highly productive separation for a hydroxy acid [118]. Many vendors provide equipment for preparative chromatographic purifications.

Chromatography can offer practical means to purify polar materials. For instance, Amberlite XAD-4 was used to separate a zwitterion from the more polar inorganic impurities (Figure 11.16). Water was used to elute the inorganic impurities, essentially de-salting the product, and EtOH was used to elute the product zwitterion. Amberlite XAD-4, stable at pH 1–14, was chosen in part because it was less expensive than reverse-phase C-8 silica gel [119].

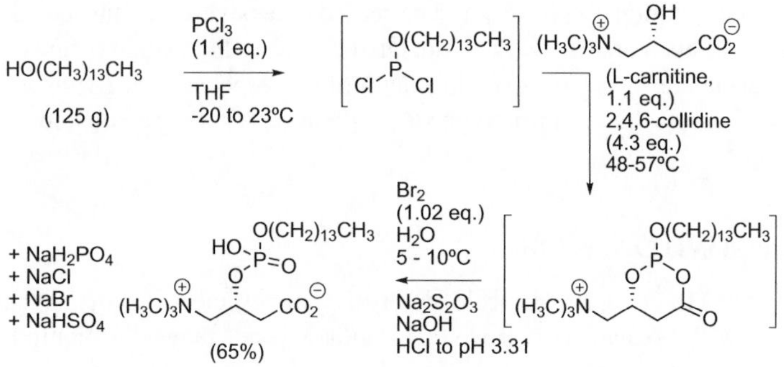

FIGURE 11.16   De-salting using Amberlite XAD-4.

Impurities can leach from polymers used for purification and then contaminate products. Dowex 1 × 2 resin, a quaternary ammonium resin, is prepared by copoly-merization of styrene with p-divinylbenzene, followed by treatments with chloromethyl methyl ether and trimethylamine, and then washing with water. Monsanto researchers extracted Dowex 1 × 2 resin with dichloromethane, and identified the major impurity as the benzylic alcohol in Figure 11.17. This impurity was probably derived from mono-meric styrene that was not removed from the polymer and hence was subjected to subsequent treatments. Such impurities, if present in polymeric resins, may be removed by washing the resins before applying material to be purified, or by extracting the eluted products with an organic solvent [120].

The cost of preparative chromatography for routine manufacturing was estimated at $1200–$5000/kg (one step, 50 kg/batch, 20 batches/year) [121]. This estimate consid-ered how much crude material could be loaded onto the stationary phase (3.5 or 7 wt%), the costs of solvents, and assumed that 90% of the solvents could be recovered by wiped film evaporation and reused. Also considered were the costs of labor and waste disposal, and the cost and amortization of the stationary phase at catalog prices, chromatography equipment and equipment used to strip solvents from the collected fractions. If no inorganic buffers are used in the eluents the cost of these materials and waste disposal decrease considerably. If almost all of the solvent can be recovered and reused, the cost of solvents drops to almost nothing. If the stationary phase can be protected by

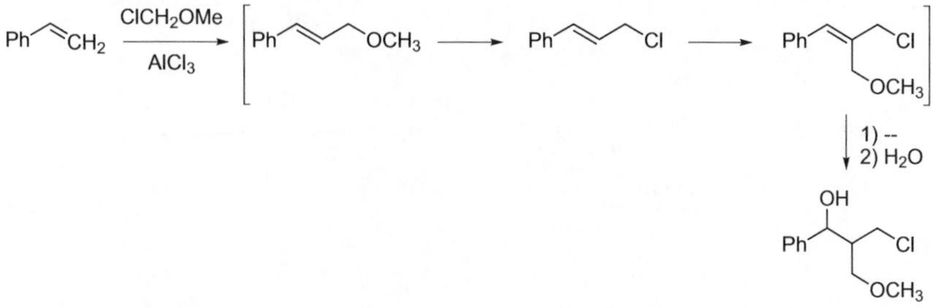

FIGURE 11.17   An impurity extracted from a styrene-based resin.

a prefilter, then the stationary phase could be reused extensively, and this cost falls. Once the equipment has been amortized the associated costs are due to maintenance and labor. CMOs that specialize in preparative chromatography, especially SMB, have manufactured drugs economically, and this type of purification should be considered today in route scouting.

IX. PERSPECTIVE ON WORKUP

Workups must be effective, and on scale ideally they are efficient. A workup that requires 12 h of stirring with a reagent to remove Pd to suitable levels may tie up equipment for an extended time, but if operations are simple this workup may be preferred over shorter procedures with more operations. Perhaps workups that entail more than two sequential treatments with the same reagent could be further optimized, or another reagent could be considered. In general procedures requiring fewer pieces of equipment are preferred, provided that operations are not extended, problematic, or require dilute conditions. For instance, an aqueous extraction would probably be preferred over filtration to remove $Et_3N \cdot HCl$ generated in toluene, as the volume of water needed would probably be small due to the high solubility of $Et_3N \cdot HCl$ in water. On the other hand, for removing poorly soluble impurities such as MgO one might consider filtration, or substituting another reagent.

The techniques and reagents discussed in this chapter could be applied to other problems, through reversed thinking. For instance, ethylenediamine has been used to remove Zn species from reaction streams (Table 11.7). GSK researchers found that the generation of a Grignard reagent was sluggish due to contamination by an ethylenediamine impurity (Figure 11.18). Treating a solution of the aryl chloride with aqueous $ZnCl_2$ precipitated the troublesome ethylenediamine impurity, and allowed Grignard formation to proceed in an acceptable manner [122].

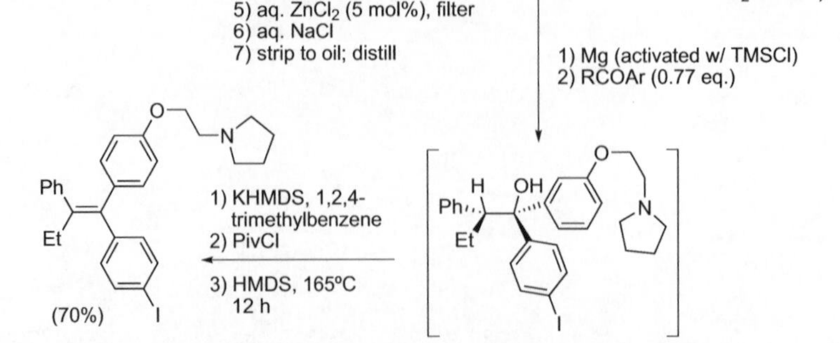

FIGURE 11.18 Treatment with $ZnCl_2$ to remove an ethylenediamine impurity.

REFERENCES

1. Zurer, P. *Chem. Eng. News* **2000,** *78*(1), 26.
2. Laird, T. *Chem. Britain* **1996** (August), 43.
3. Matlack, A. S. *Introduction to Green Chemistry*, 2nd ed.; CRC Press: Boca Raton, 2010; pp 187–214.
4. Kennedy, J. H.; Biggs, W. S. *Org. Process Res. Dev.* **1997,** *1*, 68; Misner, J. W.; Kennedy, J. H.; Biggs, W. S. *Org. Process Res. Dev.* **1997,** *1*, 77.
5. Gallou, F.; Eriksson, M.; Zeng, X.; Senanayake, C.; Farina, V. *J. Org. Chem.* **2005,** *70*, 6960. The product is analogous to doramapimod, BIRD-0796. The authors noted that the presence of alcohols, generated from urea formation using alkyl chloroformates, led to the formation of symmetrical ureas, complicating isolation.
6. Thayer, A. M. *Chem. Eng. News* **2011,** *89*(22), 21.
7. Anderson, N. G. *Org. Process Res. Dev.* **2004,** *8*, 260.
8. Chen, C.-K.; Singh, A. *Org. Process Res. Dev.* **2001,** *5*, 509.
9. http://www.jtbaker.com/msds/englishhtml/d2868.htm
10. Anderson, N. G. Unpublished. Acetic acid has also been used to quench DCC: Anderson, G. W.; Callahan, F. M. *J. Am. Chem. Soc.* **1958,** *80*, 2902.
11. Conrow, R. E.; Dean, W. D. *Org. Process Res. Dev.* **2008,** *12*, 1285.
12. The presence of hydrazoic acid in the headspace of a reactor was monitored using FTIR: Chen, C.; Frey, L. F.; Shultz, S.; Wallace, D. J.; Marcantonio, K.; Payack, J. F.; Vazquez, E.; Springfield, S. A.; Zhou, G.; Liu, P.; Kieczykowski, G. R.; Chen, A. M.; Phenix, B. D.; Singh, U.; Strine, J.; Izzo, B.; Krska, S. W. *Org. Process Res. Dev.* **2007,** *11*, 616.
13. Cason, J.; Rapoport, H. *Basic Experimental Organic Chemistry*; Prentice-Hall: Englewood Cliffs, New Jersey, 1962; pp 33–38.
14. Spagnuolo, C.; Wang, S. S. Y. *Chem. Eng. News* **1992,** *70*(22), 2; Wang, S. S. Y.; Kiang, S.; Merkl, W. *Process Safety Prog.* **1994,** *13*(3), 153.
15. Grimm, J. S.; Maryanoff, C. A.; Patel, M.; Palmer, D. C.; Sorgi, K. L.; Stefanik, S.; Webster, R. R. H.; Zhang, X. *Org. Process Res. Dev.* **2002,** *6*, 938.
16. Achmatowicz, M. M.; Thiel, O. R.; Colyer, J. T.; Hu, J.; Elipe, M. V. S.; Tomaskevitch, J.; Tedrow, J. S.; Larsen, R. D. *Org. Process Res. Dev.* **2010,** *14*, 1490.
17. Brands, K. M. J.; Payack, J. F.; Rosen, J. D.; Nelson, T. D.; Candelario, A.; Huffman, M. A.; Zhao, M. M.; Li, J.; Craig, B.; Song, Z. J.; Tschaen, D. M.; Hansen, K.; Devine, P. N.; Pye, P. J.; Rossen, K.; Dormer, P. G.; Reamer, R. A.; Welch, C. J.; Mathre, D. J.; Tsou, N. N.; McNamara, J. M.; Reider, P. J. *J. Am. Chem. Soc.* **2003,** *125*, 2129.
18. Nicola, T.; Brenner, M.; Donsbach, K.; Kreye, P. *Org. Process Res. Dev.* **2005,** *9*(4), 513.
19. Brenner, M.; Meineck, S.; Wirth, T. U.S. Patent 7,183,374 B2, 2007 (to Boehringer Ingelheim).
20. Yee, N. K.; Farina, V.; Houpis, I. N.; Haddad, N.; Frutos, R. P.; Gallou, F.; Wang, X.; Wei, X.; Simpson, R. D.; Feng, X.; Fuchs, V.; Xu, Y.; Tan, J.; Zhang, L.; Xu, J.; Smith-Keenan, L. L.; Vitous, J.; Ridges, M. D.; Spinelli, E. M.; Johnson, M. *J. Org. Chem.* **2006,** *71*, 7133.
21. Liu, W.; Nichols, P. J.; Smith, N. *Tetrahedron Lett.* **2009,** *50*, 6103.
22. Perrault, W. R.; Shephard, K. P.; LaPean, L. A.; Krook, M. A.; Dobrowolski, P. J.; Lyster, M. A.; McMillan, M. W.; Knoechel, D. J.; Evenson, G. N.; Watt, W.; Pearlman, B. A. *Org. Process Res. Dev.* **1997,** *1*, 106.
23. Prashad, M.; Har, D.; Chen, L.; Kim, H.-Y.; Repic, O.; Blacklock, T. J. *J. Org. Chem.* **2002,** *67*, 6612.
24. Modi, S. P.; Gardner, J. O.; Milowsky, A.; Wierzba, M.; Forgione, L.; Mazur, P.; Solo, A. J. *J. Org. Chem.* **1989,** *54*, 2317.
25. Hansen, K. B.; Hsiao, Y.; Xu, F.; Rivera, N.; Clausen, A.; Kubryk, M.; Krska, S.; Rosner, T.; Simmons, B.; Balsells, J.; Ikemoto, N.; Sun, Y.; Spindler, F.; Malan, C.; Grabowski, E. J. J.; Armstrong, J. D., III. *J. Am. Chem. Soc.* **2009,** *131*, 8798.

26. Benoit, G.-E.; Carey, J. S.; Chapman, A. M.; Chima, R.; Hussain, N.; Popkin, M. E.; Roux, G.; Tavassouli, B.; Vaxelaire, C.; Webb, M. R.; Whatrup, D. *Org. Process Res. Dev.* **2008,** *12,* 88.

27. Jin, F.; Wang, D.; Confalone, P. N.; Pierce, M. E.; Wang, Z.; Xu, G.; Choudhury, A.; Nguyen, D. *Tetrahedron Lett.* **2001,** *42,* 4787.

28. http://onlinelibrary.wiley.com/book/10.1002/047084289X

29. Caron, S., Ed. *Practical Synthetic Organic Chemistry: Reactions, Principles, and Techniques*; Wiley, 2011.

30. Fieser, L. F.; Fieser, M. *Reagents for Organic Synthesis,* Vol. 1; Wiley: New York, 1967; pp 581–595.

31. Fieser, L. F.; Fieser, M. *Reagents for Organic Synthesis,* Vol. 1; Wiley: New York, 1967; p 584.

32. Bellingham, R.; Buswell, A. M.; Choudary, B. M.; Gordon, A. H.; Moore, S. O.; Peterson, M.; Sasse, M.; Shamji, A.; Urquhart, M. W. J. *Org. Process Res. Dev.* **2010,** *14,* 1254.

33. Fleck, T. J.; McWhorter, W. W.; DeKam, R. N.; Pearlman, B. A. *J. Org. Chem.* **2003,** *68,* 9612.

34. Mickel, S. J.; Sedelmeier, G. H.; Niederer, D.; Daeffler, R.; Osmani, A.; Schreiner, K.; Seeger-Weibel, M.; Bérod, B.; Schaer, K.; Gamboni, R.; Chen, S.; Chen, W.; Jagoe, C. T.; Kinder, F. R., Jr.; Loo, M.; Prasad, K.; Repic, O.; Shieh, W.-C.; Wang, R.-M.; Waykole, L.; Xu, D. D.; Xue, S. *Org. Process Res. Dev.* **2004,** *8,* 92.

35. htpp://curlyarrow.blogspot.com/2009_07_01_archive.html

36. Fujino, K.; Takami, H.; Atsumi, T.; Ogasa, T.; Mohri, S.; Kasai, M. *Org. Process Res. Dev.* **2001,** *5,* 426.

37. Reducing an imide to an amine: Couturier, M.; Andresen, B. M.; Jorgensen, J. B.; Tucker, J. L.; Busch, F. R.; Brenek, S. J.; Dubé, P.; Am Ende, D. J.; Negri, J. T. *Org. Process Res. Dev.* **2002,** *6,* 42.

38. Kopach, M. E.; Singh, U. K.; Kobierski, M. E.; Trankle, W. G.; Murray, M. M.; Pietz, M. A.; Forst, M. B.; Stephenson, G. A. *Org. Process Res. Dev.* **2009,** *13,* 209.

39. Davies, I. W.; Marcoux, J.-F.; Corley, E. G.; Journet, M.; Cai, D.-W.; Palucki, M.; Wu, J.; Larsen, R. D.; Rossen, K.; Pye, P. J.; DiMichele, L.; Dormer, P.; Reider, P. J. *J. Org. Chem.* **2000,** *65,* 6415.

40. McCoy, M. *Chem. Eng. News* **2011,** *89*(4), 12.

41. http://en.wikipedia.org/wiki/Struvite (accessed 11/24/10).

42. Journet, M.; Cai, D.; Hughes, D. L.; Kowal, J. J.; Larsen, R. D.; Reider, P. J. *Org. Process Res. Dev.* **2005,** *9,* 490.

43. Conrow, R. A. *Personal communication.*

44. Li, Y.-E.; Yang, Y.; Kalthod, V.; Tyler, S. M. *Org. Process Res. Dev.* **2009,** *13,* 73.

45. Anderson, N. G.; Lust, D. A.; Colapret, K. A.; Simpson, J. H.; Malley, M. F.; Gougoutas, J. Z. *J. Org. Chem.* **1996,** *61,* 7955.

46. Anderson, N. G. Unpublished.

47. Zegar, S.; Tokar, C.; Enache, L. A.; Rajagopol, V.; Zeller, W.; O'Connell, M.; Singh, J.; Muellner, F. W.; Zembower, D. E. *Org. Process Res. Dev.* **2007,** *11,* 747.

48. Yee, N. K.; Farina, V.; Houpis, I. N.; Haddad, N.; Frutos, R. P.; Gallou, F.; Wang, X.; Wei, X.; Simpson, R. D.; Feng, X.; Fuchs, V.; Xu, Y.; Tan, J.; Zhang, L.; Xu, J.; Smith-Keenan, L. L.; Vitous, J.; Ridges, M. D.; Spinelli, E. M.; Johnson, M. *J. Org. Chem.* **2006,** *71,* 7133.

49. Isola, A. M.; Holman, N. J.; Tometzki, G. B.; Watts, J. P.; Koser, S.; Klintz, R.; Munster, P. U.S. Patent 6,101,181, 1997 (to BASF).

50. Fukumoto, T.; Yamamoto, A. U.S. 5,292,973, 1994 (to Shin-Etsu Chemical Co., Ltd.).

51. Anderson, N. G.; Coradetti, M. L.; Cronin, J. A.; Davies, M. L.; Gardineer, M. B.; Kotnis, A. S.; Lust, D. A.; Palaniswamy, V. A. *Org. Process Res. Dev.* **1997,** *1,* 315.

52. Jiang, X.; Lee, G. T.; Villhauer, E. B.; Prasad, K.; Prashad, M. *Org. Process Res. Dev.* **2010,** *14,* 883.

53. Challenger, S.; Dessi, Y.; Fox, D. E.; Hesmondhalgh, L. C.; Pascal, P.; Pettman, A. J.; Smith, J. D. *Org. Process Res. Dev.* **2008,** *12,* 575.

54. Hilborn, J. W.; Lu, Z.-H.; Jurgens, A. R.; Fang, Q. K.; Byers, P.; Wald, S. A.; Senenayake, C. H. *Tetrahedron Lett.* **2001,** *42,* 8919.

55. As seen for acid-mediated hydrolysis of an aminoquinazoline: Ripin, D. H. B.; Bourassa, D. E.; Brandt, T.; Castaldi, M. J.; Frost, H. N.; Hawkins, J.; Johnson, P. J.; Massett, S. S.; Neumann, K.;

Phillips, J.; Raggon, J. W.; Rose, P. R.; Rutherford, J. L.; Sitter, B.; Stewart, A. M., III; Vetelino, M. G.; Wei, L. *Org. Process Res. Dev.* **2005**, *9*, 440.

56. Lowenthal, H. J. E.; Zass, E. *A Guide for the Perplexed Organic Experimentalist*, 2nd ed.; Wiley, 1992; p 124.

57. Delhaye, L.; Ceccato, A.; Jacobs, P.; Köttgen, C.; Merschaert, A. *Org. Process Res. Dev.* **2007**, *11*, 160.

58. Braden, T. M.; Coffey, S. D.; Doecke, C. W.; LeTourneau, M. E.; Martinelli, M. J.; Meyer, C. L.; Miller, R. D.; Pawlak, J. M.; Pedersen, S. W.; Schmid, C. R.; Shaw, B. W.; Staszak, M. A.; Vicenzi, J. T. *Org. Process Res. Dev.* **2007**, *11*, 431.

59. Anderson, N. G.; Carson, J. R. U.S. 4,284,562, 1981 (to McNeilab, Inc.).

60. Dias, E. L.; Hettenbach, K. W.; Am Ende, D. *J. Org. Process Res. Dev.* **2005**, *9*, 39.

61. Yang, R. T. *Adsorbents: Fundamentals and Applications*; Wiley: Hoboken, NJ, 2003.

62. Silverberg, P. *Chem. Eng.* **1999**, *106*(3), 33.

63. Hogue, C. *Chem. Eng. News* **2009**, *87*(42), 27.

64. Marsh, H.; Rodríguez-Reinoso, F. *Activated Carbon*; Elsevier Science: London, 2006.

65. http://www.donau-carbon.com

66. http://www.sucrose.com/bonechar.html

67. Herr, R. J.; Meckler, H.; Scuderi, F., Jr. *Org. Process Res. Dev.* **2000**, *4*, 43.

68. Roy, G. M. *Activated Carbon Application in the Food and Pharmaceutical Industries*. Technomic Publishing Co.: Lancaster, PA, 1995.

69. http://www.gravertech.com

70. Furniss, B. S.; Hannaford, A. J.; Smith, P. W. G.; Tatchell, A. R., Eds. *Vogel's Handbook of Practical Organic Chemistry*; Addison Wesley Longman Ltd.: Harlow, England, 1985; p 140.

71. Welch, C. J.; Albanese-Walker, J.; Leonard, W. R.; Biba, M.; DaSilva, J.; Henderson, D.; Laing, B.; Mathre, D. J.; Spencer, S.; Bu, X.; Wang, T. *Org. Process Res. Dev.* **2005**, *9*, 198. In the removal of Ru high selectivity was found when samples were dissolved in MeOH and treated with various absorbents; the selectivity was greatly reduced when 1:1 MeOH: H_2O was used.

72. Stewart, G. W.; Brands, K. J. M.; Brewer, S. E.; Cowden, C. J.; Davies, A. J.; Edwards, J. S.; Gibson, A. W.; Hamilton, S. E.; Katz, J. D.; Keen, S. P.; Mullens, P. R.; Scott, J. P.; Wallace, D. J.; Wise, C. S. *Org. Process Res. Dev.* **2010**, *14*, 849.

73. Welch, C. J.; Shalmi, M.; Biba, M.; Chilenski, J. R.; Szumigala, R. H., Jr.; Dolling, U.; Mathre, D. J.; Reider, P. J. *J. Sep. Sci.* **2002**, *25*, 847.

74. Welch, C. J.; Leonard, W. R.; Henderson, D. W.; Dorner, B.; Childers, K. G.; Chung, J. Y. L.; Hartner, F. W.; Albanese-Walker, J.; Sajonz, P. *Org. Process Res. Dev.* **2008**, *12*, 81.

75. Welch, C. J.; Pollard, S. D.; Mathre, D. J.; Reider, P. J. *Org. Lett.* **2001**, *3*, 95.

76. Hough, J. S.; Briggs, D. E.; Stevens, R.; Young, T. W. *Malting and Brewing Science*, Vol. 2; Chapman and Hall, 1990.

77. Matlack, A. S. *Introduction to Green Chemistry*, 2nd ed.; CRC Press: Boca Raton, 2010; pp 197–205.

78. Kemsley, J. *Chem. Eng. News* **2008**, *86*(49), 32.

79. http://www.epa.gov/greenchemistry/pubs/pgcc/winners/grca10.html

80. Savile, C. K.; Janey, J. M.; Mundroff, E. C.; Moore, J. C.; Tam, S.; Jarvis, W. R.; Colbeck, J. C.; Krebber, A.; Fleitz, F. J.; Brands, J.; Devine, P. N.; Huisman, G. W.; Hughes, G. J. *Science* **2010**, *329*, 305.

81. Jenkins, S. *Chem. Eng.* **2010**, *117*(10), 17.

82. Gray, T. *The Elements;* Black Dog & Leventhal: New York, 2009.

83. Hanson, D. J. *Chem. Eng. News* **2011**, *89*(20), 28.

84. Everts, S. *Chem. Eng. News* **2010**, *88*(41), 11.

85. Bien, J. T.; Lane, G. C.; Oberholzer, M. R. *Topics Organomet. Chem.* **2004**, *6*, 263.

86. Thayer, A. *Chem. Eng. News* **2005**, *83*(36), 55.

87. deVries, A. H. M.; Mulders, J. M. C. A.; Mommers, J. H. M.; Henderickx, H. J. W.; deVries, J. G. *Org. Lett.* **2003**, *5*, 3285.

88. Weissman, S. A.; Zewge, D.; Chen, C. *J. Org. Chem.* **2005**, *70*, 1508.

89. Rosso, V. W.; Lust, D. A.; Bernot, P. J.; Grosso, J. A.; Modi, S. P.; Rusowicz, A.; Sedergran, T. C.; Simpson, J. H.; Srivastava, S. K.; Humora, M. J.; Anderson, N. G. *Org. Process Res. Dev.* **1997**, *1*, 311.

90. Garrett, C.; Prasad, K. *Adv. Synth. Catal.* **2004**, *346*, 889.

91. Menzel, K.; Machrouhi, F.; Bodenstein, M.; Alorati, A.; Cowden, C.; Gibson, A. W.; Bishop, B.; Ikemoto, N.; Nelson, T. D.; Kress, M. H.; Frantz, D. E. *Org. Process Res. Dev.* **2009**, *13*, 519.

92. Williams, J. M.; Brands, K. M. J.; Skerlj, R. T.; Jobson, R. B.; Marchesini, G.; Conrad, K. M.; Pipik, B.; Savary, K. A.; Tsay, F.-R.; Houghton, P. G.; Sidler, D. R.; Dolling, U.-H.; DiMichele, L. M.; Novak, T. J. *J. Org. Chem.* **2005**, *70*, 7479.

93. Tehim, A. *Personal communication.*

94. Conlon, D.; Pipik, B.; Ferdinand, S.; LeBlond, C.; Sowa, J.; Izzo, B.; Collins, P.; Ho, G.-J.; Williams, J.; Shi, Y.-J.; Sun, Y. *Adv. Synth. Catal.* **2003**, *345*, 931.

95. Kaitaisto, E.; Polomski, R. E. *Personal communication.*

96. Huang, Q.; Richardson, P. F.; Sach, N. W.; Zhu, J.; Liu, K. K.-C.; Smith, G. L.; Bowles, D. M. *Org. Process Res. Dev.* **2011**, *15*, 556.

97. Flahive, E. J.; Brigitte, L.; Ewanicki, B. L.; Sach, N. W.; O'Neill-Slawecki, S. A.; Stankovic, N. S.; Shu Yu, S.; Guinness, S. S.; Dunn, J. *Org. Process Res. Dev.* **2008**, *12*, 637.

98. Wang, L.; Green, L.; Li, Z.; Dunn, J. M.; Bu, X.; Welch, C. J.; Li, C.; Wang, T.; Tu, Q.; Bekos, E.; Richardson, D.; Eckert, J.; Cui, J. *Org. Process Res. Dev.* **2011**, *15*, 1371.

99. Miller, D. *Dave Miller's Homebrewing Guide,* Storey Communications, Inc.: Pownal, Vermont, 1995; p 132.

100. Liang, J.; Lalonde, J.; Borup, B.; Mitchell, V.; Mundorff, E.; Trinh, N.; Kochrekar, D. A.; Cherat, R. N.; Pai, G. G. *Org. Process Res. Dev.* **2010**, *14*, 193.

101. Gooding, O. W.; Voladri, R.; Bautista, A.; Hopkins, T.; Huisman, G.; Jenne, S.; Ma, S.; Mundorff, E. C.; Savile, M. *Org. Process Res. Dev.* **2010**, *14*, 119.

102. Liang, J.; Mundorff, E.; Volari, R.; Jenne, S.; Gilson, L.; Conway, A.; Krebber, A.; Wong, J.; Huisman, G.; Truesdell, S.; Lalonde, J. *Org. Process Res. Dev.* **2010**, *14*, 188.

103. Burdick, D. C.; Egger, H.; Gum, A. G.; Koschinski, I.; Muelchi, E.; Prevolt-Halter, I. U.S. Patent 7,312,199 B2, 2007, (to DSM).

104. Mountford, P. G. In *Green Chemistry in the Pharmaceutical Industry*; Dunn, P., Wells, A.; Williams, M. T., Eds.; Wiley-VCH, 2010; p 145.

105. Bernasconi, B.; Lee, J.; Roletto, J.; Sogli, L.; Walker, D. *Org. Process Res. Dev.* **2002**, *6*, 152.

106. McCoy, M. *Chem. Eng. News* **2000**, *78*(25), 17.

107. Strube, J.; Haumreisser, S.; Schmidt-Traub, H.; Schulte, M.; Ditz, R. *Org. Process Res. Dev.* **1998**, *2*, 305.

108. http://www.amberlyst.com/glycerol.htm

109. Mulllin, R. *Chem. Eng. News* **2007**, *85*(39), 49.

110. McCoy, M. *Chem. Eng. News* **2004**, *82*(41), 20.

111. Hawkins, J. M.; Watson, T. J. N. *Angew. Chem. Int. Ed. Eng.* **2004**, *43*, 3224.

112. Deshmukh, R. R.; Miller, J. E.; De Leon, P.; Leitch, W. W., II; Cole, D. L.; Sanghvi, Y. S. *Org. Process Res. Dev.* **2000**, *4*, 205.

113. Lajmi, A. R.; Schwartz, L.; Sanghvi, Y. S. *Org. Process Res. Dev.* **2004**, *8*, 651.

114. Fisher, D.; Garrard, I. J.; san den Heuvel, R.; Sutherland, I. A.; Chou, F. E.; Fahey, J. W. *J. Liq. Chromatogr.* **2005**, *28*, 1913.

115. Moriarty, R. M.; Rani, N.; Enache, L. A.; Rao, M. S.; Batra, H.; Guo, L.; Penmasta, R. A.; Staszewski, J. P.; Tuladhar, S. M.; Prakash, O.; Crich, D.; Hirtopeanu, A.; Gilardi, R. *J. Org. Chem.* **2004**, *69*(6), 1890.

116. Mickel, S. J.; Niederer, D.; Daeffler, R.; Osmani, A.; Kuesters, E.; Schmid, E.; Schaer, K.; Gambon, R.; Chen, W.; Loeser, E.; Kinder, F. R., Jr.; Konigsberger, K.; Prasad, K.; Ramsey, T. M.; Repic, O.; Wang, R.-W. *Org. Process Res. Dev.* **2004**, *8*, 122.

117. Welch, C. J.; Fleitz, F.; Antia, F.; Yehl, P.; Waters, R.; Ikemoto, N.; Armstrong, J. D., III; Mathre, D. J. *Org. Process Res. Dev.* **2004**, *8*, 186.

118. Thayer, A. *Chem. Eng. News* **2005**, *83*(36), 49.

119. Welch, C. J.; Henderson, D. W.; Tschaen, D. W.; Miller, R. A. *Org. Process Res. Dev.* **2009,** *13*, 621.

120. Prashad, M.; Amedio, J. C.; Ciszewski, L.; Lee, G.; Villa, C.; Chen, K.-M.; Wambser, D.; Prasad, K.; Repic, O. *Org. Process Res. Dev.* **2002,** *6*, 773.

121. Tou, J. S.; Kostelc, J. G.; Schlittler, M. R.; Violand, B. N. *Org. Process Res. Dev.* **2003,** *7*, 750.

122. Anderson, N. G. Chapter 2 In *The Art of Drug Synthesis*; Johnson, D. S.; Li, J. J., Eds.; Wiley: New York, 2007; pp 11–28.

123. Ace, K. W.; Armitage, M. A.; Bellingham, R. K.; Blackler, P. D.; Ennis, D. S.; Hussain, N.; Lathbury, D. C.; Morgan, D. O.; O'Connor, N.; Oakes, G. H.; Passey, S. C.; Powling, L. C. *Org. Process Res. Dev.* **2001,** *5*, 479.

Crystallization and Purification

"Most of the problems we saw in moving products to manufacturing were in the crystallization stage."

– Mary Shire, Solid State Pharmaceutical Cluster, Ireland [1]

I. INTRODUCTION

Crystallization of products often drives reactions to completion. A striking example is shown in Figure 12.1, with the three-component condensation of a chiral diol, glyoxal, and 4-fluorophenylboronic acid. Five lactol diastereomers were detected in solution in a ratio of 50 (*trans*):10:<2 (*cis*):10:20, all probably in equilibrium through the initially formed hydroxyaldehyde. After an extractive workup this aprepitant precursor [2] was crystallized from methylcyclohexane–EtOAc in 86% yield. Instead of the desired *trans*-diastereomer, which was the major diastereomer in solution, the crystallized product was identified as the *cis*-isomer, which had been the minor product in solution [3]. In this powerful transformation the chirality of the secondary alcohol controlled the stereochemistry of two chiral centers that formed. This is an example of a

Practical Process Research and Development. DOI: 10.1016/B978-0-12-386537-3.00012-5

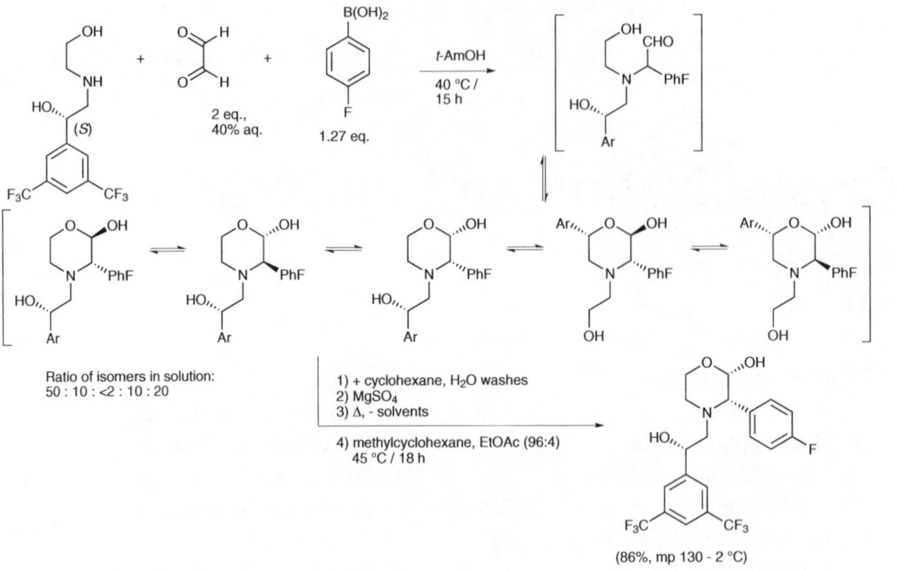

FIGURE 12.1 Formation of an aprepitant intermediate driven by crystallization (CIsAT).

Solubilities greatly influence processing and determine operations. For instance, the monosodium salt of a dicarboxylic acid may crystallize from a solution of the disodium salt before two equivalents of HCl has been added; to avoid contamination with the monosodium salt an inverse addition may be helpful. Crystals that form initially may be the kinetic product, not the thermodynamically favored polymorph; extended stirring of a suspension may lead to the crystallization of a more stable polymorph.

crystallization-induced asymmetric transformation (CIAT), which is discussed with crystallization-induced dynamic resolutions (CIDRs) in Section VI. In general attention on crystallization is focused on isolating an API or a key intermediate, to investigate whether multiple crystalline forms may be produced, and to select the optimal form. Different crystal forms may arise as the product of any reaction, and by taking advantage of the different crystal forms improved processes may result.

There are many options for purifying and isolating APIs and intermediates. Chromatography may be used early in the development of a drug candidate, and can be cost-effective if significant amounts of time are needed to develop a process to remove impurities [4]. Distillation is time-consuming and labor-intensive, and most APIs and intermediates decompose with extended heating. Wiped film evaporators (WFEs) are used on scale to distill heat-sensitive materials. Decomposition is minimized by continuous processing under vacuum with short contact time through the heated zone [5]. Lilly researchers described using a WFE to remove heptane from a less-volatile product, a benzyl azide [6]. Such specialized equipment may have to be rented.

Purifying and isolating solids by sublimation may be appropriate for preparing reference standards of some small molecules, but is too inefficient for scale-up. When solids are formed quickly from solution often they are precipitated, usually by rapid addition of an antisolvent, and generate amorphous or poorly crystalline materials of low quality; slow filtrations can be avoided by adding a diatomaceous earth admix or by increasing the surface area of the filter. Precipitation is useful for removing small amounts of impurities and isolating large peptides [7]. Crystallization in general is preferred for APIs and key intermediates, as it can be used to purge impurities. Batch crystallization, as usually employed in the pharmaceutical industry, can be labor-intensive and equipment-intensive, but once understood can routinely provide high-quality products. Another option, sometimes effective but usually unfamiliar to academic researchers, is to reslurry solids in a solvent; even though the solids don't completely dissolve batches can be easily upgraded if impurities adsorbed to surface of crystals are dissolved in the solvent. When a wet cake shrank on the filter so much that effective washing was difficult, GSK researchers resorted to reslurrying the wet cake [8]. (Reslurrying is sometimes referred to as trituration.) Reslurrying can produce a solvate or another polymorph. Crystallization is the option usually preferred in the pharmaceutical industry.

> Too often people precipitate solids, and forego the benefits of crystallization.

The goals of crystallization are to purify the solids by removing impurities, and to produce material in the desired physicochemical state. In the initial stages of developing an API, when delivering enough material by a timeline is crucial, any crystallization may be satisfactory provided that impurities are effectively removed and the product can be scraped out of the crystallizer. The process scientist wants to develop crystallizations that are practicable, i.e., those that use SAFE solvents under relatively concentrated conditions with operations that can be performed SAFELY, easily, and reproducibly on scale. Practicable crystallizations require relatively little time, making efficient use of equipment. Such crystallizations are different from techniques to prepare crystals for a single-crystal X-ray analysis [9], and are likely to be different from crystallizations performed by chemists in academia. Slurries must be transferred readily from the crystallizer to the filter, and solids must be transferred readily to and from the dryer. Hygroscopic solids can pose challenges in transfer, storage, analyses, and formulation (Chapter 13.) For routine operations the crystallization and isolation must reliably produce high-quality material that meets specifications for both chemical and physicochemical characteristics. Chemical quality includes potency of the compound and suitable levels of impurities such as H_2O, residual solvents, and heavy metals. Physicochemical quality includes considerations such as the desired polymorph, particle size distribution, and perhaps crystal habit. The process scientist designs crystallizations to meet these goals in addition to the primary goal of upgrading the quality of the compounds.

Crystallization on an industrial scale for the pharmaceutical industry has been thoroughly discussed in an excellent monograph [10], in a recent book chapter [11], and in other texts [12–15].

Practicable crystallization — Imagine a crystallization from diisopropyl ether with isolation after holding at −20 °C for 72 hours — this could be one of the least practicable crystallizations ever. Although (*i*-Pr)$_2$O can crystallize some materials very well, it is prone to forming peroxides; solid peroxides may adhere to the cap of the container (SAFETY issue). An ageing period of 72 hours would probably use equipment inefficiently on scale. The temperature of −20 °C is the temperature of a freezer; if a suspension was held in a refrigerator freezer over a weekend, then it was not stirred. Solids that crystallize upon the wall of a flask can be dislodged by a large spatula in the kilo lab, but on scale access to the inside surfaces of vessels is minimized due to SAFETY and cGMP considerations. Suspensions should be stirred for other practical reasons. Consider a layer of NaCl in water that was high enough to cover the still agitator of a vessel: the heavy layer of salt might stop the agitator from moving if it was turned on! By gently agitating a suspension as it forms crystals seed crystals are dispersed, increasing the opportunities for nucleation and more rapid crystal growth. Gentle agitation may afford a more predictable crystallization, but extended and/or rapid agitation may fragment solids into particles that plug the pores of filter media ("blinding the filter"). Fracturing the crystals may be helpful to transfer a suspension of extremely large crystals to the filter. Gentle agitation allows the suspension to drain from the crystallizer without leaving solids stuck high on the walls. On scale the mother liquor is often sprayed into the crystallizer to dislodge crystals for transfer to the filter. But the mother liquor must be returned to the crystallizer through lines that cannot be adequately cleaned, such reuse of the mother liquor would not be consistent with cGMP practices. These and other considerations may make up a practicable crystallization process.

Crystallizations should progress from a solution to a suspension of crystals, without passing through an oil dispersion. Oils can cling to the walls of vessels and dryers, and to crystals, impeding transfers and filtrations. Even small amounts of oils in a suspension of crystals can entrap significant amounts of impurities. Crystallizations that proceed through oil dispersions are likely to be unreliable, even though the process may be reproducible over many batches. The example shown in Figure 12.2 serves to illustrate the potential difficulties involved. In the initial process to manufacture the phosphinic acid from the modified Arbuzov reaction, acetonitrile and hexamethyldisiloxane were removed by azeotropic concentration, and upon cooling the concentrate produced crystals. Despite early concerns about the oil droplets in laboratory crystallizations seen under the microscope, the product crystallized in 85% yield with 97–98 wt% purity, and operations were simple. (Perhaps adding a small amount of an organic co-solvent might have avoided oiling by dissolving the product in the mixture above the crystallization temperature. A polar solvent such as acetonitrile or a hindered alcohol such as *i*-PrOH might have been effective. The polarity of hexamethyldisiloxane is similar to that of heptanes, and crystallization from the hydrolysate without removing acetonitrile and hexamethyldisiloxane also produced an oil dispersion.) This process was successfully demonstrated in eight runs in the pilot plant and in five runs introduced to manufacturing. Unfortunately there were problems later with the manufacturing campaign: batches of the phosphinic acid were slowly converted to next fosinopril sodium intermediate in only 82% yield, a disappointing yield for a simple esterification. The crystallization of the phosphinic acid in the presence of oil droplets probably occluded impurities detrimental to that acid-catalyzed process. Recrystallization from

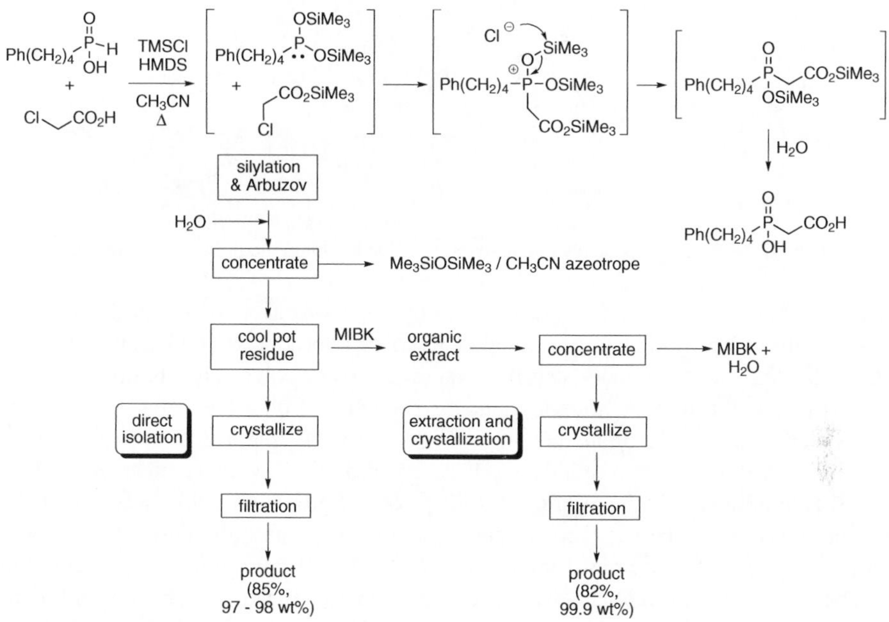

FIGURE 12.2 Direct isolation or extraction and crystallization of a phosphinic acid from a modified Arbuzov reaction.

MIBK was used to upgrade about one of every three batches of the phosphinic acid, producing batches of the benzyl ester in 92% yield. After a recurring need for reworking the phosphinic acid batches, the modified Arbuzov process was

> Oils may be detected readily by viewing a suspension under a light microscope.

redeveloped by extracting the product into MIBK, followed by concentrating and cooling the extract to crystallize the phosphinic acid. This isolation reliably produced high-quality phosphinic acid in manufacturing, at 82% yield and 99.9 wt% purity, and esterification to the next intermediate was rapid [16]. In the initial isolation from water the oiling seen was caused by too rapidly applying the crystallization pressure, producing an unreliable crystallization process. The revised process proceeded from a solution to a suspension of crystals without passing through an oil dispersion, and proved to be very rugged. The development time for this intermediate was extended by not controlling the crystallization.

Chen and Singh have outlined a screening process to identify solvents to crystallize reaction products directly from reactions [18], and Amgen workers have described a

> In general a purity of 97% is acceptable for an intermediate, even as low as 94% [17]. Acceptable purities are determined by the usefulness of the product produced. If an intermediate is processed unacceptably in the next step, a decision must be made whether to rework the intermediate, redevelop the process to prepare the intermediate, or to redevelop the downstream process. Sometimes politics affect this decision.

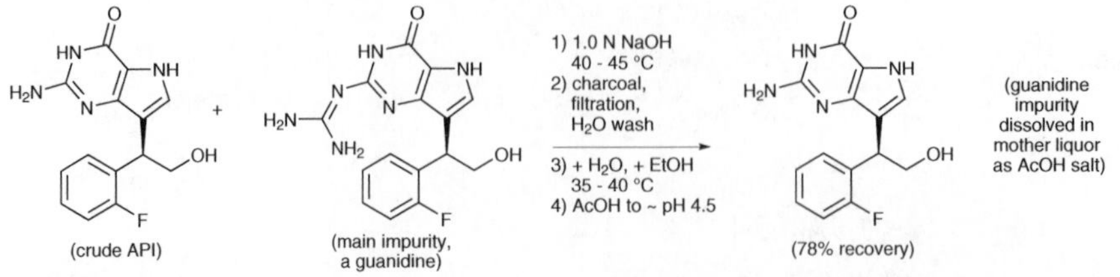

FIGURE 12.3 Purging a guanidine impurity by recrystallization in the presence of acetic acid.

high-throughput screening for purification by crystallization and trituration [19]. Crystallizations can be further designed to dissolve byproducts. For instance, moderately polar organic compounds can be crystallized from polar solvents or from lipophilic solvents; the mother liquors would be expected to dissolve impurities that are more polar or less polar than the product, respectively. Impurities with additional, ionizable functional groups may be selectively removed by pH adjustment before crystallization, as shown in Figure 12.3 [20]. Recrystallization using solvents at the extreme ends of the polarity spectrum, such as H_2O or heptanes, often gives high weight returns but with poor purging of impurities. Non-polar, highly lipophilic products can often be crystallized from polar solvents, e.g., MeOH, EtOH, and acetonitrile, and small amounts of protic solvents can change crystal morphology and size. In small amounts, highly lipophilic solvents such as heptanes and iso-octane may be useful co-solvents.

The progress of crystallization can be monitored by several approaches. The onset of crystallization can be seen as solutions turn opaque. Raman spectroscopy can be used to follow crystallizations, even for enantiotropic polymorphs [21]. Infrared spectroscopy and HPLC can be used to monitor the amount of product still in solution. For instance, if the amount of product in solution after ageing for 30 min is the same as the amount initially in solution, then further ageing is unlikely to be beneficial, so filtration may be timely.

Insoluble impurities should be removed by polish filtration before initiating crystallization, to raise the quality of the isolated product by removing these impurities, to prevent premature crystallization, and to prevent crystallization of an unwanted polymorph. Polish filtrations are simply good housekeeping.

APIs should be clarified by polish filtration of solutions before the final isolation. Hot polish filtrations are often used. For effective scale-up a warm solution should be stable for about an hour without significant degradation or premature crystallization, in order to accommodate possible delays prior to polish filtration. On scale the rapid passage of the hot API solution may adequately warm transfer lines and preclude premature crystallization in the lines. Polish filtration may be more difficult to carry out in the kilo lab than on scale. To preclude premature crystallization the hot polish filtration should be carried out at least 10 °C above the temperature at which solids begin to crystallize.

Washing a product wet cake on the filter displaces the mother liquors. Viscous solvents, such as *i*-PrOH and *t*-BuOH, may give slow filtrations. Solvents that are immiscible with the mother liquor and washes may be applied to the wet cakes (provided of course that the product does not dissolve). For instance, heptanes may be applied to a wet cake to displace water, if heptanes are more readily removed by drying.

In the laboratory, wet cakes are routinely dried in crystallizing dishes placed in drying ovens. Tray dryers may be used for pilot plant batches, but require too much labor for routine drying on scale. Agitated dryers are used on scale to decrease the drying time. Care must be taken when initially stirring the wet cake, or the product can solidify to a taffy-like mass [22].

Modes of crystallization most typically include cooling and concentration, followed by antisolvent addition, and reactive crystallization [10]. Antisolvent addition and reactive crystallization are more likely to produce metastable polymorphs [23]. Reactive crystallization can include forming a salt, adjusting the pH to neutralize or ionize a substance, and crystallization-driven transformations. These modes of crystallization are discussed in following sections of this chapter, along with guidelines for classical resolutions by crystallization and considerations for polymorphs. In early investigations in developing processes crystallization pressure should be applied gradually to avoid difficulties in processing. Details on volumes used for effective and efficient crystallization and isolation can be found in discussions related to Figure 8.12. Considerations focusing on the API, including salt selection, are discussed in Chapter 13.

II. SLOW APPLICATION OF CRYSTALLIZATION PRESSURE, AND SEEDING

As crystallization occurs molecules are arranged in an orderly fashion within the crystal lattice. The energy given off is sometimes recognized as a rise in temperature of as much as 2 °C during crystallization on scale. Figure 12.4 shows the results of applying

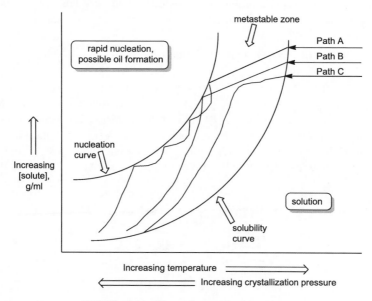

FIGURE 12.4 Effects of crystallization pressure.

a crystallization pressure, with temperature (or crystallization pressure) on the X-axis and solubility on the Y-axis. The nucleation curve, or the conditions at which crystallization occurs, roughly parallels the solubility curve. Oiling occurs beyond the nucleation curve, in the upper left corner of this figure. Between the two curves is the metastable zone. The metastable zone can be characterized simply by adding seeds at various times: if added seeds dissolve or promote oiling, then conditions are not in the metastable zone. Conductivity measurements [24] and focused beam reflectance measurement (FBRM™, also known as Lasentec™) [25,26] have also been used to characterize a metastable zone. With rapid cooling (Path A) oiling may take place, or multiple nucleation events occur and produce crystals with a relatively wide particle size distribution (PSD). More gradual cooling may lead to one nucleation event, producing crystals with a smaller PSD (Path B). (Such crystallization often occurs in campaigns of a product run in successive batches in dedicated equipment, as cleaning between batches may not be thorough enough to remove crystals absorbed to the dome or other parts of the crystallizer.) Gradual cooling into the metastable zone followed by seeding (Path C) may produce large crystals. (In this discussion cooling applies the crystallization pressure. Other ways to apply crystallization pressure include evaporating solvents, adding antisolvents, adjusting pH, and adding another component to afford a reactive crystallization.)

Effective seeding can be carried out with 0.5 wt% seeds, and cGMP seeds can be prepared. (For high-quality crystals seeding should take place after polish filtration, with high-quality seeds of the desired polymorph. Lower-quality seeds may lead to slower crystallizations, possibly allowing metastable polymorphs to form [27].) Heavy seeding, e.g., 3–10 wt%, introduces a large number of seeds and produces smaller crystals [11]. This is especially true for needles and plates. Equation 12.1 details an equation that has been developed for seeding to crystallize product of a desired size [28]. Crystallizations on scale may not proceed according to this model if the crystals are fractured by agitation or if a high crystallization pressure induces secondary nucleation [11].

$$\text{Mass of seeds} = \text{mass of crystals} \times \left(\frac{\text{size of seeds}}{\text{size of crystals}} \right)^{i} \qquad (12.1)$$

where $i = 1$ (needles), 2 (plates) or 3 (cubes or spheres).

Spontaneous crystallization is likely to produce a metastable form before a more stable form crystallizes, according to Ostwald's rule of stages [27] (see Section VIII). Seeding can help initiate not only the formation of the desired polymorph, but can also guide how the crystals form within a reactor. The phosphate salt in Figure 12.5 was poorly soluble under the crystallization conditions, and tended to crystallize on

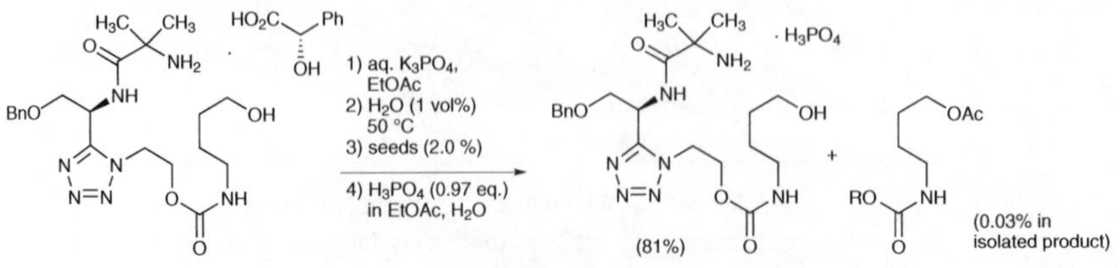

FIGURE 12.5 Controlled crystallization of a phosphate salt.

the reactor walls. (Such encrustation, or "fouling," often occurs near the heat transfer surfaces, where the temperature gradient is greater [29]. Fouling prevents efficient heat transfer in heat exchangers [30].) The mandelate salt, which was not a pharmaceutically preferred salt, was crystallized to upgrade the API. After splitting the mandelate salt, seeds of the phosphate salt were added to the solution of the free base, followed by the addition of diluted H_3PO_4 to apply a more gradual crystallization pressure. By such seeding the salt crystallized freely and could be readily transferred to the filter. By undercharging the H_3PO_4 the amount of the O-acetate byproduct was held to 0.03% [31].

III. CRYSTALLIZATION BY COOLING

Cooling a solution is probably the most common way to apply crystallization pressure and thus crystallize compounds. Various profiles of cooling are shown in Figure 12.6 [32]. With the advent of programmable units, temperatures and devices to monitor particle size, crystallizations can be controlled in the laboratory and on scale. Uncontrolled ("natural") cooling occurs when coolant is allowed to pass rapidly through the external jacket of a crystallizer; cooling is rapid at first, when the difference between the temperatures of the coolant and the reactor contents is greatest. Linear cooling can be programmed, and may be more popular now. Cubic cooling is slow initially, then more rapid; this cooling profile minimizes nucleation and encourages crystal growth, producing batches with fewer fines and a narrower particle PSD relative to linear and uncontrolled cooling [33]. FBRM can be used to determine the appropriate cooling profile, as clearly explained by researchers from Mettler–Toledo: slow cooling should be applied initially, when the concentration of crystals and hence surface area for crystal growth are low. When the rate of increase in the concentration and size of crystals has slowed, more cooling can be applied to generate supersaturation and encourage further crystallization [25]. Cubic cooling can be used to control PSD and avoid milling a final product. Effective crystallizations can be performed without programmed cooling by seeding a solution a few degrees above the cloud point and holding at that temperature range for about 0.5 h to develop a bed of crystals. During this time a temperature rise of

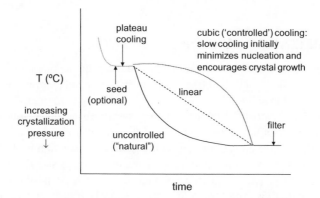

FIGURE 12.6 Profiles of cooling to control crystallizations.

about 2 °C may be observed, as the molecules give up energy to crystallize. After such a "plateau" holding period the suspension may be cooled as rapidly as possible (uncontrolled cooling). The phosphinic acid in Figure 12.2 was crystallized from MIBK by seeding at 68–70 °C and ageing at that temperature range for half an hour, followed by uncontrolled cooling to 0–5 °C and ageing at that temperature range for 1 hour [16].

To anticipate the crystallization of a solution it is also helpful to follow the temperature of the coolant in the external jackets of the crystallizer

For a temperature-controlled crystallization with multiple solvents, choose solvents with similar boiling points for ready scale-up. Negative-boiling azeotropes may not allow the higher temperatures desired for dissolution.

(Further details can be found in the discussion with Figure 8.12.) This combination of plateau cooling, with optional seeding, and uncontrolled cooling has reliably produced high-purity crystals with easy handling characteristics products on a large scale. Such "poor man's cubic cooling" [34] is easier to effect than cubic cooling if programmable temperature controllers are not available.

Fines can extend processing by plugging the pores of filters. Fines can be converted to larger crystals by Ostwald ripening, in which a slurry of crystals is subjected to cycles of moderate heating and cooling before isolation. For example, a slurry at 35 °C may be heated to 40 °C, held at that temperature for 30 minutes, cooled back to 35 °C, and held at that temperature for 30 minutes. At the higher temperatures the crystalline fines dissolve, and upon cooling the existing crystals grow bigger. With five cycles of Ostwald ripening Pfizer workers achieved API crystals of 100–200 mm for formulation, instead of 10–15 mm crystals that had been routinely encountered [35]. Fines of pentaerythritol have also been converted to larger crystals through segregation and warming, essentially a chemical engineering approach [36]. Large crystals of RDX with few fines were grown using temperature cycling [37]. To control fines batch crystallization is preferred over continuous crystallization, in which a stream of the slurry is continuously removed [38]. Conceptually the best approach to control fines is to encourage crystal growth over nucleation [29].

IV. CRYSTALLIZATION BY CONCENTRATING

Crystallizing by concentrating an extract and controlling the temperature is frequently used in the pharmaceutical and fine chemicals industry. An example was discussed in the section above. Rapid evaporation of solvent may cause solids to crystallize above the solvent line, and for successful control of the crystallization these solids should be redissolved before crystallization is initiated. Ideally solvent vapors condensing on the reactor walls will dissolve any solids and return them to the solution in the crystallizer.

Evaporation using nitrogen gas may be slow but effective, as shown for the CIAT in Figure 12.7 [39]. In general removing solvents by distillation is quicker and more convenient. Concentrating a solution by evaporation may be useful for preparing crystals for single-crystals X-ray analysis [9].

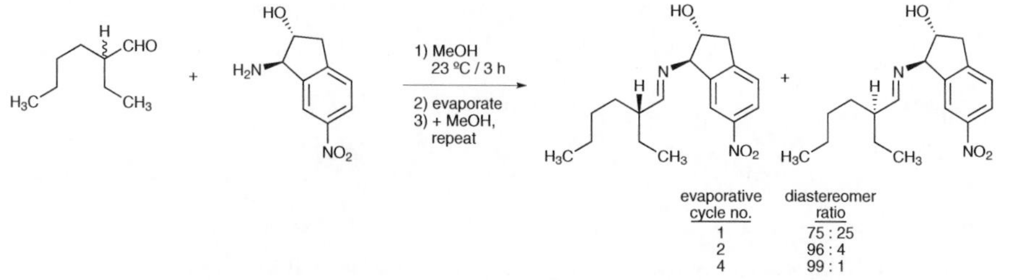

FIGURE 12.7 CIAT crystallization driven by evaporation with N_2.

V. CRYSTALLIZATION BY ADDING AN ANTISOLVENT

Adding an antisolvent to crystallize a product is sometimes referred to as a drown-out process, and may produce a metastable polymorph if the antisolvent is added rapidly [23]. Perhaps the most common use of antisolvents is adding water to a reaction run in a polar aprotic solvent such as DMF (see Figure 12.8 [40]). In this dicyclohexylcarbodiimide-mediated (DCC) coupling the byproduct dicyclohexylurea (DCU) was removed by filtration, and the filtrate was diluted with MeOH and water to crystallize the product. Slow addition is usually necessary to avoid oiling, and close examination will generally reveal some oiling where an antisolvent enters the crystallizer. Provided that the antisolvent can be transferred from a drum or storage vessel, adding the antisolvent to the rich solution requires only one reactor.

Occasionally a rich solution is added to an antisolvent to initiate crystallization, usually as a last resort, but this approach can be very effective. Since the crystallization pressure is highest at the beginning of addition, this "shock" crystallization promotes nucleation with minimal crystal growth, and is likely to produce small crystals or unstable polymorphs. This technique also risks forming oils or precipitates. Adding the rich solution to the antisolvent requires a second reactor, which can be an issue for scale-up if equipment and space are concerns. Once the antisolvent and solution of product are combined, heating may not allow dissolution of the product. Abbott researchers described a successful example of such a crystallization (Figure 12.9). When the rich NMP solution of the vinylogous amides was diluted with water the filtration on a 10 kg scale was very slow, and the dried product contained about 5 wt% NMP. When the rich NMP solution was added to water, agglomerates formed, which were readily filtered and dried on a 125 kg scale. For each addition protocol the yields and qualities were comparable, but the inverse addition was clearly preferred on scale [41].

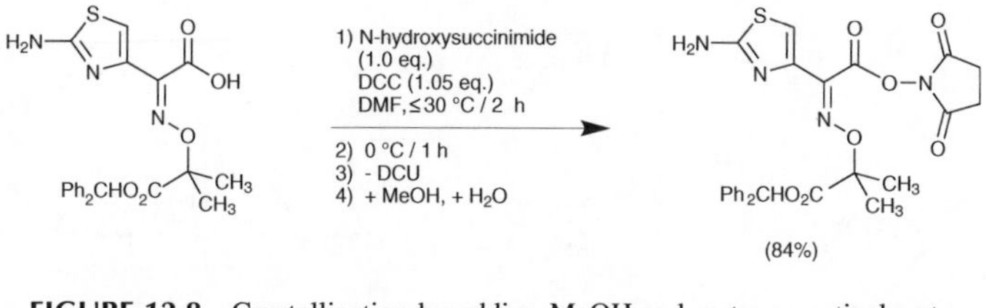

FIGURE 12.8 Crystallization by adding MeOH and water as antisolvents.

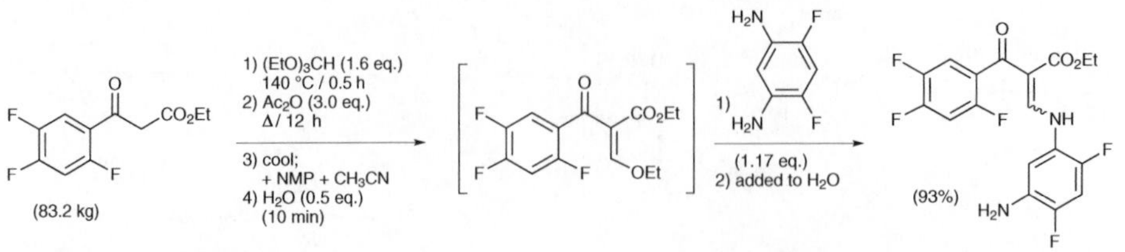

FIGURE 12.9 Crystallizations by adding product solution to an antisolvent.

VI. REACTIVE CRYSTALLIZATION

In a reactive crystallization the product crystallizes as it is formed. Reactive crystallizations can include transformations that form or break covalent bonds, salt formation, neutralizing or ionizing a substance by adjusting pH, and crystallization-driven transformations. Some reactive crystallizations take place very quickly, and others more deliberately. Rapid crystallizations, while they risk forming small particles or ill-defined precipitates [42], can be very effective. Slow crystallizations may be needed to ensure that crystallizations do not oil out, or to provide sufficient time for equilibrium processes to occur.

Salt formations are commonly mentioned as reactive crystallizations, with impinging jet technologies being employed (Chapter 14). The hydrochloride salts of two tertiary amines (structures not disclosed) were successfully isolated through reactive crystallization (Figure 12.10). Each of these amine hydrochlorides was highly crystalline, which probably encouraged rapid crystallization. The HCl salts were chosen due to the high solubilities found in water, and the higher bioavailabilities anticipated. Solvent screening showed superior stabilities of the polymorphs generated from EtOH and MeOH; unfortunately crystallizing the HCl salts from EtOH or MeOH also generated the genotoxins ethyl chloride and methyl chloride. The presence of the latter

> SAFETY – HCl will react with MeOH and EtOH to produce Me_2O (bp −25 °C), MeCl (bp −24 °C), Et_2O (bp 35 °C), and EtCl (bp 12 °C). These highly volatile compounds have flash points below −40 °C. Care must be taken to avoid generating flammable mixtures, and to scrub MeCl and EtCl. The presence of H_2O should reduce the amount of MeCl and EtCl generated.

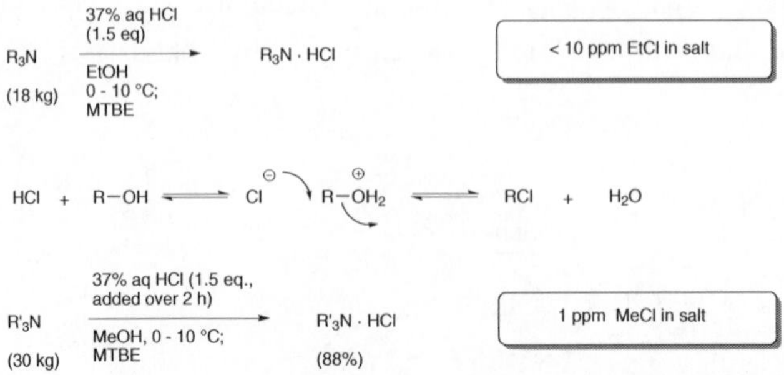

FIGURE 12.10 Formation of tertiary amine hydrochlorides from aqueous alcohol solutions.

impurities necessitated monitoring and controlling the concentrations of EtCl and MeCl in these two salts [43]. (See Chapter 13 for further discussion on genotoxic impurities.)

The first salt was crystallized from EtOH. Concentrated aqueous HCl was a convenient source of HCl, and MTBE was added as an antisolvent (Figure 12.10). The H_2O in the 37% HCl reagent was anticipated to decrease the amount of EtCl in the product, as increased amounts of H_2O were expected to shift the equilibrium away from EtCl (Figure 12.10). This salt was crystallized on an 18 kg scale from EtOH, and EtCl, as detected by headspace GC, was found in the product salt at no more than 10 ppm (the limit suitable for that stage of development) [43].

The second API candidate was crystallized from MeOH instead of EtOH, again using 37% aqueous HCl (12 M HCl). Keys to this process were the presence of H_2O, a low temperature for the HCl addition, and the crystallization of the salt as the HCl was added. The HCl salt was contaminated with the greatest amount of MeCl, 88 ppm, when anhydrous HCl was used. Extensive drying at 85 °C did not effectively expel residual MeCl found in one batch of salt, even after granulation; however, the salt could be readily recrystallized to remove the MeCl. The importance of the reactive crystallization was realized when salt formation from MeOH was carried out using 6 M HCl: the product salt stayed dissolved until the antisolvent MTBE was added, and two batches of this salt prepared with 6 M HCl contained high amounts of EtCl (17 and 44 ppm). Using 37% aqueous HCl at the conditions described above crystallization was carried out on a 30 kg scale, with residual MeCl present at less than 1 ppm [43]. The unanticipated, rapid reactive crystallizations of these two salts were exploited in the preparation of these API candidates.

Another reactive crystallization was deliberately carried out slowly. An optimized process for the preparation of captopril, one of the first blockbuster drugs is shown in Figure 12.11. In a process developed in the early 1980s the thioester was hydrolyzed in 4 M NaOH, acidified with HCl, and treated with Zn in the presence of dichloromethane to reduce any disulfide that had formed. (Classically disulfides are formed by exposing mercaptans to basic conditions.) The triphasic mixture was filtered, the rich organic extract was concentrated, and the dichloromethane was chased with water.

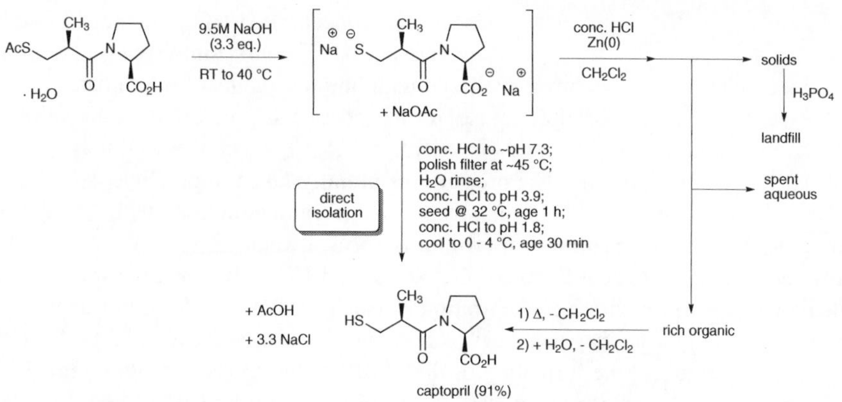

FIGURE 12.11 Processes to isolate captopril from hydrolysis of the thioester precursor.

Captopril was crystallized by cooling the aqueous concentrate. It was clear that if disulfide formation could be minimized and controlled, then captopril could be isolated directly from the hydrolysate, eliminating several operations and increasing productivity. Several chemists had tried to crystallize captopril directly from the hydrolysate and unfortunately produced oily products; these results led to the lore that such a direct isolation was not feasible.

The first step in developing the direct isolation of captopril was conducting the hydrolysis in more concentrated NaOH instead of 4 M NaOH, to minimize the loss of the highly water-soluble product to the mother liquor. The solid thioester was added to 9.5 M NaOH at room temperature, and the temperature rose to about 40 °C. (Since no external cooling was applied during this addition the temperature rise was expected to be the same on scale.) As expected, hydrolysis under these concentrated conditions (1.3 L of 9.5 M NaOH/kg starting material) was extremely rapid. The solution was acidified to about pH 7 for SAFE handling during the polish filtration. Slow acidification of the filtrate led to a cloud point, which could be clarified by adding a small amount of aqueous NaOH but not by moderate heating. The oily dispersion was dissolved by adding a small amount of aqueous NaOH, and at about pH 3.9 the solution was seeded with captopril and cooled slightly. After holding at about 32 °C for about 15 min crystallization began, and the pH rose to about 4.0. By adding a small volume of HCl the slurry was acidified by 0.1–0.2 pH units to preclude oiling, and aged for about 15 min. This cycle of acidifying and ageing was repeated over several hours, producing an oil-free slurry at about pH 1.8. (The slurry was not acidified further due to concerns that HCl might corrode metal vessels.) The product crystals were washed with cold water and dried [44].

Subsequent laboratory runs showed that cloud points occurred at pH 3.5–4, a range that was too wide for trouble-free crystallization on scale. By tightly controlling the amount of water used for washing the filter during the polish filtration, which was the only other source of variability in this process, the cloud point became reproducible, and crystallizations were initiated reliably at a pH above the cloud point. The mother liquor, containing one equivalent of acetic acid, was saturated with the main byproduct, NaCl. For operations on scale the concentrated HCl was steadily added over a period of hours, simplifying operations. This direct isolation process proved reliable with one-ton inputs of the thioester.

> Too great a crystallization pressure leads to supersaturation and oiling (the cloud point). Control conditions to crystallize *near* the cloud point.

The success of the captopril direct isolation process was predicated on the ability to control the formation of the disulfide impurity in captopril, which was routinely found at about 0.3% using this process. The precautions initially taken to preclude oxidation to the disulfide (operating under N_2 atmosphere and purging solutions with N_2) did not prove to be critical. The disulfide was primarily formed when the wet cake was washed with water. The high concentrations of the NaOH and HCl solutions that were charged created a mother liquor with a concentration of NaCl greater than 5 M. Fortunately the high ionic strength of the reagents and reaction media essentially excluded O_2 from the reaction, and minimized the formation of the disulfide during processing. Figure 12.12 shows that at high concentrations of NaCl the solubility of O_2 in water is very low, even at low temperatures [45].

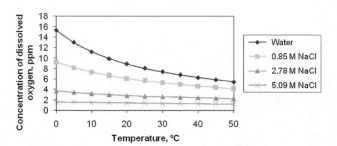

FIGURE 12.12 Solubility of oxygen in water as functions of dissolved NaCl and temperature. *Abstracted with permission from Cramer, S. D. Ind. Eng. Chem. Process Des. Dev. **1984**, 23, 618. Copyright 1984 American Chemical Society.*

In another example of a slow reactive crystallization, controlled addition of H_2SO_4 to a solution of an amine over 6–8 hours generated an amine hydrosulfate, the structure of which was not revealed. By charging H_2SO_4 gradually at first and then more rapidly, crystal growth was favored over nucleation, and larger crystals were formed. Essentially this was a "cubic addition." The filter cake was less compressible, resulting in faster deliquoring and washing, and the product had excellent flow and bulk handling characteristics [33].

Other types of slow reactive crystallizations are CIAT (Figure 12.1) and CIDR processes [46,47]. The latter processes are types of dynamic resolutions, which include processes that are solutions throughout. Figure 12.13 shows a general approach to dynamic resolutions. Through CIDR essentially one enantiomer is produced, as a salt with chiral counterpart; CIAT refers to the crystallization of one diastereomer, i.e., a compound that has at least two chiral centers, or to the crystallization of one of two olefin isomers. CIDR is a type of CIAT, which is less commonly known. CIAT and CIDR processes can be very cost-effective; for instance, the yield of a resolution through a CIDR process is theoretically 100%, instead of the 50% yield theoretically available through a classical resolution.

Probably the most familiar CIDR process is the resolution of an amino-benzodiazepine with dichlorosalicylaldehyde [48] (Figure 12.14). In this case the intermediate imine from condensation with dichlorosalicylaldehyde is racemized by the small excess of the amine, and crystallization of the desired product drives the reaction. Another CIDR process shown in Figure 12.14 involves a Strecker reaction [49]; racemic aminonitrile is generated by addition of cyanide to the imine, probably in its protonated form. The process is driven by the crystallization of the salt of L-tartaric acid with the S-enantiomer of the aminonitrile. (A critical IPC for this process is controlling the

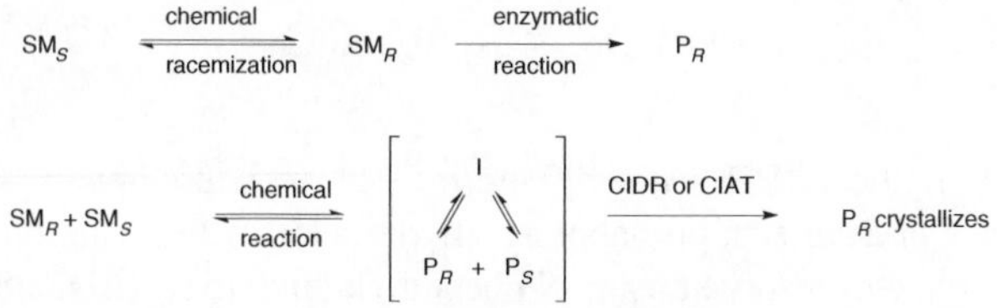

FIGURE 12.13 Dynamic processes. SM_R and SM_S and P_R and P_S refer to the R- and S-enantiomers of the starting material and the product, respectively, and I refers to an intermediate. CIAT processes produce diastereomers. Since the purpose of CIDR operations is to isolate enantiomerically pure product, the diastereomeric nature of the CIDR products is overlooked for these discussions.

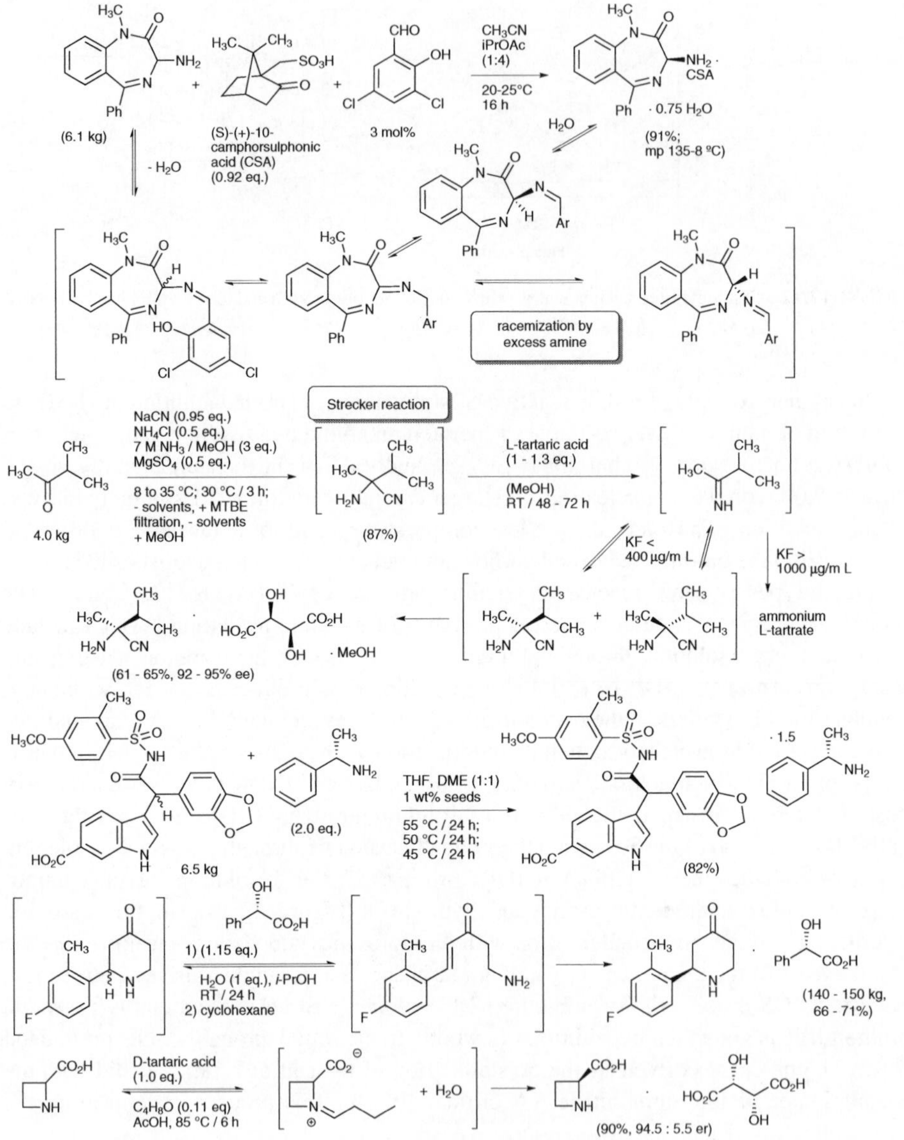

FIGURE 12.14 Some CIDR processes.

amount of H_2O in the CIDR to <400 μg/mL; at moisture levels >1000 μg/mL ammonium tartrate crystallizes, probably as a hydrate [50].) In an unoptimized process an *N*-sulfonyl amide was resolved using *S*-phenethylamine [51]. CIDR through a retro-Michael–Michael process has also been described in detail [8]. Racemic azetidine carboxylic acid was converted to L-azetidine carboxylic acid through treatment with butyraldehyde and D-tartaric acid in AcOH [52]. The reactions shown in Figure 12.14 represent some different functional groups involved in CIDR processes.

In Figure 12.15 are shown some CIAT processes. Condensation of D-tryptophan methyl ester hydrochloride with piperonal (classified by the DEA as a List I controlled substance [53]) by a Pictet–Spengler reaction produced the *cis*-product in high yield [54]. In this reaction the hydrochloride salt of the desired diastereomer, a penultimate intermediate of tadalafil, was less soluble than the hydrochloride salt of the unproductive diastereomer. Olefins can also be preferentially formed by CIAT processes; condensation of the aryl aldehyde with the diarylmethylene component provided the *trans*-olefin

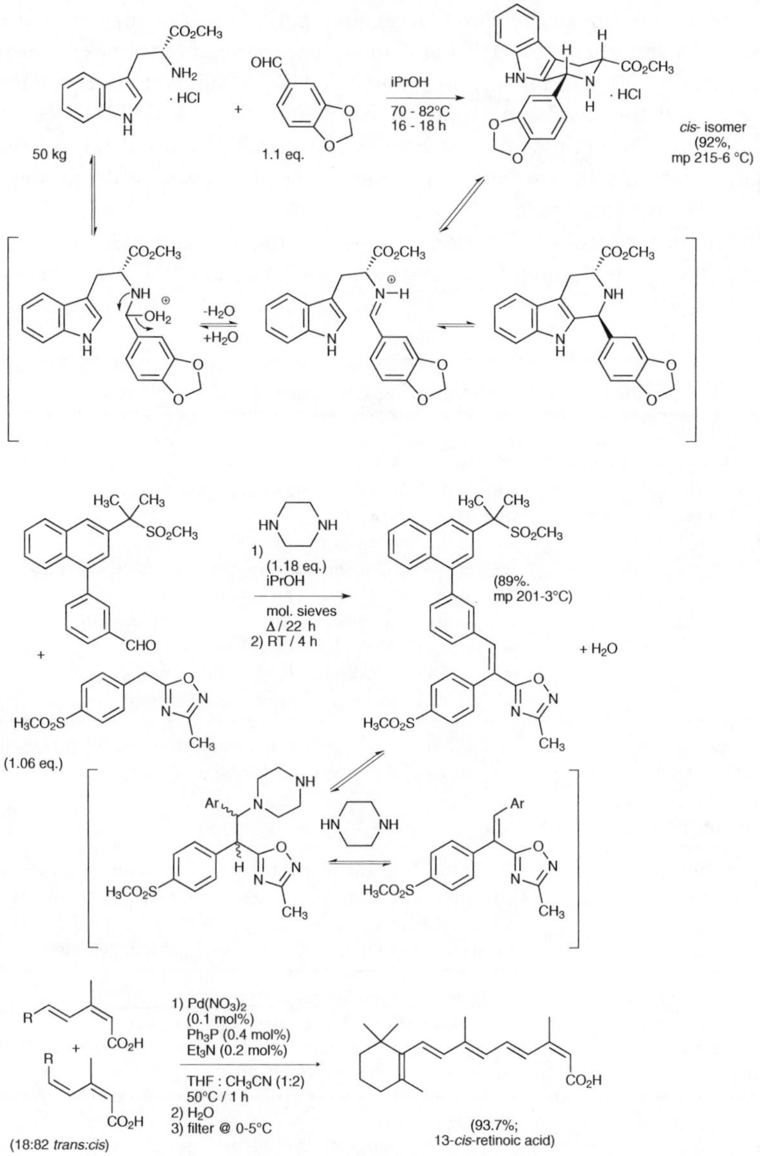

FIGURE 12.15 Some CIAT processes.

in high yield by a Knoevenagel process [55]. Piperidine probably added reversibly in solution to the newly formed double bond molecules, and crystallization of the less-soluble *trans*-isomer led to the accumulation of the product. This process was a slurry-to-slurry transformation. A striking example of a CIAT process through crystallization of an olefin is found in the Hoffmann-LaRoche patent for a palladium-mediated isomerization [56].

As is evident from Figures 12.14 and 12.15, CIDR and CIAT processes may require as ageing periods as long as three days, which ordinarily is considered a long time to tie up equipment. An ageing period of one day is more than compensated by the improved yields and simpler workups, in processes that require only filtration. In 2005, Merck was granted a US Presidential Green Chemistry Challenge Award for developing routes to aprepitant through CIAT processes (see Figure 3.2) [57]. The optimized route using CIAT required 20% of the solvents and raw materials used by the original route, suggesting the power of decreasing waste through CIAT processes [58].

Developing a successful CIDR or CIAT process requires considerations shown in Table 12.1. Often clues leading to development of CIAT and CIDR are noticed by the

TABLE 12.1 Considerations for Developing CIDR and CIAT Processes (1)

Key Factor	Comment
Finding conditions for facile racemization or epimerization	Moderate heating may accelerate racemization or epimerization by ● enolization by acids or bases ● retro-Michael eliminations, then Michael additions ● elimination to form carbonyls or imines and addition ● addition to carbonyls or imines and elimination ● use of catalysts such as aldehydes ● retro-Diels—Alder to Diels—Alder sequences ● additives to form co-crystals or clathrates
Finding a suitable resolving agent that is effective and stable during racemization or epimerization	A stable resolving agent will afford more reliable processing, and may allow recovery and recycle of that component
Developing effective crystallization conditions	Select highly crystalline compounds that melt significantly above the temperature of crystallization. (For 25 examples in the reference below, the average mp was 155 °C, and compounds melted on average 115 °C above the crystallization temperature)
Isolating and handling the product under conditions that do not lead to degradation	Dissolution in a different solvent, e.g., one used for NMR, may convert the product to a mixture of enantiomers or diastereomers. High temperatures during drying have also been shown to racemize products (2)

(1) Anderson, N. G. Org. Process Res. Dev. **2005**, 9, 800.
(2) Bravo, F.; Cimarosti, Z.; Tinazzi, F.; Smith, G. E.; Castoldi, D.; Provera, S.; Westerduin, P. Org. Process Res. Dev. **2010**, 14, 1162.

prepared mind. Conditions to screen for a dynamic resolution have been detailed [59]. CIAT and CIDR processes may be slurry-to-slurry transformations, and due to the extended ageing times small particles may result. Thus CIDR and CIAT processes may require more cautious scale-ups than other processes [46].

VII. HANDLING AND ISOLATING WATER-SOLUBLE PRODUCTS

Crystallization of water-soluble products from water can be problematic and require extensive operations, especially if water-soluble byproducts are present. For instance, to isolate a highly water-soluble zwitterion Merck workers hydrolyzed an ester with 50% NaOH, added 1.0 equivalent of concentrated HCl, and filtered off the NaCl byproduct; the filtrate was azeotroped dry with i-PrOH and the product was crystallized by the addition of hexane [60]. One way to circumvent extensive processing is to generate byproducts that are soluble in organic solvents, and crystallize the product under these conditions. For instance, hydrolyzing the benzoate in Figure 12.16 would generate a benzoate salt or benzoic acid along with the water-soluble d4T. Reacting the benzoate with n-butylamine generated n-butyl benzamide, which was soluble in the mother liquors and washes when d4T was crystallized from i-PrOH [61]. If the amidation could be conducted in i-PrOH or if the benzoyl group could be removed by transesterification with i-PrOH, then perhaps a reactive crystallization process could be developed.

When hydrophilic products are crystallized from water, controlling the pH and selecting the counterions are often keys for a successful process. Unionized acids and bases can be crystallized from water, as described with Figure 12.11, and increasing the ionic strength of an aqueous medium can "salt out" products. For zwitterionic species, such as those shown in Figure 3.10, crystallizing from water at the isoelectric point (pI) can afford convenient and high-yielding processes. Zwitterionic species are generally the least soluble at the pI. Acidic products can be isolated as crystalline salts with amines, such as dicyclohexylamine (DCHA), metal cations, or quaternary ammonium cations. Screening counterions can reveal surprising differences in solubilities and crystallinity. The sodium salt precursors to aztreonam and carumonam can be isolated in good yield from water [62] or water–MeOH mixtures [63], respectively (Figure 12.17). The potassium salt of the aztreonam precursor is highly water-soluble, and the tetrabutylammonium salt can be extracted into dichloromethane [64] and crystallized from organic solvents. Alcoholic solvents such as MeOH, EtOH or i-PrOH, sometimes with some water present, can be useful to crystallize water-soluble products.

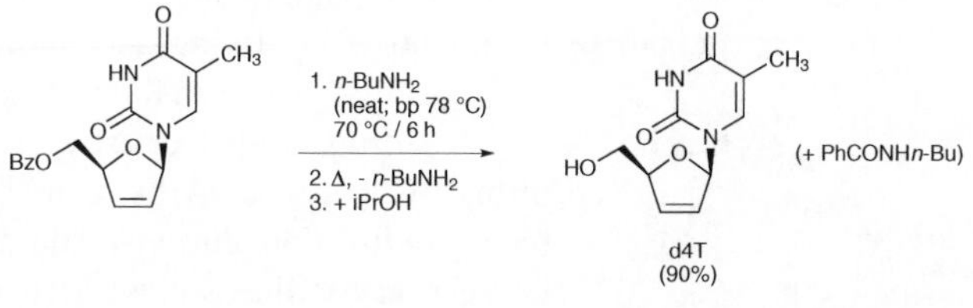

FIGURE 12.16 Crystallization of d4T from an anhydrous reaction medium.

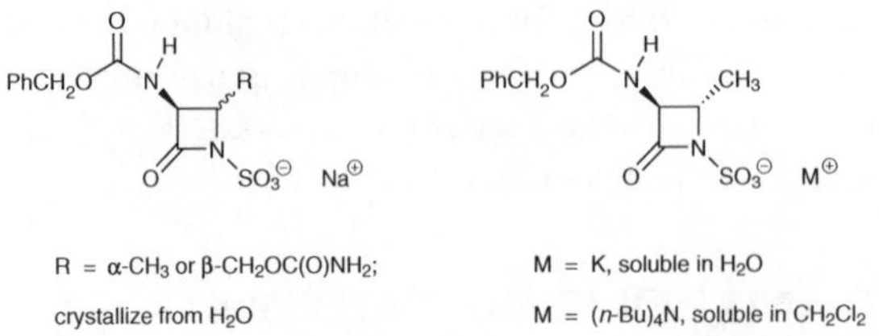

R = α-CH₃ or β-CH₂OC(O)NH₂;
crystallize from H₂O

M = K, soluble in H₂O
M = (n-Bu)₄N, soluble in CH₂Cl₂

FIGURE 12.17 Solubilities of sulfonic acid salts.

Just as solubilities of a sodium carboxylate and a potassium carboxylate can be very different, salts present during crystallization can have noticeable effects. For instance, acidification of a sodium carboxylate from hydrolysis of an ester with NaOH precipitated a gummy solid, while acidification of the potassium carboxylate consistently crystallized a carboxylic acid with good handling characteristics [65]. Oiling and precipitation can occur when compounds with very low solubility are generated, and the insoluble nature of such compounds may make them inaccessible to further changes in pH.

VIII. POLYMORPHS

Polymorphs are different crystal forms of the same compound, differing due to the arrangements of molecules within the unit cells of the crystalline lattice of each crystal form. Overall the contents of the crystals are the same, as can be judged by HPLC, GC (for any solvated solvents), elemental analysis, KF titration, and other assays. Different polymorphs may display different physical properties, such as solubility, melting points, heats of fusion upon melting, density, hardness, crystal shape, and stability to moisture and light [66]. Crystals containing a solvated solvent are sometimes referred to as polymorphs, but are more appropriately termed pseudopolymorphs. The presence of solvents can be inferred by thermogravimetric analysis (TGA) as small weight losses near the boiling points of the solvent, and can be confirmed by GC, NMR, and other analyses. (Solvents may be lost from solids at temperatures above the boiling points: the strength of hydrogen bonds between a solvent and an API or excipient may be different than that of hydrogen bonding of one solvent molecule to another [67].) Solvents that are stoichiometric with the API will be present at concentrations higher than those allowed for residual solvents, unless the solvent in question is H₂O (as a hydrate). Hydrates are crystals containing water of solvation. If an anhydrous form is present, it may be referred to as an anhydrate. Other types of solids are desolvated solvates, brittle solids from which residual solvents in crystal lattice were removed by drying, and amorphous solids, which display no crystalline content [68–71].

The FDA definition of polymorphs includes not only crystalline anhydrates, hydrates, and solvates, but also amorphous forms [72].

Most drug substances are crystalline solids, primarily because APIs can be upgraded by crystallization during manufacturing, and because crystalline solids often display bulk stability superior to that of amorphous solids.

Good bulk stability of the drug substance will usually confer good shelf life of the drug product (see Chapter 13). In a drug product, a change in the morphic state of the API can alter the ease of formulation, the rate of tablet dissolution, the solubility and bioavailability of API, and the shelf life. Changes in the appearance of the drug product, such as tablets that disintegrate upon storage or crystals growing through walls of gel-caps, can affect patient compliance with the medication. In 1969 Haleblian and McCrone discussed the impact of polymorphs on various formulations, and recommended that pharmaceutical companies thoroughly investigate polymorphism in APIs. They also recommended that the most stable polymorph be developed [66].

Two examples serve to illustrate the importance of understanding polymorphism. Crystallization of abecarnil (Figure 12.18) became a problem when Form B, the least-stable of three known polymorphs, could no longer be crystallized on scale. Process improvements had generated API batches of higher potency, and key impurities were no longer present to help template the formation of Form B. The goal became to make Form A without contamination by Form C, the most stable polymorph. One scale-up solution was to generate Form B by spray drying [73], which affords another example of Ostwald's rule of crystallization by stages. Subsequent laboratory developments found the optimal processing was to use clean A as seeds, activate the seeds by slurrying in solvent before charging to solution, charge 3 wt% seeds at proper concentration and temperature, and cool at 0.5–1.0 °C/minute (rapid cooling on scale) to prevent crystallization of Form C. Critical to the success of these operations on scale was rapidly removing the mother liquor by loading a centrifuge to a wet cake thickness of only 1 cm and centrifuging; these steps prevented rapid solvent-mediated transformation to Form C [74]. The detail required for these procedures, which are certainly creative and feasible, point out the desirability of selecting the most stable polymorph as the API.

Probably the most renowned (and painful) case study for an unexpected polymorph is that of the AIDS drug ritonavir (Figure 12.19). In late 1992 ritonavir was discovered, and after rapid development Abbott filed an NDA in late 1995 and began commercial manufacturing in early 1996. In March 1996 the FDA approved sales of the drug. About two years later the drug product failed dissolution tests, and another formulation was necessary. Sales of ritonavir were interrupted, prompting high-pressure investigations to control the crystallization of the API. Examination by polarized light microscopy revealed a new crystal form, and this polymorph was thoroughly characterized as Form II (Table 12.2). The density of Form II was slightly less than the density of Form I, which was unusual for the more stable polymorph. XRPD and near-IR could detect 3% and 1%, respectively, of Form II in bulk API. Ultimately a seed detection test was demonstrated to be able to detect 1 ppm of Form II in the bulk material. (For the latter test the solvents of formulation, which were ~92% EtOH and some proprietary agents, including

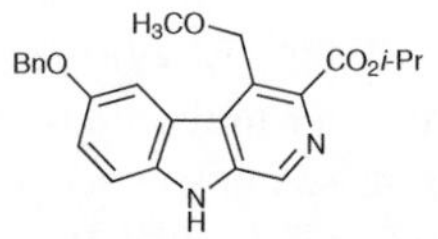

FIGURE 12.18 Abecarnil.

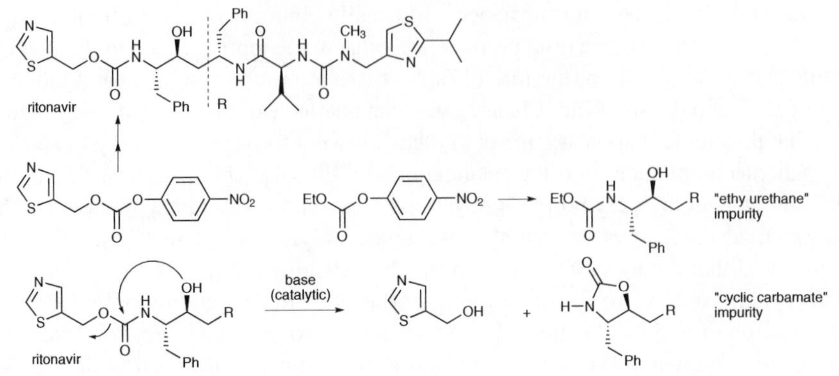

FIGURE 12.19 Impurities found in batches of ritonavir.

TABLE 12.2 Characteristics of Ritonavir Polymorphs Forms I and II

Test	Form I	Form II
Light microscopy	Rods	Needles
Melting point	~122 °C	~125 °C
Solubility (5 °C)	Standard	Much less
Bulk density	1.28 g/cm^3	1.25 g/cm^3
XRPD	Standard	Very different
Near-IR	Standard	Very different
Solid ^{13}C NMR	Standard	Very different

saturated fatty acids, were seeded with bulk ritonavir at a concentration about 110% of the saturation of Form II, but below the saturation level of Form I. Hence any crystals that formed were due to the presence of Form II.) No control of the crystal form had been required for the initial formulation, in accordance with the ICH guidelines current at the time of filing [75].

Once Form II had appeared the crystallization of Form I could be controlled in laboratories, but not in manufacturing equipment. (The inability to crystallize Form I after Form II had appeared which may have been due to crystals of Form II residing in process equipment in locations that were difficult to clean. Routine manufacturing of a metastable polymorph of another API in another facility was carried out through meticulous cleaning between batches [76].) The formation of Form II was attributed to the presence of the ethyl urethane impurity in the API (Figure 12.19); the ethyl urethane was derived from ethyl p-nitrophenylcarbonate, an impurity in the thiazolylmethyl p-nitrophenylcarbonate intermediate [77]. In another paper, it was noted that seeding solutions of ritonavir with the cyclic carbamate degradent shown in Figure 12.19 led to the crystallization of Form II. Some of the cyclic carbamate impurity was invariably present in bulk ritonavir, and was found to be less soluble than ritonavir. The hydrogen bonding through the cyclic carbamate (the "U"-orientation of H–N–C–O) was similar to the orientation of the cisoid carbamate portion of Form II, and may have prompted crystallization of Form II [75]. Curiously, one processing change associated with batches containing Form II was the application of an additional basic wash of the ritonavir crystals [77].

Ostwald's rule (or Ostwald's rule of stages or Ostwald's progression) states that the first polymorph to crystallize is the one closest to the orientation in solution, and hence has the least amount of energy to give up in crystallizing. Hence ritonavir Form I would be expected to crystallize first [78,27]. Said another way, the more stable of two polymorphs will probably crystallize second.

One Abbott researcher told me informally in 2000 that Abbott felt they had thoroughly researched polymorphs associated with ritonavir, and thought that there was a 10,000–1 chance against their being "burned" by the appearance of another polymorph. Abilities to search for polymorphs have greatly expanded since 1995. Using high-throughput screening (HTS), Transform Pharmaceutical identified Forms I, II, and three new forms of ritonavir in 2000 crystallization experiments in four weeks [79]. Abbott's experience with ritonavir underscores the importance of searching for and controlling polymorphs of APIs.

Different morphic forms are commonly encountered [80], usually for final products since the economic impact of different forms is more pressing. The more a compound is examined and processed, the more likely polymorphs, pseudopolymorphs and hydrates will be discovered or created; 56 crystalline forms of atorvastatin calcium have been reported in patents [81]. More than 100 solvates of sulfathiazole have been characterized [82]. Even simple compounds may form several polymorphs; for instance, six polymorphs were found for triacetone triperoxide (TATP) [83]. Data from routine screening by SSCI (Table 12.3) showed that over 80% of solids exhibited multiple crystalline forms, almost half of the solids displayed polymorphism, and almost half of the salts screened were hydrates [84].

TABLE 12.3 Prevalence of Different Solid Forms

	All Compounds Screened by SSCI (1)	Salts Screened	Non-Salts Screened
Total	245 (2)	95	128
Multiple crystalline forms (3)	82%	81%	82%
Polymorphs	48%	39%	55%
Solvates	32%	36%	28%
Hydrates	38%	48%	30%
Amorphous	48%	54%	43%

(1) Stahly, G. P. Cryst. Growth Des. **2007,** *7, 1007.*
(2) Structures of 22 compounds were unknown to SSCI.
(3) Including polymorphs, solvates, and hydrates.

HTS approaches for potential polymorphs and salts have been described in detail [85–90]. HTS for both crystallization and crystal structure analysis has been described for orcinol (3,5-dihydroxytoluene) and salts [91]. Crystallization experiments are usually carried out by reslurrying, cooling, adding antisolvent, and evaporation. Grinding solids may produce polymorphs [92] and was found to be effective in generating co-crystals of highly insoluble compounds [93]. A wide range of solvents may be employed to attempt to generate polymorphs, even solvents that are unlikely to be used on scale, such as Et_2O, hexafluoroisopropanol, or pyridine [94]. Tests for morphic forms can include X-ray powder diffraction, DSC, mp, microscopy, IR (primarily diffuse reflectance), Raman spectroscopy, solid-state NMR, solution NMR and GC (for solvates), and powder density (for desolvated solvates). Attempts are made to prepare different morphic forms by varying solvents (especially considering solvent polarity, temperature, concentration, pH, and H_2O contents).

> To develop crystals of a compound that had not crystallized previously, sonicate an oil dispersion of this material. Alternatively, a solution of this material may be seeded with a crystalline compound that has a high amount of structural similarity to it.

Screening for polymorphs (and solvates and hydrates) generally occurs after screening to select the desired salt. After different solid forms have been identified studies are undertaken to identify the most stable form. Researchers from Pfizer and the University of Minnesota found that about 25% (of 43) of early discovery compounds screened were transformed to more stable polymorphs. Their approach to identify the stable polymorphs was to reslurry solids at room temperature for up to two weeks in solvents (or solvent mixtures) that partially dissolved the solids. Transformation of ritonavir Form I into Form II took place fairly readily in solvents that produced solubilities of about 5 mg/mL (about 7 mM), with transformation being slower in solvents with lower solubility. In screening a compound under development at Pfizer they confirmed that polymorph transformation was faster in solvents that conferred solubilities of at least 3.4 mg/mL (about 8 mM). Using this approach they identified conditions for converting the undisclosed compound to the desired polymorph over 36 hours in high yield [95].

> If desired conversion occurs overnight, try to develop a practicable conversion with that solvent. For example, if 5 mg of solids/mL dissolves at RT, then heat 100 mg of solid/mL of solvent (10% w/v) to dissolve and cool. Crystallization should return 90–95 wt%, depending on initial purity.

Close attention to detail may be required to crystallize the desired polymorph. Solvent composition, concentration of the solute, rate of cooling, and type of agitation may be important. For instance, crystallization of one polymorph of 2,6-dihydroxybenzoic acid was slow when solutions were mixed using a turbine agitator, but rapid when stirred magnetically: the grinding effect of the magnetic stirbar produced secondary nucleation and accelerated crystallization [96]. For crystal forms that are very similar, very slow cooling or other application of crystallization pressure may be necessary to crystallize the desired form. For example, storage of some samples of treprostinil diethanolamine (Figure 12.20) produced a second form, as evidenced by

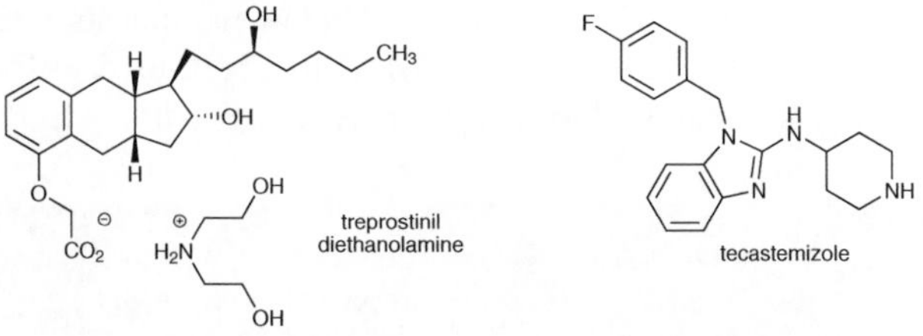

FIGURE 12.20 APIs crystallized as desired polymorphs through slow cooling.

DSC and XRPD. Researchers identified polymorph Form A (mp 103 °C, heat of fusion 109.0 J/g) and Form B (mp 107 °C, heat of fusion 109.2 J/g), indicating that Form B was only slightly more stable than Form A. There were no noticeable differences in solubility between these two forms, and extensive studies were undertaken to find conditions to crystallize the more stable form. For example, crystallization from EtOH–acetone at ratios of 1:5, 1:6, or 1:8 produced a mixture of polymorphs, while crystallization from EtOH–acetone (1:7) *with seeding* produced Form B. Form B was optimally crystallized from EtOH–acetone (1:7) through seeding with 1% Form B at 40 °C, holding at that temperature for 1 hour to develop a good seed bed, and cooling to ambient temperature at the rate of 1 °C/h. This protocol produced Form B, but as rod-like crystals that were deemed unsuitable for formulation. By reslurrying in heptane the product was converted to agglomerates of Form B with few fines, more suitable for formulation [97]. In a second example using the help of SSCI, Sepracor researchers identified conditions to isolate the desired of two polymorphs from tecastemizole. Form A was the less-stable form, and crystallized before Form B. By slurrying at temperatures above 70 °C the solvent-mediated conversion of Form A to Form B was rapid, and the slurry was cooled slowly. Below 70 °C cooling was progressively more rapid. The rate of cooling was similar to cubic cooling, and reliably produced Form B. No seed crystals were necessary. Unreliable production of Form B using linear cooling had prompted these investigations [98].

The polymorph that crystallizes above room temperature may not be the most stable polymorph at room temperature. For a monotropic pair of polymorphs, the more stable polymorph has the higher-melting point, a higher heat of fusion (as measured by DSC), lower solubility, and higher bulk density, and is more stable at all temperatures, as described by Novartis for a drug candidate [99] and by Matrix Laboratories for temazepam [100]. With an enantiotropic pair of polymorphs, the relative stabilities change above a transition temperature; by cooling a solution of irbesartan below the transition temperature irbesartan Form I was crystallized reproducibly from a solvent at whichForm I was less soluble than Form II [101]. One experiment to determine if a pair is enantiotropic is to suspend the lower-melting polymorph in a solvent that partially dissolves the solids, and add small amount of the higher-melting polymorph. After slurrying for 4–7 days the solids are isolated by filtration and characterized; if the lower-melting form predominates, the pair is enantiotropic [23]. (The latter manuscript also clearly described a process to crystallize the desired polymorph of an enantiotropic pair.) Reslurry experiments at elevated temperatures may serve to identify the transition temperature. If the product is crystallized below the transition temperature, the isolated product should be the polymorph that is more stable at room temperature. For instance, Pfizer workers crystallized a drug candidate at 45 °C, about 5 °C below the transition temperature for a pair of enantiomorphic polymorphs [102]. FBRM has been used to follow the polymorph transformations for acitretin [26] and for pravastatin sodium [103].

> Search for the most stable polymorph early in development of a drug candidate. Check for the presence of polymorphs using all solvents used to prepare and formulate an API. Screen in the presence and absence of water. Impure samples of the API may contain impurities that help to template the crystallization of another polymorph, so at some point it may be wise to screen for polymorphs using both acceptable and highly purified material. The form that is most stable at room temperature will probably provide the best bulk stability for the API (and the drug product). Identifying the desired form early in development may avoid the need to change formulation or initiate a new tox study. New forms are patentable intellectual property, and can extend the patent life of a drug.

Disappearing polymorphs have been at best curiosities, and at other times have interrupted the supply of drug product, as described for ritonavir. Impurities have been found to accelerate or inhibit the transformation of one polymorph to another [104]. The influence of impurities or additives in changing crystal forms may be difficult to predict, as was found for crystallization of vanillin [105]. Novartis workers found for an API candidate that they were unable to cleanly crystallize Forms A and C once Form D had been isolated from a pilot plant batch, and attributed the formation of Form D to either the presence of a new impurity or the absence of impurities generated from earlier processes. They concluded that polymorph screening should also take place on the material to be generated using the procedures designed for scale-up [99].

IX. PROCESSING ENANTIOMERIC CRYSTALS

Characteristics of enantiomeric crystals are summarized in Table 12.4. True racemates are crystals in which each enantiomer occurs in the unit cell of the crystalline lattice, and are the most prevalent. The melting point of a true racemate may be above, below, or at the same temperature as that of either enantiomer. Only one enantiomer crystallizes in the unit cell of conglomerates (sometimes called a racemic mixture), and the shape of the crystals may be visually different. (The classic example of conglomerates is told with Pasteur's physically separating crystals of sodium ammonium tartrate.) The melting point of a conglomerate mixture is lower than that of either enantiomer. Conglomerates are found no more than 10% of the time. The enantiomers of pseudoracemates have little affinity toward each other, and are sometimes called racemic solid solutions. Phase diagrams for true racemates, conglomerates, and pseudoracemates may be found in Testa's excellent book [106].

The solvent most often used for classical resolutions is 95% EtOH [107,108], probably because so many salts crystallize as hydrates. Hence solvents containing water should be considered when screening for resolution conditions. This guideline is not always true; for instance, Merck researchers found that water was deleterious to a resolution, while the desired salt crystallized in the presence of i-PrOH [60].

Controlling cooling to control the rate of crystal growth can be important to effect good resolution. For instance, rapid cooling to the isolation temperature and immediate

TABLE 12.4 Characteristics of Crystals of Enantiomeric Compounds

Characteristic	Conglomerate	True Racemate	Pseudoracemate
Contents of crystal unit cell for racemic mixture	Only 1 enantiomer/ unit cell of a crystal; enantiomers found in different crystals	Both enantiomers in unit cell	Both enantiomers in unit cell
Occurrence	≤10%	Most common	Very rare
Characteristics of racemic crystals relative to a pure enantiomer	Decreased mp Increased solubility	Increased mp Decreased solubility	mp similar to either enantiomer, similar solubility
Approaches for resolution by crystallizing (1)	Entrainment: seed with one enantiomer, less prevalent enantiomer stays dissolved	Remove undesired enantiomer first by filtering off less-soluble racemate, then crystallize product; or make diastereomers by salt formation or covalent modification	NA
Effect of racemic switch	Increased mp Decreased solubility	Decreased mp Increased solubility	Very little change

(1) Resolution can also be accomplished through kinetic resolution and second-order asymmetric transformations.

filtration raised the er to 99:1 as compared to 85:15 after slow cooling [109]. Merck researchers detailed how a desired diastereomeric salt crystallized faster than the undesired diastereomeric salt [60].

Conglomerates can be resolved by crystallization by entrainment, while true racemates must be derivatized with chiral agents to effect resolution. With entrainment a solution is seeded with one enantiomer, and crystallization of the enriched solution produces a mother liquor enriched with the other enantiomer. Seeding that mother liquor with the other enantiomer prompts the crystallization of the second enantiomer, with that mother liquor enriched in the first enantiomer. This process continues until the level of impurities built up in the mother liquors impedes crystallization. Monosodium glutamate (MSG) and BINOL dimethyl ether have been resolved by entrainment [110]. In the crystallization of L-asparagine monohydrate by entrainment from aqueous solutions of racemic asparagine, the rate cooling was found to have the biggest effect on crystallization, while particle size, stirring rate, and the mass of seed crystals were found to have negligible to only small effects [111]. Crystallization of enantiomeric salts of imeglimin by entrainment has also been studied [112].

The purity of a chiral compound may be upgraded by crystallization, whether a racemic mixture crystallizes as a true racemate or as a conglomerate. For instance, the racemic free base of sitagliptin was found to be a conglomerate; by crystallizing the free base derived from chiral hydrogenation the undesired enantiomer was purged because it was solubilized in the mother liquor as the conglomerate. The chiral purity of the API could not be upgraded by crystallizing the H_3PO_4 salt [113]. (The biocatalytic route to sitagliptin was stereospecific, eliminating the operations to crystallize the free base and raising the yield [114].) If the undesired enantiomer crystallizes as a true racemate, crystals of the latter should form before the desired chiral product crystallizes. For example, a racemic carboxylic acid (also derived from chiral hydrogenation) crystallized from an i-PrOAc concentrate, thus removing the undesired enantiomer and producing higher quality acid (as the DCHA salt) [115].

X. GUIDELINES FOR A PRACTICAL CRYSTALLIZATION

Some practical guidelines for assessing a good crystallization are shown in Table 12.5. Thoroughly removing the mother liquor may be necessary to prevent co-crystallization

TABLE 12.5 Laboratory Assessment of Crystals Suitable for Scale-Up

Characteristic	Comment
Slurry can be readily suspended by increased agitation	Dense solids may be difficult to remove completely from the bottom of crystallizer
No oil seen in slurry (microscopic examination)	Excessive crystallization pressure may form oils. Oils may plug filters and slow operations
Slurry can be readily poured into filter	Transferability
Filter cake does not crack or pull away from walls	Breaks in cake lead to poor washing and incomplete removal of impurities. Consider reslurrying (1)
Little compression of solids upon filtration	Compression will plug filters in thick cakes. Use filters with greater surface area
Mother liquor or wash readily suspends solids left in crystallizer	Leave material in kettle for next unit operation in a campaign
Good filtration rate: 500 mL slurry dewatered by filtration through 4 cm high cake in Buchner funnel in ≤5 min	Best with filter cloth used for scale-up. If slow, use filter press or centrifuge with small cake thickness
Washing removes mother liquor and impurities	Adjust solvent content/temperature. Applying a solvent to chase the mother liquor may be helpful, but can precipitate impurities. Gradually changing the solvent composition of the wash solvent may be necessary (1)
Washing does not dissolve cake	Adjust solvent content/temperature
Dried product not hygroscopic	Hygroscopicity will cause formulation difficulties

(1) Bravo, F.; Cimarosti, Z.; Tinazzi, F.; Smith, G. E.; Castoldi, D.; Provera, S.; Westerduin, P. Org. Process Res. Dev. **2010**, 14, 1162.

of an undesired polymorph [74] or an undesired diastereomeric salt [8]. For a CIDR process (Figure 12.14) GSK workers considered hydrocarbon solvents to efficiently displace the mother liquor, and chose cyclohexane because its viscosity was most similar to that of the *i*-PrOH-containing mother liquor. To avoid precipitating compounds dissolved in the mother liquor by adding cyclohexane directly to the wet cake, they applied four washes in a gradient fashion: *i*-PrOH, 1:1 *i*-PrOH:cyclohexane, 1:3 *i*-PrOH: cyclohexane, and finally cyclohexane [8]. Long drying times on scale can create bottlenecks in routine operations, and thoroughly washing a wet cake on the filter may be key to shorten drying times. Considerations from chemical engineers are especially valuable in this regard.

XI. PERSPECTIVE

Many factors are considered in developing rugged crystallizations. Physical factors commonly considered are cooling, agitation, and seeding. Sonochemistry [116,117] or polarized light [118] may also be applied to initiate crystallization. In the pharmaceutical industry usually crystallizations are carried out in batch processing, but continuous operations may be viable. Critical considerations, which may not become obvious until trouble arises, are the presence or absence of impurities, as discussed in Chapter 13.

The crystals encountered by organic process scientists are all too often needles. Needle crystals can pose processing difficulties because they can fragment with excessive agitation, with smaller crystals potentially blinding the filter elements. Needles also can form relatively impenetrable mats in a filter, making washing difficult. "Crystal engineering" has had some success in preparing crystals that are more amenable to processing. In some cases small amounts of impurities have been added, usually to inhibit the growth of needle-like crystals along the long axis; a drawback of this approach is that the additional chemical components must be considered as potential impurities in an API [119]. For instance, crystallization of the *t*-butyl ester acetonide (an atorvastatin precursor) from *i*-PrOH produced larger crystals that filtered faster when the crystallization was carried out in the presence of 0.5 wt% of cetyltrimethylammonium bromide (CTAB; Figure 12.21) [120,121]. (Processing an API or late-stage intermediate in the presence of a chemical such as CTAB would require investigations into the residual level of such a compound in the API. Although CTAB has been used as a preservative, it is toxic systemically [122].) Crystallizing a cephalosporin in the presence of 0.2% of hydroxypropylcellulose type-M generated the desired blade-like habit [123]; hydroxypropylcellulose is GRAS and is used in formulating APIs into drug

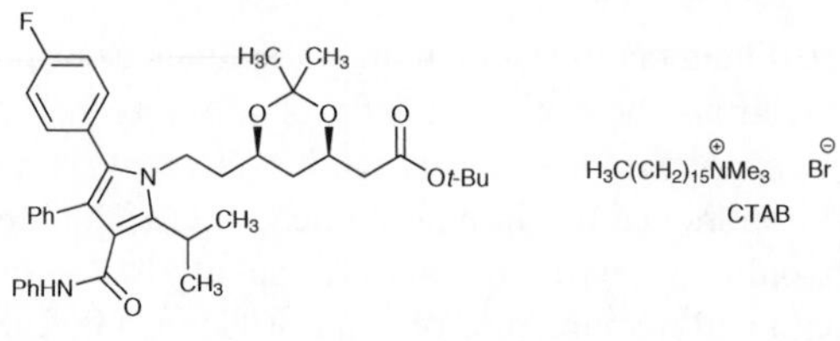

FIGURE 12.21 An atorvastatin precursor crystallized in the presence of CTAB.

products. Sonication has been applied to convert irbesartan needles into "bricks" that filter and handle better on scale [116,124]; such treatment is best carried out by circulating the crystal slurry continuously past or through [125] the source of the sonic energy, and not every facility is familiar with continuous operations. (Benefits and applications of continuous processing are discussed in Chapter 14.)

Oiling is another problem frequently encountered, indicating a narrow range of supersaturation. Controlling the physical aspects of the processing may be the preferred approach to develop a reliable crystallization on scale. Merck engineers were faced with developing a reliable crystallization of an API that tended to oil or produce small crystals that filtered very slowly. Slow addition of the antisolvent to a solution of the API in the presence of seeds, followed by ageing, produced crystals with a broad particle size distribution (PSD). By simultaneously adding a solution of API and a solution of antisolvent to a slurry containing up to 10 wt% seeds the filtration time was markedly improved. Many variables were systematically examined for the crystallization, and the primary factors for rapidly filtering cakes were low power input during the agitation, and charging milled seeds. A higher solvent/antisolvent ratio, higher temperature for the additions, and longer charging times also produced crystals that filtered more easily. Under the optimized conditions in manufacturing the API was centrifuged in about a quarter of the best time previously available [126].

Controlling particle size and PSD is essential for producing effective drug products, as discussed in Chapter 13. Dry milling generates dust and coats equipment with the NCE; the physical losses of up to 5% of an NCE can be critical in the early stage of development. Wet milling is safer for reducing the particle size of highly potent compounds and does not have physical losses [127], but contact with solvents can produce another polymorph, solvate, or hydrate if the optimal final form has not been selected. Micronization may be necessary to formulate inhalable drugs and drug products requiring even distribution of the API, such as suspensions, poorly soluble compounds, and highly potent compounds [128]. Small particles can be produced by wet milling with high shear [129] or crystallizing using impinging jet technology [10]. Particle size can also be controlled by sonication under continuous processing conditions [116,130], dispersion using supercritical CO_2 (sc CO_2), Solution Enhanced Dispersion of Solids (SEDS) [131], or by spray drying. Techniques that stress crystals mechanically and chemically may produce metastable polymorphs. To develop a process to generate small crystals, less time is needed for micronization than for developing particle engineering, so micronization may be employed early in the development of an NCE. Micronization increases the number of unit operations, so particle engineering may ultimately be preferred to manufacture an API. [128]. Process modeling tools can be used to scale-up crystallizations [132]. Controlling particle size is the subject of a recent book [13].

Engineers more often than chemists appreciate the nuances of physical parameters; after successful investigations the optimal conditions may seem deceptively obvious. For instance, engineers appreciate how excessive power input on large scale can fragment crystals, and the importance of eliminating heat gradients to avoid fouling [133].

Polymorphism undoubtedly will continue to be an issue for developing drug substances and for continued manufacture of drug products. GSK researchers recently described developing an API as a mixture of polymorphs; they determined that the presence of up to 27% of the minor polymorph was not a critical quality attribute (CQA) for the API [134]. Solid-state transformations should not be overlooked. Hydrates may be overdried; rehydration of a dried phosphate salt was carried out by passing moist nitrogen through the dryer [135]. Water activity calculations may be helpful in predicting suitable solvent mixtures to crystallize a desired hydrate (Chapter 6). The methanol solvate of an API could be dried and converted to the desired hydrate by exposure to moist air; attempts at crystallizing the hydrate directly failed [94]. Polymorph conversion in the solid state has been demonstrated by exposure to CO_2 [136]. (Perhaps API final form conversions will be carried out in facilities that prepare decaffeinated coffee using supercritical CO_2.)

Attention to details is heightened as operations converge closer to making a marketable product. Being able to understand and control crystallizations is thus extremely important, as the quality of the crystals can determine whether a batch can be used to create drug product. Poorly behaved crystallizations can tie up equipment, lead to process bottlenecks, and lower productivity. Analyzing reworked batches also increases the efforts and costs of quality control departments. Thus crystallizations can have a significant financial impact.

To develop a crystallization process for routine manufacturing it is necessary to consider various "what if" scenarios; for example, what if the solution of the product had to be held at a warm temperature for an extended time prior to polish filtration? (This could occur if processing were delayed due to various equipment difficulties. Such failure modes and effects analyses (FMEA) considerations are discussed in Chapter 16.) Would the product decompose in solution? If volatilized solvents were not returned to the crystallizer would the product crystallize before the solution was clarified by polish filtration? Will

> Develop a rework procedure before it is needed. Often a low-quality batch can be subjected to the original crystallization conditions, perhaps with more care to remove insoluble impurities or to coax the crystallization along.

byproducts co-crystallize if the slurry is held while the first part of the batch is being centrifuged? If crystallization occurs prematurely can the suspension be warmed to redissolve the product? Such considerations are useful to avoid processing difficulties, and to minimize the number of batches that need to be reworked.

REFERENCES

1. Mullin, R. *Chem. Eng. News* **2010,** *88*(29), 20.
2. Nelson, T. D. The product that crystallized was an intermediate in one route explored to aprepitant. In *Strategies and Tactics in Organic Synthesis*, Vol. 6; Harmata, M., Ed. Elsevier: Amsterdam, 2005; p 321.
3. Pye, P. J.; Rossen, K.; Weissman, S. A.; Malialkal, A.; Reamer, R. A.; Ball, R.; Tsou, N. N.; Volante, R. P.; Reider, P. J. *Chem.–Eur. J.* **2002,** *8*, 1372.
4. Welch, C. J.; Henderson, D. W.; Tschaen, D. W.; Miller, R. A. *Org. Process Res. Dev.* **2009,** *13*, 621. And references therein.
5. http://www.pfaudler.com/wiped_film_evaporators.php?gclid=CL69yarHqakCFQELbAodyiHUKQ
6. Kopach, M. E.; Murray, M. M.; Braden, T. M.; Kobierski, M. E.; Williams, O. L. *Org. Process Res. Dev.* **2009,** *13*, 152.
7. Jarvis, L. M. *Chem. Eng. News* **2006,** *84*(29), 23; Thayer, A. M. *Chem. Eng. News* **2011,** *89*(22), 21.

8. Bravo, F.; Cimarosti, Z.; Tinazzi, F.; Smith, G. E.; Castoldi, D.; Provera, S.; Westerduin, P. *Org. Process Res. Dev.* **2010,** *14,* 1162. See footnote 15.

9. Etter, M. C.; Jahn, D. A.; Donahue, B. S.; Johnson, R. B.; Ojala, C. *J. Cryst. Growth* **1986,** *76,* 64.

10. Tung, H.-H.; Paul, E. L.; Midler, M.; McCauley, J. A. *Crystallization of Organic Compounds, an Industrial Perspective.* Wiley: New York; 2009.

11. Wieckhusen, D. In *Process Chemistry in the Pharmaceutical Industry,* Gadamasetti; Braish, Eds.; 2nd ed.; CRC Press: Boca Raton, Florida, 2008; pp 295–312.

12. Davey, R.; Garside, J. *From Molecules to Crystallizers; an Introduction to Crystallization.* Oxford University Press: Oxford, 2000.

13. Leubner, I. *Precision Crystallization: Theory and Practice of Controlling Particle Size*; CRC Press: Boca Raton, Florida, 2009.

14. Mullin, J. W. *Crystallization,* 4th ed.; Butterworth-Heinemann, 2001.

15. Braatz, R. D. *Annu. Rev. Control* **2002,** *26,* 87.

16. Anderson, N. G.; Ciaramella, B. M.; Feldman, A. F.; Lust, D. A.; Moniot, J. L.; Moran, L.; Polomski, R. E.; Wang, S. S. Y. *Org. Process Res. Dev.* **1997,** *1,* 211.

17. Belecki, K.; Berliner, M.; Bibart, R. T.; Meltz, C.; Ng, K.; Phillips, J.; Ripin, D. H. B.; Vetelino, M. *Org. Process Res. Dev.* **2007,** *11,* 754.

18. Chen, C.-K.; Singh, A. K. *Org. Process Res. Dev.* **2001,** *5,* 508.

19. Tan, H.; Reed, M.; Gahm, K. H.; King, T.; Seran, M. D.; Bostick, T.; Luu, V.; Semin, D.; Cheetham, J.; Larsen, R.; Martinelli, M.; Reider, P. *Org. Process Res. Dev.* **2008,** *12,* 58.

20. Prashad, M.; Har, D.; Chen, L.; Kim, H.-Y.; Repic, O.; Blacklock, T. J. *J. Org. Chem.* **2002,** *67,* 6612.

21. Hu, Y.; Liang, J. K.; Myerson, A. S.; Taylor, L. S. *Ind. Eng. Chem. Res.* **2005,** *44,* 1233.

22. Anderson, N. G. Personal experience.

23. Müller, M.; Meier, U.; Wieckhusen, D.; Beck, R.; Pfeffer-Hennig, S.; Schneeberger, R. *Cryst. Growth Des.* **2006,** *6,* 946.

24. He, G.; Tjahjono, M.; Chow, P. S.; Tan, R. B. H.; Garland, M. *Org. Process Res. Dev.* **2010,** *14,* 1469.

25. Barrett, P.; Smith, B.; Worlitschek, J.; Bracken, V.; O'Sullivan, B.; O'Grady, D. *Org. Process Res. Dev.* **2005,** *9,* 348.

26. Sathe, D.; Sawant, K.; Mondkar, H.; Naik, T.; Deshpande, M. *Org. Process Res. Dev.* **2010,** *14,* 1373.

27. Beckmann, W.; Otto, W.; Budde, U. *Org. Process Res. Dev.* **2001,** *5,* 387.

28. Cook, D. M.; Jones, R. H.; Kabir, H.; Lythgoe, D. J.; McFarlane, I. M.; Pemberton, C.; Thatcher, D. M.; Walton, J. B. *Org. Process Res. Dev.* **1998,** *2,* 157.

29. Wey, J. S. In *Handbook of Industrial Crystallization*; Myerson, A. S., Ed.; Butterworth-Heinemann: Boston, 1993; p 209.

30. Kerner, J. *Chem. Eng.* **2001,** *118*(6), 35.

31. Davulcu, A. H.; McLeod, D. D.; Li, J.; Katipally, K.; Littke, A.; Doubleday, W.; Xu, Z.; McConlogue, C. W.; Lai, C. J.; Gleeson, M.; Schwinden, M.; Parsons, R. L., Jr. *J. Org. Chem.* **2009,** *74,* 4068.

32. Davey, R.; Garside, S. *From Molecules to Crystallizers,* Oxford University Press: New York, 2002; p 70.

33. Kim, S.; Lotz, B.; Lindrud, M.; Girard, K.; Moore, T.; Nagarajan, K.; Alvarez, M.; Lee, T.; Nikfar, F.; Davidovich, M.; Srivastava, S.; Kiang, S. *Org. Process Res. Dev.* **2005,** *9*(6), 894.

34. Myerson, A. Personal communication.

35. Ripin, D. H. B.; Bourassa, D. E.; Brandt, T.; Castaldi, M. J.; Frost, H. N.; Hawkins, J.; Johnson, P. J.; Massett, S. S.; Neumann, K.; Phillips, J.; Raggon, J. W.; Rose, P. R.; Rutherford, J. L.; Sitter, B.; Stewart, A. M., III; Vetelino, M. G.; Wei, L. *Org. Process Res. Dev.* **2005,** *9,* 440.

36. Wey, J. S. In *Handbook of Industrial Crystallization*; Myerson, A. S., Ed.; Butterworth-Heinemann: oston, 1993; p 209; reference to Zipp, G. L.; Randolph, A. D. *Ind. Eng. Chem. Res.* **1989,** *28,* 1446

37. Kim, J.-W.; Kim, J.-K.; Kim, H.-S.; Koo, K.-K. *Org. Process Res. Dev.* **2011,** *15,* 602.

38. For a recent article on continuous crystallization on an industrial scale, see Genck, W. *Chem. Eng.* **2011,** *118*(7), 28.

39. Kosmrlj, J.; Weigel, L. O.; Evans, D. A.; Downey, C. W.; Wu, J. *J. Am. Chem. Soc.* **2003,** *125,* 3208.

40. Singh, J.; Denzel, T. W.; Fox, R.; Kissick, T. P.; Herter, R.; Wurdinger, J.; Schierling, P.; Papaioannou, C. G.; Moniot, J. L.; Mueller, R. H.; Cimarusti, C. M. *Org. Process Res. Dev.* **2002**, *6*, 863.
41. Haight, A. R.; Ariman, S. Z.; Barnes, D. N.; Benz, N. J.; Gueffier, F. X.; Henry, R. F.; Hsu, M. C.; Lee, E. C.; Morin, L.; Pearl, K. B.; Peterson, M. J.; Plata, D. J.; Willcox, D. R. *Org. Process Res. Dev.* **2006**, *10*, 751.
42. See Reference 10, Chapter 10.
43. Yang, Q.; Haney, B. P.; Vaux, A.; Riley, D. A.; Heidrich, L.; He, P.; Mason, P.; Tehim, A.; Fisher, L. E.; Maag, H.; Anderson, N. G. *Org. Process Res. Dev.* **2009**, *13*, 786.
44. Anderson, N. G.; Bennett, B. J.; Feldman, A. F.; Lust, D. A.; Polomski, R. E. U.S. Patent 5,026,873, 1991 (to E. R. Squibb & Sons.)
45. Cramer, S. D. *Ind. Eng. Chem. Process Des. Dev.* **1984**, *23*, 618.
46. Anderson, N. G. *Org. Process Res. Dev.* **2005**, *9*, 800.
47. Brands, K. M. J.; Davies, A. J. *Chem. Rev.* **2006**, *106*, 2711.
48. Reider, P. J.; Davis, P.; Hughes, D. L.; Grabowski, E. J. J. *J. Org. Chem.* **1987**, *52*, 957.
49. For a review of enantioselective Strecker reactions, see Gröger, H. *Chem. Rev.* **2003**, *103*, 2795.
50. Kuethe, J. T.; Gauthier, D. R., Jr.; Beutner, G. L.; Yasuda, N. *J. Org. Chem.* **2007**, *72*, 7469.
51. Ashcroft, C. P.; Challenger, S.; Clifford, D.; Derrick, A. M.; Hajikarimian, Y.; Slucock, K.; Silk, T. V.; Thomson, N. M.; Williams, J. R. *Org. Process Res. Dev.* **2005**, *9*, 663.
52. Barth, P.; Pfenninger, A. WIPO Patent WO/1997/041084A1 (to AstraZeneca).
53. http://www.deadiversion.usdoj.gov/schedules/index.html
54. Orme, M. W.; Martinelli, M. J.; Doecke, C. W.; Pawlak, J. M.; Chelius, E. C. WO 04/011463, 2004 (to Lilly); also EP 1546149, 2005.
55. Conlon, D. A.; Drahus-Paone, A.; Ho, G.-J.; Pipik, B.; Helmy, R.; McNamara, J. M.; Shi, Y.-J.; Williams, J. M.; Macdonald, D.; Deschênes, D.; Gallant, M.; Mastracchio, A.; Roy, B.; Scheigetz, J. *Org. Process Res. Dev.* **2006**, *10*, 30.
56. Lucci, R. U.S. 4,556,518, 1985 (to Hoffmann-La Roche).
57. Nelson, T. D. In *Strategies and Tactics in Organic Synthesis*, Vol. 6; Harmata, M., Ed.; Elsevier: Amsterdam, 2005; p 321.
58. http://www.epa.gov/greenchemistry/pubs/docs/award_recipients_1996_2007.pdf
59. Kiau, S.; Discordia, R. P.; Madding, G.; Okuniewicz, F. J.; Rosso, V.; Venit, J. J. *J. Org. Chem.* **2004**, *69*, 4256. See supporting information.
60. Chung, J. Y. L.; Hughes, D. L.; Zhao, D.; Song, Z.; Mathre, D. J.; Ho, G.-J.; McNamara, J. M.; Douglas, A. W.; Reamer, R. A.; Tsay, F.-R.; Varsolona, R.; McCauley, J.; Grabowski, E. J. J.; Reider, P. J. *J. Org. Chem.* **1996**, *61*, 215.
61. Chen, B.-C.; Quinlan, S. L.; Stark, D. R.; Reid, J. G.; Audia, V. H.; George, J. G.; Eisenreich, E.; Brundidge, S. P.; Racha, S.; Spector, R. H. *Tetrahedron Lett.* **1995**, *36*, 7957; Reddy, J. R.; Reid, J. G.; Renner, D. A.; Quinlan, S. L.; Weaver, D. G.; Chen, B.-C.; Stark, D. R.; Gao, Q. *Org. Process Res. Dev.* **1998**, *2*, 203.
62. Denzel, T.; Cimarusti, C. M.; Singh, J.; Mueller, R. H. U.S. 5,194,604, 1993 (to E. R. Squibb & Sons).
63. Wei, C. C.; Weigele, M. U.S. 4,502,994, 1985 (to Hoffmann-LaRoche).
64. Floyd, D. M.; Fritz, A. W.; Cimarusti, C. M. *J. Org. Chem.* **1982**, *47*, 176.
65. Bullock, K. M.; Burton, D.; Corona, J.; Diederich, A.; Glover, B.; Harvey, K.; Mitchell, M. B.; Trone, M. D.; Yule, R.; Zhang, Y.; Toczko, J. F. *Org. Process Res. Dev.* **2009**, *13*, 303.
66. Haleblian, J.; McCrone, W. *J. Pharm. Sci.* **1969**, *58*, 911.
67. Thanks to Ann Newman for this point.
68. Brittain, H. G., Ed.; *Polymorphism in Pharmaceutical Solids*, 2nd ed.; Informa Healthcare: New York, 2009.
69. Storey, R., Ed.; *Solid State Characterization of Pharmaceuticals*; Wiley-Blackwell, 2008.
70. Zakrzewski, M.; Zakrzewski, K., Eds.; *Solid State Characterization of Pharmaceuticals*; ASSA International, 2006.

71. Thompson, M. D.; Authelin, J.-R. Chemical development of the drug substance solid form, In *Process Chemistry in the Pharmaceutical Industry*; Gadamasetti, K., Ed.; Dekker: New York, 1999; pp 371–388.

72. ICH Q6A (2000), http://www.ema.europa.eu/docs/en_GB/document_library/Scientific_guideline/2009/09/WC500002823.pdf

73. Beckmann, W.; Otto, W. H. *Chem. Eng. Res. Dev.* **1996,** *74A,* 750.

74. Beckmann, W.; Nickisch, K.; Budde, U. *Org. Process Res. Dev.* **1998,** *2,* 298.

75. Bauer, J.; Spanton, S.; Henry, R.; Quick, J.; Dziki, W.; Porter, W.; Morris, J. *Pharm. Res.* **2001,** *18,* 859.

76. Moniot, J. L. Personal communication.

77. Chembukar, S. R.; Bauer, J.; Deming, K.; Spiwek, H.; Patel, K.; Morris, J.; Henry, R.; Spanton, S.; Dziki, W.; Porter, W.; Quick, J.; Bauer, P.; Donaubauer, J.; Narayanan, B. A.; Soldani, M.; Riley, D.; McFarland, K. *Org. Process Res. Dev.* **2000,** *4,* 413.

78. Ostwalds' rule is nicely stated in Bauer, J.; Spanton, S.; Henry, R.; Quick, J.; Dziki, W.; Porter, W.; Morris. *J. Pharm. Res.* **2001,** *18,* 859.

79. Morissette, S. L.; Soukasene, S.; Levinson, D.; Cima, M. J.; Almarsson, O. *Proc. Natl. Acad. Sci.* **2003,** *100,* 2180.

80. Mangin, D.; Puel, F.; Veesler, S. *Org. Process Res. Dev.* **2009,** *13,* 1241.

81. An, S.-G.; Sohn, Y.-T. *Arch. Pharm. Res.* **2009,** *32,* 933.

82. Bingham, A.L.; Hughes, D.S.; Hursthouse, M.B.; Lancaster, R.W.; Tavener, S.; Threlfall, T.L. *Chem. Commun.* **2001,** 603.

83. Reany, O.; Kapon, M.; Botoshansky, M.; Keinan, E. *Cryst. Growth Des.* **2009,** *9,* 3661.

84. Stahly, G. P. *Cryst. Growth Des.* **2007,** *7,* 1007.

85. Detoisien, T.; Forite, M.; Taulelle, P.; Teston, J.; Colson, D.; Klein, J. P.; Veesler, S. *Org. Process Res. Dev.* **2009,** *13,* 1338.

86. Rubin, A. E.; Tummala, S.; Both, D. A.; Wang, C.; Delaney, E. J. *Chem. Rev.* **2006,** *106,* 2794.

87. Remenar, J. F.; MacPhee, J. M.; Larson, B. K.; Tyagi, V. A.; Ho, J. H.; McIlroy, D. A.; Hickey, M. B.; Shaw, P. B.; Almarsson, O. *Org. Process Res. Dev.* **2003,** *7,* 990.

88. Almarsson, O.; Hickey, M. B.; Peterson, M. L.; Morissette, S. L.; Soukasene, S.; McNulty, C.; Tawa, M.; MacPhee, J. M.; Remenar, J. F. *Cryst. Growth Des.* **2003,** *3,* 927.

89. Peterson, M. L.; Morissette, S. L.; McNulty, C.; Goldsweig, A.; Shaw, P.; LeQuesne, M.; Monagle, J.; Encina, N.; Marchionna, J.; Johnson, A.; Gonzalez-Zugasti, J.; Lemmo, A. V.; Ellis, S. J.; Cima, M. J.; Almarsson, Ö *J. Am. Chem. Soc.* **2002,** *124,* 10958.

90. Morissette, S. L.; Almarsson, Ö; Peterson, M. L.; Remenar, J. F.; Read, M. J.; Lemmo, A. V.; Ellis, S.; Cima, M. J.; Gardner, C. R. *Adv. Drug. Deliv. Rev.* **2004,** *56.* 275.

91. Mukherjee, A.; Grobelny, P.; Thakur, T. S.; Desiraju, G. R. *Cryst. Growth Des.* **2011,** *11,* 2637.

92. Etter, M. C.; Frankenbach, G. M.; Bernstein, J. *Tetrahedron Lett.* **1989,** *30,* 3617.

93. André, V.; Fernandes, A.; Santos, P. P.; Duarte, M. T. *Cryst. Growth Des.* **2011,** *11,* 2325.

94. Shiraki, M. *J. Pharm. Sci.* **2010,** *99,* 3986.

95. Miller, J. M.; Collman, B. M.; Greene, L. R.; Grant, D. J. W.; Blackburn, A. C. *Pharm. Dev. Technol.* **2005,** *10,* 291.

96. Davey, R. J.; Blagden, N.; Righini, S.; Alison, H.; Ferrari, E. S. *J. Phys. Chem. B* **2002,** *106,* 1954.

97. Batra, H.; Penmasta, R.; Phares, K.; Staszewski, J.; Tuladhar, S. M.; Walsh, D. A. *Org. Process Res. Dev.* **2009,** *13,* 242.

98. Saranteas, K.; Bakale, R.; Hong, Y.; Luong, H.; Foroughi, R.; Wald, S. *Org. Process Res. Dev.* **2005,** *9,* 911.

99. Prashad, M.; Sutton, P.; Wu, R.; Hu, B.; Vivelo, J.; Carosi, J.; Prasad, K.; Liang, J. *Org. Process Res. Dev.* **2010,** *14,* 878.

100. Jetti, R. K. R.; Bhogala, B. R.; Gorantla, A. R.; Karusula, N. R.; Datta, D. *Cryst. Growth Des.* **2011,** *11,* 2039.

101. Veesler, S.; Lafferrère, L.; Garcia, E.; Hoff, C. *Org. Process Res. Dev.* **2003,** *7,* 983.

102. Dugger, R., contributed talk to Current Process Chemistry Symposium, Philadelphia, June 2–3, 2011.
103. Jia, C.-Y.; Yin, Q.-X.; Zhang, M.-J.; Wang, J.-J.; Shen, Z.-H. *Org. Process Res. Dev.* **2008**, *12*, 1223.
104. Mukuta, T.; Lee, A. Y.; Kawakami, T.; Myerson, A. S. *Cryst. Growth Des.* **2005**, *5*, 1429.
105. Pino-García, O.; Rasmuson, Å. C. *Cryst. Growth Des.* **2004**, *4*, 1025.
106. Testa, B. *Principles of Organic Stereochemistry;* Marcel Dekker: Basel, 1979; pp 181–184.
107. Jaques, J.; Collet, A.; Wilen, S. H. *Enantiomers, Racemates, and Resolutions;* John Wiley & Sons: New York, 1981; pp 381–395.
108. Kozma, D., Ed.; *CRC Handbook of Optical Resolutions via Diastereomeric Salt Formation*; CRC Press, 2001.
109. Federsel, H.-J. *Org. Process Res. Dev.* **2000**, *4*, 352.
110. Crosby, J. In *Chirality in Industry*; Collins, A. N.; Sheldrake, G. N.; Crosby J., Eds.; Wiley: New York, 1992; pp 24–26
111. Petrusevska-Seebach, K.; Seidel-Morgenstern, A.; Elsner, M. P. *Cryst. Growth Des.* **2011**, *11*, 2149.
112. Wacharine-Antar, S.; Levilain, G.; Dupray, V.; Coquerel, G. *Org. Process Res. Dev.* **2010**, *14*, 1358.
113. Hansen, K. B.; Hsiao, Y.; Xu, F.; Rivera, N.; Clausen, A.; Kubryk, M.; Krska, S.; Rosner, T.; Simmons, B.; Balsells, J.; Ikemoto, N.; Sun, Y.; Spindler, F.; Malan, C.; Grabowski, E. J. J.; Armstrong, J. D., III. *J. Am. Chem. Soc.* **2009**, *131*, 8798.
114. Savile, C. K.; Janey, J. M.; Mundorff, E. C.; Moore, J. C.; Tam, S.; Jarvis, W. R.; Colbeck, J. C.; Krebber, A.; Fleitz, F. J.; Brands, J.; Devine, P. N.; Huisman, G. W.; Hughes, G. J. *Science* **2010**, *329*, 305.
115. Lorenz, J. Z.; Busacca, C. A.; Feng, X.; Grinberg, N.; Haddad, N.; Johnson, J.; Kapadia, S.; Lee, H.; Saha, A.; Sarvestani, M.; Spinelli, E. M.; Varsolona, R.; Wei, X.; Zeng, X.; Senanayake, C. H. *J. Org. Chem.* **2010**, *75*, 1155. Supporting information for this manuscript describes the isolation of the DCHA salt.
116. Kim, S.; Wei, C.; Kiang, S. *Org. Process Res. Dev.* **2003**, *7*, 997.
117. C3 technology to control nucleation, deagglomeration, and crystal size distribution, http://www. prosonix.co.uk
118. Myerson, A. S.; Garetz, B. A. U.S. 6,426,406, 2002.
119. Reference 10, p217.
120. Kljajic, A.; Zupet, R. WO2010069593, 2010 (to Krka, d.d., Slovenia).
121. Bowles, D. M.; Boyles, D. C.; Choi, C.; Pfefferkorn, J. A.; Schuyler, S. *Org. Process Res. Dev.* **2011**, *15*, 148. See footnote 23.
122. Ash, M.; Ash, I. *Handbook of Preservatives*; Synapse Information Resources, Inc., 2004; pp 323–324.
123. Masui, Y.; Kitaura, Y.; Kobayashi, T.; Goto, Y.; Ando, S.; Okuyama, A.; Takahashi, H. *Org. Process Res. Dev.* **2003**, *7*, 334.
124. Franc, B.; Hoff, C.; Kiang, S.; Lindrud, M. D.; Monnier, O.; Wei, C. U.S. 7,008,959, 2006 (to Sanofi-Aventis).
125. Prosonix offers annular sonic sources bonded to the outside of pipes for continuous flow and reproducible crystallization, including small particles for inhalation, http://www.chemsourcesymposia.org.uk/abstracts.htm
126. Reference 10, pp 188–192.
127. Dhodapkar, S.; Theuerkauf, J. *Chem. Eng.* **2001**, *118*(6), 45.
128. am Ende, D. J.; Rose, P. R. In *Fundamentals of Early Clinical Development: From Synthesis Design to Formulation*; Abdel-Magid, A.; Caron, S., Eds.; Wiley, 2006; pp 247–267.
129. Kamahara, T.; Takasuga, M.; Tung, H. H.; Hanaki, K.; Fukunaka, T.; Izzo, B.; Nakada, J.; Yabuki, Y.; Kato, Y. *Org. Process Res. Dev.* **2007**, *11*, 699.
130. Accentus: http://www.accentus.co.uk
131. http://www.supercriticalfluids.com
132. Schmidt, B.; Patel, J.; Ricard, F. X.; Brechtelsbauer, C. M.; Lewis, N. *Org. Process Res. Dev.* **2004**, *8*, 998.
133. Jenkins, S. *Chem. Eng.* **2010**, *117*(10), 31.

134. Cimarosti, Z.; Castagnoli, C.; Rossetti, M.; Scarati, M.; Day, C.; Johnson, B.; Westerduin, P. *Org. Process Res. Dev.* **2010,** *14,* 1337.

135. Rassias, G.; Hermitage, S. A.; Sanganee, M. J.; Kincey, P. M.; Smith, N. M.; Andrews, I. P.; Borrett, G. T.; Slater, G. R. *Org. Process Res. Dev.* **2009,** *13,* 774.

136. Tian, J.; Dalgarno, S. J.; Atwood, J. L. *J. Am. Chem. Soc.* **2011,** *133,* 1399.

Final Product Form and Impurities

"…98.5% purity doesn't mean very much when we are talking about biologically active contaminants that, in varying minuscule amounts, can create powerful effects…."

P. Raphals [1]

I. INTRODUCTION

For approval to sell drugs pharmaceutical and biotechnology companies must adhere to the standards posed by various regulatory authorities. The qualities considered in APIs include potencies, identity and levels of impurities, and physicochemical characteristics. In the United States the Food and Drug Administration (FDA) approves new drug applications (NDAs). The FDA enforces testing methodology that has been established by the U.S. Pharmacopeia (USP), a non-profit, non-governmental organization [2]. Regulatory authorities for the European Union (the European Medical Association, EMA, formerly known as EMEA), Japan, and other countries are similarly concerned about the SAFETY and quality of drug products, but may require different data for environmental impact and other studies. The International Conference on Harmonization (ICH) strives to make guidelines consistent worldwide, or at least for the major economic markets.

Modes of delivery of drug substance include primarily ingestion, followed by injection, topical and mucosal application, and inhalation. Most drug substances are formulated into tablets, followed by capsules, lyophiles, creams, microsuspensions, and sustained release forms [3]. Tablets are the preferred formulation, because there is high

Practical Process Research and Development. DOI: 10.1016/B978-0-12-386537-3.00013-7

patience compliance, and formulation is rapid and economical. Furthermore, tablets generally have good stability, probably because the outer molecules in the crystal lattice of the API and the surrounding solids protect the API from degradation [4]. Many compounds, such as aspirin, decompose readily in solution.

The primary tasks of the various regulatory agencies are to ensure that medicines are efficacious and not harmful. If the concentration of a drug substance within a drug product is variable, the drug product may be ineffective or even toxic if the margin of safety is low; therefore, one responsibility of those marketing drugs is to ensure uniformity of a drug substance within a pill, a bottle, a batch, and finally within all batches. Process chemists attempt to minimize heterogeneity within a vessel, and so operate in solutions whenever possible. Except for lyophilized formulations, pharmaceutical scientists don't have the luxury of stirring solutions, and converting solids and liquids into homogeneous mixtures is very challenging. (Anyone who has made a vinaigrette or baked a cake "from scratch" knows the importance of mixing materials in the right order and at the right speed.) In the process of formulating an API into a tablet, pharmaceutical scientists study the mixing of a drug substance with excipients, therapeutically inactive components. Operations involved in formulation may include screening, blending, milling with or without liquids added, drying, compression, and film coating. Excipients may include fillers, disintegrants, sweeteners, flavors, colorants, and lubricants; excipients that could affect the stability of the API are preferably not chosen. Various aspects of material science are important to the pharmaceutical scientist [5]. Triboelectrification, or the ability to pick up a charge from contact with other materials, can cause particles, especially small particles, to clump together and mix poorly [6–8]. By heating to the glass transition temperature (T_g) a compound melts and becomes fluid; with rapid cooling it may become amorphous, mobile, and less stable [4,9]. Any solids that adhere to the punches or dyes of a tabletting press will decrease productivity [10]. Interactions of APIs with excipients can degrade APIs, and hydration encountered during mixing or energy-intensive operations such as milling can promote the formation of a hydrate or polymorph [11]. Pharmaceutical scientists can control the characteristics of excipients by specifications or processing. For reproducible formulation and stability of a drug product it is necessary to control the physical qualities of the API, such as the polymorph, hygroscopicity, particle size and particle size distribution (PSD), and shape, all of which may affect the flow characteristics of solids during formulation. The Quality Analysis department (QA) assesses the physical and chemical characteristics of excipients and chemicals used to make APIs, and the Quality Control department (QC) monitors the physical and chemical characteristics of drug substances. (QA and QC may be merged into one unit called Quality.) It became the job of QC units to assess the uniformity of ampoules, tablets, capsules, and other formulations, and to determine how to obtain statistically significant samples from such batches [12]. The cost of formulation is a significant part of the cost of converting a drug substance into a drug product [13]. A huge cost may be incurred if the physical form of an API is not controlled and a dosage form fails specifications, thus interrupting clinical trials or discontinuing sales of a drug. Hence process scientists pay a great deal of attention in preparing APIs with consistent physicochemical characteristics.

Most small molecule APIs are usually designed to be solids that can be crystallized and recrystallized, to control the quality of both the drug substance and the drug product. As a rule, crystalline materials are more stable than amorphous solids, so crystalline drug substances are more likely to provide an acceptable shelf life (preferably two years) for a drug product. Some drug substances are formulated as amorphous solids, such as quinapril hydrochloride [14] and nelfinavir mesylate [15]. Amorphous solids may be preferred because other forms crystallize as solvates with unacceptably high levels of residual solvents, or as attempts to skirt patents on crystalline final forms [16], or due to higher solubilities of amorphous solids in water and biological fluids. (As an exception to this generality, Amgen researchers found that an amorphous form of a calcium salt was less soluble in water than was a dihydrate form of that calcium salt. They attributed the lower solubility of the amorphous material to its poor wettability [17].) A compound with low solubility may be rendered amorphous by mixing a solution with polymers and spray-drying: the addition of the polymers and perhaps dilution stabilize the API in the spray-dried dispersion, and the bioavailability may greatly increase. The ability to formulate such poorly soluble APIs greatly expands the potential of drug substances: 70–90% of compounds in development may have low solubility [18]. Pharmaceutical scientists and process chemists and engineers must continue to work together to produce high-quality, stable drug products. For the process scientist any process to crystallize an API must be practicable, that is, able to be performed on scale [19].

> Pharmaceutical chemists may be accustomed to producing drug product from micronized drug substance, but micronization may not be necessary. Micronization and milling can expose personnel to potent, toxic compounds, and these operations invariably lead to loss of API. If there is doubt that micronization is needed to provide a suitable drug product, formulation with API from routine production should be examined.

We have a moral responsibility to understand the nature of impurities in substances to which humans are exposed. Unfortunately we cannot predict the toxicity of the compounds we prepare, much less the toxicity of related impurities. The quote beginning this chapter was directed toward impurities in L-tryptophan dietary supplements (Figure 13.1) that lead to the painful and sometimes fatal eosinophilia myalgia syndrome [1]. An impurity that may have been responsible for the syndrome is the bis-tryptophan derivative [20,21]. European women who ingested thalidomide during pregnancy found that some of their children had deformed limbs (phocomelia) due to the presence of the (S)-enantiomer of thalidomide; FDA inspector Frances Oldham Kelsey denied approval for the use of thalidomide in 1960 [22]. Facile racemization of thalidomide occurs *in vivo*. An unintended result of preparing the isostere of meperidine led to the generation of 1-methyl-4-phenyl-tetrahydropyridine (MPTP); in the brain this impurity is probably converted to the cation MPP^+, and led to the permanent symptoms similar to Parkinson's disease in a recreational drug user who prepared and ingested the meperidine isostere [23]. The United States dropped the defoliant Agent Orange onto the Vietnam countryside during the Vietnam war. The components of Agent Orange, a 50:50 mixture of 2,4-D and 2,4,5-T, were rapidly metabolized; unfortunately impurities present in Agent Orange, especially 2,3,7,8-tetrachlorodibenzo-*p*-dioxin (TCDD), proved to be carcinogenic and teratogenic, even at extremely low levels [24]. Genotoxic impurities are also compounds that are active at extremely low levels. We must understand the impurities associated with the compounds we prepare.

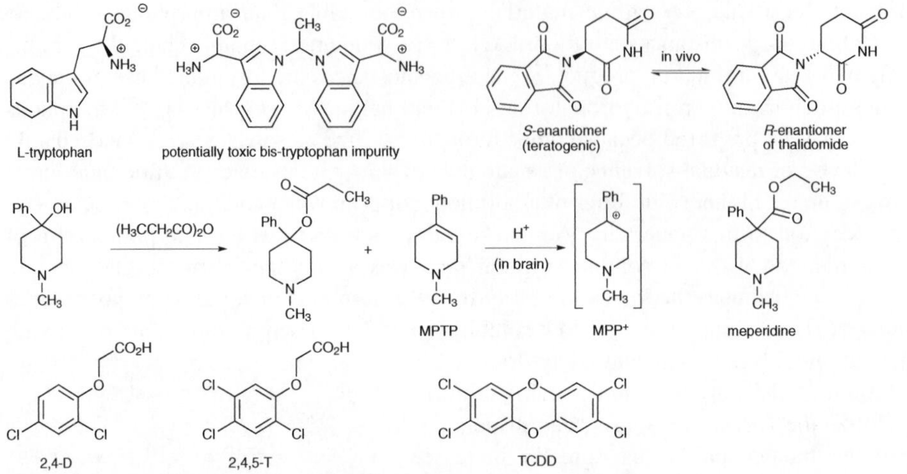

FIGURE 13.1   Toxic impurities found in some chemicals.

Sections in this chapter discuss final form selection, limits for impurities in APIs, controlling genotoxic impurities, and impurities that may be formed after storage of bulk API and drug products.

II. PHYSICOCHEMICAL CONSIDERATIONS IN APIs AND FINAL FORM SELECTION

Different salts and polymorphs have different stabilities, solubilities, and various physicochemical characteristics, such as melting points, density, particle shape, and flowability [25]. Scientists are not able to predict the optimal salt and polymorphs for a new chemical entity (NCE), and so screening is necessary [26–32]. The data derived from salt and polymorph screening guide selection of the final form of an API. Compared to the neutral compounds, salts often display higher melting points, higher solubility in water, and higher bioavailability. For instance, salts of codeine are more soluble in water than the free base (Figure 13.2) [33].

In addition to good aqueous solubility, key physicochemical properties of solid APIs include no or negligible hygroscopicity, good stability vs. relative humidity, and good chemical stability vs. heat and light. Hygroscopic solids may be difficult to transfer from the filter to the dryer to storage containers. In formulating a drug product hygroscopic solids may clump together and not be mixed evenly. If an extremely hygroscopic

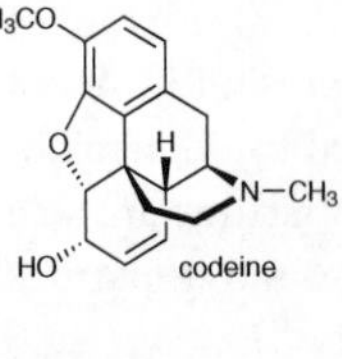

solid form of codeine	solubility in H₂O at RT, mg/mL
amine · 0.5 H₂O	8
· H₂SO₄	33
· HCl	50
· H₃PO₄ · 1.5 H₂O	435

FIGURE 13.2   Solubilities of codeine and salts in water.

material adsorbs a significant amount of moisture as milligram amounts are being weighed for analysis, a low potency will be assayed. Melting points should be a primary factor in selecting a final form, beyond the fact that low-melting compounds can require additional care to crystallize. Formulating a physiologically active compound that melts below 60 °C may be a challenge, as compression during tabletting typically produces localized temperature increases of 3–30 °C [34]. If the formulation mixture sticks to the punches and dyes of the tabletting presses, the productivity of the formulation will be slowed. Low-melting compounds may exhibit poor chemical stability [4], and the stress of the bulk mass of a tablet during long-term storage may lead to plastic deformation of a low-melting API, resulting in bulk aggregation [28]. To avoid difficulties in formulating as tablets, low-melting compounds may be formulated in capsules. Compounds with high melting points may not compact readily during tabletting, and plasiticizers, such as avicel, may be added [35]. If the crystalline lattice is too strong, poor solubility and hence poor bioavailability may result. (High-melting salts would be expected to have better solubility in aqueous systems than high-melting non-ionic compounds.) The melting point (or DSC data) of a potential final form should be one of the first physicochemical characteristics considered.

If the API is ionizable screening for a salt form may take place, and later the selected chemical form may be screened for optimal final form (polymorph, pseudopolymorph, hydrate, or amorphous) [25]. Screening for the preferred salt may occur early in development of a drug candidate, or as late as prior to formulation for Phase 2 studies. Full screening efforts may be postponed to minimize costs, and undertaken only after proof of concept and initial safety studies have been demonstrated [26]. Salt forms may be viewed as intellectual property; for instance, sitagliptin phosphate hydrate [36] was patented after the main patent for sitagliptin [37]. (During development Merck may have discovered that sitagliptin phosphate hydrate was the most stable form, and the easiest to process on scale [38].) Polymorph identification would be carried out after any screening for salt forms, followed by studies to identify the most stable polymorph [39]. Different dosage forms may have different requirements for the API. An overview of the final form selection process for a drug product is shown in Table 13.1.

Residual impurities can impede salt screening, and this is especially true for oils, which can trap impurities. Removing residual volatile impurities, such as trifluoroacetic acid and dichloromethane can be difficult (see discussion of Raoult's law with Figure 11.3). Volatile impurities such as TFA and CH_2Cl_2 may be removed by lyophilization. Sonication can help to crystallize oils during screening studies [41].

If a molecule contains a basic group, acids will be screened. Acids routinely screened include HCl, H_2SO_4, H_3PO_4, CH_3SO_3H, and others (Table 13.2). A wider diversity of salts has been approved in recent years [42]. When acids having more than one ionizable proton are considered, screening routinely is carried out for non-stoichiometric salts. For example, H_3PO_4 could form salts with one or two equivalents of an amine. If the molecule contains an acidic group, bases will be screened. For stable salt formation Bastin et al. have proposed that there should be a difference of at least 3 pH units between the pK_a of the neutral compound and that of the salt-forming counterpart [26].

Not all acids are created equal in making salts of basic molecules. HCl salts are very common, so manufacturers of the API and formulators are familiar with the characteristics of HCl and handling limitations. However, HCl is corrosive to some equipment that

TABLE 13.1 Outline of Final Form Selection Process

1. Salt screening to identify the optimal salt if drug candidate has ionizable groups
   A. Series of compounds is selected to form salts.
   B. Series of solvents is selected. ICH Class 3 solvents are preferred to crystallize an API, but a wide selection of solvents may be considered in an effort to generate solid forms.
   C. Crystallization experiments are undertaken. Variables considered include slow or rapid cooling, short or extended contact with solvent, mixtures of solvents, and moisture content.
   D. Crystalline compounds are identified and characterized.
   E. Accelerated stability tests may be used to select the most stable forms.

2. Polymorph screening to identify solid forms, salt or non-salt
   A. Steps (1) B–E are carried out. Solid samples may also be ground in an effort to generate other solid forms.
   B. Steps A–C may be repeated using a wider range of solvents if results are not satisfactory.
   C. Stability tests may be undertaken to identify most stable form.
   D. Accelerated stability tests may be used to identify degradents.

3. Co-crystal screening to identify other solid forms. This screening may also be carried out simultaneously with polymorph screening. Rarely used before 2003
   A. Series of GRAS additives [40] is selected for crystallization studies.
   B. Crystallization experiments as described for (2) above.

TABLE 13.2 Acids Used to Form Salts in Marketed Drugs

Acid	Total Distribution in 2006 (1)	Distribution of APIs Approved in 2002–2006 (1)	pK_a, in H_2O (2–5)	Comments, Examples of Marketed Salts
HCl	53.4%	38.9%	−6.10	Salts may decompose by loss of HCl on storage Ciprofloxacin hydrochloride
H_2SO_4	7.5%	5.6%	−3.00, 1.92	Abacavir sulfate
HBr	4.6%	8.3%	−9	Dextromethorphan hydrobromide
Methanesulfonic	4.2%	8.3%	−1.20	Nelfinavir mesylate
Maleic	4.2%	5.6%	1.92	May form impurities from Michael addition (6) Rosiglitazone maleate
L-Tartaric	3.8%	8.3%	3.02, 4.36	Levalbuterol tartrate
AcOH	3.3%	2.8%	4.76	Leuprolide acetate
HNO_3	1.7%	2.8%	−1.32	Miconazole nitrate
L-Malic	2.8%	0.4%	3.46, 5.10	Sunitinib malate

TABLE 13.2 Acids Used to Form Salts in Marketed Drugs—cont'd

Acid	Total Distribution in 2006 (1)	Distribution of APIs Approved in 2002–2006 (1)	pK_a, in H$_2$O (2–5)	Comments, Examples of Marketed Salts
H$_3$PO$_4$	2.7%	5.6%	1.96, 7.12, 12.32	Often forms hydrates (7) Sitagliptin phosphate hydrate
Citric	2.7%	2.8%	3.13, 4.76, 6.40	Relatively high molecular weight means less active ingredient by weight basis Sildenafil citrate
Fumaric	1.7%	0.0%	3.03, 4.38	May form impurities from Michael addition (6). Quetiapine fumarate
Succinic	1.2%	2.8%	4.21, 5.64	Metoprolol succinate
p-Toluenesulfonic	0.4%	2.8%	−1.34	Sorafenib tosylate
Oxalic	0.2%	2.8%	1.25, 4.14	Escitalopram oxalate
Benzenesulfonic	0.2%	0.0%	0.70	Samples of PhSO$_3$H may be contaminated by benzene (8) Amlodipine besylate
D-Glucuronic	0.2%	0.0%	3.18	Trimetrexate glucuronate
Hippuric	0.2%	0.0%	3.55	Methenamine hippurate
Benzoic	0.2%	0.0%	4.2	Rizatriptan benzoate
Glutamic			2.19, 4.25	
L-Ascorbic			4.17, 11.57	
Glutaric			4.31, 5.41	
Adipic			4.44, 5.44	
Total drugs	523	36		

(1) Paulekuhn, G. S.; Dressman, J. B.; Saal, C. J. Med. Chem. **2007,** 50, 6665. Source: the FDA Orange Book, which lists approved generic drugs.

(2) http://www.zirchrom.com/organic.htm

(3) Gordon, A. J.; Ford, R. A. The Chemist's Companion; Wiley: New York, 1972; pp 58–59.

(4) Brittain, H. G. Am. Pharm. Rev. **2009,** November/December, 62.

(5) For values on acidities in DMSO, see Bordwell, F. G. Acc. Chem. Res. **1988,** 29, 456.

(6) Abramowitz, R.; O'Donoghue, D.; Prabhakaran, P.; Mathew, R.; Suda, R. F.; Jain, N. B., poster presented at AAPS Meeting, Boston, 1997; Chatterjee, S.; Pedireddi, V. S.; Rao, C. N. R. Tetrahedron Lett. **1998,** 39, 2843; see also Giron, D.; Grant, D. J. W., in Handbook of Pharmaceutical Salts: Properties, Selection and Use; Stahl, P. H.; Wermuth, C. G., Eds.; Wiley-VCH: Weinheim, 2002; p 41.

(7) Handbook of Pharmaceutical Salts: Properties, Selection, and Use; Stahl, P. H.; Vermuth, C. G., Eds.; Wiley-VCH: Weinheim, 2002; pp 213, 302.

(8) Gross, T. D.; Schaab, K.; Ouellette, M.; Zook, S.; Reddy, J. P.; Shurtleff, A.; Sacaan, A. I.; Alebic-Kolbah, T.; Bozigian, H. Org. Process Res. Dev. **2007,** 11, 365.

is used to crystallize, isolate, and dry APIs, and can destroy valuable equipment and contaminate the API with heavy metal salts. Equipment such as polypropylene filters and Hastelloy reactors resist corrosion by HCl. Some HCl salts have been shown to be relatively unstable in the solid state, as with the volatilization of free HCl from amino-salicylic acid and caffeine salts [43], as shown in Figure 13.3. Abbott researchers found that salts of an azetidine with either HCl or TFA were unstable, but the tosic acid salt was stable [44], probably due to the non-volatile nature of tosic acid. H_3PO_4 is not volatile, and hence is not lost to the atmosphere due to the equilibrium between the salt vs. the free base and unionized acid. In addition, H_3PO_4 is essentially not corrosive to equipment used to crystallize, isolate, and dry APIs, and often is preferred for that reason. H_3PO_4 usually forms hydrates [45], and H_2O in a crystal may lead to decomposition of the API. Large excesses of H_3PO_4 in salt formations are avoided due to environmental impact on waste streams. Acetic acid may not be a strong enough acid to form stable salts with weak amines. Citric acid (FW 198) will contribute a great deal of mass to a salt, seeming like ballast, but may generate salts with the most desired characteristics. Succinic acid may be a good substitute for maleic and fumaric acids, which can form impurities by Michael addition of the amine to the unsaturated acids [46,47]. Since maleic acid can be

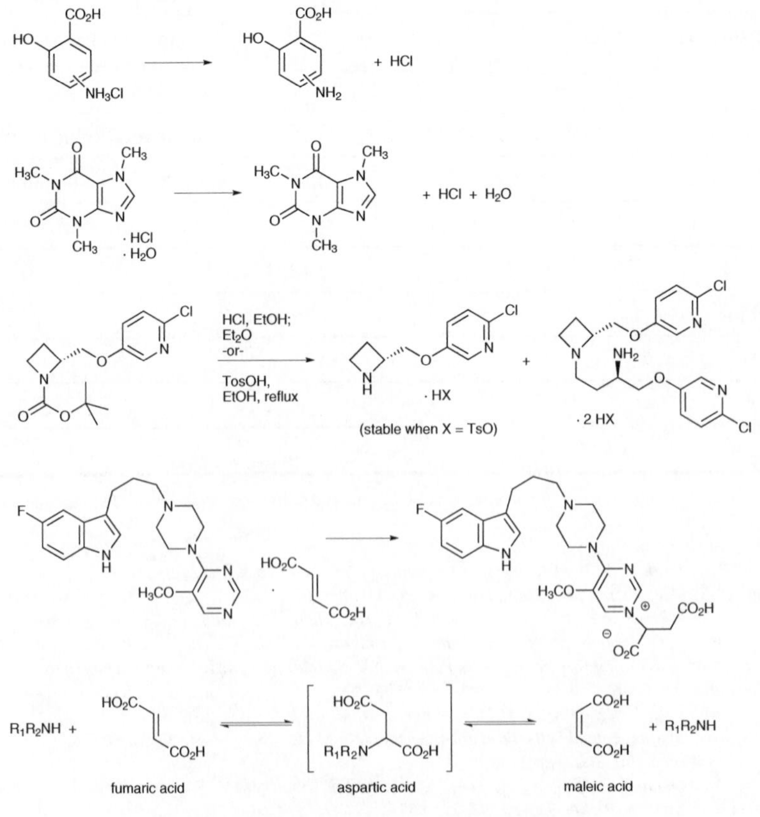

FIGURE 13.3 Side reactions of salts.

isomerized by heating in the presence of amines such as 4,4'-bipyridine [48], amine salts of maleic acid and fumaric acid could be in equilibrium in the solid state. Oxalic acid can readily form crystalline salts, but has been shunned because oxalic acid is known to be toxic. Lundbeck's escitalopram oxalate displays stability required for a drug product, and at the daily dosage of escitalopram (as high as 20 mg) the amount of oxalic acid consumed is much less than the toxic level (600 mg/kg for humans) [49,50].

If a product is acidic, salt formation may increase the solubility in water and bioavailability. Cations routinely screened include Na^+, K^+, Ca^{++}, Mg^{++}, organic amines such as diethanolamine and tromethamine (tris-(hydroxymethyl)-amino-methane, or THAM), and basic amino acids such as lysine and arginine (Table 13.3) [51]. In the acidic environment of the gut salts may precipitate, but as amorphous solids they may become more bioavailable [52].

Co-crystals and inclusion compounds may be investigated in the search for the optimal final form, especially if only amorphous forms of an API were found earlier. In co-crystals the partner molecules are held together by strong hydrogen bonds within the

TABLE 13.3 Cations Used to Form Salts in Anionic Marketed Drugs (1)

Counterion	Total Distribution in 2006 (1)	Distribution of APIs approved in 2002−2006 (1)	Comments, Examples of Marketed Salts
Na^+	75.3%	62.5%	May be hygroscopic
Ca^{++}	6.9%	18.8%	Low solubility, often small particles Atorvastatin calcium (2)
K^+	6.3%	6.3%	Solubility in water may be higher than Na salt Losartan potassium
Meglumine−H^+	2.9%	0%	Diatrizoate meglumine
Tromethamine−H^+ (tris, THAM)	1.7%	0%	tris-(hydroxymethyl)-aminomethane May be irritating to bowels
Mg^{++}	1.2%	6.3%	Low solubility, often small particles, slimy Esomeprazole magnesium
Zn^{++}	1.2%	0%	Insulin zinc
Lysine−H^+	0.6%	6.3%	Ibuprofen lysine
Total	174	16	

(1) Paulekuhn, G. S.; Dressman, J. B.; Saal, C. J. Med. Chem. **2007**, 50, 6665. Source: the FDA Orange Book, which lists approved generic drugs.
(2) Atorvastatin calcium can be recrystallized from EtOAc/hexane: Roth, B. D. US Patent 5,273,995, 1993 (to Warner−Lambert Company).

crystals; many definitions have been offered. Co-crystals are unique forms, as can be judged by XRPD, IR, solid-state NMR, and melting point/DSC [53]. Inclusion compounds are crystals in which guest molecules are held inside the cavities formed by the host molecule, and weak hydrogen bonding may or may not be involved. Inclusion complexes may be clathrates, wherein the spaces are channels or tunnels, or cage compounds, where the guest molecules are enclosed [54]. Clathrates may accommodate various molecules of similar size, such as MeOH, EtOH or *n*-PrOH [55], or host larger molecules, as do cyclodextrin–API complexes [56]. Co-crystals and clathrates have been discovered by accident and by screening. For instance, the co-crystal of ethyl *p*-hydroxybenzoate ("ethyl paraben," a preservative used in cosmetics) with loracarbef was discovered during formulation studies (Figure 13.4) [57]. In principle useful co-crystals and clathrates can be formed from any non-toxic compound, such as compounds from the GRAS list [40].

Co-crystals and clathrates have provided an array of advantages over previously existing materials, depending on the uses (Figure 13.4). Co-crystals can increase the solubility of neutral compounds. For instance, the rate of dissolution of a co-crystal (mp 142 °C) with glutaric acid (mp 97.5 °C) was 18 times faster than that of the extremely weakly basic pyrimidine alone (mp 206 °C) [58]. Melting points are generally changed; a survey showed that 51% of co-crystals had melting points between those of the API and the coformer, while 39% had lower melting points [53]. Merck prepared a chiral

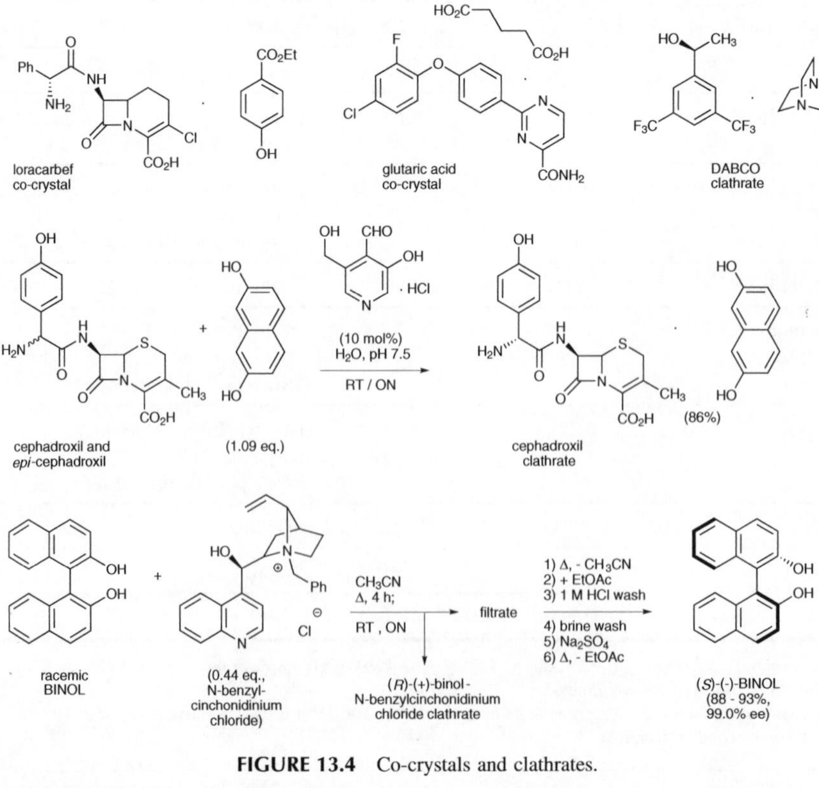

FIGURE 13.4 Co-crystals and clathrates.

alcohol (an aprepitant precursor) as an inclusion complex with DABCO; the complex was beneficial because it had a higher melting point and was more conveniently handled [59]. A CIAT process was used to prepare cefadroxil in high yield as a clathrate [60]. BINOL has been resolved by clathrate formation, and both the clathrate and the free BINOL can be purified [61]. Racemic lamivudine has been resolved through co-crystal formation with (S)-(−)-BINOL [62]. Boehringer-Ingelheim has reported screening for co-crystals using glutaric acid [63]. The pharmaceutical industry and CMOs will continue to develop and exploit co-crystals and clathrates [64,65].

Amgen researchers have described in a detailed and chronological manner the screening to find the most acceptable form of a carboxylic acid NCE. The free acid was found to be poorly soluble, and from results of early screening a sodium salt form was selected because it displayed the highest melting point. For long-term development this salt was abandoned in favor of forms that were more highly crystalline, less hygroscopic, and did not require storage at 2–8 °C. Further screening identified a calcium salt dihydrate as the preferred form. Because the aqueous solubility of the calcium salt was so much less than that of the sodium salt (0.003 mg/mL vs. >100 mg/mL), studies in primates were conducted. The bioavailability–pharmacokinetics profiles of a suspension of the calcium salt (1 mg/kg) and a solution of the sodium salt (1 mg/kg) were similar, indicating that aqueous solubility was not a good predictor of bioavailability. This contribution graphically describes the use of DSC, XRDP, and vapor sorption in assaying potential final forms [17].

III. IMPURITY CONSIDERATIONS IN APIs

Low levels of impurities can interrupt manufacturing and sales of drug products. Humans can detect malodorous impurities at very low levels, and smells are subjective. Genotoxins may need to be controlled at the ppm level (or even ppb), as discussed in the next section. Since impurities can be purged by attrition in downstream processing, sometimes by conversion to other impurities [66], process chemists may position reactions that generate troublesome impurities early in the route.

> Significant levels of impurities can form from what organic chemists might consider a "poor" reaction. Often it is prudent to verify whether an impurity is present.

To determine whether batches of an API contain harmful impurities, a batch of the API candidate (the "tox batch") is subjected to animal testing. If acceptable levels of toxicity are found in the animal studies regulatory authorities may grant permission for Phase 1 trials in humans, and the impurities in the tox batch are said to be qualified. (For batches of NCEs and APIs destined for human use, the impurity profiles must be no worse than that of the impurities in the qualifying tox batch, that is, no new impurities can be present, and the levels of the qualified impurities can be no higher than those found in the tox batch.) A tox batch with extremely high-quality, e.g., 99.8% AUC, can pose problems for manufacturing if other impurities arise through routes or operations developed after the tox batch was prepared. Different impurity profiles may trigger a bridging tox study.

According to ICH Guidance Q3A(R2) [67], applicants must summarize the "*actual and potential impurities* that are most likely to arise during the *synthesis, purification,* and *storage* of the new drug substance. This summary should be based on sound scientific appraisal of the *chemical reactions involved in the synthesis, impurities associated with raw materials* that could contribute to the impurity profile of the new drug substance, and *possible degradation products. This discussion can be limited to those impurities that might reasonably be expected* based on knowledge of the chemical reactions and conditions involved." Emphasis by italics was added to indicate that known and reasonable impurities must be considered; for example, degradents of impurities are generally not considered PGIs.

Standard limits for impurities, based on the dosage of API or drug candidate, are described by the ICH in guidance Q3A. Table 13.4 shows the limits for reporting, identifying, and qualifying impurities. It should be noted that while higher reporting thresholds could be justified, lower thresholds may be applied if impurities are highly toxic.

Limits in APIs have been set for some common impurities, such as residual solvents, based on toxicological data. Limits shown in Table 13.5 are based on Option 1, with dosages of 10 g/day. These limits are acceptable for all drug substances, excipients, and drug products, and are used when dosages have not been defined. Limits for drug products with dosages greater than 10 g/day should be calculated by Option 2 [68]. Higher limits are permissible using Option 2 if the daily dosage is less than 10 g/day.

Calculations for the permissible daily exposure (PDE) are based on Equation 13.1:

$$PDE = (NOEL \times weight\ adjustment)/(F1 \times F2 \times F3 \times F4 \times F5) \qquad (13.1)$$

where

NOEL is the no observed effect limit,
$F1$ is the extrapolation between species ($F1 = 12$ for mice to humans),
$F2$ is the individual variation,
$F3$ is the length of tox study,
$F4$ is the factor for severe toxicity, and
$F5$ is the factor if only LOEL (lowest observed effect limit) is available.

Specifications for residual metals have recently come under scrutiny, for several reasons. The original method for determining the concentration of "heavy metals" involved precipitating metals as sulfides from aqueous solutions and visually comparing

TABLE 13.4 Q3A Impurity Limits in New Drug Substances (1)

Maximum Daily Dose of Drug Substance	Reporting Threshold (2–3)	Identification Threshold (2)	Qualification Threshold (2)
≤2 g/day	0.05%	0.10% or 1.0 mg/day intake (whichever is lower)	0.15% or 1.0 mg/day intake (whichever is lower)
≥2 g/day	0.03%	0.05%	0.05%

(1) http://www.fda.gov/RegulatoryInformation/Guidances/ucm127942.htm#i
(2) Higher reporting thresholds should be scientifically justified.
(3) Lower thresholds can be appropriate if the impurity is unusually toxic.

TABLE 13.5 Limits for Some Residual Solvents in APIs (1)

	ICH Solvent Class	Limit (2) (ppm)
1,2-Dichloroethane (DCE)	1	4
1,2-Dimethoxyethane (DME)	2	100
EtOH	3	5000 ppm (0.5%)

(1) ICH Q3C: Tables and list: http://www.fda.gov/RegulatoryInformation/Guidances/ ucm128223.htm
(2) Based on dosage of 10 g/day (Option 1).

the colors to that of a standard lead sulfide standard. This method was known to underestimate the content of heavy metals, and to be non-specific. Current methodology, primarily atomic absorption and inductively coupled plasma-mass spectroscopy, is more accurate and specific for metal species. (Unfortunately such instrumentation is not inexpensive [2].) Limits for "elemental metals" are shown in Table 13.6. As of 2009 a USP advisory panel was also considering limits on antimony, boron, chromium III/VI, cobalt, mercury(I) as MeHg, mercury(II), tin, and tungsten [69]. The ICH is expected to release guidance on limits of heavy metals as Q3D [70].

Q: What does 10 ppm look like? What does 10 ppb look like?
A: Consider a million or a billion marbles of half-inch diameter. After tapping to allow settling, random packing of solid spheres fills about 60% of the total volume of a container, i.e., about 40% of the total volume is empty space [71]. Ten parts per million would be equivalent to 10 half-inch diameter marbles distributed in slightly more than 5 drums of such marbles (5.15 drums at 55 US gal, or 1.4018 yd^3 or 1.0717 m^3 or 1072 L). Ten parts per billion would be equivalent to 10 half-inch diameter marbles distributed in 234 dump trucks with 10 yd^3 capacity each.

As mentioned in Chapter 12, solutions of APIs should be polish-filtered before the API is isolated. In some cases it may not be possible to hold a compound in solution long enough to effect a polish filtration, forcing polish filtration at an earlier stage and cGMP handling up to the preparation of the final form. For instance, the beta form of aztreonam would dissolve briefly in warm EtOH, and crystallize [72], but subsequent warming did not redissolve the API. To circumvent this inconvenience aztreonam was dissolved in water–EtOH, polish-filtered, and crystallized as the alpha form, then converted to the more stable beta form under GMP conditions. Also removed by polish filtration are endotoxins, lipopolysaccharide complexes from the cell membrane of Gram-negative bacteria, which can cause pyrogenic reactions

The best strategy to control impurities in final product is to first control the impurities in the penultimate intermediate. Then conduct tolerance tests to determine the extent to which the final processing will remove these impurities.

TABLE 13.6 Permitted Daily Exposure (PDE) for Elemental Metals in APIs (1–3)

Element (4)		PDE (μg/day) (Oral Dosage (5–6))	wt% (In 1 g/day Oral Dosage (5–6))
Arsenic (inorganic)	As	15	0.0015%
Cadmium	Cd	5	0.0005%
Chromium	Cr	250	0.025%
Copper	Cu	2500	0.25%
Iron (7)	Fe	13,000	1.30%
Lead	Pb	10	0.001%
Manganese	Mn	2500	0.25%
Mercury (inorganic)	Hg	15	0.0015%
Molybdenum	Mo	250	0.025%
Nickel	Ni	250	0.025%
Palladium	Pd	100	0.01%
Platinum	Pt	100	0.01%
Vanadium	V	250	0.025%
Zinc (7)	Zn	13,000	1.30%
Iridium Osmium Rhodium Ruthenium	Ir Os Rh Ru	100 (for sum of these metals)	0.01% (for sum of these metals)

(1) EMA Guideline of the Specification Limits for Residues of Metal Catalysts or Metal Reagents, 2008, http://www.ema.europa.eu/docs/en_GB/document_library/Scientific_guideline/2009/09/WC500003586.pdf;

(2) Pharmacopeial Forum **2010**, 36, 1: http://www.usp.org/pdf/EN/hottopics/232_ElementalImpuritiesLimits.pdf

(3) http://www.usp.org/pdf/EN/hottopics/2009-04-22MetalImpuritiesToxChart.pdf

(4) USP advisory panel was also considering limits on antimony, boron, chromium III/VI, cobalt, mercury(I) as MeHg, mercury(II), tin, and tungsten: http://www.usp.org/pdf/EN/hottopics/2009-04-22MetalImpuritiesToxChart.pdf

(5) Limits for parenteral or inhalation exposure are one-tenth the limit for oral, mucosal, and topical and dermal administration (Ref. 2 above).

(6) ICH Q3D guidance may change values.

(7) EMA Guideline of the Specification Limits for Residues of Metal Catalysts or Metal Reagents, 2008, http://www.ema.europa.eu/docs/en_GB/document_library/Scientific_guideline/2009/09/WC500003586.pdf

or even death when injected into the bloodstream. These impurities can arise from insufficiently purified water for injection (WFI) or from reaction streams derived from bacterial biocatalysts. Endotoxins are difficult to destroy by heat or pH extremes, and may be removed by extraction due to their hydrophobic nature, or by filtration through a 0.22 μm filter [73].

IV. GENOTOXIC IMPURITIES

Over the past 11 years concern about genotoxic impurities has grown considerably within the pharmaceutical and CMO industries. Genotoxic and carcinogenic impurities are included in the category of unusually toxic impurities described in ICH Q3A, and so are subject to lower limits [67]. Assaying and controlling impurities at the ppm or even ppb level are significant challenges, and can delay the development of an NCE [74–77]. Teasdale has recently written an excellent book on genotoxins [78] and an extensive overview was published in *Chemical & Engineering News* [79].

Genotoxins, or mutagens, are potentially carcinogenic to humans through damage to DNA [80]. (Rarely is a compound identified as genotoxic to humans, due to the ethical considerations of running such trials on humans.) On a molecular level, the DNA may be broken or covalently modified, leading to problems in transcription and ultimately translation to proteins. Terms that unfortunately have been used interchangeably in the literature are PGIs (potentially genotoxic impurities), GTIs (compounds determined by toxicological testing to be genotoxic impurities), and GIs (genotoxic impurities). Some carcinogens have been tested on animals to determine the TD_{50}, the dosage at which half of the animals tested develop tumors, and such data is available through the Carcinogenic Potency Database [81], the National Toxicology Program Report on Carcinogens [82] and other sources. By using the TD_{50} data it is possible to calculate a limit for a GTI in a drug product, and this limit may be higher than the default limit of an untested impurity (1.5 µg/day for a clinical trial of 12 months or longer; see discussion with Table 13.7). For instance, ethyl chloride has been shown to be carcinogenic in mice, but at low potency (TD_{50} of 1810 mg/kg/day). The California EPA assigned a limit of 150 µg/day for EtCl [83]. Using the approach of some FDA researchers [84,85], a limit of 1.81 mg/day may be permissible [86]. Whenever possible, data should be used to set limits on impurities, in this case treating EtCl as a GTI [83,87].

Some functional groups (structural alerts) flagged as potentially conferring genotoxic activity to molecules are shown in Figure 13.5 [88]. Alkylating agents are generally the first group to be thought of as genotoxins [89]. Aldehydes may exert genotoxic activity by reacting at N_1 and the exocyclic amine of deoxyguanosine residues, as demonstrated for the reaction of acetaldehyde with human cells [90]. Polycyclic aromatic hydrocarbons, such as the extremely potent mutagen aflatoxin B_1, may be reactive after metabolic epoxidation [91,92]. Aryl boronic acids and boronic esters have been shown to be mutagenic in bacterial screens; data from animal testing is not available [93]. Extremely reactive functional groups, e.g., RCOCl, are missing from Figure 13.5, probably due to

TABLE 13.7 Staged TTC Proposed by FDA

Duration of Exposure	<14 days	14 day– 1 mo	1–3 mos	3–6 mos	6–12 mos	>12 mos
Allowable intake (mg/day) of genotoxic and carcinogenic impurities	120	60	20	10	5	1.5

http://www.fda.gov/downloads/Drugs/GuidanceComplianceRegulatoryInformation/Guidances/ucm079235.pdf

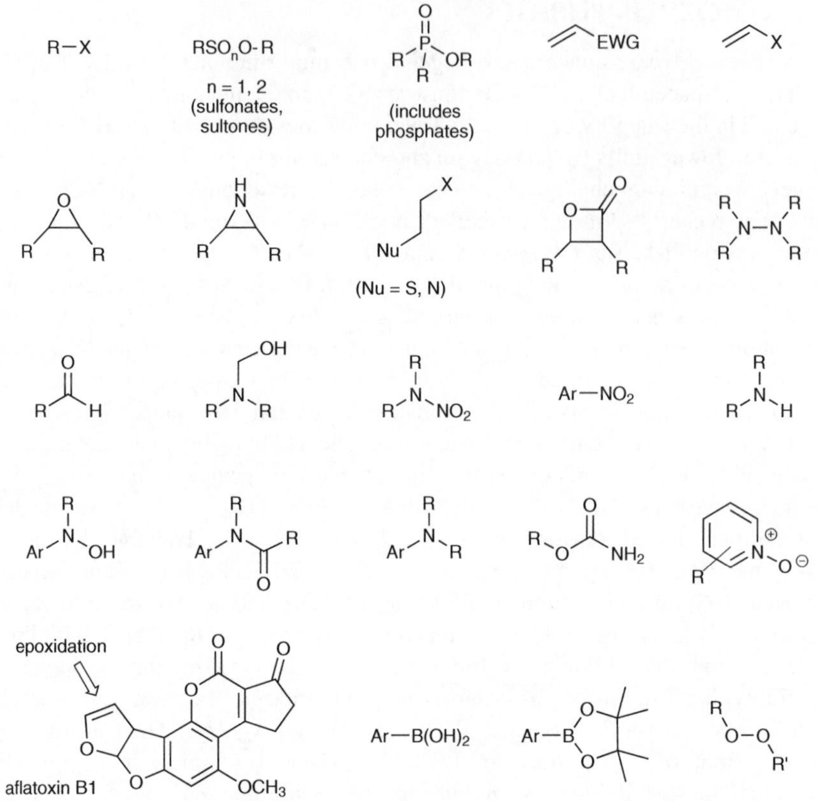

(R = aryl, alkyl, H; EWG = electron-withdrawing group; X = leaving group, such as Br)

FIGURE 13.5   Structural alerts, or functional groups flagged as potentially genotoxic.

hydrolysis *in vitro* (and *in vivo*). Some programs, such as ToxTree, consider such reactive functional groups as PGIs [94]. A number of software programs are acceptable to regulatory authorities to predict the carcinogenicity of compounds, including MultiCASE (MCase), Oncologic, DEREK, TOPKAT, RTECS (Registry of Toxic Effects of Chemical Substances), and LeadScope. The FDA also has some databases on toxic substances that may be present as impurities in APIs. Software, such as Pharma D$_3$, may overestimate the number of PGIs generated as metabolites [95]. Researchers from the FDA recently critiqued the MCase and Oncologic programs [96]. The EMA has stated that "the absence of a structural alert based on a well-performed [software] assessment…will be sufficient to conclude that the impurity is of no concern with respect to genotoxicity…" [97]. Further draft guidelines have recently been proposed by the FDA [98]. It should be noted that regulatory authorities in some countries use structural alerts, but not software packages [18]. Furthermore, software packages are self-learning tools, and hence the quality of predictions will be based on the quality of the data that was inputted.

The accepted sequence of dealing with genotoxic impurities is (1) desktop screening (*in silico*); (2) microbial screening, such as the Ames test; and (3) animal testing, often mice and rats. The goal in testing is to determine a threshold dosage, the dosage at which there is considered no risk for causing cancer. The rationale for this sequence has been described in detail [99]. In the early stages of developing a drug, PGIs often are not tested in animals, because PGIs (often in small amounts) may be difficult to isolate, identify, and prepare, and because toxicology testing requires time and money.

PGIs by definition are *potentially* genotoxic impurities, that is, impurities without sufficient toxicological data to establish an acceptable exposure limit. The EMA stated that if the presence of such impurities cannot be avoided, they must be removed down to levels as low as reasonably practicable (ALARP) [99]. By default PGIs are subject to the "threshold of toxicological concern" (TTC) limit for clinical trials (Table 13.7) [100]; the limit of 1.5 µg/day for clinical trials longer than 12 months also applies to drugs administered chronically. These limits were based on an acceptable risk of 1 in 100,000 of contracting cancer by exposure to a compound at a given dosage. Higher limits were allowed for shorter clinical studies (a staged TTC, Table 13.7), which were nonetheless lower than the limits proposed by researchers in the pharmaceutical community [101]. Bruce Ames and Lois Gold [102], Roche researchers [103] and others have questioned the validity of the linear extrapolation used to define acceptably safe limits.

Biological systems contain many repair mechanisms when confronted with foreign chemicals. For instance, chemicals can be excreted unchanged, intercepted by endogenous nucleophiles such as glutathione, or otherwise metabolized. Enzymes can repair damaged genetic material. Damaged cells may be programmed to die (apoptosis), or the organism can die, perhaps due to cancers. If cellular damage is less extensive, the organism may survive, with harmful genes being passed on to successive generations. Lack of mutagenicity in whole animal tests indicates that protective mechanisms are at work to deactivate PGIs. For example, acrylic acid, acrylates and methacrylates produced DNA damage in mouse lymphoma cells *in vitro*, but when tested in whole animals there was no evidence of mutagenic effects [104]. Snodin questioned the true impact of alkyl mesylates as genotoxins in 2006 [89]. Data from toxicology testing of PGIs is more definitive than *in silico* and *in vitro* tests, as was evidenced with the contamination of nelfinavir mesylate by ethyl mesylate (EtOMs).

The concern of genotoxins was aired in June 2007, when Hoffmann-La Roche withdrew from the European market all batches of nelfinavir mesylate (Viracept™) that had been recently made in Switzerland. Patient complaints of odor had caused Roche to investigate, and Roche found that batches of drug product produced during March–June 2007 contained high levels of ethyl mesylate. EtOMs has been shown to be carcinogenic, genotoxic, and teratogenic, and to alkylate DNA primarily at N7 of guanidine [105]. The later batches of API used to make these batches of drug product contained up to about 1100 ppm of EtOMs. At dosages of 2.5 g/day, up to about 2.75 mg of EtOMs was being consumed daily by patients [106,107].

Viracept™ was made by spray-drying a mixture of the free base and MsOH in EtOH (Figure 13.6). Roche identified the storage of MsOH in a tank dedicated to manufacturing nelfinavir mesylate as the primary root cause of high levels of EtOMs. This holding tank had been cleaned by draining and rinsing with EtOH, but unfortunately not all of the EtOH had been removed from the tank before fresh

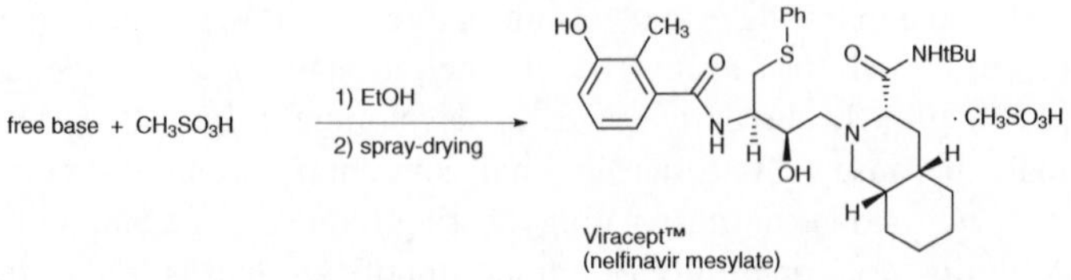

free base + CH₃SO₃H

1) EtOH
2) spray-drying

Viracept™
(nelfinavir mesylate)

FIGURE 13.6 Preparation of nelfinavir mesylate.

MsOH was added. Under these anhydrous conditions the concentration of EtOMs in the MsOH gradually rose over months of storage; hence EtOMs was being applied with the MsOH to the free base during salt formation. Batches of the API made in the October 2006 and the January–February campaigns contained progressively higher levels of EtOMs. Inspectors for the EMA and the Spanish agency Agemed identified another root cause as the piping for EtOH and N_2 in the plant, which may have returned volatilized EtOH from the drying process to the MsOH holding tank. The inspectors identified another possible cause as the presence of methyl methanesulfonate (MeOMs) in the MsOH, which could form EtOMs by acid-catalyzed transesterification. The EMA's overall criticism was that Hoffmann-La Roche did not fully understand the process that they had purchased [107].

Roche addressed the contamination issue in many ways. A toxicology study on EtOMs was carried out, and chromosomal damage was seen at 25 mg/kg/day (mice). Calculations showed that an exposure of about 370 times the greatest level of EtOMs that had been ingested in contaminated batches would be needed to impede mammalian DNA repair mechanisms [108]. The EMA accepted that 2 mg/kg/day of EtOMs would be SAFE (for a 2.5 g dosage of Viracept™ in a 50 kg man, this is equivalent to 4 wt% of EtOMs in Viracept™) [109]. Roche set a specification for no more than 1 ppm of EtOMs and MeOMs total in MsOH, to limit the formation of EtOMs by transesterification, and set a specification of NMT 0.01 ppm of EtOMs and MeOMs combined in recycled EtOH. Roche changed details of the salt formation; for instance, MsOH was charged directly from a disposable container, eliminating the holding tank. The pipe for charging MsOH to the suspension of free base in EtOH was optimized, perhaps by substituting a pipe with a smaller diameter to decrease mesomixing. Only 0.97–0.995 equivalents of MsOH was to be added, to limit acid-catalyzed esterification (see discussion concerning Figure 13.7). In the process of charging the MsOH about 90% of the required amount was added, and then the mixture was adjusted to pH 3.5–3.0 by adding more MsOH, a pH range that was less acidic than previously

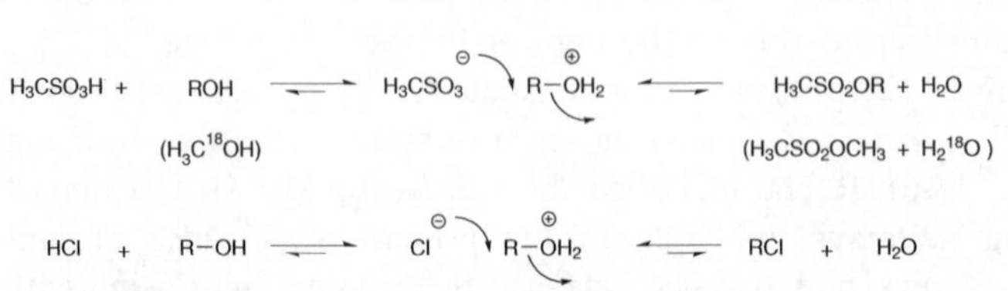

FIGURE 13.7 Equilibria involved in the formation of alkyl mesylates and alkyl chlorides.

allowed. These actions produced nelfinavir mesylate that routinely met the specification in place of NMT 0.5 ppm of EtOMs and MeOMs combined. In September 2007 the EMA lifted restrictions of Viracept™ sales [107,109–111].

More important than theory are the data on mutagenicity of GTI and the data on the levels of PGI/GTI found in batches of an API.

As mentioned regarding Viracept®, alkyl mesylates and alkyl chlorides can form as side products while crystallizing amine salts from alcoholic solvents, and strategies for controlling alkyl chlorides have been reviewed [112]. The obvious solutions to this concern are to avoid using alcoholic solvents to generate an API salt (and the penultimate), to select acids that do not form as byproducts esters that are alkylating agents, such as tartaric and oxalic acid, or even to develop the free base as an API. Unfortunately the best polymorph of a salt may crystallize from an alcoholic solvent, using an undesirable acid. The formation of alkyl mesylates is acid-catalyzed, as PQRI researchers demonstrated for MsOH and ^{18}O-labeled methanol (Figure 13.7). The PQRI researchers showed that the reaction is slow (0.2% MeOMs formed at 50 °C after 50 hours when 0.06 wt% H_2O was present) and that adding the weak base 2,6-lutidine slowed the reaction further. Adding H_2O pushed the equilibria to the left, away from MeOMs. To further decrease the formation of alkyl sulfonates the PQRI researchers recommended undercharging the sulfonic acid and minimizing the contact time of the sulfonic acid and the alcohol [113,114]. Similar considerations apply for the formation of alkyl chlorides from alcohols and HCl.

The straight-forward approach of adding water to decompose impurities has been quite effective. The crude tosylate salt from deprotection of a Boc-amine in i-PrOH contained 20–60 ppm of isopropyl tosylate, which was completely hydrolyzed during recrystallization after 1 hour at 72 °C (Figure 13.8) [115,116]. PTC-mediated alkylation of a bis-mesylate with a phenol occurred under weakly basic conditions at 102 °C. An excess of the bis-mesylate was required for alkylation, but this intermediate was a PGI and so needed to be controlled in the API. Hydrolysis of the alkyl mesylate was shown to depend on the concentration of water, while hydrolysis of the alkyl carboxylate depended on the concentration of hydroxide. The conditions described were developed to effect a high level of alkylation with decomposition of the excess alkyl mesylate *without* hydrolyzing the ester [117]. In another example the THP group on an adenine derivative was removed by treatment with TFA in i-PrOH, and the concern was that isopropyl trifluoroacetate, a PGI, would be present. Accordingly TFA-mediated deprotection was carried out in aqueous i-PrOH to minimize the formation of isopropyl trifluoroacetate [118]. Alkyl sulfonates have also been decomposed by reactive resins [119]. Dimethylcarbamoyl chloride (DMCC), a side product from acid chloride formation using $SOCl_2$ and DMF, is a known animal carcinogen at ppb levels [120]. (A proposed mechanism for formation of DMCC is shown in Figure 13.8. The presence of pyridine in this reaction increased the amount of DMCC formed [121].) Acid chloride formation is catalyzed by the Vilsmeier reagent shown. Experiments showed that the hydrolysis of DMCC was fairly rapid: under the quenched reaction conditions the half-life ($t_{1/2}$) was 4.13 min at 80 °C. Using the optimized reaction conditions, about 8 ppm of DMCC was formed in the initial reaction to generate the Vilsmeier reagent, and by holding at 70 °C for 30 min after

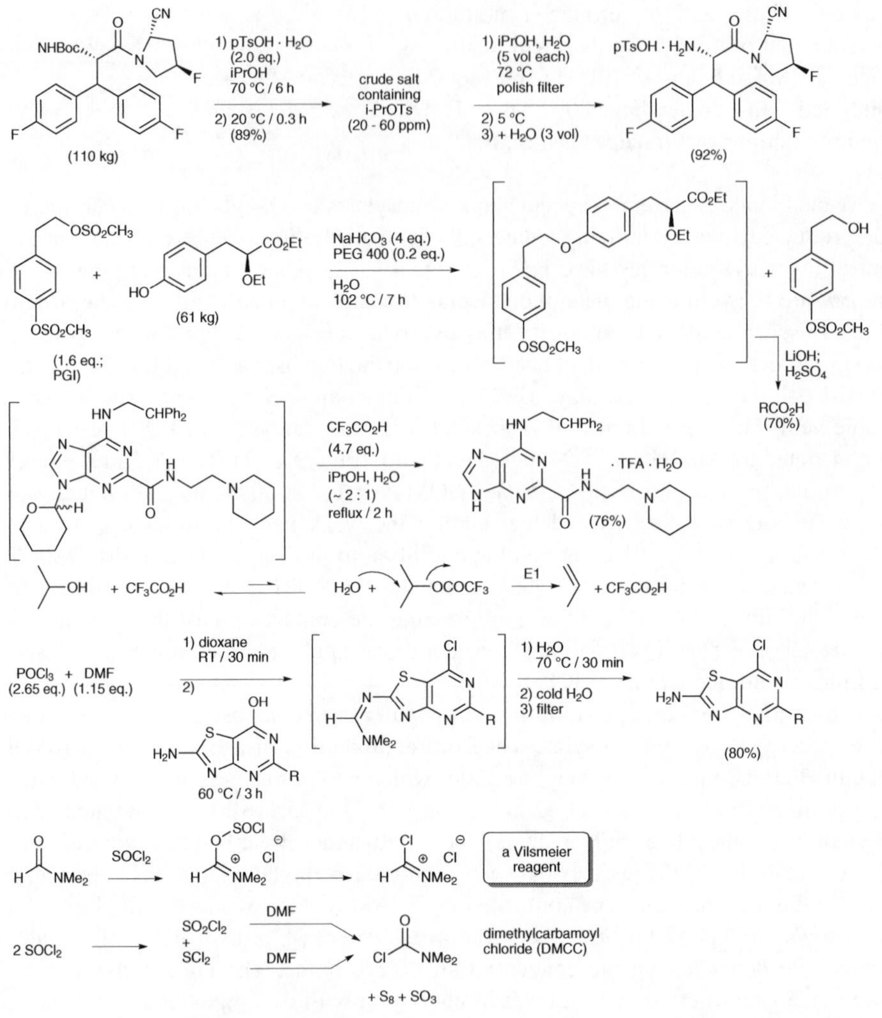

FIGURE 13.8 Use of water to decompose troublesome impurities.

quenching with water the concentration of DMCC remaining would be calculated at less than 1 ppm [121]. Other chloropyrimidines would decompose under these conditions used to decompose DMCC [122]. Aryl boronic acids and esters have been shown to be mutagenic in bacteria [93]; perhaps treatment with water will be used to decompose this class of compounds through protodeboronation (see Chapter 10). Figure 13.8 shows some excellent examples of adding H_2O reaction to decompose reactive impurities *en route* to isolating the reaction products in good yields.

Factors that can influence ability to purge impurities include reactivity, solubility, ionizability, volatility, and chromatography [123]. Approaches to control impurities in APIs can be summarized in the acronym ACE, for *Avoid*, *Control*, and *Expel* (Table 13.8). The examples in Figure 13.8 *control* the levels of impurities by

TABLE 13.8 ACE Approach to Minimizing Impurities, Including Genotoxic Impurities

Technique	Examples and Comments
Avoid use or formation	Avoid ICH Class 1 solvents Choose different starting materials Change order of steps in routes
Control formation or decomposition	Mechanistic studies on chemical fate
Expel from product	Chemical fate arguments used in QbD
Extractive workup	Exploit solubility differences
Absorption	Activated carbon, resins
Evaporation	Removing solvents; azeotroping
Selective crystallization	Gradually apply crystallization pressure Exploit solubility differences
Washing product cake	Displace impurities, e.g., high-bp solvents
Drying	Remove VOCs, low-MW compounds

decomposing them. DMF can be *avoided* in forming chloropyrimidines [122], but without DMF the chlorination in Figure 13.8 may have been unacceptably slow. DBU has been substituted for DMF in chlorinations [124], thus *avoiding* the use of DMF. Examples on *expelling* impurities from the product can be found in the chapters on workup and crystallization. The reactive crystallizations discussed with Figure 12.10 were employed in part because drying did not *expel* the alkyl chlorides [125].

Examples on *avoiding* the use or formation of troublesome materials can be found in the chapters on solvents, reagents, and route selection. Two additional examples shown in Figure 13.9 were developed to avoid generating a carcinogenic byproduct and a PGI. In the first example Zeneca researchers wanted to append an aminoethyl group onto a phenol, but realized that the obvious two-carbon units shown had disadvantages. They developed an oxathiazolidine-*S*-oxide reagent for the transformation [126]. In the second example a mesylate intermediate was used in the second route to an API; to avoid generating this PGI the manufacturing route employed the alcohol shown, with the coupling going through the trifluoroacetate ester formed *in situ* [127].

One aspect of *controlling* genotoxic impurities is to positioning PGIs early in the route, thus creating more opportunities to purge (*expel*) them from the APIs. Pierson described Lilly's three-step strategy of (1) positioning a PGI at least four steps from final step, and at each stage analyze potential for processing to remove it; (2) stressing purifications by spiking into process streams additional PGI prior to isolations (process tolerance studies); and (3) monitoring the concentration of the PGI in the penultimate, and setting limit tests. By validating the results a QbD approach is possible [128,129]. Vertex has described how risk assessment for controlling genotoxins leads to QbD [130]. The spiking experiments to determine how readily acetamide was removed from process streams are good examples

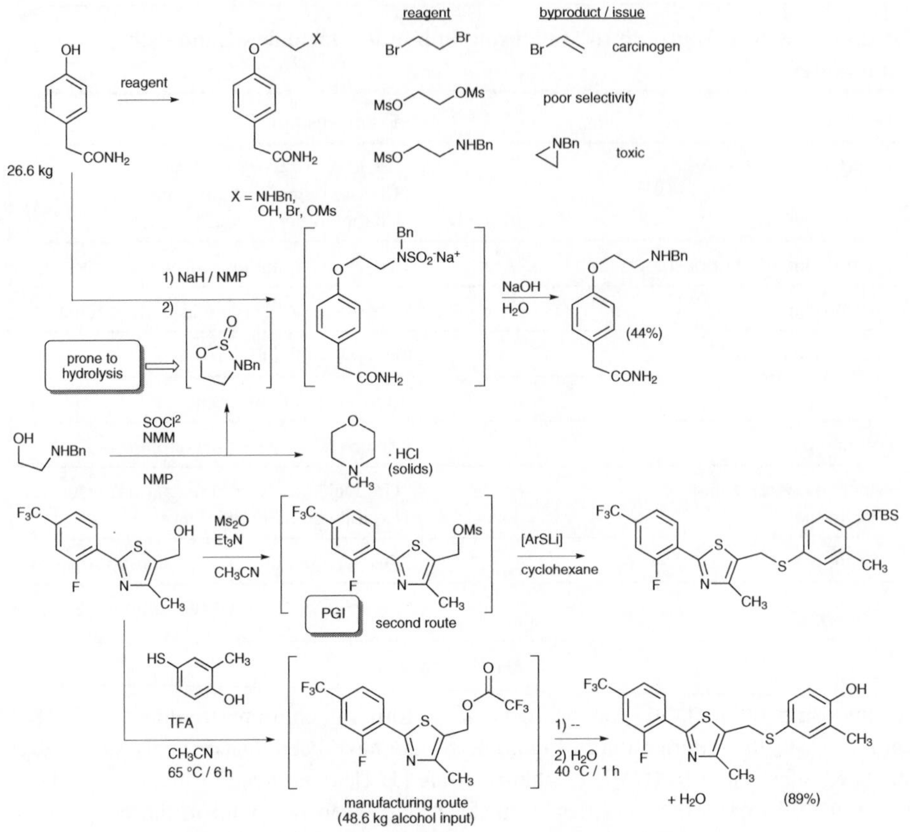

FIGURE 13.9 Avoiding troublesome intermediates or byproducts.

The best approach to control GTIs is to use data, not theory. Understand the mutagenicity of a GTI. Know the levels of PGI/GTI found in the API. Develop control points, chemical fate arguments, and PARs. Ultimately thorough understanding and QbD should simplify operations.

of process tolerance studies [131], and Lilly researchers have detailed the detection of formaldehyde in an intermediate and acceptable process tolerance that led to the removal of formaldehyde from a drug candidate [132]. Figure 11.10 shows how a PGI was controlled by positioning its use earlier in the route, and then by developing an extractive workup to *expel* it from the product [133].

V. STABILITY OF DRUG SUBSTANCE

Bulk stability studies are required by FDA on the API (for the NDA) and for the drug product. Guidance provides that stability testing should be carried out on at least three batches of drug substance, which were produced by what is intended to be the routine manufacturing method; accelerated stability testing can be carried out on one batch [134]. Standard conditions for photostability testing are described in ICH Q1B [135]. Guidance on the shelf life of a drug product has been described in ICH Q6A [136] and ICH Q6B [137], with guidance on degradents found separately in ICH Q3B(R2) [138]. The stability of drug products is tightly monitored; interactions of excipients with drug substance can produce impurities not found in the drug substance [138]. Specifications of APIs should be set so that batches of drug product will stay within specification limits throughout the shelf life, provided that the drug product is stored under the defined conditions [136,140]. Storage conditions and recommended frequencies for stability testing are shown in Tables 13.9 and 13.10.

The primary routes for degradation of APIs in the solid state and for drug products are through oxidation and hydrolysis [4]. To a lesser extent degradation occurs through dehydration, photolytic reactions, rearrangements, and loss of volatile components, such as loss of HCl from a hydrochloride salt (see Figure 13.3). Dehydration of gabapentin to the lactam would be expected upon bulk storage (Figure 13.10); perhaps surprisingly, the lactam (LD_{50} 300 mg/kg for mice) is much more toxic than gabapentin (LD_{50} 8000 mg/kg for mice). Cyclization occurred under acid, alkaline, or dry conditions, and could be controlled to the limit of 0.5% by controlling the acidity at less than 20 ppm [141]. A degradent of the macrocylic lactone ABT-773 was formed by β-elimination of the sugar moiety, followed by a heterocyclic Diels–Alder reaction [142]. Some oxidative impurities identified in olanzapine are also shown in Figure 13.10 [143].

In anticipating degradation of APIs, consider functional groups in molecule as nucleophiles and electrophiles. Neighboring group participation, intermolecular reactions, and possible interactions with O_2, solvents, and other reagents may degrade the API. Any cyclization to form a five-membered byproduct is extremely likely (Illuminati, G.; Mandolini, L. *Acc. Chem. Res.* **1981,** *14,* 95). Excessive byproduct formation can necessitate changing how the API is isolated, or even prompt abandoning the API.

VI. PERSPECTIVE AND SUMMARY

The fate of a drug product is closely affected by the expectations of regulators and patients. To avoid surprises the guidance of regulatory authorities can be and should be consulted frequently. Consumers may become more aware of quality issues and expenses associated with drugs; for instance, *The Wall Street Journal* reported in 2000 that many drugs remain potent years after the expiration dates, and this article has been widely circulated. This study was carried out by the US Army on drugs that had been stored, unopened, in ideal conditions [144]. Litigation may accelerate or slow changes.

Greater clarity regarding genotoxins has been published [97,98] and more is expected. The EMA requested that manufacturers consider the risk of sulfonate esters contaminating their marketed products [145,146]. There are some inconsistencies in guidelines; for instance, the FDA listed acetamide as a food additive [147] and yet has regarded acetamide as a genotoxin [148–150]. (The carcinogenic activity of acetamide in rodents may be due to formation of hydroxylamine [151].) Ethyl chloride is considered a genotoxin, and yet is marketed as a topical anesthetic [152]. Jacobson-Kram and Jacobs have discussed the positions of the FDA on genotoxins, and pointed

TABLE 13.9 Stability Studies for Drug Substances (1)

Study	Storage Condition	Minimum Time Period Covered by Data at Submission
General Stability Studies for Drug Substances		
Long-term (2)	25 °C ± 2 °C/60% RH ± 5% RH or 30 °C ± 2 °C/65% RH ± 5% RH	12 months
Intermediate (3)	30 °C ± 2 °C/65% RH ± 5% RH	6 months
Accelerated	40 °C ± 2 °C/75% RH ± 5%	6 months
Drug Substances Intended for Storage in a Refrigerator		
Long-term	5 °C ± 3 °C	12 months
Accelerated	25 °C ± 2 °C/60% RH ± 5% RH	6 months
Drug Substances Intended for Storage in a Freezer		
Long-term	−20 °C ± 5 °C	12 months

(1) Q1A(R2) Stability Testing of New Drug Substances and Products, 2003: http://www.fda.gov/downloads/Drugs/GuidanceComplianceRegulatoryInformation/Guidances/UCM073369.pdf
(2) It is up to the applicant to decide whether long-term stability studies are performed at 25 °C ± 2 °C/60% RH ± 5% RH or 30 °C ± 2 °C/65% RH ± 5% RH.
(3) If 30 °C ± 2 °C/65% RH ± 5% RH is the long-term condition, there is no intermediate condition.

TABLE 13.10 Recommended Testing Frequencies for Long-Term Stability Testing (1)

	Year One	Year Two	Thereafter Through Proposed Shelf Life
Long-term (shelf life of at least 12 mos)	0, 3, 6, 9, 12 mos	18, 24 mos	Annually
Accelerated (6 mos)	0, 3, 6 mos		
Intermediate (12 mos) (2)	0, 6, 9, 12 mos (at least 4 points)		

(1) Q1A(R2) Stability Testing of New Drug Substances and Products, 2003: http://www.fda.gov/downloads/Drugs/GuidanceComplianceRegulatoryInformation/Guidances/UCM073369.pdf
(2) Intermediate storage may be required if accelerated stability studies show significant changes.

out that impurities in general don't have beneficial impacts [153]. Snodin [89] and Elder and Harvey [154] have tried to put GTIs in perspective regarding the risk vs. benefits. Perhaps contributions from pharmaceutical industries will quantitate the costs of controlling gentoxins at ppm levels when developing drugs. Despite toxicological data,

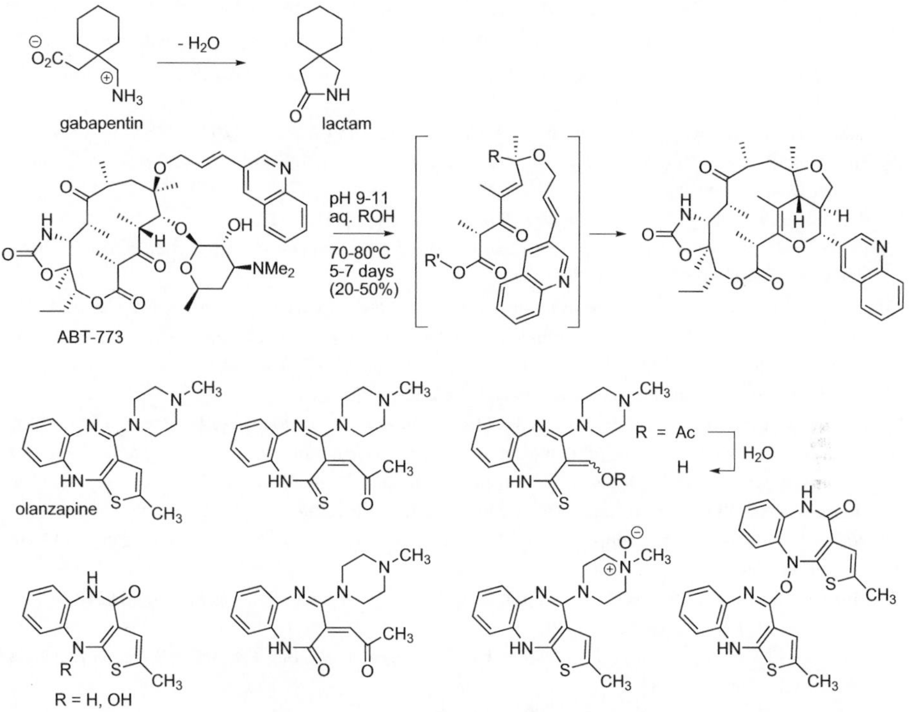

FIGURE 13.10 Some examples of degradents found in APIs.

limits for GTIs may be continue to be set as though these impurities are PGIs, following the ALARP principle. The ICH is expected to release guidance on limits of genotoxins as M7 [155].

The economics of failing to control the physical form and impurities in APIs provide good reasons to monitor operations and scrutinize data. Furthermore, from a moral standpoint we must control the impurities in APIs.

Presenting data in tables may be the best means to convince a scientist of the validity of an argument. Organizing data in the appropriate columns and rows gets the points across quickly. Policy-makers may base decisions on more than hard data, but clear presentations can minimize discussions.

REFERENCES

1. Raphals, P. *Science* **1990,** *249,* 619.
2. Kemsley, J. *Chem. Eng. News* **2008,** *86*(49), 32.
3. Kleinman, M. H.; Lee, B. In *Fundamentals of Early Clinical Development: From Synthesis Design to Formulation*; Abdel-Magid, A.; Caron, S., Eds.; Wiley, 2006; pp 269–294.
4. Byrn, S. R.; Xu, W.; Newman, A. W. *Adv. Drug. Deliv. Rev.* **2001,** *48,* 115. APIs can react with excipients; this reference described the reaction of fluoxetine hydrochloride with lactose to form brown impurities by the Maillard reaction.
5. Brittain, H. G.; Boganowich, S. J.; Bugay, D. E.; DeVincentis, J.; Lewen, G.; Newman, A. W. *Pharm. Res.* **1991,** *8,* 963.

6. Karner, S.; Urbanetz, N. A. *J. Aerosol Sci.* **2011,** *42*, 428.

7. Meier, M.; John, E.; Wieckhusen, D.; Wirth, W.; Peukert, W. *Eur. J. Pharm. Sci.* **2008,** *34*, 45.

8. Watanabe, H.; Ghadiri, M.; Matsuyama, T.; Ding, Y. L.; Pitt, K. G.; Maruyama, H.; Matsusaka, S.; Masuda, H. *Int. J. Pharm.* **2007,** *334*, 14.

9. Jadhav, N. R.; Gaikwad, V. L.; Nair, K. J.; Kadam, H. M. *Asian J. Pharm.* **2009,** *3*, 82.

10. Zanowiak, P. In *Encyclopedia of Pharmaceutical Technology,* Vol. 4, Swarbrick, J.; Boylan, J. C., Eds.; Marcel Dekker: New York, 1990; pp 209–229.

11. Beckmann, W.; Otto, W.; Budde, U. *Org. Process Res. Dev.* **2001,** *5*, 387.

12. Repic, O. *Principles of Process Research and Chemical Development in the Pharmaceutical Industry*; Wiley: New York, 1998; pp 179–194.

13. Detailed costs of formulations are difficult to assess. For one approach, see Butters, M.; Catterick, D.; Craig, A.; Curzons, A.; Dale, D.; Gillmore, A.; Green, S. P.; Marziano, I.; Sherlock, J.-P.; White, W. *Chem. Rev.* **2006,** *106*, 3002.

14. http://www.fingertipformulary.com/?action=label&drugId=750

15. *Physician's Desk Reference*, 53rd ed.; Medical Economics Co., Inc.: Montvale, New Jersey, 1999; p 484.

16. Regarding amorphous forms of atorvastatin calcium, Harrington, P. J. *Pharmaceutical Process Chemistry for Synthesis; Rethinking the Routes to Scale-Up*; Wiley-VCH, 2011; p 341. The preferred crystalline form of atorvastatin calcium is the trihydrate: Briggs, C. A.; Jennings, R. A.; Wade, R.; Harasawa, K.; Ichikawa, S.; Minohara, K.; Nakagawa, S. U.S. 5,969,156, 1999 (to Warner Lambert).

17. Morrison, H.; Jona, J.; Walker, S. D.; Woo, J. C. S.; Li, L.; Fang, J. *Org. Process Res. Dev.* **2011,** *15*, 104.

18. Thayer, A. M. *Chem. Eng. News* **2010,** *88*(22), 13.

19. Stahl, P. H.; Vermuth, C. G., Eds.; *Handbook of Pharmaceutical Salts: Properties Selection and Use*; Wiley-VCH: Zurich, 2008.

20. Swinbanks, D.; Anderson, C. *Nature* **1992,** *358*, 96.

21. Mayeno, A. N.; Lin, F.; Foote, C. S.; Loegering, D. A.; Ames, M. M.; Hedberg, C. W.; Gleich, G. J. *Science* **1990,** *250*, 1707.

22. http://www.nytimes.com/2010/09/14/health/14kelsey.html

23. Abell, C. W.; Shen, R. S.; Gessner, W.; Brossi, A. *Science* **1984,** *224*, 405.

24. Pearn, J. H. *J. Paediatr. Child Health* **1985,** *21*, 237.

25. Bowker, M. J. In *Handbook of Pharmaceutical Salts: Properties, Selection, and Use*; Stahl, P. H.; Wermuth, C. G., Eds.; Wiley-VCH: Weinheim, 2002, pp 163–164.

26. Bastin, R. J.; Bowker, B. J.; Slater, M. J. *Org. Process Res. Dev.* **2000,** *4*, 427.

27. Brittain, H. G. *Am. Pharm. Rev.* **2009,** 62 (November/December).

28. Gould, P. *Int. J. Pharm.* **1986,** *31*, 201.

29. Morris, K. R.; Fakes, M. G.; Thakur, A. B.; Newman, A. W.; Singh, A. K.; Venit, J. J.; Spagnuolo, C. J.; Serajuddin, A. T. M. *Int. J. Pharm.* **1994,** *105*, 209.

30. Antonucci, V.; Yen, D.; Kelly, J.; Crocker, L.; Dienemann, E.; Miller, R.; Almarrsson, O. *J. Pharm. Sci.* **2002,** *91*, 923.

31. Remenar, J. F.; MacPhee, J. M.; Larson, B. K.; Tyagi, V. A.; Ho, J. H.; McIlroy, D. A.; Hickey, M. B.; Shaw, P. B.; Almarsson, O. *Org. Process Res. Dev.* **2003,** *7*, 990.

32. Serajuddin, A. T. M.; Pudipeddi, M. In *Handbook of Pharmaceutical Salts: Properties, Selection, and Use*; Stahl, P. H.; Wermuth, C. G., Eds.; Wiley-VCH: Weinheim, 2002; pp 149–150.

33. *The Merck Index*, 13th ed.; Merck & Co., Inc.: Whitehouse Station, NJ, 2001; p 431.

34. Bogda, M. J. Tablet compression: machine theory, design, and process troubleshooting. In *Encyclopedia of Pharmaceutical Technology*, Vol. 18; Suppl. 1; Swarbrick, J.; Boylan, J. C., Eds.; Marcel Dekker: New York, 1999; p 252

35. Morris, K. Personal communication.

36. Wenslow, R. M.; Armstrong, J. D., III; Chen, A.; Cypes, S.; Ferlita, R. R.; Hansen, K.; Lindemann, C.; Spartalis, E. WO 2005/020920 A2; (to Merck & Co., Inc.).

37. Edmondson, S. D.; Fisher, M. H.; Kim, D.; MacCoss, M.; Parmee, E. R.; Weber, A. E.; Xu, J. WO 03/004498, 2003 (to Merck & Co., Inc.).

38. Hansen, K. B.; Hsiao, Y.; Xu, F.; Rivera, N.; Clausen, A.; Kubryk, M.; Krska, S.; Rosner, T.; Simmons, B.; Balsells, J.; Ikemoto, N.; Sun, Y.; Spindler, F.; Malan, C.; Grabowski, E. J. J.; Armstrong, J. D., III, *J. Am. Chem. Soc.* **2009,** *131,* 8798.

39. Miller, J. M.; Collman, B. M.; Greene, L. R.; Grant, D. J. W.; Blackburn, A. C. *Pharm. Dev. Technol.* **2005,** *10,* 291.

40. The FDA has established an EAFUS (Everything Added to Food in the United States) list of acceptable additives. These are generally regarded as safe (GRAS): http://www.cfsan.fda.gov/~dms/eafus.html

41. Newman, A. Personal communication.

42. Paulekuhn, G. S.; Dressman, J. B.; Saal, C. *J. Med. Chem.* **2007,** *50,* 6665. The source for these data is the FDA Orange Book, which lists approved generic drugs. Acids in this reference that were used less frequently used to form marketed salts were not included in Table 13.2.

43. Byrn, S. R. *Solid-State Chemistry of Drugs; Academic,* 1982; p 22.

44. Lynch, J. K.; Holladay, M. W.; Ryther, K. B.; Bai, H.; Hsiao, C.-N.; Morton, H. E.; Dickman, D. A.; Arnold, W.; King, S. A. *Tetrahedron Asymmetry* **1998,** *9,* 2791.

45. Lee, S.; Hoff, C. In *Handbook of Pharmaceutical Salts: Properties, Selection, and Use*; Stahl, P. H.; Vermuth, C. G., Eds.; Wiley-VCH: Weinheim, 2002; p 213.

46. Abramowitz, R.; O'Donoghue, D.; Prabhakaran, P.; Mathew, R.; Suda, R.F.; Jain, N.B. Poster presented at AAPS Meeting, Boston, 1997.

47. Giron, D.; Grant, D. J. W. In *Handbook of Pharmaceutical Salts: Properties, Selection and Use*; Stahl, P. H.; Wermuth, C. G., Eds.; Wiley-VCH: Weinheim, 2002; p 41.

48. Chatterjee, S.; Pedireddi, V. S.; Rao, C. N. R. *Tetrahedron Lett.* **1998,** *39,* 2843.

49. http://en.wikipedia.org/wiki/Oxalic_acid

50. Rhubarb is known for having high levels of oxalic acid in the leaves, and deaths occurred during World War I when rhubarb leaves were eaten as vegetables. For an interesting discussion, see http://www.rhubarbinfo.com/poison

51. Paulekuhn, G. S.; Dressman, J. B.; Saal, C. *J. Med. Chem.* **2007,** *50,* 6665. Source: the FDA Orange Book, which lists approved generic drugs.

52. Thanks to Elizabeth Vadas for this comment.

53. Schultheiss, N.; Newman, A. *Cryst. Growth Des.* **2009,** *9,* 2950.

54. Smith, M. B.; March, J. *March's Advanced Organic Chemistry,* 6th ed.; Wiley, 2007; pp 126–127.

55. Bingham, A. L.; Hughes, D. S.; Hursthouse, M. B.; Lancaster, R. W.; Tavener, S.; Threlfall, T. L. *Chem. Commun.* **2001,** 603.

56. Uekama, K.; Hirayama, F.; Irie, T. *Chem. Rev.* **1998,** *98,* 2045.

57. Amos, J. G.; Indelicato, J. M.; Pasini, C. E.; Reutzel, S. M. U.S. 6,001,996, 1999 (to Eli Lilly).

58. McNamara, D. P.; Childs, S. L.; Giordano, J.; Iarriccio, A.; Cassidy, J.; Shet, M. S.; Mannion, R.; O'Donnell, E.; Park, A. *Pharm. Res.* **2006,** *23,* 1888.

59. Hansen, K. B.; Chilenski, J. R.; Desmond, R.; Devine, P. N.; Grabowski, E. J. J.; Heid, R.; Kubryk, M.; Mathre, D. J.; Varsolona, R. *Tetrahedron: Asymmetry* **2003,** *14,* 3581.

60. Kemperman, G. J.; Zhu, J.; Klunder, A. J. H.; Zwanenburg, B. *Org. Lett.* **2000,** *2,* 2829.

61. Cai, D.; Hughes, D. L.; Verhoeven, T. R.; Reider, P. J. *Org. Synth. Coll. Vol. 10* **2004,** 93.

62. Roy, B. N.; Singh, G. P.; Srivastava, D.; Jadhav, H. S.; Saini, M. B.; Aher, U. P. *Org. Process Res. Dev.* **2009,** *13,* 450.

63. Li, Z.; Yang, B.-S.; Jiang, M.; Eriksson, M.; Spinelli, E.; Yee, N.; Senanayake, C. *Org. Process Res. Dev.* **2009,** *13,* 1307.

64. Vishweshwar, P.; McMahon, J. A.; Bis, J. A.; Zaworotko, M. J. *J. Pharm. Sci.* **2006,** *95*(3), 499.

65. Childs, S. L.; Rodríguez-Hornedo, N.; Reddy, L. S.; Jayasankar, A.; Maheshwari, C.; McCausland, L.; Shipplett, R.; Stahly, B. C. *Cryst. Eng. Comm.* **2008,** *10,* 856.

66. Anderson, N. G.; Coradetti, M. L.; Cronin, J. A.; Davies, M. L.; Gardineer, M. B.; Kotnis, A. S.; Lust, D. A.; Palaniswamy, V. A. *Org. Process Res. Dev.* **1997**, *1*, 315.

67. Impurities in new drug substances, Q3A(R2), *Step 4* version (25.10.06): http://www.fda.gov/RegulatoryInformation/Guidances/ucm127942.htm#i

68. Guidance for Industry Q3C Impurities: Residual Solvents http://www.fda.gov/downloads/Drugs/GuidanceComplianceRegulatoryInformation/Guidances/ucm073394.pdf

69. http://www.usp.org/pdf/EN/hottopics/2009-04-22MetalImpuritiesToxChart.pdf

70. Q3D guidance on elemental metal limits had not been released by the end of 2011.

71. http://books.google.com/books?id=GC85CKs7URkC&pg=PA265&lpg=PA265&dq=tap+density+spheres&source=web&ots=MYqy6iVAn0&sig=8Uwf30msc2R3sJq20_J3VKngInk&hl=en&sa=X&oi=book_result&resnum=9&ct=result#PPA265, M1.

72. Floyd, D.; Kocy, O.; Monkhouse, D. C.; Pipkin, J. D. U.S. 4,946,838, 1990 (to E. R. Squibb & Sons).

73. Robinson, R. *Org. Process Res. Dev.* **2010**, *14*, 323.

74. Delaney, E. J. *Regul. Toxicol. Pharmacol.* **2007**, *49*(2), 107.

75. Humfrey, C. D. N. *Toxicol. Sci.* **2007**, *100*(1), 24.

76. Bouder, F. *Expert Rev. Clin. Pharmacol.* **2008**, *1*, 241.

77. Elder, D. P.; Teasdale, A.; Lipczynski, A. M. *J. Pharm. Biomed. Anal.* **2008**, *46*, 1.

78. Teasdale, A., Ed. *Genotoxic Impurities: Strategies for Identification and Control*; Wiley Interscience, 2010.

79. Thayer, A. *Chem. Eng. News* **2010**, *88*(39), 16.

80. Kruhlak, N. L.; Contrera, J. F.; Benz, R. D.; Matthews, E. J. *Adv. Drug Delivery Rev.* **2007**, *59*, 43.

81. http://potency.berkeley.edu/cpdb.html

82. http://ntp.niehs.nih.gov/index.cfm?objectid=72016262-BDB7-CEBA-FA60E922B18C2540

83. Bercu, J. P.; Callis, C. M. *Org. Process Res. Dev.* **2009**, *13*, 938.

84. Cheeseman, M. A.; Machuga, E. J.; Bailey, A. B. *Food Chem. Toxicol.* **1999**, *37*, 387.

85. Kroes, R.; Renwick, A. G.; Cheeseman, M.; Kleiner, J.; Mangelsdorf, I.; Piersma, A.; Schilter, B.; Schlatter, J.; van Schothorst, F.; Vos, J. G.; Würtzen, G. *Food Chem. Toxicol.* **2004**, *42*, 65.

86. An acceptable exposure presents a risk of getting cancer of 1 in 100,000. The TD_{50} represents a risk of 1 in 2 (50%). To reach a risk of 1 in 100,000 the TD_{50} for mice (1810 mg/kg/day) is divided by 50,000, producing a value of 0.0362 mg/kg/day. For a human weighing 50 kg, limiting dosage would be 1.81 mg/day, or 1810 µg/day. This approach ignores species differences.

87. Haney, B. P.; Mason, P.; Anderson, N. G. *Org. Process Res. Dev.* **2009**, *13*, 939.

88. Abstracted from Snodin, D. J.. *Org. Process Res. Dev.* **2010**, *14*, 960.

89. Snodin, D. J. *Regul. Toxicol. Pharmacol.* **2006**, *45*, 79.

90. Garcia, C. C. M.; Angeli, J. P. F.; Freitas, F. P.; Gomes, O. F.; de Oliveira, T. F.; Loureiro, A. P. M.; Di Mascio, P.; Medeiros, M. H. G. *J. Am. Chem. Soc.* **2011**, *133*, 9140.

91. Baertschi, S. W.; Raney, K. D.; Stone, M. P.; Harris, T. M. *J. Am. Chem. Soc.* **1988**, *110*, 7929.

92. Dai, Q.; Xu, D.; Lim, K.; Harvey, R. G. *J. Org. Chem.* **2007**, *72*, 4856.

93. O'Donovan, M. R.; Mee, C. D.; Fenner, S.; Teasdale, A.; Phillips, D. H. *Mutat. Res.* **2011**, *724*(1–2), 1.

94. ToxTree software: toxtree.sourceforge.net/

95. Raillard, S. P.; Bercu, J.; Baertschi, S. W.; Riley, C. M. *Org. Process Res. Dev.* **2010**, *14*, 1015.

96. Mayer, J.; Cheeseman, M. A.; Twaroski, M. L. *Regul. Toxicol. Pharmacol.* **2008**, *50*, 50.

97. http://www.ema.europa.eu/docs/en_GB/document_library/Scientific_guideline/2009/09/WC500002907.pdf

98. Guidance for Industry: Genotoxic and Carcinogenic Impurities in Drug Substances and Products: Recommended Approaches (December 2008): http://www.fda.gov/downloads/Drugs/GuidanceComplianceRegulatoryInformation/Guidances/ucm079235.pdf

99. Guidelines for genotoxicity testing and data interpretation have been described in ICH Guidance S2: http://www.ema.europa.eu/docs/en_GB/document_library/Scientific_guideline/2011/12/WC500119604.pdf

100. *Guideline on the limits of genotoxic impurities.* CPMP/SWP/5199/02European Medicines Evaluation Agency, Committee for Medicinal Products for Human Use (CHMP), http://www.ema. europa.eu/docs/en_GB/document_library/Scientific_guideline/2009/09/WC500002903.pdf; June 28, 2006, London.

101. Müller, L.; Mauthe, R. J.; Riley, C. M.; Andino, M. M.; De Antonis, D.; Beels, C.; DeGeorge, J.; De Knaep, A. G. M.; Ellison, D.; Fagerland, J. A.; Frank, R.; Fritschel, B.; Galloway, S.; Harpur, E.; Humfrey, C. D. N.; Jacks, A. S.; Jagota, N.; Mackinnon, J.; Mohan, G.; Ness, D. K.; O'Donovan, M. R.; Smith, M. D.; Vudathala, G.; Yotti, L. *Regul. Toxicol. Pharmacol.* **2006,** *44*(3), 198.

102. Ames, B. N.; Gold, L. S. *Mutat. Res.* **2000,** *447*, 3.

103. Brigo, A.; Müller, L. In *Genotoxic Impurities: Strategies for Identification and Control*; Teasdale, A., Ed.; Wiley Interscience, 2010; p 27.

104. Robinson, D. I. *Org. Process Res. Dev.* **2010,** *14*, 946.

105. Gocke, E.; Lave, T.; Müller, L.; Pfister, T. http://www.gta-us.org/2008Meeting/2008Presentations/ Gocke.pdf

106. Mueller, L.; Gocke, E.; Larson, P.; Lave, T.; Pfister, T. Poster LBPE1157, XVII International AIDS Conference, 3–8 August 2008, Mexico City, Mexico; http://www.aids2008.org/Pag/Abstracts.aspx? AID=16184

107. http://www.ema.europa.eu/docs/en_GB/document_library/EPAR_-_Assessment_Report_-_Variation/ human/000164/WC500050681.pdf

108. Müller, L.; Gocke, E.; Lavé, T.; Pfiste, T. *Toxicol. Lett.* **2009,** *190*, 31.

109. http://www.ema.europa.eu/docs/en_GB/document_library/Medicine_QA/2009/11/WC500015048.pdf

110. Müller, L.; Singer, T. *Toxicol. Lett.* **2009,** *190*, 243.

111. Müller, L.; Gocke, E. *Toxicol. Lett.* **2009,** *190*, 330.

112. Elder, D. P.; Lipczynski, A. M.; Teasdale, A. *J. Pharm. Biomed. Anal.* **2008,** *48*, 497.

113. Teasdale, A.; Eyley, S. C.; Delaney, E.; Jacq, K.; Taylor-Worth, K.; Lipczynski, A.; Reif, V.; Elder, D. P.; Facchine, K. L.; Golec, S.; Oestrich, R. S.; Sandra, P.; David, F. *Org. Process Res. Dev.* **2009,** *13*, 429.

114. Teasdale, A.; Delaney, E. J.; Eyley, S. C.; Jacq, K.; Taylor-Worth, K.; Lipczynski, A.; Hoffmann, W.; Reif, V.; Elder, D. P.; Facchine, K. L.; Golec, S.; Oestrich, R. S.; Sandra, P.; David, F. *Org. Process Res. Dev.* **2010,** *14*, 999.

115. Patterson, D. E.; Powers, J. D.; LeBlanc, M.; Sharkey, T.; Boehler, E.; Irdam, I.; Osterhaut, M. H. *Org. Process Res. Dev.* **2009,** *13*, 900.

116. Patterson, D. E. Personal communication.

117. Chan, L. C.; Cox, B. G.; Sinclair, R. S. *Org. Process Res. Dev.* **2008,** *12*, 213.

118. Challenger, S.; Dessi, Y.; Fox, D. E.; Hesmondhalgh, L. C.; Pascal, P.; Pettman, A. J.; Smith, J. D. *Org. Process Res. Dev.* **2008,** *12*, 575.

119. Lee, C.; Helmy, R.; Strulson, C.; Plewa, J.; Kolodziej, E.; Antonucci, V.; Mao, B.; Welch, C. J.; Ge, Z.; Al-Sayah, M. A. *Org. Process Res. Dev.* **2010,** *14*, 1021.

120. Levin, D. *Org. Process Res. Dev.* **1997,** *1*, 182.

121. Stare, M.; Laniewski, K.; Westermark, W.; Sjögren, M.; Tian, W. *Org. Process Res. Dev.* **2009,** *13*, 857.

122. Anderson, N. G.; Ary, T. D.; Berg, J. L.; Bernot, P. J.; Chan, Y. Y.; Chen, C.-K.; Davies, M. L.; DiMarco, J. D.; Dennis, R. D.; Deshpande, R. P.; Do, H. D.; Droghini, R.; Early, W. A.; Gougoutas, J. Z.; Grosso, J. A.; Harris, J. C.; Haas, O. W.; Jass, P. A.; Kim, D. H.; Kodersha, G. A.; Kotnis, A. S.; LaJeunesse, J.; Lust, D. A.; Madding, G. D.; Modi, S. P.; Moniot, J. L.; Nguyen, A.; Palaniswamy, V.; Phillipson, D. W.; Simpson, J. H.; Thoraval, D.; Thurston, D. A.; Tse, K.; Polomski, R. E.; Wedding, D. L.; Winter, W. J. *Org. Process Res. Dev.* **1997,** *1*, 300.

123. Teasdale, A.; Fenner, S.; Ray, A.; Ford, A.; Phillips, A. *Org. Process Res. Dev.* **2010,** *14*, 943.

124. Frampton, G. A.; Hannah, D. R.; Henderson, N.; Katz, R. B.; Smith, I. H.; Tremayne, N.; Watson, R. J.; Woollam, I. *Org. Process Res. Dev.* **2004,** *8*, 415.

125. Yang, Q.; Haney, B. P.; Vaux, A.; Riley, D. A.; Heidrich, L.; He, P.; Mason, P.; Tehim, A.; Fisher, L. E.; Maag, H.; Anderson, N. G. *Org. Process Res. Dev.* **2009,** *13*, 786.

126. Barker, A. C.; Boardman, K. A.; Broady, S. D.; Moss, W. O.; Patel, B.; Senior, M. W.; Warren, K. E. H. *Org. Process Res. Dev.* **1999**, *3*, 253.

127. Brown, A. D.; Davis, R. D.; Fitzgerald, R. N.; Glover, B. N.; Harvey, K. A.; Jones, L. A.; Liu, B.; Patterson, D. E.; Sharp, M. J. *Org. Process Res. Dev.* **2009**, *13*, 297.

128. Pierson, D. A.; Olsen, B. A.; Robbins, D. K.; DeVries, K. M.; Varie, D. L. *Org. Process Res. Dev.* **2009**, *13*, 285.

129. For further commentaries on Pierson's article, see Snodin, D. A. *Org. Process Res. Dev.* **2009**, *13*, 409; Pierson, D. *Org. Process Res. Dev.* **2009**, *13*, 410.

130. Looker, A. R.; Ryan, M. P.; Neubert-Langille, B. J.; Naji, R. *Org. Process Res. Dev.* **2010**, *14*, 1032.

131. Schülé, A.; Ates, C.; Palacio, M.; Stofferis, J.; Delatinne, J.-P.; Martin, B.; Lloyd, S. *Org. Process Res. Dev.* **2010**, *14*, 1008.

132. Argentine, M. D.; Owens, P. K.; Olsen, B. A. *Adv. Drug Delivery Rev.* **2007**, *59*, 12.

133. Zegar, S.; Tokar, C.; Enache, L. A.; Rajagopol, V.; Zeller, W.; O'Connell, M.; Singh, J.; Muellner, F. W.; Zembower, D. E. *Org. Process Res. Dev.* **2007**, *11*, 747.

134. Q1A(R2) Stability Testing of New Drug Substances and Products, 2003, http://www.fda.gov/downloads/Drugs/GuidanceComplianceRegulatoryInformation/Guidances/UCM073369.pdf

135. http://www.fda.gov/downloads/Drugs/GuidanceComplianceRegulatoryInformation/Guidances/UCM073373.pdf

136. ICH Q6A Specifications: Test Procedures and Acceptance Criteria for New Drug Substances and New Drug Products: Chemical Substances (2000), http://www.ema.europa.eu/docs/en_GB/document_library/Scientific_guideline/2009/09/WC500002823.pdf

137. http://www.fda.gov/downloads/Drugs/GuidanceComplianceRegulatoryInformation/Guidances/ucm073488.pdf

138. http://www.fda.gov/downloads/Drugs/GuidanceComplianceRegulatoryInformation/Guidances/UCM073389.pdf

139. Jacobsen, T. M.; Wertheimer, A. I. *Modern Pharmaceutical Industry: A Primer*; Jones & Bartlett: Sudbury, Massachusetts, 2010.

140. Beaman, J. V. In *Pharmaceutical Testing to Support Global Markets*; Huynh-Ba, K., Ed.; Springer, 2009; p 211.

141. Augart, H.; Gebhardt, W.; Herrmann, W. U.S. 6,054,482, 2000 (to Gödecke AG).

142. Stoner, E. J.; Allen, M. S.; Christesen, A. C.; Henry, R. F.; Hollis, L. S.; Keyes, R.; Marsden, I.; Rehm, T. C.; Shiroor, S. G.; Soni, N. B.; Stewart, K. D. *J. Org. Chem.* **2005**, *70*, 3332.

143. Rao, P. S.; Ray, U. K.; Hiriyanna, S. G.; Rao, S. V.; Sharma, H. K.; Handa, V. K.; Mukkanti, K. *J. Pharm. Biomed. Anal.* **2011**, *56*, 413.

144. Cohen, L. P. This article is readily available from secondary sources on the internet, such as. *The Wall Street J.*, http://articles.mercola.com/sites/articles/archive/2001/02/07/drug-expiration-part-two.aspx; March 2000.

145. http://www.ema.europa.eu/docs/en_GB/document_library/Other/2009/11/WC500015375.pdf

146. After reviewing the toxicological data on ethyl mesylate, the EMA concluded that monitoring patients who had been inadvertently exposed to high levels of EtOMs in these batches of nelfinavir mesylate was not necessary. http://www.emea.europa.eu/humandocs/PDFs/EPAR/Viracept/38225608en.pdf

147. In the FDA's EAFUS (Everything Added to Food in the U.S.) database (http://www.cfsan.fda.gov/~dms/eafus.html), acetamide "has not yet been assigned for toxicology literature search." The Flavor and Extract Manufacturers Association Expert Panel, an independent organization, included acetamide in the GRAS list, stating an average maximum use level of 5 ppm for baked goods. http://members.ift.org/NR/rdonlyres/B9066605-9760-4E67-82BA-B2AA8D343020/0/0805gras22_complete.pdf

148. Acetamide is negative in the Ames assay and thus is not a mutagen or genotoxin. Based on rodent toxicity data the International Agency for Research on Cancer (IARC) has classified acetamide as possibly carcinogenic to humans (Group 2B) and recommended a limit of NMT 5 µg/day. http://www.inchem.org/documents/iarc/vol71/053-acetamide.html

149. Personal communications with clients and consultants.

150. For spiking studies to purge acetamide from process streams, see Reference 131.

151. Snodin, D. J. *Org. Process Res. Dev.* **2010,** *14,* 960.

152. Gebauer's Ethyl Chloride®: http://www.gebauerco.com/Products/Gebauer-s-Ethyl-Chloride-%281%29. aspx. Thanks to John Knight for mentioning this point.

153. Jacobson-Kram, D.; Jacobs, A. *Int. J. Toxicol.* **2005,** *24,* 129. McGovern, T.; Jacobson-Kram, D. *Trends Anal. Chem.* **2006,** *25,* 790.

154. Elder, D.; Harvey, J. In *Genotoxic Impurities: Strategies for Identification and Control*; Teasdale, A., Ed.; Wiley Interscience, 2010; p 193. See references therein.

155. M7 guidance on genotoxins had not been released by the end of 2011.

Continuous Operations

"The idea is primarily to switch from batch to continuous processing…. Sometimes our clients come because they are looking for a cheaper process, but mostly because they have SAFETY or quality issues. The new systems increase the homogeneity of mixtures and control reactions better, while producing fewer impurities so you can concentrate on the target molecule."

– Laurent Pichon, European Institute for Innovative Processes (MEPI) [1]

I. INTRODUCTION

A chemist needing to make 10 kg of a product might find that the reaction that worked well on a 10 g scale in a 200 mL round-bottomed flask did not work well on a 100 g scale. Unless this problem can be outsourced, the chemist is presented with at least three possibilities: to use the current chemistry on a scale calculated to make somewhat more than 10 kg in hopes of making that much material in the end; to develop conditions that would ensure successful scale-up in batches of 3–10 kg; or to run 1000 reactions on a 10 g scale in the equipment used for the successful 10 g run. Is it better to change the chemistry or the operations? The third option of running 1000 reactions is probably not an efficient use of time, but the chemist might be confident of success in each individual run, and no further process development might be necessary. *If* the reaction could be run in the equipment used for the 10 g batch operations *and* the reactor could be automatically filled and emptied at the end of the reaction, then making more material using the small reactor could be effective and efficient. This train of thought highlights some attractive features of continuous operations.

Practical Process Research and Development. DOI: 10.1016/B978-0-12-386537-3.00014-9

Continuous operations, or flow operations, allow products to be made in small reactors with efficient use of energy and generation of less waste, and such operations have been referred to as process intensification [2]. Many high-volume compounds, such as cyclopropylamine [3], sucrose, and table salt, have been manufactured for decades using continuous operations. Recently the pharmaceutical industry has been applying continuous operations to manufacture APIs.

Continuous operations are developed for processing primarily in five areas. First of all, continuous reactors can provide efficient mixing and tight temperature control, thus minimizing side products caused by micromixing and temperature excursions. Reactors such as static mixers create turbulent (efficient) mixing by pumping the reaction streams past internal elements in the tube. If high dilution is not needed to control micromixing, then solvent usage may be minimized. By circulating heat exchange fluids past the surface of a small-diameter tube or channel, temperatures can be controlled within narrow temperature ranges, due to the high surface area-to-volume ratios. More reliable processing through better temperature control and mass transport can minimize the number of rejected batches [4]. A small reactor such as a static mixer requires only a small footprint on a plant floor, and focuses the energy supplied to the reactor on a small volume [5]. Secondly, continuous operations can afford benefits for both cryogenic and high-temperature processes. Some processes requiring cryogenic temperatures in batch operations can be carried out at higher temperatures and in short reaction times through continuous operations. This approach may eliminate the need for Hastelloy reactors used for cryogenic batch operations. The high temperature–short time (HTST) regimen has been well established to pasteurize milk and prepare other food products [6]. Processes conducted at high temperatures, such as 140–300 °C, can be carried out in a coiled metal tube immersed in a high-temperature heating fluid or in a GC oven; fittings on each end of the tube allow the flow of process streams. In this fashion a high-temperature process can be constrained to a small area and better controlled to minimize SAFETY concerns from leaks of hot liquids. Third, intermediates that are unstable at the reaction conditions can be rapidly transferred from reactors to other vessels for subsequent reactions or workup. This approach can minimize decomposition of unstable intermediates by physically separating reactive intermediates and products. Fourth, continuous flow operations can ensure close contact of reaction streams with specialized areas of a reactor, such as light sources, sonic horns, beds of immobilized catalysts such as palladium on carbon, and reservoirs containing enzymes. The high localized concentrations afforded by these conditions can produce rapid reactions, and continuous operations may be the only practical way to make large amounts of material from photochemical, sonochemical, and microwave processes. Last but not least, continuous operations may be instituted for SAFETY reasons. Highly reactive compounds, such as phosgene and diazomethane, can be generated and used immediately, so that large amounts do not accumulate [7]. Continuous flow operation temperatures can also allow for SAFE control of reactions that require high temperatures and yet are exothermic slightly above the desired reaction temperatures [8,9]. By heating only a small amount of a reaction mixture at any moment to high temperatures the potential damage is lessened if problems should arise with controlling reaction temperatures. Hence for SAFETY reasons continuous operations may be preferred over batch or semi-batch operations. Continuous operations have been reviewed [10–18].

An organic chemist may first think of microreactors when the topic of continuous operations is raised [10,19]. Microreactors are produced by precision engineering, with the diameter of fluid channels ranging from sub-micrometers to sub-millimeters [19]. Through these small channels mass transport and heat exchange become more efficient [19]. Microreactors can include static mixers, areas with immobilized reagents, separators, and more [20]. Microreactors have been used in peptide synthesis [21], polymerase chain reactions [22], and even manufacturing monomers. Many microreactors have become commercially available over the past 10 years. Some of the benefits and drawbacks of microreactors are discussed in Section II.

Besides microreactors, other types of continuous reactors include simple tubing, continuously stirred tank reactors (CSTR, Figure 1.7), spinning discs [23], spinning tube-in-a-tube reactors [24,25], and plug flow reactors (PFRs). With the spinning disc reactor liquids are pumped onto the center of the disc, and dispersed to the edges by centrifugal forces; contact of the liquids with the disc is in the order of a few seconds. Spinning disc reactors were found to be effective with an intrinsic PTC reaction and to crystallize an API with small particle size and narrow PSD [26]. Static mixers are a type of PFR, with stationary internal elements that split and twist the flow of the stream. Some of the mixing elements may be seen for the sanitary static mixer in Figure 14.1. Specialized static mixers are also available, such as the static ozone mixer used to introduce ozone as micronized bubbles when sterilizing water in water treatment plants [27,28]. Diverse applications are possible with these energy-saving devices, mixing fluids ranging from non-viscous liquids to petroleum extracts to tomato paste [29]. Small static mixers are portable, inexpensive, and may be used in the laboratory, pilot plant, and manufacturing settings. Static mixers may be used for most of the operations described in Section III [30].

As mentioned above, a potential benefit of developing a continuous operation is that further process development may not be necessary to make large amounts of product. The simplest approach to make more material is to run the operations longer, "scaling out." Alternatively, running operations in several small continuous reactors

FIGURE 14.1 Sanitary static mixer, showing removable internal mixing elements. *(Used by permission of Chemineer, Inc.)*

simultaneously can generate large amounts of product through "numbering up" [19]. If larger reactors are employed for continuous operations, some process development may be necessary; calculations of Reynolds numbers have been used to assist the successful scale-up of PFRs (Section III).

Slightly different approaches are necessary for developing continuous operations as compared to batch or semi-batch operations. More equipment is needed for continuous operations. In addition to the continuous reactor, at least one pump is needed to control the flow of the input streams, along with at least one vessel to contain the starting materials and a vessel to collect the product stream. Secure fittings are needed to preclude leaks, and pressure gauges, pressure relief valves, and devices to prevent back-flow are necessary. The mindset for optimizing continuous processes is also slightly different from that of batch processing, as shown in Table 14.1. For continuous

TABLE 14.1 Some Comparisons of Batch and Semi-Batch Operations to Continuous Operations

Parameter/ Consideration	Batch or Semi-Batch Operations	Continuous Operations
Beginning of reaction	When desired temperature or pressure is reached; or, end of addition of a key component	When steady state is reached; or when input streams begin passing into or through a reactor
Reaction time	When conversion is suitably complete according to IPC	τ (Residence time)
End of reaction	When conversion is suitably complete according to IPC	When all input streams have passed through a reactor under steady-state conditions; or simply when all input streams have passed through the reactor
Comparative reaction temperatures	Cryogenic	May be warmer than cryogenic temperatures (higher with shorter τ)
To increase turbulence	Increase agitation rate, use baffles	Decrease τ (increase flow rate) or add more mixing elements
Typical concentration of reaction components	Routinely >0.3 M	The concentration of each component in separate streams may be higher before combining in the continuous reactor, to increase productivity
In-process assays	Usually external, can be on-line (PAT)	Readily adaptable for PAT

operations complete reactions are generally desired in one passage of material flowing through the reactor; hence reaction times become the mean residence time (τ, "tau"), the average time required for a molecule to pass through a reactor from beginning to end. For a PFR τ can be calculated by dividing the volume of the PFR by the flow rate. Steady-state conditions, necessary for reliable processing, might be defined by reaching a target flow rate, back pressure, temperature, or other criteria. The end of a continuous reaction can be defined as the end of steady-state conditions, or when all inputs have passed through the reactor. An advantage of using continuous operations for manufacturing APIs and drug products is the ease with which continuous, on-line or in-line analyses are possible; through such process analytical technology (PAT [31]) and feedback control of conditions, quality by design (QbD [32]) objectives may be readily met (see Chapters 7 and 16).

The following sections are based roughly on the diameter of the continuous processing channels or tubes. Section II discusses microreactors and reactors using small-diameter tubing. Operations conducted in reactors larger than the channels of microreactors may be called mesofluidics. Section III discusses continuous operations using larger diameter tubing or reactors, equipment that often has been used to prepare larger quantities of material.

II. MICROREACTORS AND SMALL-DIAMETER TUBING FOR SCALE-UP

Microreactors characteristically are made of stacks of channels with depths of <1 μm to <1 mm. Components include mixers, splitters, heat exchangers, and separators. Usually microreactors are made of metal or glass. They offer enhanced selectivity through improved mixing and heat transfer, and SAFETY [19,33].

Improved yields may result from shorter reaction times (τ) vs. semi-batch reaction times; cryogenic temperatures often are not needed [15,17]. Microreactors may be considered cGMP-friendly: inexpensive, virgin microreactors could be purchased for each new campaign, and continuous processing fits well with the FDA's PAT initiative [31].

Many uses have been demonstrated and promised for microfluidics. Microreactors have been used for on-demand production of hazardous materials, such as H_2O_2 [34,35]. Biodiesel has been prepared using microreactors [36,37]. Trimethyl orthoacetate has been manufactured using tubes of 1 mm diameter [38]. Drug discovery and analyses have been carried out on a "lab on a chip" [39]. Rapid process optimization has been touted [40–43]. Sigma-Aldrich discussed using microreactors for reaction screening and production in 2004 [44]. Prof. Seeburger was quoted as saying that "[m]icroreactors will become the round-bottom flasks of the 21st century" [41]. Using glass microreactors, Corning and DSM carried out a hazardous nitration under cGMP conditions, manufacturing more than 25 metric tons in four weeks [45].

In one of the most striking examples of using continuous operations for economic reasons, DSM developed continuous operations with custom metal microreactors to decrease the CoG of a monomer (Figure 14.2). Through continuous operations manufacturing personnel were able to control a very exothermic reaction and contain the volatile and toxic acrylonitrile. Rapid processing minimized decomposition of both starting materials under the reaction conditions. Impressive results were obtained: waste

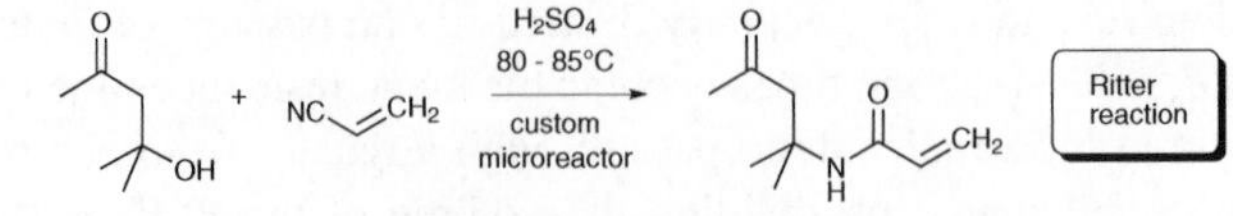

FIGURE 14.2 Continuous Ritter reaction to make a monomer.

was decreased by 15%, the yield was increased by 15%, and 40 tons/day of this monomer could be produced [46,47].

A number of suppliers have recently made small-scale equipment available for continuous operations in the laboratory. The oxadiazoles in Figure 14.3 were prepared as part of a library for discovery compounds [48]. Flow hydrogenolysis using an H-Cube® Continuous-flow Hydrogenation Reactor [49] was used to remove a benzyl group [50]. Equipment for operating SAFELY in continuous mode up to 200 bar and 350 °C is also available commercially [51]. An exothermic Paal–Knorr reaction was run in microreactors at 20 °C, generating the pyrrole in essentially quantitative yield. When four units were run in parallel, 55.8 g of the pyrrole was produced in an hour [52].

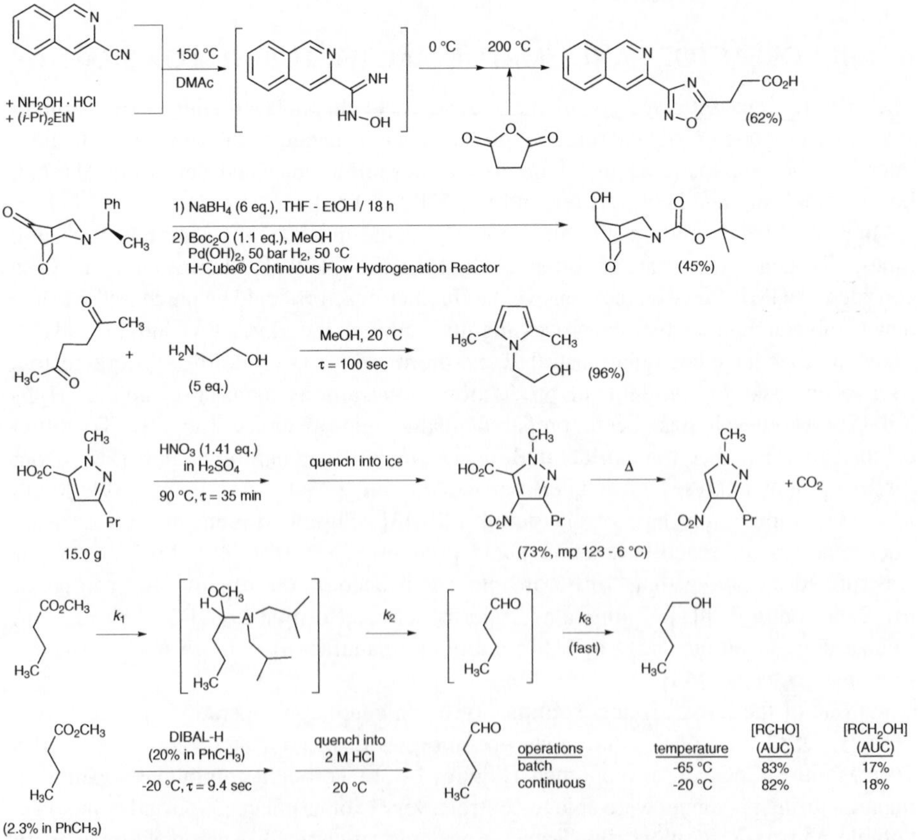

FIGURE 14.3 Reactions carried out in commercially available microfluidic reactors.

A feasibility study was carried out for a continuous nitration of a sildenafil intermediate. Temperatures of about 90 °C were favored for rapid reaction, but reactions run above 100 °C led to exothermic decomposition with CO_2 evolution. The reaction temperature was readily controlled due to efficient heat exchange in the microreactor, with channel widths less than 100 μm [9]. In the diisobutylaluminum hydride-mediated reduction of esters to aldehydes, over-reduction can readily occur if the first intermediate rapidly breaks down in situ to the desired aldehyde. The limitations inherent to external heat removal with batch operations can extend operations and increase the amount of the over-reduced side product. For instance, reduction of an ester at −78 °C generated the aldehyde at 95% AUC in the laboratory, but about 80% AUC in the pilot plant. The Lonza researchers showed that using continuous operations equivalent results were achieved at non-cryogenic temperatures [53]. The Buchwald and Jensen groups have shown that phenols can be converted to the corresponding triflates and on to biaryls through Suzuki–Miyaura couplings using a packed-bed reactor; key to this process was the microfluidic liquid–liquid extraction unit for removing water-soluble impurities [54,55]. A Swern oxidation in a microreactor at 20 °C has been described [56]. Many applications are possible.

Coiled tubing has been used for continuous processing, some using a commercially available unit [57]. McQuade's group has prepared ibuprofen in a three-stage sequence of acid-catalyzed acylation, oxidative rearrangement, and hydrolysis [58] (Figure 14.4). A hydroxamic acid was prepared in coiled tubing, and the data in Figure 14.4 show how the yield was influenced by τ and the temperature of the coil [59]. [3 + 2] Dipolar cycloadditions were carried out in tubing of 1 mm internal diameter (ID), producing the pyrrolidine shown in good yield [60]. A library of indazoles was also prepared quickly at 150–250 °C; by using the coiled tubing reactor reactions with methylhydrazine and hydrazine were carried out SAFELY [61]. A Newman–Kwart rearrangement was carried out in DME at 300 °C and 1000–1000 psi; under these conditions the solvent became supercritical and doubled in volume. At this temperature the starting material and product were highly soluble, and the solvent could be readily removed after the reaction, facilitating isolation. A syringe pump that operated at these pressures was used to charge the solution of the starting material [62]. The exothermic nature of a halogen–metal exchange (and the stability of the lithiated thiophene) limited the ability to scale-up a reaction using batch operations, so continuous operations were investigated. A 92% conversion was seen using 0.25 inch coiled tubing, and optimization was rapid [63]. Researchers from the Stahl group and Lilly showed that O_2-mediated oxidation of 1-phenylethanol with a soluble palladium catalyst could generate the ketone in quantitative yield. The reactor was made of 0.375 in outside diameter (OD) stainless steel tubing, 7 L in volume, and for SAFE operations the concentration of the oxidant was 8% O_2 in N_2 [64]. Ozonolysis of 1-decene has been carried out in the Vapourtech R4 unit (1 mm ID tubing), with productivity equal to that of batch ozonolysis of the same olefin. In contrast to batch operations, with continuous operations the reaction temperature was significantly higher (−10 °C vs. −78 °C for batch operations) and the space–time yields from the small reactor were higher [65–67]. Formation of a benzyl azide was also carried out in a hot tube reactor using 2.16 mm ID stainless steel tubing [68]. These processes indicate the success possible with simple equipment.

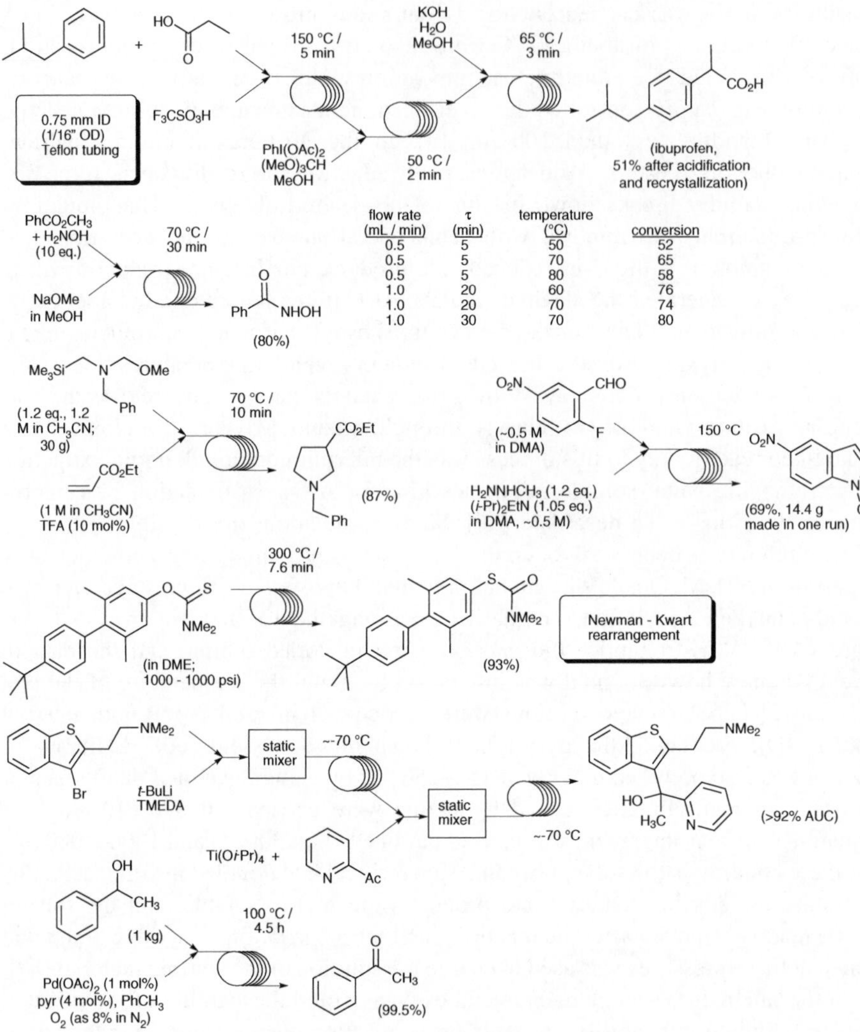

flow rate (mL / min)	τ (min)	temperature (°C)	conversion
0.5	5	50	52
0.5	5	70	65
0.5	5	80	58
1.0	20	60	76
1.0	20	70	74
1.0	30	70	80

FIGURE 14.4 Reactions carried out in coiled tubing.

III. CONTINUOUS OPERATIONS USING TUBING OR REACTORS OF LARGER DIAMETER

This section focuses preparing relatively large quantities of material using equipment larger than microreactors and small-diameter coiled tubing.

PFRs are often used for making larger amounts of material through continuous operations. Mixing in these metal or plastic tubes occurs primarily radially, not axially [69]; one might imagine a thin disc traveling down a tube, with the composition changing within the disc, until the disc emerges from the tube as a plug of product [70]. Primarily there are two types of flow considered in PFRs: laminar flow and turbulent flow. Laminar flow has no mixing perpendicular to the axis of the tube or vessel, and is relatively inefficient. Turbulent flow comprises random flow, and with effective turbulent mixing heat transfer and mass transfer occur efficiently. For turbulent flow in a tube the Reynolds number (*Re*) should be greater than 3500–4000, as calculated in Equation 14.1 [71,72]. (For calculating this dimensionless number keeping the units consistent is important.) This equation could be used to scale-up a continuous process using PFRs of increasing diameter. For instance, to reach the same *Re* with twice the diameter the linear flow rate would be halved, but the volume passing through the reactor per time (m^3/s) would double.

$$Re = du\rho/\mu \qquad (14.1)$$

where d = tube diameter (m), u = mean velocity (m/s), ρ = bulk density of fluid (kg/m^3), and μ = dynamic viscosity (N-s/m^2).

Continuous operations may be used to separate reactive intermediates and products. A continuous chlorination was carried out when there were unacceptable yields on a 100 g scale in batch mode (Figure 14.5). Close examination showed that the

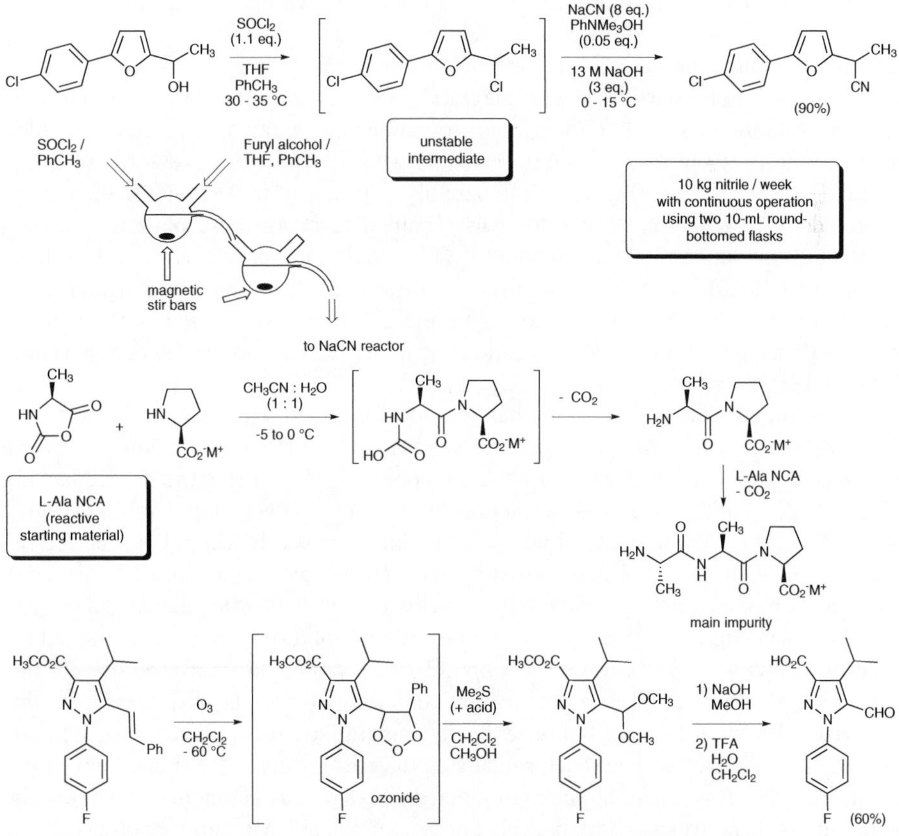

FIGURE 14.5   Continuous reactions with reactive intermediate and product.

intermediate chloride had a half-life of only 20 min (by NMR analysis in CDCl$_3$). Converting the unstable chloride as it formed to the nitrile product was necessary to make more nitrile efficiently. CSTRs were used, with ganged syringe pumps delivering an output flow rate of 10 mL/min. Using continuous reaction conditions the yield was 90%, and *10 kg of nitrile could be made in a week using two 10-*mL *round-bottomed flasks* [73]. This is an excellent example of both scaling out and converting an unstable intermediate to the desired product as it formed. As another example, Merck researchers made L-alanyl–L-proline, an enalapril intermediate, using the *N*-carbonic anhydride (NCA) of L-alanine. This approach saved two steps by avoiding the protection and deprotection of L-alanine. In a batch mode with an addition time for the NCA of less than 5 seconds, the yield was 90%. Unfortunately a 50-fold scale-up gave reduced yields. When the addition time was extended to only about 3 minutes the yield dropped to about 65% [74,75]. An engineer explained in another publication that the problem was the generation of the tripeptide impurity L-alanyl–L-alanyl–L-proline, formed by reaction of the desired dipeptide with the NCA. In this case the product is reactive toward the starting material, and using continuous operations the product was physically separated from the NCA. An in-line static mixer with a one-second residence time reduced the formation of the tripeptide. The operations were scaled up from 0.8 cm diameter in the laboratory to 1 inch in the plant with no change in selectivity [76]. Ozonolysis of a styrene was carried out using counter-current flow through a chromatography column packed with glass beads to ensure better gas–liquid contact. The output solution of ozonide was quenched in a solution of dimethyl sulfide as it emerged from the column [77,78]. Abbott researchers developed an ozonolysis under CSTR conditions: the reaction mixture overflowed into a quench vessel containing dimethyl sulfide. About 90 g/day of the ozonolysis product, an API intermediate, was generated. A larger ozone generator was obtained to make more material, and an output of 80 g/hour resulted, again under CSTR conditions. The researchers described the extensive SAFETY precautions they employed for these reactions, which were carried out in a fume hood [79]. Scaling up the CSTR ozonolysis was more deemed practical than numbering up, which would have increased the complexity of generating ozone and monitoring for leaks of ozone [80].

Successful continuous processing has been demonstrated with contained enzymatic and inorganic catalysts. Tao and co-workers developed a continuous enzyme-mediated reduction to prepare a chiral alcohol, with ammonium formate being the bulk reductant (Figure 14.6). The process was accelerated by forced contact of the substrate with the enzymes: the starting material was pumped into a reservoir where D-lactate dehydrogenase (D-LDH) and formate dehydrogenase (FDH) were constrained by a nanofiltration membrane, through which the stream of chiral product exited [81]. This procedure is analogous to an enzyme immobilized to a solid support and constrained by a filter to one vessel. Merck researchers optimized a dynamic kinetic resolution for the preparation of a chiral ethyl ester precursor to odanacatib. Under the basic conditions the two azlactones were in equilibrium, and the immobilized *Candida antarctica* lipase B (CAL-B) in the packed-bed reactor mediated the ethanolysis of the desired azlactone enantiomer [82]. By containing the immobilized enzyme in a column the enzyme is not subjected to shear from agitation [83]. Successful Suzuki–Miyaura couplings using a continuous-flow fixed-bed reactor and flow-through conditions were demonstrated

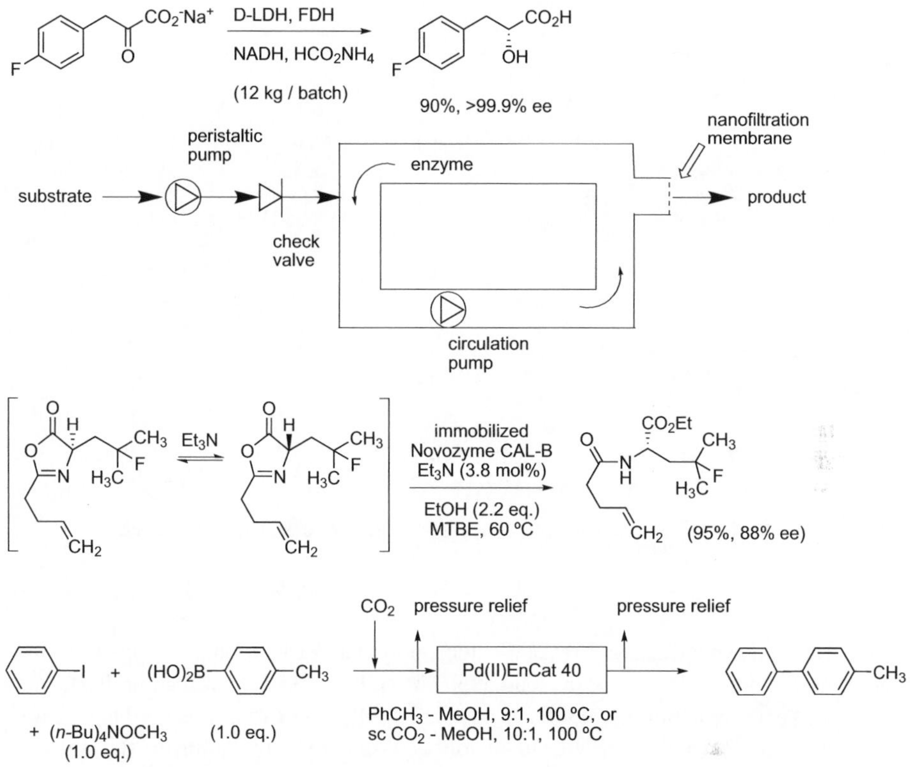

FIGURE 14.6 Continuous catalytic reactions.

using either conventional solvents (toluene–MeOH) or sc CO_2–MeOH. Using the latter solvent the conversion to the biphenyl was quantitative after two passages through the column containing the Pd(II)EnCat 40 catalyst, as long as the appropriate amount of the base (n-Bu)$_4$NOMe was present [84]. In another example, a column was packed with raw almond meal, a source of oxynitrilases. Through the column was passed a solution of HCN and aromatic aldehydes in isopropyl ether, producing cyanohydrins with up to 99% ee [85]. The latter example demonstrates the utility of a continuous-flow reactor for crude catalyst preparations. Benzylic alcohols have been oxidized to either aldehydes or acids using silica-supported Jones reagent [86]. The Poliakoff group has used fixed catalyst beds for continuous operations using sc CO_2, as in the methylation of alcohols, carboxylic acids and amines by dimethyl carbonate [87,88] and hydrogenation of iso-phorone to the saturated ketone [89]. By passing reaction streams through catalysts reaction rates may be increased due to the forced contact.

Continuous operations can make large amounts of material using processes that require localized sources of energy, such as photolysis [90,91], sonochemistry [92–94], microwave chemistry, and heat. The intensity of energy radiating from a point is inversely proportional to the square of the distance from the energy source [95], hence process streams are directed to flow near the energy source for efficient operations. The photolyses in Figure 14.7 were carried out by circulating the reaction stream

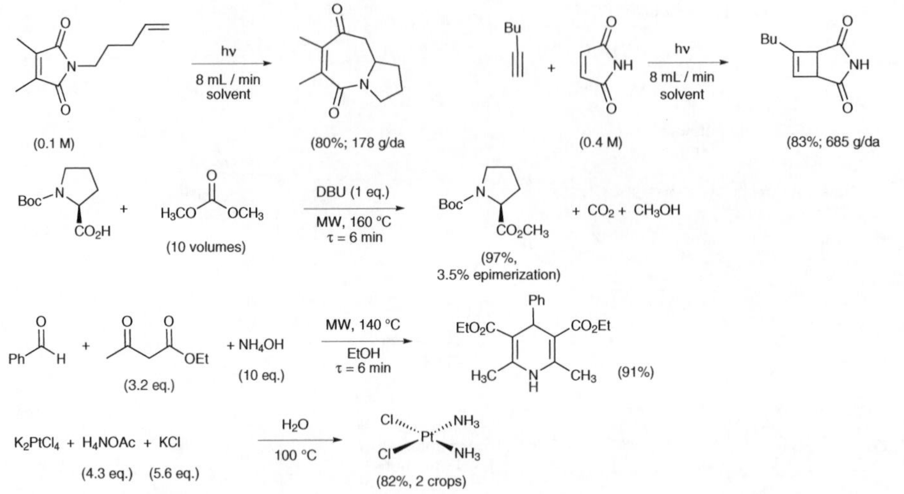

Practical Process Research and Development

FIGURE 14.7 Some reactions benefiting from close contact with energy sources.

through UV-transparent, solvent-resistant fluoropolymer (FEP) tubing wrapped around a lamp [96]. Photolysis has been conducted through a glass microreactor [97], and a continuous photoreactor is commercially available [98]. Sonication was used to fracture large crystals and make filtration and handling easier, by circulating crystal slurries externally past a sonic probe. Subsequent Ostwald ripening removed the fines and made the crystals more uniform [99]. Sonication with temperature cycling is used for the manufacture of irbesartan [100]. Annular sonic sources bonded to the outside of pipes have also provided reproducible crystallization, even generating small particles for inhalation [101]. Microwave esterification of Boc-L-proline with dimethyl carbonate was found to be about 80 times faster than thermal esterification (90 °C), and esterification could be carried out in continuous mode [102]. (Esterification with dimethyl carbonate does not need microwave conditions; such esterification has also been carried out using DABCO [103], K_2CO_3/DMF [104], and acid catalysis [105].) The Leadbeater group has described approaches for scale-up of microwave-promoted reactions. Under flow conditions the 1,4-dihydropyridine was generated in high yield [106], and large-scale reactors are available for batch reactions [107]. Kappe's group has shown that high temperatures may be the critical part of microwave reactions [108]. Conditions developed for microwave heating have been applied to heat reaction mixtures conventionally; using continuous flow the streams exiting the reactor can be diluted with organic solvents, thus avoiding clogging from solid products [109]. Cisplatin has been prepared through conventional heating using continuous processing [110]. These technologies may rapidly provide large amounts of material for initial scale-up, and for routine manufacturing.

Solids can plug microchannel reactors, tubing, and static mixers, causing researchers to develop homogeneous solutions for continuous processing. Nonetheless clogging can occur [47]. Eliminating constrictions and applying sonication can help avoid plugging [111].

IV. SUMMARY AND PERSPECTIVE

Continuous processing is often the only way to make large amounts of material using processes that require rapid heat transfer (for low- and high-temperature processes), efficient mixing, or contact with localized energy sources (photolyses, sonolyses, and thermolyses). Continuous reactors may be employed for the SAFE scale-up of some reactions. Despite the pressure generated under high temperatures continuous reactions can be carried out safely using back-pressure regulators and pressure relief valves; for instance, a Heck reaction was conducted in DMF or EtOH at 130 °C under continuous conditions [112], aminolyses in EtOH were conducted in a microreactor at 245 °C [113], and a Newman–Kwart rearrangement was conducted in glyme at 300 °C [62; Figure 14.4]. Using microreactors [113] and PFRs volatile components may be constrained to the liquid phase, not volatilized into the headspace of a reactor. Some reactions may benefit from continuous operations by physically separating reactive intermediates and products, hence minimizing impurity formation and easing scale-up. Other types of continuous processing, such as continuous separations, continuous concentration using wiped film evaporators [114] and continuous crystallization using CSTRs [115] or impinging jets [116,117], cannot be covered in this brief review. The equipment used for many continuous processes is simple and inexpensive. A chemist may find that his or her chemical engineer counterpart has studied continuous operations, and would be eager to investigate such optimization.

Concerns about material produced during non-steady-state operations or material that was out of specification have discouraged some from considering continuous operations for manufacturing APIs. Steady-state conditions can be reached by controlling parameters such as the flows of process streams and heat transfer fluids. Ideally non-steady-state conditions would be limited to perhaps two residence times at both the beginning and end of operations. At least two approaches for cGMP manufacturing using continuous operations are feasible and practical. The most direct option is to separate substandard portions of the product stream from the main portion of the batch, and subject these portions to rework or disposal. Under some circumstances the sub-par portions could be recirculated through the continuous reactor with additional starting materials or reagents if the impurities and subsequent processing were understood well enough to validate the overall process. Another approach is to combine the entire process stream in one vessel and process further to ensure high quality and homogeneity; recrystallization is an obvious option. In this fashion the operations collectively become semi-continuous or semi-batch. Blending *batches* that fail specifications with acceptable batches in order to meet specifications is not acceptable [118]. Successful process validation, necessary to make cGMP material for human use, depends in part on understanding the ability of processing to tolerate levels of impurities that are routinely generated [119].

Not all processes and reactions are suitable for continuous processing. For batch operations that routinely perform well, such as the crystallization of an API, there is little impetus to introduce continuous operations. Reactions that occur over several hours would need extremely large reactors, which might negate the advantages of continuous operations. Some fermentations, multi-component reactions, and CIAT/CIDR processes would be included in this category. However, reactions that occur in milliseconds and where micromixing is an issue (Table 8.3) may be good candidates for continuous operations. Some specialized reactions may only be amenable to continuous operations to make more material, as discussed in this chapter, and continuous operations should be considered as possible approaches.

The continuing pressure to speed process development may discourage some scientists from investing time and capital into new directions such as continuous processing, but various CROs offer those services [120]. In 2011 worldwide 20–30 plants were using microstructured reactors for manufacturing [47]. Industry and various academic groups are developing practical approaches to continuous processing. The push toward process intensification as cost-effective manufacturing operations will continue.

REFERENCES

1. Mullin, R. *Chem. Eng. News* **2011,** *89*(30), 18.
2. Lapkin, A. A.; Plucinski, P. K. RSC Green Chemistry No. 5, In *Chemical Reactions and Processes under Flow Conditions*; Luis, S. V.; Garcia-Verdugo, Eds.; The Royal Society of Chemistry, 2010; pp 18–24.
3. Diehl, H.; Blank, H. U.; Ritzer, E. U.S. 5,032,687, 1991 (to Kernforschunganlage Julich Gesellschaft); Kleemiss, W.; Kalz, T. U.S. 5,728,873, 1998 (to Huels Aktiengesellschaft). See discussion with Figure 1.6.
4. Understanding mass transport and heat transfer is one of the principles of green chemistry: Winterton, N. *Green Chem.* December **2001,** G73.
5. Maximizing the space–time efficiencies of operations is a principle of green engineering: Anastas, P. T.; Zimmerman, J. B. *Environ. Sci. Technol.* **2003,** *37*(5), 95.
6. See discussion with the continuous Swern oxidation in Figure 1.6.
7. Proctor, L. D.; Warr, A. J. *Org. Process Res. Dev.* **2002,** 6, 884; Struempel, M.; Ondruschka, B.; Daute, R.; Stark, A. *Green Chem.* **2008,** *10,* 41. Diazomethane was generated in these procedures from Diazald®, KOH, and Carbitol® (diethylene glycol monoethyl ether).
8. Bogaert-Alvarez, R. J.; Demena, P.; Kodersha, G.; Polomski, R. E.; Soundararajan, N.; Wang, S. S. Y. *Org. Process Res. Dev.* **2001,** *5*, 636.
9. Panke, G.; Schwalbe, T.; Stirner, W.; Taghavi-Moghadam, S.; Wille, G. *Synthesis* **2003,** *18*, 2827.
10. Wiles, C.; Watts, P. *Micro Reaction Technology in Organic Synthesis.* CRC Press: Boca Raton, FL, 2011.
11. RSC Green Chemistry No. 5; Luis, S. V.; Garcia-Verdugo, Eds.; *Chemical Reactions and Processes under Flow Conditions*; The Royal Society of Chemistry, 2010.
12. Roberge, D. M.; Zimmermann, B.; Rainone, F.; Gottsponer, M.; Eyholzer, M.; Kockmann, N. *Org. Process Res. Dev.* **2008,** *12*, 905.
13. Ley, S. V.; Baxendale, I. R. *Chimia* **2008,** *62*, 162.
14. Mason, B. P.; Price, K. E.; Steinbacher, J. L.; Bogdan, A. R.; McQuade, D. T. *Chem. Rev.* **2007,** *107*, 2300.
15. Watts, P.; Wiles, C. *Org. Biomol. Chem.* **2007,** *5*, 727.
16. Roberge, D. M.; Ducry, L.; Bieler, N.; Cretton, P.; Zimmermann, B. *Chem. Eng. Technol.* **2005,** *28*, 318.
17. Penneman, H.; Watts, P.; Haswell, S. J.; Hessel, V.; Loewe, H. *Org. Process Res. Dev.* **2004,** *8*, 422.
18. Anderson, N. G. *Org. Process Res. Dev.* **2001,** *5*, 613.
19. Ehrfeld, W.; Hessel, V.; Löwe, H. *Microreactors: New Technology for Modern Chemistry*, Wiley-VCH: Weinheim, 2000.
20. Ondrey, G. *Chem. Eng.* **2001,** *108*(7), 27.
21. Watts, P.; Wiles, C.; Haswell, S. J.; Pombo-Villar, E.; Styring, P. *Chem. Commun.* **2001,** p 990.
22. Kopp, M. U.; de Mello, A. J.; Manz, A. *Science* **1998,** *280*, 1046.
23. Ramshaw, C. In *Re-Engineering the Chemical Processing Plant: Process Intensification*; Moulijn, J. A., Ed.; Dekker, 2007; p 69.

24. Gonzalez, M. A.; Ciszewski, J. T. *Org. Process Res. Dev.* **2009,** *13*, 64.

25. Hampton, P. D.; Whealon, M. D.; Roberts, L. M.; Yaeger, A. A.; Boydson, R. *Org. Process Res. Dev.* **2008,** *12*, 946.

26. Brechtelsbauer, C.; Lewis, N.; Oxley, P.; Ricard, F. *Org. Process Res. Dev.* **2001,** *5*, 65.

27. http://www.komax.com/Static-Mixer-Ozone.html

28. For an example of an ozonolysis carried out in water, see Fleck, T. J.; McWhorter, W. W.; DeKam, R. N.; Pearlman, B. A. *J. Org. Chem.* **2003,** *68,* 9612.

29. http://www.komax.com/pdf_files/Tomato_Paste_Juice_Case_Study.pdf

30. More information and equipment is available from http://www.komax.com, http://www.chemineer.com, http://www.kenics.com, http://www.staticmixer.com, http://www.koflo.com, and others.

31. Guidance for Industry: PAT – A Framework for Innovative Pharmaceutical Development, Manufacturing, and Quality Assurance; September 2004, http://www.fda.gov/downloads/Drugs/GuidanceComplianceRegulatoryInformation/Guidances/UCM070305.pdf

32. ICH Q8(R2) Pharmaceutical Development, November 2009, http://www.fda.gov/downloads/Drugs/GuidanceComplianceRegulatoryInformation/Guidances/ucm073507.pdf

33. Thomas, P. http://www.pharmamanufacturing.com/articles/2004/237.html

34. Freemantle, M. *Chem. Eng. News* **2004,** *82*(41), 39.

35. *Chem. Eng.* **2006,** *114*(5), 18.

36. Petkewich, R. *Chem. Eng. News* **2006,** *84*(18), 48.

37. *Chim. Oggi* **2006,** *24*(3), 6.

38. *Chem. Eng.* **2006,** *113*(5), 13.

39. Hrusovsky, K.; Roskey, M. *Mod. Drug Disc.* **2004,** *7*(12), 24.

40. Gray, R. *Curr. Drug Disc.* **2004,** 9 [July].

41. Freemantle, M. *Chem. Eng. News* **2005,** *83*(6), 11.

42. Rattner, D. M.; Murphy, E. R.; Jhunjhunwala, M.; Snyder, D. A.; Jensen, K. F.; Seeburger, P. H. *Chem. Commun.* **2005,** 578.

43. Schwalbe, T.; Simons, K. *Chim. oggi* **2006,** *24*(2), 56.

44. Rouhi, A. M. *Chem. Eng. News* **2004,** *82*(27), 18.

45. Short, L. *Chem. Eng. News* **2008,** *86*(42), 37.

46. Bohn, L.; Braune, S.; Kotthaus, M.; Kraut, M.; Pöchlauer, P.; Vorbach, M.; Wenka, A.; Schubert, K. poster presented at IMRET-9, 9th International Conference on Microreaction Technology, Potsdam Germany, 6–8 Sept. 2006; Pöchlauer, P.; Kotthaus, M.; Vorbach, M.; Deak, M.; Zich, T.; Marr, R. US 2008/0306300 A1.

47. Ondrey, G. *Chem. Eng.* **2011,** *118*(4), 17.

48. Grant, D.; Dahl, R.; Cosford, N. D. P. *J. Org. Chem.* **2008,** *73*, 7219.

49. Thales Nanotechnology's H-Cube® Continuous-flow Hydrogenation Reactor generates H_2 by electrolysis, and uses disposable packed catalyst cartridges. Faster reductions are afforded by forcing contact of the reaction stream with the catalyst. This innovative equipment has been the equipment routinely used in drug discovery laboratories for hydrogenations and hydrogenolyses, http://www.thalesnano.com/products/h-cube

50. Darout, E.; Basak, A. *Tetrahedron Lett.* **2010,** *51*, 2998.

51. Thales Nanotechnology's X-Cube Flash™, http://www.thalesnano.com/node/130

52. Nieuwland, P. J.; Segers, R.; Koch, K.; van Hest, J. C. M.; Rutjes, F. P. J. T. *Org. Process Res. Dev.* **2011,** *15*, 783.

53. Ducry, L.; Roberge, D. M. *Org. Process Res. Dev.* **2008,** *12*, 163.

54. Noël, T.; Kuhn, S.; Mushio, A. J.; Jensen, K. F.; Buchwald, S. L. *Angew. Chem. Int. Ed.* **2011,** *50*, 5943.

55. Ritter, S. K. *Chem. Eng. News* **2011,** *89*(23), 39.

56. Kawaguchi, T.; Miyata, H.; Ataka, K.; Mae, K.; Yoshida, J. *Angew. Chem. Int. Ed.* **2005,** *44*, 2413.

57. http://www.vapourtec.co.uk/products/reactors

58. Bogdan, A. R.; Poe, S. L.; Kubis, D. C.; Broadwater, S. J.; McQuade, D. T. *Angew. Chem. Int. Ed.* **2009**, *48*, 8547.

59. Riva, E.; Gagliardi, S.; Mazzoni, C.; Passarella, D.; Rencurosi, A.; Vigo, D.; Martinelli, M. *J. Org. Chem.* **2009**, *74*, 3540.

60. Grafton, M.; Mansfield, A. C.; Fray, M. J. *Tetrahedron Lett.* **2010**, *51*, 1026.

61. Wheeler, R. C.; Baxter, E.; Campbell, I. B.; Macdonald, S. J. F. *Org. Process Res. Dev.* **2011**, *15*, 565.

62. Tilstam, U.; Defrance, T.; Giard, T.; Johnson, M. D. *Org. Process Res. Dev.* **2009**, *13*, 321.

63. Gross, T. D.; Chou, S.; Bonneville, D.; Gross, R. S.; Wang, P.; Campopiano, O.; Ouellette, M. A.; Zook, S. E.; Reddy, J. P.; Moree, W. J.; Jovic, F.; Chopade, S. *Org. Process Res. Dev.* **2008**, *12*, 929.

64. Ye, X.; Johnson, M. D.; Diao, T.; Yates, M. H.; Stahl, S. S. *Green Chem.* **2010**, *12*, 1180.

65. Roydhouse, M. D.; Ghaini, A.; Constantinou, A.; Cantu-Perez, A.; Motherwell, W. B.; Gavriilidis, A. *Org. Process Res. Dev.* **2011**, *15*, 989.

66. Ozonolysis using a falling film microreactor: Steinfeld, N.; Abdallah, R.; Dingerdissen, U.; Jähnisch, K. *Org. Process Res. Dev.* **2007**, *11*, 1025.

67. Ozonolysis using structured microreactors: Hübner, S.; Bentrup, U.; Budde, U.; Lovis, K.; Dietrich, T.; Freitag, A.; Küpper, L.; Jähnisch, K. *Org. Process Res. Dev.* **2009**, *13*, 952.

68. Kopach, M. E.; Murray, M. M.; Braden, T. M.; Kobierski, M. E.; William, O. L. *Org. Process Res. Dev.* **2009**, *13*, 152.

69. Michel, B. J. *Chem. Processing* **1989**, July, 24.

70. Roberts, G. W. *Chemical Reactions and Chemical Reactors*. Wiley, 2009.

71. Atherton, J. H.; Carpenter, K. J. *Process Development: Physicochemical Concepts*; Oxford University Press: Oxford, 1999; pp 37–38.

72. http://www.engineeringtoolbox.com/reynolds-number-d_237.html

73. Foulkes, J. A.; Hutton, J. *Synth. Commun.* **1979**, *9*, 625.

74. Blacklock, T. J.; Shuman, R. F.; Butcher, J. W.; Shearin, W. E., Jr.; Budavari, J.; Grenda, V. J. *J. Org. Chem.* **1988**, *53*, 836.

75. NCAs of amino acids are used to make poly-amino acids. The reactivity of *N*-alkyl NCAs is probably much lower. A quinapril precursor was prepared on scale using 2 mol% of acetic acid, perhaps as a catalyst for reacting with the NCA. The *t*-butyl ester of (*S*)-1,2,3,4-tetrahydro-3-isoquinolinecarboxylic acid (from 30 kg of the tosic acid salt) in toluene was added over about 5 min to a solution of the ethyl ester of (2*S*,4*S*)-2-(4-methyl-2,5-dioxo-oxazolidin-3-yl)4-phenyl-butyric acid (23.5 kg) and 90 g of AcOH in toluene. No temperature was mentioned and no yield was given, but the in-process HPLC data indicated the conversion was essentially quantitative. Jennings, S. M. U.S. 6,858,735 B2, 2005 (to Warner–Lambert).

76. Paul, E. L. *Chem. Eng. Sci.* **1988**, *43*, 1773.

77. Bowles, D. M.; Boyles, D. C.; Choi, C.; Pfefferkorn, J. A.; Schuyler, S. *Org. Process Res. Dev.* **2011**, *15*, 148. Some residual acid must have been present during the quench, as the dimethyl acetal was formed. The acetal ester was converted to the desired aldehyde ester by treatment with NaOH and TFA, as shown in the figure.

78. For a description of a similar column, see Pelletier, M. J.; Barilli, M. L.; Moon, B. *Appl. Spectrosc.* **2007**, *61*, 1107.

79. Allian, A. D.; Richter, S. M.; Kallemeyn, J. M.; Robbins, T. A.; Kishore, V. *Org. Process Res. Dev.* **2011**, *15*, 91.

80. Allian, A. D. Personal communication.

81. Tao, J.; McGee, K. *Org. Process Res. Dev.* **2002**, *6*, 520; Tao, J.; McGee, K. In *Asymmetric Catalysis on Industrial Scale*; Blaser, H. U.; Schmidt, E., Eds.; Wiley-VCH: Weinheim, 2004; pp 321–334.

82. Truppo, M. D.; Hughes, G. *Org. Process Res. Dev.* **2011**, *15*, 1033.

83. Yu, R. H.; Polniaszek, R. P.; Becker, M. W.; Cook, C. M.; Yu, L. H. L. *Org. Process Res. Dev.* **2007**, *11*, 972.

84. Leeke, G. A.; Santos, R. C. D.; Al-Duri, B.; Seville, J. P. K.; Smith, C. J.; Lee, C. K. Y.; Holmes, A. B.; McConvey, I. F. *Org. Process Res. Dev.* **2007**, *11*, 144. The authors noted that some Pd was leached from the catalyst used.

85. Chen, P.; Han, S.; Lin, G.; Li, Z. *J. Org. Chem.* **2002,** *67,* 8251.

86. Wiles, C.; Watts, P.; Haswell, S. J. *Tetrahedron Lett.* **2006,** *47,* 5261.

87. Gooden, P. N.; Bourne, R. A.; Parrott, A. J.; Bevinakatti, H. S.; Irvine, D. J.; Poliakoff, M. *Org. Process Res. Dev.* **2010,** *14,* 411.

88. Parrott, A. J.; Bourne, R. A.; Gooden, P. N.; Bevinakatti, H. S.; Poliakoff, M.; Irvine, D. J. *Org. Process Res. Dev.* **2010,** *14,* 1420.

89. Stevens, J. G.; Gómez, P.; Bourne, R. A.; Drage, T. C.; George, M. W.; Poliakoff, M. *Green Chem.* **2011,** *13,* 2727.

90. Hoffmann, N. *Chem. Rev.* **2008,** *108,* 1052.

91. Van Gerven, T.; Mul, G.; Moulijn, J.; Stankiewicz, A. *Chem. Eng. Process.: Process Intensification* **2007,** *46,* 78.

92. Nowak, F. M., Ed. *Sonochemistry: Theory, Reactions and Syntheses, and Applications*; Nova Science, 2010.

93. Mason, T. J.; Peters, D. *Practical Sonochemistry: Power Ultrasound Uses and Applications*, 2nd ed.; Woodhead Publishing, 2002.

94. Mason, T. J. *Chem. Soc. Rev.* **1997,** *26,* 443.

95. The inverse square law: http://en.wikipedia.org/wiki/Inverse-square_law#Light_and_other_electromagnetic_radiation

96. Hook, B. D. A.; Dohle, W.; Hirst, P. R.; Pickworth, M.; Berry, M. B.; Booker-Milburn, K. I. *J. Org. Chem.* **2005,** *70,* 7558.

97. Wootton, R. C. R.; Fortt, R.; de Mello, A. J. *Org. Process Res. Dev.* **2002,** *6,* 187.

98. From mikroglas chemtech GmbH: http://www.mikroglas.com

99. Kim, S.; Wei, C.; Kiang, S. *Org Process Res. Dev.* **2003,** *7,* 997.

100. Franc, B.; Hoff, C.; Kiang, S.; Lindrud, M. D.; Monnier, O.; Wei, C. U.S. 7,008,959, 2006, to Sanofi-Aventis.

101. Prosonix: http://www.chemsourcesymposia.org.uk/abstracts.htm

102. Shieh, W.-C.; Dell, S.; Repic, O. *Tetrahedron Lett.* **2002,** *43,* 5607.

103. Shieh, W.-C.; Dell, S.; Bach, A.; Repic, O.; Blacklock, T. J. *J. Org. Chem.* **2003,** *68,* 1954.

104. Jiang, X.; Tiwari, A.; Thompson, M.; Chen, Z.; Cleary, T. P.; Lee, T. B. K. *Org. Process Res. Dev.* **2001,** *5,* 604.

105. Rekha, V. V.; Ramani, M. V.; Ratnamala, A.; Rupakalpana, V.; Subbaraju, G. V.; Satyanarayana, C.; Rao, C. S. *Org. Process Res. Dev.* **2009,** *13,* 769.

106. Bowman, M. D.; Holcomb, J. L.; Kormos, C. M.; Leadbeater, N. E.; Williams, V. A. *Org. Process Res. Dev.* **2008,** *12,* 41.

107. Schmink, J. R.; Kormos, C. M.; Devine, W. G.; Leadbeater, N. E. *Org. Process Res. Dev.* **2010,** *14,* 205.

108. Damm, M.; Glasnov, T. N.; Kappe, C. O. *Org. Process Res. Dev.* **2010,** *14,* 215.

109. Kelly, C. B.; Lee, C. X.; Leadbeater, N. E. *Tetrahedron Lett.* **2011,** *52,* 263.

110. Pedrick, E. A.; Leadbeater, N. E. *Inorg. Chem. Commun.* **2011,** *14,* 481.

111. Hartman, R. L.; Naber, J. R.; Zaborenko, N.; Buchwald, S. L.; Jensen, K. F. *Org. Process Res. Dev.* **2010,** *14,* 1347.

112. Nickbin, N.; Ladlow, M.; Ley, S. M. *Org. Process Res. Dev.* **2007,** *11,* 458.

113. Bedore, M. W.; Zaborenko, N.; Jensen, K. F.; Jaimson, T. F. *Org. Process Res. Dev.* **2010,** *14,* 432.

114. Kopach, M. E.; Murray, M. M.; Braden, T. M.; Kobierski, M. E.; Williams, O. L. *Org. Process Res. Dev.* **2009,** *13,* 152.

115. Genck, W. *Chem. Eng.* **2011,** *118*(7), 28.

116. Tung, H.-H.; Paul, E. L.; Midler, M.; McCauley, J. A. *Crystallization of Organic Compounds, an Industrial Perspective.* Wiley: New York, 2009.

117. Lindrud, M. D.; Kim, S.; Wei, C. US 6,302,958 B1, 2001 (to Bristol-Myers Squibb).

Practical Process Research and Development

118. ICH Q7A Good Manufacturing Practice Guidance for Active Pharmaceutical Ingredients, August 2001, http://www.fda.gov/downloads/Drugs/GuidanceComplianceRegulatoryInformation/Guidances/ucm073497.pdf
119. Anderson, N. G.; Burdick, D. C.; Reeve, M. M. *Org. Process Res. Dev.* **2011,** *15*, 162.
120. For instance, Thomas Swan offers continuous supercritical fluid processing, http://www.thomas-swan.co.uk/ASP/ProductsAndServices/ProductsAndServices.asp?Type=SupercriticalProcessing. AMPAC, Carbogen Amcis and Lonza also offer continuous reaction services.

Chapter 15

Refining the Process for Simplicity and Ruggedness

"…in industry, and especially in process chemistry, the accurate analysis of yield and selectivity is common, and procedures for ensuring the accuracy of data abound. Process chemists, even if they carry out the initial analysis such as HPLC themselves, are usually aware of the limitations of their methodology…"

– Trevor Laird [1]

I. INTRODUCTION

The process chemist knows the limitations of a process through observation and thoughtful experiments. Anticipating and avoiding problems is the reward, so that problems do not need to be solved on the floor of the pilot plant or manufacturing unit.

Thorough process understanding is gained in the laboratory by investigating the critical processing parameters (CPPs), process inputs that directly influence the critical quality attributes (CQAs) of intermediates and final products. CPPs can include reaction temperature, pH, pressure of H_2, residual moisture prior to crystallization, rate of cooling a slurry during crystallization, and so on. CPPs and CQAs are formally parts of process validation and implementation (Chapter 16), and rarely is there enough time to determine all the optimal processing conditions prior to going into the pilot plant. The laboratory experiments in Table 15.1 may be undertaken to optimize operations and develop robust processes, and as many of these experiments as possible should be considered before implementing a process on scale.

Practical Process Research and Development. DOI: 10.1016/B978-0-12-386537-3.00015-0

TABLE 15.1 Critical Laboratory Investigations before Implementing a Process on Scale (1)

Experiment	Type	Purpose
Optimization	Range-finding	Identify normal operating range (NOR) (2)
Stress tests/abuse tests	Range-finding	Identify proven acceptable range (PAR) (2)
Extended runs	Mimic times needed for scale-up runs (3)	Identify side reactions, degradents, and physical difficulties that may occur on scale
Calorimetry	Qualitative Quantitative	SAFETY (avoiding runaway reactions) Predict processing times for exothermic operations when the cooling capability of scale-up equipment is known
Tolerance studies	Conduct processing with increased inputs of selected impurities under PAR conditions	Establish permissible levels of impurities in inputs or process streams to produce APIs and intermediates with acceptable CQAs
Use-test	Subject all inputs to processing conditions	Predict success for scale-up operations; identify whether any problems of scale-up are related to inputs and/or operations

(1) Rarely is there time before pilot plant introduction for extensive optimization. These tests may be more fully considered for validating and implementing a process for manufacturing.
(2) NOR and PAR are validation terms referring to conditions within a design space.
*(3) For batch processing a 10-fold scale-up can be expected to at least double the processing time: Anderson, N. G. Org. Process Res. Dev. **2004**, 8, 260.*

The process chemist is familiar with designing laboratory experiments optimize conditions to make a final product, but may be less familiar with the other five types of experiments in Table 15.1. With stress tests reaction mixtures are purposely exposed to more extreme conditions to assess the potential impact. For instance, a key reagent may be added in portions to determine whether higher levels of impurities will form due to micromixing. Another example of a stress test is shown in Figure 15.1. Merck

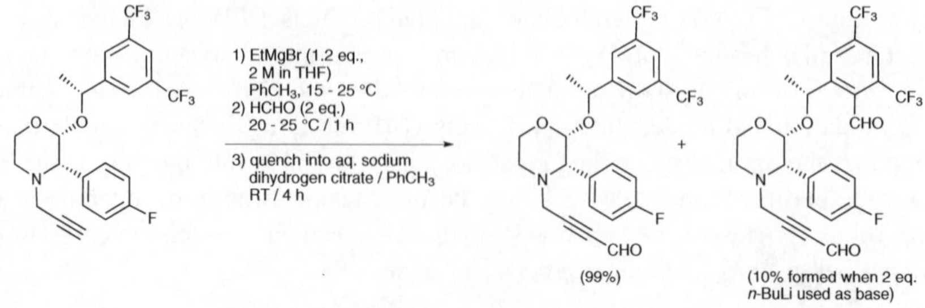

FIGURE 15.1   Selection of an optimal base after a stress test.

researchers anticipated that by using n-BuLi as base small amounts of excess base would lead to metalation in one aryl ring, leading to a bis-aldehyde. When two equivalents of n-BuLi was charged in a stress test, 10% of this side product was indeed formed. Further screening showed that EtMgBr could be used as a base, even in 20% excess, avoiding the need to titrate and control the addition of the n-BuLi solution [2]. An abuse test is an extreme stress test; instead of holding a reaction at $-40\,^\circ$C a reaction may be warmed gradually to $0\,^\circ$C and analyzed by HPLC. Calorimetry experiments are usually conducted on specialized equipment by trained personnel, perhaps on a contractual basis. For tolerance tests impurities may be added to a process stream, such as during a crystallization, to determine how readily these impurities can be purged by routine processing. Use-tests should be carried out in advance of scale-up runs to ensure that processing will proceed as anticipated before resources are committed to the scale-up.

Operations should be extended in the laboratory to judge stability and anticipate physical and chemical issues on scale. For instance, if a suspension of crystals is to be filtered on a centrifuge in a second load, then the remaining product slurry will probably continue to be agitated in the crystallizer while the centrifuge is in use. Extended agitation may decrease the particle size, allow the co-crystallization of an unwanted impurity, or lead to the formation of another polymorph that affects subsequent handling and quality. The stability of the suspension could be examined in the laboratory by storing a small portion of the product slurry in a refrigerator overnight and analyzing both solid and liquid phases. Such extended holding periods can be anticipated by comparing the capacities of equipment with the volumes of the process streams.

Knowing the chemical stability of intermediates can also be important for effective scale-up, and potential difficulties uncovered by extended stability testing may be addressed by modifying workups or by telescoping. For instance, carboxylation of a vinyl Grignard intermediate proved troublesome (Figure 15.2). Increased ageing of the preparation in the presence of CO_2 reduced the yield of the acetate. FTIR showed that the concentration of free CO_2 in solution was initially about 0.25 M after the vinyl Grignard had been consumed, and fell with ageing. The carbonate, implicated by experiments, was not effectively acetylated, and formed the undesired lactone upon workup. KOt-Bu was charged effectively to remove CO_2 from the reaction mixture, as other methods to remove CO_2 failed [3]. As another example, the aldehyde in Figure 15.2 was too unstable to isolate, as complete racemization at $0\,^\circ$C was seen after storage over two days, and so this intermediate was telescoped [4].

The chloropyridone in Figure 15.2 readily decomposed under basic conditions during the workup: if the reaction was quenched with aqueous $KHCO_3$ the product decomposed at the rate of 6%/hour, but quenching with aqueous NaCl gave acceptable stability [5]. Compounds with some functional groups may exhibit poor stability and should not be good candidates for extended storage or shipping. For instance, aromatic aldehydes are susceptible to

Don't discard a "ruined" reaction too soon, as you may be able to extract some valuable information from it. For instance, a reaction that unfortunately warmed to room temperature over lunch may provide information on a byproduct that could be troublesome [6].

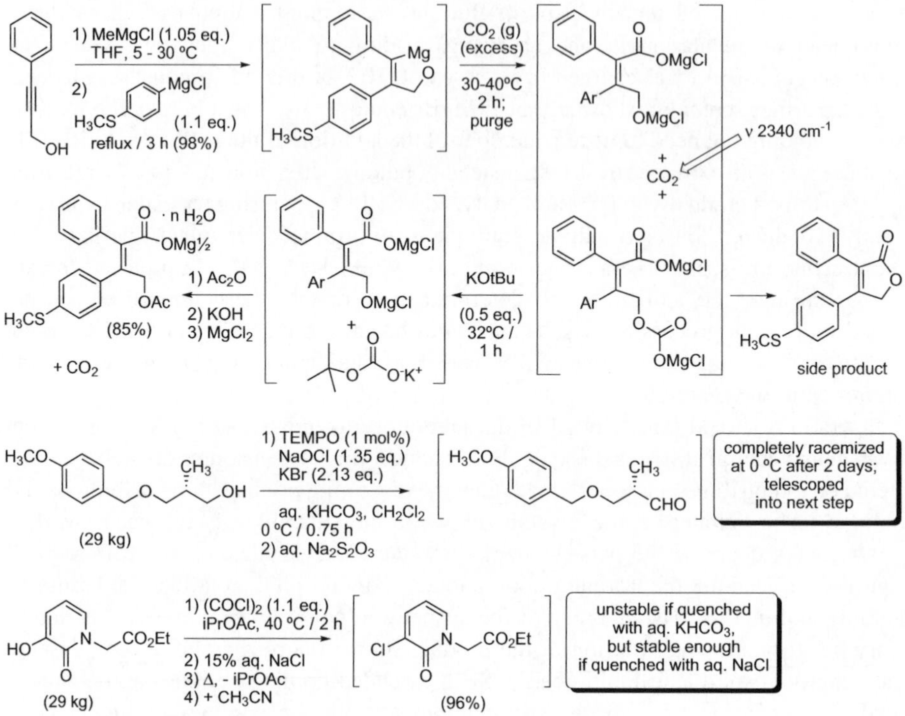

FIGURE 15.2 Understanding and controlling the stability of intermediate streams.

Choose data over assumptions when possible. For instance, a brine wash may not be necessary. Furthermore, NaCl entrained in the organic phase may lead to a high ROI or formation of a hydrochloride salt. Confirm that the brine wash removes H_2O or the desired impurity from the extract, or avoid the brine wash altogether.

air oxidation; cyanohydrins and cyanoamines (the products from Strecker reactions) may be prone to extruding HCN. Comparative stabilities can be assessed by holding a reaction stream or isolated product for several days and analyzing by a suitable, convenient assay, such as HPLC.

Table 15.2 is a checklist of considerations for productive, economic scale-ups. Some of these points were mentioned in Chapter 8.

II. THE IMPORTANCE OF MASS BALANCES

A mass balance approaching 100% indicates that the reaction product and byproducts have been satisfactorily quantified, and the disposition of the reaction product and byproducts is well understood.

What defines a satisfactory mass balance depends on how much time is available to improve the mass balance. When a low mass balance occurs, it is possible that process streams have been incorrectly quantified for product and byproduct; or some sources of physical losses have not been identified; or some byproducts have not been identified.

TABLE 15.2 Considerations for Productive and Economic Scale-Ups

☐ Use minimal volumes of solvents for reaction that allow smooth control of processing. Example: try 0.3–0.4 M reactions in the laboratory, but increase the concentration if the reaction mass is fluid and no appreciable exotherm is noticed

☐ Develop process conditions where all streams are mobile

☐ Use exothermic additions to warm mixture to desired temperature, as long as dose-controlled addition can be throttled and sufficient cooling is available to maintain desired temperature range for reaction

☐ Develop means to recharge reagents if the reaction becomes stalled

☐ Minimize V_{max}, the maximum volume/kg of starting material, to increase productivity in selected equipment. Usually V_{max} occurs during quench or extractions

☐ If $V_{max} \gg V_{min}$, sizing kettles may be difficult. Keep $V_{max}:V_{min} \leq 10:1$ ($\leq 5:1$ better)

☐ Develop a direct isolation if possible, or telescope into the next step

☐ Use dilute, not saturated solutions for extractions

☐ Minimize number and volumes of extractions, while minimizing V_{max}

☐ Use no more than two different solvents other than water

☐ Design process to chase lower-bp solvent with higher-bp solvent

☐ Crystallize at 5–10 volumes/kg

☐ Wash wet cake with two cake volumes of solvent. If necessary apply a lower-boiling or less viscous solvent to displace solvent on cake and ease drying

☐ Fine-tune operations based on laboratory in-process assays, and IPC during scale-up

☐ Identify critical and flexible processing steps

☐ Identify potential for any extended operations

☐ Conduct extended reaction, process tolerance tests and abuse tests whenever appropriate

If the reason for the low mass balance is not due to quantitation errors, then it is likely that the operations are not understood and controlled. With a low mass balance successful scale-up is not assured, and additional time to understand operations may be warranted.

Satisfactory mass balance data may be generated using HPLC without reference standards that have been exhaustively analyzed. For a hypothetical reaction, where all starting material has been consumed and the molecular weight and absorbance of Impurity A is expected to be similar to those of the product, a semi-quantitative mass balance can be prepared as mentioned in Table 15.3. Key for this is generating a reasonable estimate of how much product is dissolved in the combined mother liquor and washes. This calculation ignores any product or impurities that may have been partitioned into any extracts.

TABLE 15.3 Example of a Semi-Quantitative Mass Balance for a Crystallization

A	B	C	D	E	F	G
	Yield ("as is")	Product AUC (HPLC)	Product purity vs. std. (HPLC)	Yield, corrected	Impurity A AUC (HPLC)	Impurity A est. M% (E × F/C)
Dry cake (9.16 g of 10.0 g theory)	91.6%	97%	95.0%	87.0 M% (B × D)	2%	1.8 M%
ML + washes (120 mL)		33%	0.002 g/mL	2.4 M% (D × 120/ 10.0 g theory)	66%	4.8 M%
Total				89.4 M%		6.6 M%
Mass balance				*96.0 M%*		

III. DOCUMENTING A PROCESS FOR TECHNOLOGY TRANSFER

The fine details of operations are included for best reproducibility. In the experimental section for the preparation of a thiophene (Figure 15.3) many details were mentioned [7], similar to those found in descriptive log sheet or an *Organic Syntheses* preparation. Some examples included the directions to jog the agitator (turn the agitator motor on for a few seconds, then turn it off) to mobilize beads of the aqueous phase adhering to the walls of the separator so that they fall down to the interface, then drain the contents of the vessel into separate containers, and wash the vessel with acetone and steam to remove traces of the odoriferous thiol. The solution of the thiol was stored in a sealed bottle, and rinsed from the bottle with a small amount of solvent. The Pharmacia authors also described adding 0.1 equivalents of

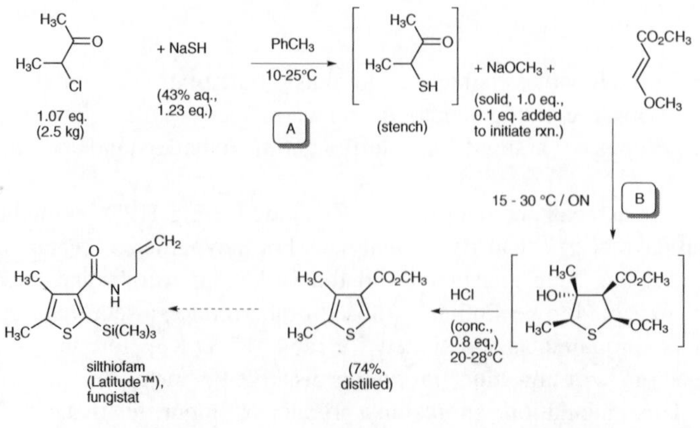

FIGURE 15.3 Details for the preparation of a thiophene.

TABLE 15.4 Guidelines for Technology Transfer

- Include reaction scheme, with balanced equation, and process flow diagram

- Include SAFETY assessment. Propose SAFETY considerations for scale-up

- Describe the processes with details necessary for reproducibility, especially observations, such as temperature rises or how a slurry thickened at a temperature. Be specific

- Highlight critical process parameters, e.g., strictly anhydrous conditions, and how to achieve these conditions, e.g., azeotropically dry at 76–80 °C/atmospheric pressure

- Describe in detail in-process assays, including sampling, sample preparation, and chromatography conditions, and IPC limits

- Describe procedures to neutralize byproducts and waste streams and how to clean equipment

- Include a rework procedure

- Include tables of yields, quality, impurities, and mass balance

- Provide history of previous process R&D

- Critique your optimal process: explain what is known and not known in current process, and what worked and did not work. Propose areas for future optimization

- Propose areas of concern for scale-up, e.g., when vigorous agitation is needed

- Emphasize openness and honesty

- Ensure that your document is reviewed and approved by management

---

the acrylate to initiate the reaction, then adding the remainder gradually to control the temperate at about 30 °C. This level of details is helpful in scale-up and in transferring processes.

A document for technology transfer should stand alone, with minimal reference to other documents. Considerations for efficient, effective technology transfer are shown in Table 15.4.

Follow up after technology transfer to learn how effectively was your process transferred, and what knowledge can be applied to future processing. The best way to assess the performance of your process is to participate on-site in the technology transfer.

IV. CASE STUDIES ON REFINED PROCESSES

Conformational changes can radically influence processing. In the first-generation ring-closing metathesis (RCM) process to provide the hepatitis C candidate BILN 2061 the concentration of the diene was optimized at 0.011 M (Figure 15.4). About 400 kg was made in batches with inputs of 20.2 kg of diene [8]. Derivatization of the nitrogen of the secondary amide led to higher yields from the RCM, by encouraging a conformation that favored cyclization [9]. In the second-generation route a Boc group was bonded to that amide nitrogen. For the RCM the concentration of that diene was raised to 0.1 M, producing the olefin in excellent yield [10].

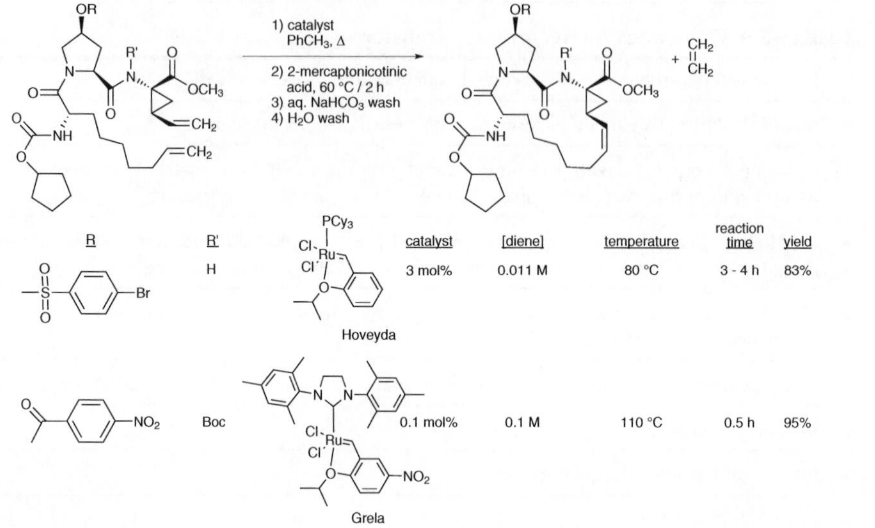

FIGURE 15.4 Conformational changes led to more concentrated processes in macrolide formation.

Merck scientists extensively optimized processes for manufacture of sitagliptin. The case study that follows reviews some of the process R&D and how these processes were carried on scale-up. Sitagliptin is an orally active treatment for type-2 diabetes, a DPP-4 inhibitor, and first in class. Merck received a Presidential Green Chemistry Award (2006) for asymmetric hydrogenation of the unprotected enamine, which avoided protection and deprotection steps in sitagliptin manufacturing. In 2009 Merck and Codexis received a US Presidential Green Chemistry Award for the biocatalytic route to sitagliptin, which decreased waste by eliminating one synthetic step and a recrystallization. The manufacturing routes are outlined in Figure 15.5.

The substituted piperazine was initially prepared from a chloropyrazine (Figure 15.6). The first step was problematic, requiring an excess of hydrazine and reaction temperatures that were close to 85 °C, at which the reaction mass became uncontrollably exothermic. The chloropyrazine decomposed under the reaction conditions, the tedious extractive workup used CH_2Cl_2, and the subsequent reaction with superphos was quite exothermic [11]. Control and destruction of the excess hydrazine were also issues. This route was abandoned in favor of a more efficient route to the piperazine that began with only one equivalent of hydrazine. Selective reaction with ethyl trifluoroacetate and acylation under Schotten–Baumann conditions afforded the mixed hydrazide. Reaction with phosphorus oxychloride afforded the chloromethyloxadiazole in a slightly lower yield than with phenylphosphoryl dichloride, but $POCl_3$ generated more benign waste streams. Reaction with ethylenediamine produced the piperazine. If less than 2.8 equivalents of ethylenediamine was charged the intermediates were contaminated by the crystalline dihydrochloride salt of ethylenediamine. Isolation of the amidine intermediate (not shown) purged excess ethylenediamine, and the salt was readily prepared by treatment with concentrated HCl. The optimized route allowed for safer handling of hydrazine, with only one isolated intermediate in five steps.

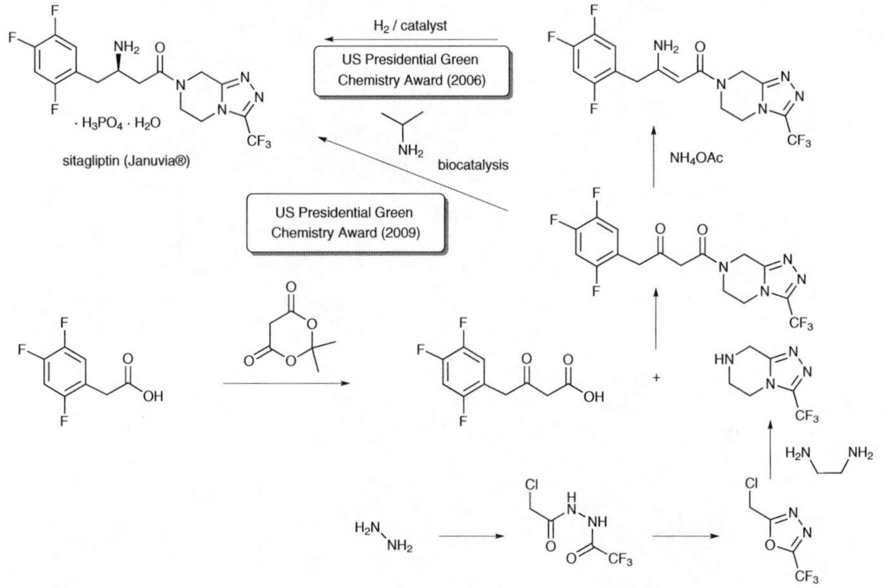

FIGURE 15.5 Manufacturing routes to sitagliptin.

Overall the yield for the optimized route was nearly twice that of the earlier, problematic route [12].

The first-generation route scaled up to make sitagliptin began with chiral hydrogenation of a β-ketoester, which was then hydrolyzed and coupled with benzyloxyamine

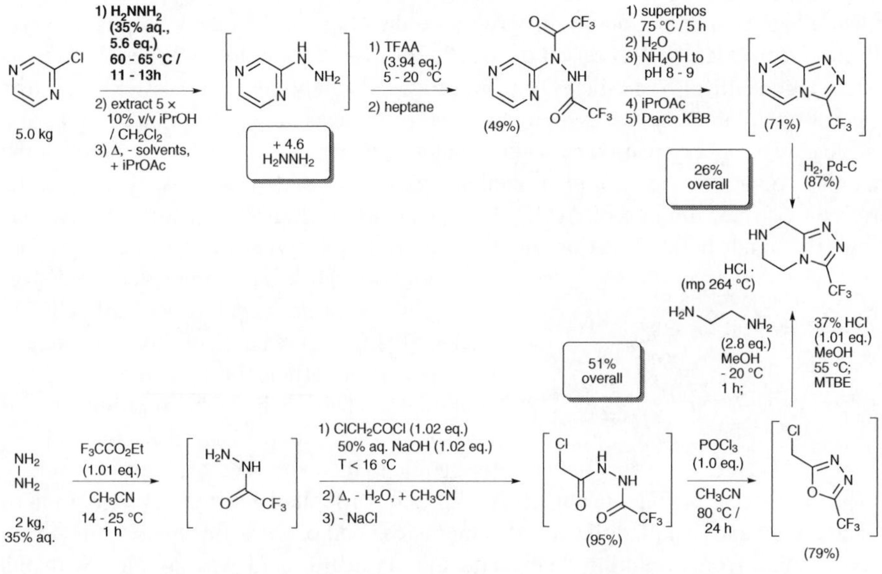

FIGURE 15.6 Preparations of the piperazine hydrochloride for sitagliptin.

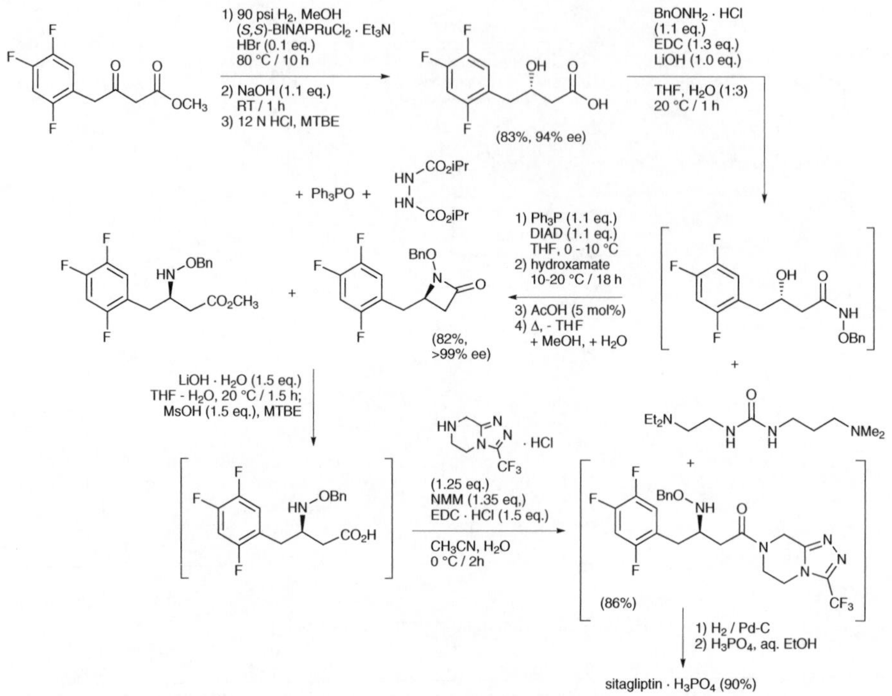

FIGURE 15.7　First-generation coupling to make sitagliptin.

using EDC in a predominantly aqueous solution (Figure 15.7). The β-lactam was formed through a Mitsunobu reaction, generating the byproducts triphenylphosphine oxide and the hydrazide from reduction of diisopropyl azodicarboxylate. In the laboratory recrystallization from MeOH/H$_2$O raised the quality of the β-lactam to >99% ee, but unfortunately recrystallization in the pilot plant produced solely the methyl ester during the solvent swap. Methanolysis was found to take place with as little as 0.1 mol% NaOH, and on scale may have been caused by residual detergents from cleaning the reactors. For the remaining batches in the pilot plant methanolysis was avoided by concentrating in the presence of a trace amount of AcOH. Hydrolysis of the β-lactam or methyl ester was followed by amide bond formation with EDC, hydrogenolysis to deprotect the amine, and

> For effective introduction on scale, processes should accommodate routine operations in existing equipment.

salt formation [11]. This route delivered over 100 kg of sitagliptin for safety and early clinical studies [13]. The drawbacks of this route include the poor atom efficiencies of the two EDC couplings and the Mitsunobu reaction, use of two protecting groups, and the high solvent usage due to the extractive workups.

The large-scale manufacturing of sitagliptin incorporated a different preparation of the ketoamide and installed the chiral center at the end of the sequence (Figure 15.8). The acetonide from Meldrum's acid and the arylacetic acid was coupled with the substituted piperazine through acid catalysis. This reaction probably proceeded

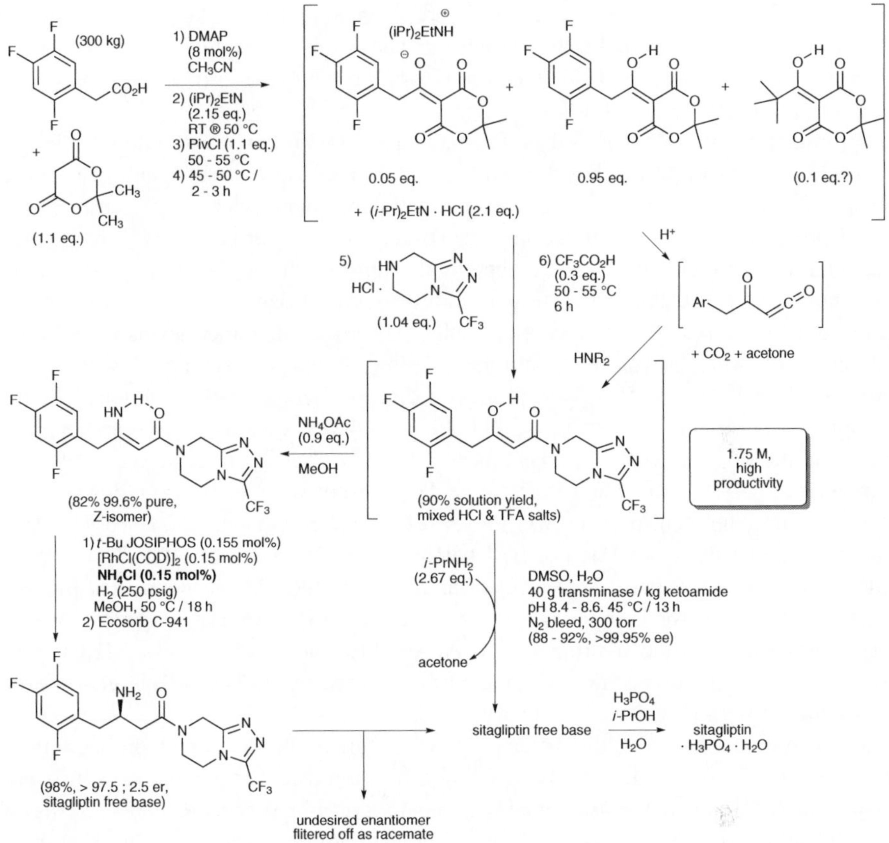

FIGURE 15.8 Optimized endplays to sitagliptin.

through formation of the acyl ketene, and the process was highly productive at 1.75 M. The β-ketoamide was telescoped into the formation of the enamine, which crystallized from the reaction as the Z-enamine [14]. Extensive hydrogenation studies were carried out to find reagents and conditions to convert the enamine into the desired chiral β-aminoamide [15,16]. Hydrogenating the enamine risked reducing the fluorine atoms on the aromatic ring. Optimization proceeded after selecting the *t*-Bu JOSIPHOS catalyst and MeOH (see discussion with Figure 10.16). Control issues included efficient de-gassing, agitation, and H_2 transfer from gas to the solution. Dry conditions may have been needed to preclude hydrolysis of enamine, and probably involved IPCs for moisture. Optimal conditions for the hydrogenation required a moderate temperature and a moderate pressure, requiring a relatively specialized reactor. At a concentration of 11 wt% (0.25 M) the productivity was reasonable; higher concentrations of the initial slurry gave slow hydrogenations due to poor mass transfer. The hydrogenation required a relatively long time, due to product inhibition. Low catalyst loading required good control of inhibitors, and the expensive rhodium was recovered by adding Ecosorb® C-941 in an atmosphere of H_2. To raise the chiral purity the free base was crystallized; the undesired enantiomer was removed by filtration as the crystalline racemate before the free base was crystallized. By placing the chiral catalyst at end of the route less catalyst needed to be purchased. The levels of residual Rh and Fe in the product were below 20 ppm [13]. This route avoided the use of protecting groups and was more atom-efficient.

Optimization on scale revealed some control issues with the chiral hydrogenation in Figure 15.8. Ketoamide batches 1–9 gave good conversion and acceptable ee, while batches 10–18 gave conversion no higher than 82% and enantiomeric excesses as low as 89%. Ironically batches 10–18 had been reworked to raise purities; rework had either added an inhibitor, or removed a beneficial impurity. Initial analyses showed no differences between batches, but deeper analyses showed that batches 1–9 were more acidic, and non-aqueous capillary electrophoresis revealed higher levels of residual ammonium chloride, a byproduct from the coupling. Batches of ketoamide with higher levels of NH_4Cl, i.e., >1500 ppm, produced dimeric byproducts (see Figure 10.16). Crossover experiments showed that a beneficial impurity was at play. Hydrogenation of ketoamide lot 1 gave 95% conversion and 95% ee, but after reslurry that lot was hydrogenated to 91% conversion and 91% ee. When ketoamide lot 10 was reduced in the mother liquor generated from reslurrying lot 1, both the conversion and ee rose to 95% from the initial values of 91%. For best performance, the range of NH_4Cl was determined to be 500–1500 ppm. That level of ammonium chloride was controlled by processing batches of ketoamide to contain no more than 0.1% of NH_4Cl. Then 0.05 wt% NH_4Cl was added to the hydrogenation. A possible role of the excess NH_4Cl is to promote imine formation critical to the addition of hydrogen. The NH_4Cl charge was stoichiometric with $[RhCl(COD)]_2$ charge [17].

Reductive amination of the ketoamide was studied for the direct production of sitagliptin free base. The feasibility of using a transaminase for this conversion is not obvious, as substrates for transaminases classically are α-keto acids. Under 435 psig of H_2 with $Ru(OAc)_2$-R-dm-segphos® (1 mol%) and ammonium salicylate as the ammonia source Merck and Takasago researchers prepared the free base in 91% yield, with 99.5% ee [18]. This work and the earlier manufacturing route were superseded by the transaminase process developed by Codexis and Merck (Figure 15.8). In this highly efficient process the transamination is driven to completion by removing the byproduct acetone as an azeotrope with i-PrNH$_2$. The benefits from this route are removing one step from the total overall, eliminating the burden of removing and recovering rhodium, and avoiding the need to build a specialized manufacturing plant [19]. The drawbacks of this substitution are the costs of developing and purchasing the biocatalyst, and the cost of refiling the sitagliptin manufacturing process. Obviously the benefits were substantial for Merck to refile the manufacturing process.

Other examples of refining processes may be seen in the direct isolation of captopril (see discussion with Figure 12.11) and in a modified Arbuzov process (see discussion with Figure 12.2).

V. SUMMARY AND PERSPECTIVE

Processes can be scaled up more efficiently if time is spent in the laboratory refining the operations first; unfortunately time is often a rare commodity. Parallel, high-throughput screening may save a considerable amount of time; for instance, using automation a group at BMS ran 116 reactions in four days to screen for optimal conditions to hydrolyze a *N*-trifluoroacetyl amino acid derivative [20]. Cary estimated that "...automated high-throughput equipment can speed up research by five to 10 times..." [21]. A reward of thorough process investigations is avoiding problems on scale-up, where there can be considerable pressure. Processes may also be refined, simplified, and further optimized in production, where there may be significant economic rewards for the resources invested in process optimization.

> Good process development is like well-installed sheetrock (wallboard) and plastering: the surface appears smooth and seamless, but all is anchored to a solid structure of support. Workers applied efforts many times to smooth out the rough edges and give a simple appearance. Only by shining a light on the surface from the right angle is it possible to see the seams.

REFERENCES

1. Laird, T. *Org. Process Res. Dev.* **2011,** *15*, 305.
2. Journet, M.; Cai, D.; Hughes, D. L.; Kowal, J. J.; Larsen, R. D.; Reider, P. J. *Org. Process Res. Dev.* **2005,** *9*, 490.
3. Engelhardt, F. C.; Shi, Y.-J.; Cowden, C. J.; Conlon, D. A.; Pipik, B.; Zhou, G.; McNamara, J. M.; Dolling, U.-H. *J. Org. Chem.* **2006,** *71*, 480.
4. Mickel, S. J.; Sedelmeier, G. H.; Niederer, D.; Daeffler, R.; Osmani, A.; Schreiner, K.; Seeger-Weibel, M.; Bérod, B.; Schaer, K.; Gamboni, R.; Chen, S.; Chen, W.; Jagoe, C. T.; Kinder, F. R., Jr.; Loo, M.; Prasad, K.; Repic, O.; Shieh, W.-C.; Wang, R.-M.; Waykole, L.; Xu, D. D.; Xue, S. *Org. Process Res. Dev.* **2004,** *8*, 92.
5. Ashwood, M. S.; Alabaster, R. J.; Cottrell, I. F.; Cowden, C. J.; Davies, A. J.; Dolling, U. F.; Emerson, K. M.; Gibb, A. D.; Hands, D.; Wallace, D. J.; Wilson, R. D. *Org. Process Res. Dev.* **2004,** *8*, 192.
6. For instance, treatment of a 5-chlorothiophene with *n*-BuLi at −20 °C instead of −40 °C led to deprotonation at the 4-position instead of chloride–lithium exchange: Conrow, R. E.; Dean, W. D.; Zinke, P. W.; Deason, M. E.; Sproull, S. J.; Dantanarayana, A. P.; DuPriest, M. T. *Org Process Res. Dev.* **1999,** *3*, 114.
7. Fevig, T. L.; Phillips, W. G.; Lau, P. H. *J. Org. Chem.* **2001,** *66*, 2493.
8. Nicola, T.; Brenner, M.; Donsbach, K.; Kreye, P. *Org. Process Res. Dev.* **2005,** *9*, 513.
9. Shu, C.; Zeng, X.; Hao, M.-H.; Wei, X.; Yee, N. K.; Busacca, C. A.; Han, Z.; Farina, V.; Senanayake, C. H. *Org. Lett.* **2008,** *10*, 1303.
10. Farina, V.; Shu, C.; Zeng, X.; Wei, X.; Han, Z.; Yee, N. K.; Senanayake, C. H. *Org. Process Res. Dev.* **2009,** *13*, 250.
11. Hansen, K. B.; Balsells, J.; Dreher, S.; Hsiao, Y.; Kubryk, M.; Palucki, M.; Rivera, N.; Steinhuebel, D.; Armstrong, J. D., III; Askin, D.; Grabowski, E. J. J. *Org. Process Res. Dev.* **2005,** *9*, 634.
12. Balsells, J.; DiMichele, L.; Liu, J.; Kubruk, M.; Hansen, K.; Armstrong, J. D. *Org. Lett.* **2005,** *7*, 1039.
13. Hansen, K. B.; Hsiao, Y.; Xu, F.; Rivera, N.; Clausen, A.; Kubryk, M.; Krska, S.; Rosner, T.; Simmons, B.; Balsells, J.; Ikemoto, N.; Sun, Y.; Spindler, F.; Malan, C.; Grabowski, E. J. J.; Armstrong., J. D., III. *J. Am. Chem. Soc.* **2009,** *131*, 8798.
14. Xu, F.; Armstrong, J. D., III; Zhou, G. X.; Simmons, B.; Hughes, D.; Ge, Z.; Grabowski, E. J. J. *J. Am. Chem. Soc.* **2004,** *126*, 13002.
15. Ikemoto, N.; Tellers, D. M.; Dreher, S. D.; Liu, J.; Huang, A.; Rivera, N. R.; Njolito, E.; Hsiao, Y.; McWilliams, J. C.; Williams, J. M.; Armstrong, J. D., III; Sun, Y.; Mathre, D. J.; Grabowski, E. J. J.; Tillyer, R. D. *J. Am. Chem. Soc.* **2004,** *126*, 3048.

16. Hsiao, Y.; Rivera, N. R.; Rosner, T.; Krska, S. W.; Njolito, E.; Wang, F.; Sun, Y.; Armstrong, J. D., III; Grabowski, E. J. J.; Tillyer, R. D.; Spindler, F.; Malan, C. *J. Am. Chem. Soc.* **2004,** *126*, 9918.

17. Clausen, A. M.; Dziadul, B.; Cappuccio, K. L.; Kaba, M.; Starbuck, C.; Hsiao, Y.; Dowling, T. M. *Org. Process Res. Dev.* **2006,** *10*, 723.

18. Steinhuebel, D.; Sun, Y.; Matsumura, K.; Sayo, N.; Saito, T. *J. Am. Chem. Soc.* **2009,** *131*, 11316.

19. Savile, C. K.; Janey, J. M.; Mundorff, E. C.; Moore, J. C.; Tam, S.; Jarvis, W. R.; Colbeck, J. C.; Krebber, A.; Fleitz, F. J.; Brands, J.; Devine, P. N.; Huisman, G. W.; Hughes, G. J. *Science* **2010,** *329*, 305.

20. Rosso, V. W.; Pazdan, J. L.; Venit, J. J. *Org. Process Res. Dev.* **2001,** *5*, 294.

21. Rouhi, A. M., Ed.; *Chem. Eng. News.* **2003,** *81*(28), 42.

Chapter 16

Process Validation
and Implementation

Chapter Outline

"…but eventually I had to go into real plants to check the willingness of Mother Nature to abide by my simulations' wishes. She was never completely compliant."

– Mike Resetarits, Fractionation Research Inc. [1]

"Test whatever delays you would reasonably expect for a plant operation, then test a delay twice as long. Before plant introduction, try to test every feature that you can in a fashion that is identical to that planned for production – materials, reaction conditions, time line, assay conditions/equipment, but especially those things that you are sure don't need any testing because you've done them successfully so many times before."

– Jeffrey L. Havens, Pharmacia & Upjohn [2]

I. INTRODUCTION

As part of the approval process to market drugs that are safe for human use, processes to manufacture APIs must be validated. Once a drug application has been approved, the API and key intermediates must be manufactured under current good manufacturing practice (cGMP) conditions. To be in compliance with cGMP the API and key

Practical Process Research and Development. DOI: 10.1016/B978-0-12-386537-3.00016-2

intermediates must be made in accordance with the validated processes, and the products of these processes must meet previously approved specifications. Process validation of APIs begins with laboratory optimization, before a process is implemented on scale [3].

The International Conference on Harmonisation of Technical Requirements for Registration of Pharmaceuticals for Human Use (ICH) is an alliance of the US FDA and the regulatory authorities of Europe and Japan, and was formed to bring SAFE, high-quality drugs efficiently to the market [4]. The ICH has issued a number of guidances to assist in the development and approval of drugs. Some of the titles of recent guidances relevant to process validation and implementation are shown in Table 16.1. Guidelines are characterized as Quality, Efficacy, SAFETY, and Multi-disciplinary [5]. Quality Control regulations have been discussed regarding the development, manufacture and regulation of APIs [6].

The guidances in Table 16.1 stress among other points that product quality cannot be tested into batches: quality is inherent in the product due to thorough process understanding and quality by design (QbD) [7–9]. According to a process analytical technology (PAT) guidance, "…a process is generally considered well understood when 1) all sources of variability are identified and explained; 2) variability is managed by the process; and 3) product quality attributes can be accurately and reliably predicted over the design space established for the materials used, process parameters, manufacturing, environmental and other conditions." [7]

In January 2011 the FDA issued guidance entitled "Process Validation: General Principles and Practices" [8], which replaced the 1987 guidance. This document stated

TABLE 16.1 Some Recent Guidances Related to Process Validation and Implementation

Guideline	Subject
Q1A–Q1F	Stability
Q3A	Impurities in drug substance
Q3B	Impurities in drug product
Q3C	Residual solvent impurities
Q3D	Residual metal impurities
Q6A	Specifications for "small molecule" APIs and drug products
Q7	Good manufacturing practice
Q8	Pharmaceutical development
Q9	Quality risk management
Q10	Pharmaceutical quality systems
Q11	Development and manufacture of drug substances
S2(R1)	Genotoxicity testing

that "The basic principle of quality assurance is that a drug should be produced that is fit for its intended use." The guidance stresses lifecycle management, with the ability to store and retrieve data being important. The document describes three basic stages of process validation: process design, process qualification, and continued process verification. Design of experiments (DoEs) and statistical process control (SPC) were urged, and process flow diagrams (PFDs) were preferred to describe operations. The document states that operations controlled by PAT may be preferred over non-PAT systems. Validation during process implementation, referred to as process performance qualification (PPQ), is not favored. The FDA urges readers to contact them with specific questions in areas not covered by this and related guidances [8].

The general sequence for process validation is shown in Table 16.2 [3].

This chapter is divided into six main sections: laboratory investigations focused on process implementation; considerations before introducing a process to the pilot plant or manufacturing plant; actions during process implementation; cleaning and removing water from scale-up equipment; follow-up activities after process introduction; and managing production through others.

> Murphy's Law states that "Whatever can go wrong, will." To minimize surprises on scale-up, plan scale-up operations as though Murphy's Law is true.

II. LABORATORY INVESTIGATIONS FOCUSED ON PROCESS IMPLEMENTATION

To develop suitable processes to manufacture materials for human usage, scientists explore the design space, which is the range and combination of process parameters in operations that convert starting materials into products [7]. The critical process parameters (CPPs) must be suitably controlled to produce material that has the desired critical quality attributes (CQAs). The conditions that are demonstrated to generate acceptable material are known as the proven acceptable range (PAR), and the preferred operating range that lies within the PAR is the normal operating range (NOR). Product generated outside the PAR may not have acceptable CQAs (Figure 16.1).

Many considerations for effective and efficient operations in pilot plant and manufacturing facilities are shown in Table 16.3. These guidelines may be used for initial process development studies and may not apply to all processes.

III. ACTIVITIES PRIOR TO INTRODUCING A PROCESS TO THE PILOT PLANT OR MANUFACTURING PLANT

A log sheet, which is essentially a permission slip allowing the use of a pilot plant or manufacturing facility, must be developed to guide operations on scale. Before preparing a log sheet, or batch record, a number of aspects of process implementation should be considered (Table 16.4). For instance, if the goal of a pilot plant run is to conduct a large-scale experiment, then less effort may be expended to isolate all of the product as compared to a run to make material needed for a clinical trial.

TABLE 16.2 General Sequence for Validating Processes (1)

	Activity	Where	Comments
1	SAFETY analyses	Desktop review	Identify and avoid known chemical and toxicological hazards
2	Range-finding experiments	Laboratory studies	Accurate in-process assays speed process development to find acceptable operating conditions
3	SAFETY analyses	Laboratory testing	Quantitate exotherms, identify toxicological hazards of intermediates in proposed process streams
4	Risk assessment (e.g., FMEA)	Multi-disciplinary review	Consider how process variations can impact product quality. Anticipate and avoid scale-up hazards for the kilo lab and pilot plant
5	Scale-up to stationary equipment	Pilot plant	Confirm operating ranges; best if laboratory people are present for introduction
6	Analyze pilot plant data	Laboratories	Confirm process understanding, or identify areas for further optimization
7	Risk assessment (e.g., FMEA)	Multi-disciplinary review	Consider how process variations can impact product quality. Anticipate and avoid hazards for further scale-up (built upon data since previous assessment)
8	Technology transfer	Manufacturing facility	Comprehensive, accurate, honest
9	Manufacturing introduction according to validation plan	Manufacturing facility	Best if those who introduced the process to the pilot plant are present for introduction
10	Analyze data from initial manufacturing batches	Laboratories	Confirm that process has been validated; identify areas for further optimization
11	Maintain validated state for registered, commercial process	Quality, regulatory, R&D	Annual reviews of production, and assess changes, e.g., in raw materials
12	Further optimization	Laboratory, pilot plant, manufacturing facility	Optimization within filed process description does not require further regulatory review or approval

(1) Abstracted with permission from Anderson, N. G.; Burdick, D. C.; Reeve, M. M. Org. Process Res. Dev. **2011**, *15, 162. Copyright 2011 American Chemical Society.*

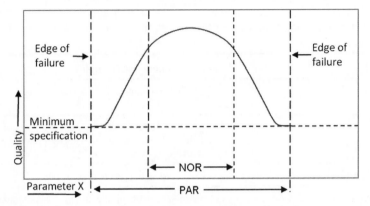

FIGURE 16.1 Relationship of proven acceptable range (PAR) and normal operating range. *Reprinted with permission from Anderson, N. G.; Burdick, D. C.; Reeve, M. M. Org. Process Res. Dev.* **2011**, *15, 162. Copyright 2011 American Chemical Society.*

TABLE 16.3 Desirable Operations for Pilot Plant and Manufacturing

Operation	Comments
● Rugged in-process assays developed	Helpful to have a spare HPLC column available
● Procedures developed to drive a stalled reaction to completion	Extend reaction or add more reagent?
● Simple, effective workups in place	Complex workups deselect processes
● Extractions generally preferred over filtrations to remove impurities	High aqueous volumes undesirable for extractions. Long filtration times undesirable
● Number and volume of extractions have been minimized	Shorter processing time, less chance of physical loss or contamination
● Emulsions, if any, can be clarified	Separation times increase with height of vessel. Include filtration as an option if it helps to clarify an emulsion
● Good stability of process streams on scale	Extend operations in the laboratory to anticipate problems
● Suspensions can be readily transferred from vessel	No big spatulas may be available on scale
● Solids can be readily transferred from filter to dryer to containers	Sticky, low-melting, or deliquescent solids cause difficulties. Best to isolate the most stable polymorph
● Every step has been considered and judged necessary or at least desirable	Unnecessary steps decrease productivity
● Rework has been developed (e.g., recrystallization or reslurry)	The product may be subjected to isolation conditions

TABLE 16.4 Some Considerations for Process Implementation and Log Sheet Preparation

Step	Comments or Examples
Identify goals	Quality, quantity, timelines; according to customer needs
SAFETY	Desirable to complete hazard assessment for the kilo lab; should be mandatory for pilot plant and manufacturing
Review background	Discuss with prior process owners
Identify critical steps	Additions, reaction, work-up, concentration, crystallization, isolation, drying
Equipment limitations	Low-temperature ability, corrosivity, centrifuge
Rugged IPC	Sampling, separations, quantitations
Contingency plans for incomplete operations	Extend processing, add more reagent, or just continue?
Contingency plans for a runaway reaction	Surge tanks, SAFETY equipment
Anticipate results of extended operations	Examine in the laboratory
Anticipate results of interrupted operations	Examine in the laboratory
Develop methods to qualify components	Simple analyses preferred; use-tests
Qualify components	Laboratory use-tests
Identify process tolerance	Solvents $\pm$ 5 vol%, reagents $\pm$ 2 wt%
Product analyses in place	Analyses; use-test of product in next step; mass balance
Identify cleaning procedure	Usually based on how to dissolve product
Identify waste disposal procedure	Usually neutralization
Abuse tests	Laboratory use-tests may consider overcharge and undercharge, pH and temperature excursions

In preparing a log sheet for the pilot plant or manufacturing facility, thoroughly describing the details of the operations is critical to avoid delays and problems. For instance, directions should describe whether any interface ("rag layer") formed during extractions is held with the top phase, lower phase, or separately. General standard operating procedures (SOPs) may be used for some operations, such as cleaning, but in general a batch record should contain all the instructions and information needed. A batch record to prepare material for a toxicology study should use good laboratory practice (GLP), and may not be as detailed as a log sheet to prepare NCE or API designated for human use [10].

Many operations considered second-nature for laboratory procedures may need to be described in detail for SAFE, efficient, and effective scale-up. The simple action of pipetting concentrated H_2SO_4 from a gallon reagent bottle into an addition funnel uses steps that a scientist very familiar with laboratory operations might neglect to mention. Such "automatic" thinking might include

1. How to safely secure the bottle of concentrated H_2SO_4 from tipping and spilling?
2. Is it safer or more convenient to remove funnel from reactor to transfer the reagent into the funnel?
3. Where to put the cap of the reagent bottle to minimize contamination? Up or down? On a paper towel, to ease clean-up?
4. Should the concentrated H_2SO_4 be measured by weight or volume?
5. If charging by weight, should the measurement be made by the difference of the material lost from the bottle or by the amount charged to a beaker or the funnel? Should the beaker be rinsed with reaction solvent to complete the transfer?
6. If charging by volume, should the container, such as a to-contain pipet, be rinsed with reaction solvent?
7. Carefully transfer concentrated H_2SO_4, noting any drips.
8. Put used pipet into a beaker filled with water.
9. Tightly secure cap on reagent bottle.
10. Temporarily sequester any equipment used for measuring and charging, then rinse.
11. Clean up any drips and spills to prevent accidents.
12. Remove reagent and used equipment from staging area.
    Such level of detail may be described in a log sheet.

For a log sheet suitable in-process controls must be in place to determine completion of reaction, pH of extraction, completion of solvent displacement, conditions for crystallization, and so on. Acceptable ranges are best.

The compatibility of equipment with reaction streams should be considered before committing resources to scale-up, to preclude contamination or physical losses. For instance, the corrosivity of aqueous HCl to steel is well known, prompting engineers to use either glass-lined vessels for strongly acidic HCl processes, or to acidify process streams in steel tanks to no more than pH 2. The corrosive nature of methanesulfonic acid to 316 stainless steel was discovered when batches run in an autoclave were found to have high contents of iron [11]. As another example, some gaskets and flanges become brittle when exposed to THF streams. McConville has prepared extensive tables describing the compatibility of various chemicals with metals, plastics, and elastomers [12]. If there are any issues that cannot be resolved through literature searching model solutions should be examined for potential issues. Pieces of the metal used in reactors (metal coupons) can be soaked in test solutions for a suitable period, dried, weighed, and examined for loss of weight and pitting of the surface.

Physicochemical properties of reagents and process streams can greatly influence the success of operations. For instance, some portion of a charge of volatile reagents, such as $SOCl_2$, may be lost to a vacuum system if charged by suction. If micromixing during an addition is known to be an issue, the initial preference may be to add the reagents more slowly, in a diluted stream, subsurface, at a lower temperature, or by spraying into the reactor. Dense slurries may not be evenly suspended by conventional impellers. Small particles may blind filters; large particles may prevent ready transfer from a crystallizer to the filter, and increased agitation may be necessary to fragment the particles. Wet cakes of crystals that don't de-liquor readily by pressure or suctioning on a filter may lose the liquids more readily by centrifuging. Filtration rates should be examined in the laboratory to select equipment suitable for scale-up. The best approaches to such physical issues may be found through discussions with the people responsible for supplying, maintaining and using the equipment [13].

> Generally equipment is not altered for pilot plant use, as some time and thought may be required to do so effectively. For instance, installing another type of agitator in a multi-purpose reactor may require time not only for swapping agitators but also for ensuring that the reactor does not leak afterward.

> All IPCs should be investigated in the laboratory, then recommended for implementation in the pilot plant or manufacturing. An assay that is expected to be routine may not perform as expected; for instance, a compound in a process stream may unexpectedly interfere with a KF titration.

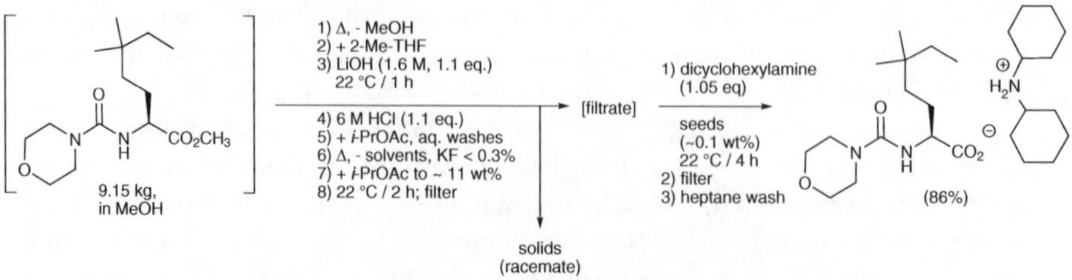

A PFD is useful to track the sequence of operations, and may be required prior to implementation in a pilot plant. A PFD is particularly useful to identify equipment needed for scale-up operations, to highlight the times for in-process assays and in-process controls, and to track process streams through equipment. In Figure 16.2 is shown a reaction sequence for the hydrolysis of an ester and subsequent workup. One notable facet of this preparation is the fact that after hydrolysis the water-soluble solvent MeOH was not removed by concentration before extractions, as might be characteristic of a workup in academia. Secondly, the quality of the chiral salt was improved by first crystallizing and filtering off the racemate of the free acid [14]. A possible PFD for this salt preparation is shown in Figure 16.3. Such PFDs could be customized by highlighting the IPCs in color, or by overlaying the operations on a spreadsheet so that inputs would be automatically calculated based on the amount of starting material.

FIGURE 16.2   Hydrolysis of an ester and DCHA salt formation.

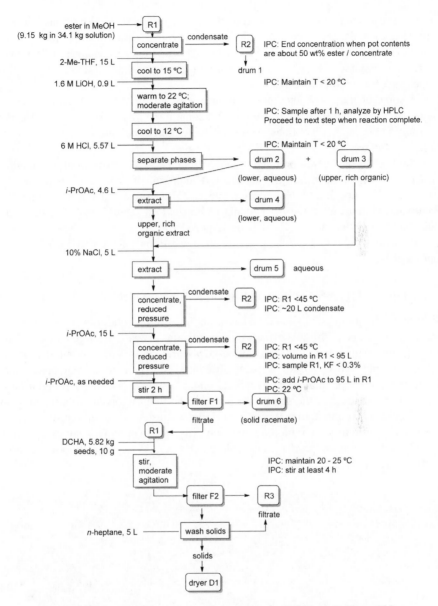

FIGURE 16.3 Process flow diagram for preparing the DCHA salt in Figure 16.2.

As noted in Figure 16.3, each "spent" phase may be drummed for storage. The streams may be analyzed for mass balance calculations and potential product recovery before disposal.

Mass balance should be carried out in the laboratory prior to pilot plant introduction, and on large-scale runs. A low mass balance suggests that a "mass hole" is present; if one does not know where the lost material is, one does not know what it is and how it is formed, hence a significant amount of product could be lost on scale. What constitutes an acceptable mass balance can change with time; for instance, with a deadline looming for delivery of material for a critical GO/NO GO study, a poor mass balance of 75% may provide an acceptable risk for scale-up to a pilot plant. Performing a mass balance also provides data for troubleshooting.

With some of the above thoughts in mind preparing the log sheet may be easier. Although the log sheet should be very specific, some flexibility should be allowed, and there should be room to record observations that are not critical but may provide useful data for the future.

As the log sheet is being prepared some actions may be underway in the laboratory. Critical reagents can be titrated, and QA approval of chemical inputs may be requested. Scale-up should begin after successful use-tests, including using the solvents and perhaps even the filter media to be used on scale. The QC department should be alerted that new IPCs will be required for the new operations, and that expedited assays from QC will help secure the operations for the second and subsequent batches.

IV. ACTIONS DURING PROCESS IMPLEMENTATION

Operations in the pilot plant should be based on the optimal conditions determined by laboratory experience. Physical effects may be greater on scale. A large batch may be cooled faster than anticipated, causing a low temperature and an emulsion. As another example, when a cake is being dried to remove a great deal of residual solvent, the initial cake may have a high amount of solvents. Mechanical agitation is often applied to turn over the solids and increase the rate of solvent volatilization. If the wet cake is agitated too strongly initially, a pasty mass may result, which increases the difficulty in removing residual solvents. Applying heat or agitation gradually at first may be prudent. On scale batch operations generally require more time than anticipated.

For efficient, effective process introduction
- Make sure operations are carried out completely and thoroughly. For instance, confirm that chemicals *from all containers* in a lot were charged.
- Make sure that physical transfers are not contaminated.
- Minimize physical losses, e.g., blow transfer lines clear with N_2.
- Get confirming sample for key steps, e.g., reaction completion for first run implemented on scale.
- Be alert for possible areas of contamination.
- Be alert for sub-optimal conditions, in order to complete process efficiently and safely. For instance, a high rate of N_2 flowing over a reaction surface with an overnight reaction could lead to the evaporation of a significant amount of solvent.

During process implementation there is usually time to focus on details for successful scale-up, for instance, while a batch is being concentrated. Some processing questions to direct actions, in descending priority, could be

1. What component or process stream is most valuable?
2. What material or operation is potentially hazardous?
3. Where can material be lost? For example, leaks, incomplete transfers through hoses, or the scrubber.
4. What are possible sources of contamination?
5. How can I make processing easier or faster?
6. How can I make clean-up easier?
7. What should I consider for the next batch or campaign?

> During startups everyone can watch for leaks and operations that are not SAFE. Most problems encountered when first introducing processes to a manufacturing facility occur after midnight. Plan to be there.

Troubleshooting will take place if difficulties arise during process implementation. Ideally any changes needed will within the processing possibilities described in the log sheet, and not be considered major changes that could jeopardize the cGMP stature of a batch [15]. On-the-job training is required for all the details of efficient process implementation.

V. CLEANING AND REMOVING WATER FROM PROCESSING EQUIPMENT

Small-scale laboratory equipment can be readily cleaned and dried in an oven. Glassware can then be quickly assembled hot and allowed to cool under nitrogen or argon, thus excluding atmospheric moisture. These operations can also be carried out in glove boxes to preclude entry of water and oxygen. But large-scale equipment cannot readily be treated in that fashion. Therefore, it is essential that water be flushed from all portions of equipment, especially stationary equipment and external sampling lines.

On scale, cleaning equipment demands thought and time. Since not all internal surfaces can be reached by a cleaning brush, residues are usually dissolved with an appropriate solvent. It may be necessary to reflux the solvent throughout the system and collect some distillate to ensure that solvent contacts all surfaces of the equipment train. Particularly important are the "dead legs," process lines that may contain volumes of liquid that are difficult to displace. (Often dead legs are found in inconvenient locations, such as pipes near the ceilings of process facilities or condenser lines that are improperly pitched. The best approaches are to insert a drain in such places, or avoid using this part of the equipment train.) To decrease the amount of time required for cleaning, the cleaning solution may be drained from a kettle while it is hot, and the vessel may then be allowed to cool under N_2. Optionally the equipment may be rinsed with water, and the water drained off. Then a water-miscible solvent may be used to rinse the equipment, and this solvent is drained off. Next the vessel is rinsed out with a small volume of solvent to be used for the reaction, and charged with the reaction solvent. Some solvent used for rinsing may be left in the lines if that solvent does not interfere with processing. A Karl Fischer titration is carried out on the reactor contents to determine if the level of residual water is acceptably low. Standard operating procedures may be established for cleaning.

Scrutinize cleaning operations before and after a run
- Minimal cleaning is often carried out between batches in a campaign of the same compound. Solids built up on the walls can harbor impurities that are deleterious to crystallizations and other operations.
- Note that acetone and MeOH, frequently used for cleaning reactors, have very low flash points.
- Consider that residual solvents may react exothermically with combinations of materials such as EtOAc, $SOCl_2$, and Zn [16]; or acetone and $POCl_3$ [17].
- Thoroughly wash reactors to remove residual soaps, which may be basic enough to cause side-reactions, as mentioned with Figure 15.7.

VI. FOLLOW-UP AFTER PROCESS INTRODUCTION

All relevant data, such as yield and quality, will be tabulated for batches prepared during the introduction and compared. Batches must meet the pre-described CQAs to be approved for human use. The first batches of product may be subjected to additional assays to ensure that nothing was overlooked in the assays that were put in place for cGMPs. For instance, batches containing fluorine might be analyzed by ^{19}F NMR, which is unlikely to be a specified test for QC. If satisfactory results are found, such confirming data would not need to be gathered in subsequent batches or reported to regulatory authorities.

After technology transfer data from routine runs may be examined using SPCs to assess robustness in manufacturing, and identify areas for further optimization [18–20]. Some process evaluation and optimization will occur in a manufacturing facility, which may be the best proving ground. Subsequent changes to manufacturing operations should be readily acceptable to regulatory authorities if those changes are within the PARs.

For successful process development after process implementation
- Identify optimal bulk storage conditions. Store in freezer as default.
- Keep retains as informal bulk stability samples.
- Pursue complete mass balance for complete understanding of operations.

VII. MANAGING OUTSOURCING

As the pharmaceutical industry is pressured to economize, managing outsourcing has become a specialized area [21]. Biotechs and small pharmaceutical companies may outsource all phases of drug development, while others prefer to keep all intellectual property in-house. Some of the bigger pharmaceutical companies outsource the initial steps and complete the end of the synthesis in-house for better control of quality and timelines. Sometimes companies wisely "outsource their problems"; for instance, a highly energetic step may be contracted to a CRO/CMO that specializes in that area. Outsourcing demands special considerations, especially for containment practices to manufacture highly potent compounds such as anti-cancer agents or anti-coagulants [22]. Patience is often needed to deal with communication that may not be clear or forthcoming, due to delays in time zones and differing social styles among cultures. Long-distance relationships require more work.

By submitting a request for quotation (RFQ) to several vendors, competitive bidding may lower the projected cost of an API or key intermediate. Scrutinizing the bids received is necessary to ensure that the bids and proposals equally cover the scope of the project. The lowest bid may not be the least expensive overall [23]. Some experienced people automatically disregard the lowest and highest bids.

Due diligence may be applied to selecting a CRO/CMO. As due diligence someone may question a CRO/CMO about their operations and policies before contracting to use the services of that organization [24–28]. In selecting a vendor people will often visit the site of a CRO or CMO as part of due diligence. In Table 16.5 are some items that may be considered for due diligence in selecting a CRO/CMO. Due diligence may also refer to the efforts a buyer undertakes to determine what is known about a compound or project before purchasing it.

TABLE 16.5 Some Due Diligence Considerations for Selecting a CRO/CMO

Consideration	Example/Comment
Process development of an API	Prepared to handle toxic chemicals Willing to perform use-tests Willing to bring up ideas, yet work on directed tasks
Physicochemical properties of an API	Familiar with polymorph issues Skilled in crystallization, does not rely on precipitation
Analytical capabilities	Suitable analytical equipment available for IPCs Able to quantitate residual metals, solvents, and PGIs in-house Equipment available to identify impurities that arise
Environmental, Health, & SAFETY	Are waste streams drummed or incinerated Are waste water streams treated microbiologically on-site? To whom are spills reported? Are respirators and suits with air sources used in event of emergencies? What is protocol for hazard evaluation before scale-up?
Manufacture of an API	Practices in place to preclude cross-contamination Will scale-up be performed on-site Suitable size of equipment available to meet demand
Regulatory	QA/QC separate from manufacturing? Will regulatory personnel help with filings?
Cost	Detail the costs for FTE, delays, cGMP operations, regulatory assistance Is an additional charge assessed to identify an impurity? Define clearly who owns IP

VIII. SUMMARY AND PERSPECTIVE

The explosion of outsourcing has changed the landscape of APIs and drug substances: in 2011, more than 80% of APIs and more than 40% of drug products were outsourced, yet the FDA inspected far fewer foreign manufacturers than manufacturers in the US [29]. Even if a company has successfully passed an FDA inspection there are no guarantees that the drug in hand was made under cGMP conditions. Many companies may be involved in the supply chain, making it difficult to trace the origin of poor-quality drugs. Counterfeit drugs have become an issue, as highlighted by the sale of heparin contaminated with over-sulfated chondroitin sulfate [30,31]. The roles of QA and QC departments have been emphasized [32]. Outsourcing will impact all phases, from the supplier to the formulator to the regulatory authorities to the consumer.

Scalability, reproducibility and ruggedness are the keys for successful technology transfer and scale-up. A scalable reaction gives the same yield and quality as scale increases, without any significant process changes. For reproducible operations paying attention to the details of all phases of processing is essential, such as monitoring and recording the volumes of aqueous extracts and filter cake washes. Rugged processes are necessary for continued productivity. Rugged processes can tolerate a wide range of process conditions, even variable purity of components, and produce consistent in-process assays, consistent cycle times, consistent yields, and consistent qualities. There are no unexpected difficulties, including extended processing time. The ruggedness of a process can only be assessed after a number of runs. The steady performance of a rugged process on scale is the result of successful technology transfer, process implementation, and the inevitable process optimization on scale.

The landscape for process validation has changed quite a bit in the past two decades. QbD guidance encourages companies to establish the NOR and PAR; a validated process should be allowed to vary within the PAR without further regulatory filing, creating more flexibility for manufacturers. Some QbD aspects have been reviewed [33], as for torcetrapib [34]. Cimarosti has recently detailed some QbD case studies carried out by GSK [35–37]. PAT has also been encouraged for current process development. PAT has been associated more with formulation (milling, blending, granulation, tabletting, and coating) than with the preparation of drug substances. For the latter PAT has been used to monitor reactions, monitor the size and shape of crystals during crystallization, filtration rates, and drying of solids [38]. PAT will probably become more important to the manufacture of APIs (Chapter 7). Thorough process validation will increase the safety of APIs, the reliability of API manufacturing, and lower the CoG. Validation will continue to evolve as science and experience evolve.

Successful process validation is the fruit of the labor of process scientists. Some people feel that the paperwork is excessive, but the penalties for failing to validate a process can be significant. On the other hand, confirming through process validation that one understands a process is extremely gratifying.

REFERENCES

1. Resetarits, M. *Chem. Eng.* **2010**, *117*(13), 22.
2. Stinson, S. C. *Chem. Eng. News* **1999**, *77*(44), 35.
3. Anderson, N. G.; Burdick, D. C.; Reeve, M. M. *Org. Process Res. Dev.* **2011**, *15*, 162.

4. http://www.ich.org/

5. These guidances may be found on websites of the FDA, ICH, and others.

6. Nusim, S., Ed.; *Active Pharmaceutical Ingredients: Development, Manufacturing, and Regulation*, 2nd ed.; Informa Healthcare; 2009.

7. Guidance for Industry: PAT – A Framework for Innovative Pharmaceutical Development, Manufacturing, and Quality Assurance: http://www.fda.gov/downloads/Drugs/GuidanceComplianceRegulatoryInformation/Guidances/ucm070305.pdf

8. Guidance for Industry. Process Validation: General Principles and Practices, issued 1/24/2011: http://www.fda.gov/downloads/Drugs/GuidanceComplianceRegulatoryInformation/Guidances/UCM070336.pdf

9. Robinson, D. Regulatory highlights. *Org. Process Res. Dev.* **2009**, *13*, 391.

10. Guidance for Industry: Good Laboratory Practices, Questions and Answers: http://www.fda.gov/downloads/ICECI/EnforcementActions/BioresearchMonitoring/UCM133748.pdf

11. Lipton, M. F.; Mauragis, M. A.; Maloney, M. T.; Veley, M. F.; VanderBor, D. W.; Newby, J. J.; Appell, R. B.; Daugs, E. D. *Org. Process Res. Dev.* **2003**, *7*, 385.

12. McConville, F. *The Pilot Plant Real Book*, 2nd ed.; FXM Engineering & Design: Worcester MA, 2007. http://www.pprbook.com

13. Chemical Engineering Buyers' Guide annually updates suppliers for equipment.

14. Lorenz, J. Z.; Busacca, C. A.; Feng, X.; Grinberg, N.; Haddad, N.; Johnson, J.; Kapadia, S.; Lee, H.; Saha, A.; Sarvestani, M.; Spinelli, E. M.; Varsolona, R.; Wei, X.; Zeng, X.; Senanayake, C. H. *J. Org. Chem.* **2010**, *75*, 1155. The details in the supporting information for this manuscript describe the preparation and isolation of the DCHA salt.

15. *ICH Q7A Good Manufacturing Practices for Active Pharmaceutical Ingredients*, 2001, http://www.fda.gov/downloads/Drugs/GuidanceComplianceRegulatoryInformation/Guidances/ucm073497.pdf See Sections 12.4 and 12.5.

16. Spagnuolo, C. J.; Wang, S. S. Y. *Chem. Eng. News*, 2. Wang, S. S. Y.; Kiang, S.; Merkl, W. *Process Safety Progr.* July 1994, 153.

17. Brenek, S. J.; am Ende, D. J.; Clifford, P. J. *Org. Process Res. Dev.* **2000**, *4*, 585.

18. Peterson, J. J.; Snee, R. D.; McAllister, P. R.; Schofield, T. L.; Carella, A. http://biometrics.com/wp-content/uploads/2009/06/gsk-bds-tr-2009-2.pdf

19. Torbeck, L. *Validation by Design: The Statistical Handbook for Pharmaceutical Process Validation*; PDA Books: Bethesda, MD; 2010.

20. Process Robustness – A PQRI White Paper. *Pharm. Eng.* **2006,** *26*(6), 1. http://www.pqri.org/pdfs/06ND-online_Glodek-PQRI.pdf

21. Ainsworth, S. *Chem. Eng. News* **2011,** *89*(19), 48.

22. Wennerberg, J.; Granquist, B.; Klingberg, D.; Lindwall, C.; Sjövall, S. *Chim. oggi (Chem. Today)*: **May 2004,** 18.

23. Resetarits, M. *Chem. Eng.* **2011,** *118*(5), 28.

24. Barnhart, R. W.; Dale, D. J.; Ironside, M. D.; Vogt, P. F. *Org. Process Res. Dev.* **2009,** *13*, 138.

25. Robinson, D. *Org. Process Res. Dev.* **2007,** *11*, 311.

26. Lu, J.; Shinkai, I. In *Process Clhemistry in the Pharmaceutical Industry, Volume 2*; Gadamasetti, K.; Braish, T., Eds.; CRC Press: Boca Raton, FL, 2008; p 465.

27. Robins, D.; Hannon, S. In *Process Chemistry in the Pharmaceutical Industry, Volume 2*; Gadamasetti, K.; Braish, T., Eds.; CRC Press: Boca Raton, FL, 2008; p 471.

28. Andler-Burzlaff, S.; Bertola, J.; Marti, R. E. In *Fundamentals of Early Clinical Drug Development*; Abdel-Magid, A.; Caron, S., Eds.; Wiley, 2006; p 113.

29. Mullin, R. *Chem. Eng. News* **2011,** *89*(20), 11.

30. www.fda.gov/MedicalDevices/DeviceRegulationandGuidance/IVDRegulatoryAssistance/ucm123775.htm

31. Kemsley, J. *Chem. Eng. News* **2008,** *89*(33), 40.

32. Ainsworth, S. J. *Chem. Eng. News* **2011,** *86*(19), 38.

33. Somma, R. *J. Pharm. Innov.* **2007,** *2,* 87.
34. am Ende, D.; Bronk, K. S.; Mustakis, J.; O'Connor, G.; Santa Maria, C. L.; Nosal, R.; Watson, T. J. N. *J. Pharm. Innov.* **2007,** *2,* 71.
35. Cimarosti, Z.; Bravo, F.; Castoldi, D.; Tinazzi, F.; Provera, S.; Perboni, A.; Papini, D.; Westerduin, P. *Org. Process Res. Dev.* **2010,** *14,* 805.
36. Cimarosti, Z.; Bravo, F.; Stonestreet, P.; Tinazzi, F.; Vecchi, O.; Camuri, G. *Org. Process Res. Dev.* **2010,** *14,* 993.
37. Cimarosti, Z.; Castagnoli, C.; Rossetti, M.; Scarati, M.; Day, C.; Johnson, B.; Westerduin, P. *Org. Process Res. Dev.* **2010,** *14,* 1337.
38. Gade, S.; Jain, D. *Chem. Eng.* **2010,** *117*(9), 47.

Troubleshooting

"Look for what's missing. Many advisers can tell a president how to improve what's proposed or what's gone amiss. Few are able to see what isn't there."

– Donald Rumsfeld, "Rumsfeld's Rules" [1]

"Troubles overcome are good to tell."

– Yiddish proverb

I. INTRODUCTION

Troubleshooting (problem-solving) goes on everyday in all scientific ventures. The impact of problems in a pilot plant or a manufacturing facility is much greater than in the laboratory. For instance, extended processing times or a decrease in quality or yield can affect productivity, and new impurities, impurities that exceed specifications, or impurities that have a strong odor or taste can cause batches to be rejected. Even thoroughly validated processes may undergo troubleshooting, because processes change as the priorities and CQAs change. For instance, the use of a new HPLC method may uncover the presence of a PGI. The keys to efficient troubleshooting are to determine the root cause of the problem, select the appropriate IPCs, and remedy the problem. Troubleshooting provides opportunities for optimization, which can lead to increased productivity, patents, and publications. Solutions to troubleshooting lie in paying attention to details and in successfully applying the scientific method.

Practical Process Research and Development. DOI: 10.1016/B978-0-12-386537-3.00017-4

The best approach in process development is to anticipate and avoid problems, especially problems that could occur on large scale. The details of two published examples that follow describe some challenges and successes of troubleshooting.

In introducing a CDI-mediated coupling to the pilot plant, Pfizer researchers found that amide formation required 50 hours, while only 12 hours was needed for laboratory couplings (Figure 17.1). In troubleshooting the extended reaction time on scale, incomplete removal of CO_2 was initially thought to be the cause. Surprisingly, laboratory investigations showed that couplings with isolated imidazolide proceeded faster when they were sparged with CO_2 before the amine was added: after 4 hours, the reaction was 88% complete, as compared to being 37% complete without CO_2 sparging. A catalytic amount of CO_2 was found to be beneficial to the amidations examined, and reactions run under 50 psi of CO_2 were as slow as those not treated with CO_2. The researchers found that stalled amidations could be "jump-started" by sparging with CO_2, a feature possibly helpful for scale-up. The authors proposed that the rate acceleration may be due to the reaction of a carbamate salt that formed. In the pilot plant, more turbulent agitation and a high N_2 sweep through the reactor had encouraged the loss of CO_2 and produced a poor coupling. A better approach on scale was to activate the acid and conduct amide formation under conditions that retained CO_2, probably by not venting the reactor [2].

The yields for an imino-Reformatsky reaction (Figure 17.2) were lower than anticipated when performed on scale by two CMOs. To accommodate processing on scale, the streams of the Reformatsky product were held in cold storage as the zinc complex, then quenched and processed through to the tosylate salt of the t-butyl ester. Unfortunately the average assay from the first campaign was only 77%, with 90% quality material being produced from the second campaign. The major impurity in the tosylate salt was identified as the amino acid, not the desired t-butyl ester. (Perhaps some HBr or $ZnBr_2$ was generated in situ, decomposing some of the t-butyl ester intermediate.) Analysis of retained in-process samples and review of the batch records showed that the level of the amino acid increased with extended holding of the Reformatsky intermediate before quenching. The quality of the batches was conveniently upgraded by reslurrying the product in acetone; this patch raised the quality from 90% to 95% [3].

The difficulties encountered in the troubleshooting examples in Figures 17.1 and 17.2 centered on how the processing differed from those reactions run in the laboratory relative to those in the pilot plant. In the first example, a beneficial byproduct was removed upon scale-up, as confirmed by laboratory use-tests mimicking the operations in the pilot plant. In the second example, extended holding of a process stream led to significant decomposition, which was confirmed by analyzing the process streams and reviewing the batch records. The best ways to discern the causes of troubles are (1) observe current processing first-hand; (2) talk with those who ran the troublesome batches, e.g., operators instead of the foreman or a bench chemist; and (3) compare batch records of troublesome batches with those of successful batches. Use-tests on-site can be extremely helpful to determine whether difficulties are associated with changes in the chemicals or the processing. Long-distance troubleshooting is difficult.

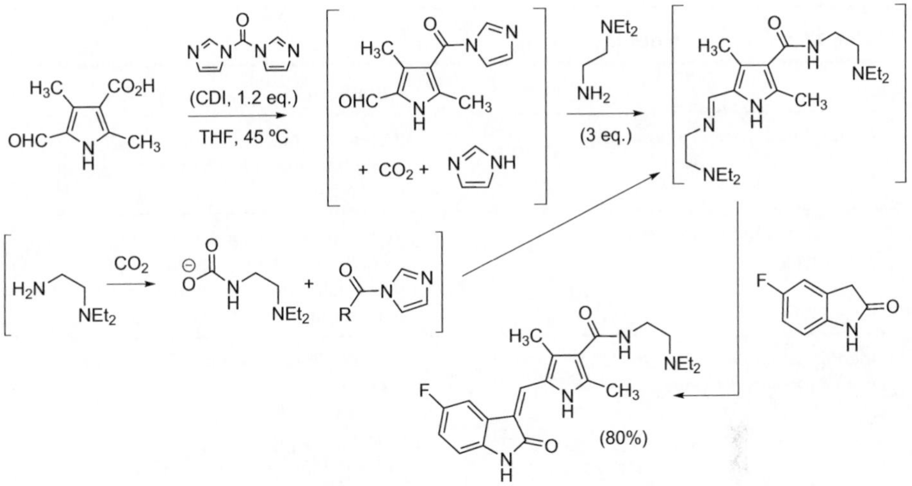

FIGURE 17.1 Troubleshooting a CDI-mediated amide bond formation.

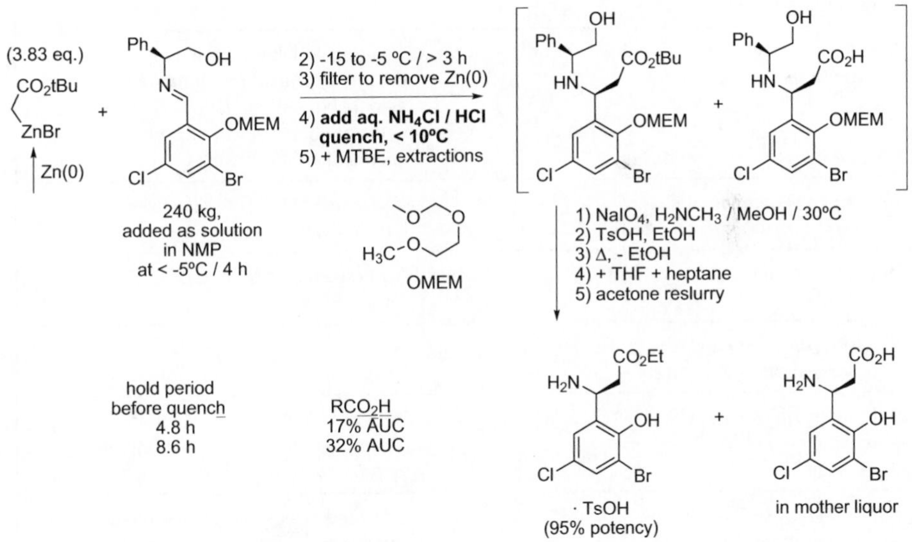

FIGURE 17.2 Troubleshooting an impurity formed during extended hold times.

Some steps involved in troubleshooting are outlined in Table 17.1.

Scrutinizing the details can be important to identify the root cause of a difficulty. For instance, the contents of solids at the top of a drum may be different from the contents at the bottom of a drum. Researchers at AstraZeneca found that scale-ups of a potassium fluoride-catalyzed Suzuki–Miyaura reaction did not proceed well, even though initial runs using the same container of KF were reproducible. They hypothesized that H_2O, a beneficial additive, had been adsorbed to the surface layer of hygroscopic KF in the bottle, while KF deeper in the container had not adsorbed H_2O [4].

TABLE 17.1 Steps in Troubleshooting

Step	Comment
1) Confirm that there is a problem	If there is a problem, is it significant enough to attempt to fix within the time allowed?
2) Establish baseline for expected performance	Retrospective data collection from batch records may be useful. Helpful to have many data points; temporarily disregard what seems to be irrelevant data
3) Identify when processing deviated from expected performance	If origin of the problem(s) is unknown, for rapid resolution consider first the causes that can be most easily fixed. Start from end of processing and work toward the beginning. Ask operators for any observations, even those they may not feel are relevant
4) Formulate feasible cause of difficulties	Intuition can be very helpful
5) Devise suitable means to confirm hypothesis	Use-tests in laboratory may be helpful to determine whether the problem(s) is associated with operations or materials. Correlate batches with key physical or chemical characteristics
6) Institute experimental changes on scale	Consider SAFETY. Obtain additional hazard evaluation if prudent
7) Analyze results of changes in operations	Accurate analyses are crucial
8) Repeat procedure from Step (4) if problem has not been solved	
9) Institute appropriate changes on scale	New procedures or chemical inputs may be required. Simplify operations whenever possible without compromising reliability
10) Analyze results of changes in operations	If needed assays can be fine-tuned and simplified
11) Continue to monitor batches after changes have been implemented	Reliable operations will confer ruggedness and determine whether true cause of problem has been identified

Identifying the cause of a problem is often the most difficult part of problem-solving. The key questions are the seven "W" questions: "Who? What? When? Where? Why? hoW? hoW much?" Often the "how much?" question provides perspective and is the most meaningful.

Identifying the cause of a problem will probably be easier if there is only one root cause. Unfortunately in trying to reduce the cause to only one source (reductivist thinking [5]) we may lose sight of the true expanse of influencing factors.

Facts always trump intuition and opinion, and end discussions. Your supervisor's experience may be very relevant to troubleshooting, but discussions may be more efficient if you present a possible solution with the problem. Experience helps in troubleshooting, but be wary of drawing conclusions too quickly from limited data. Drawing a straight line with two data points is easy, but any number of curves can pass through two points.

There are three primary sources of processing problems on scale. Mixing is perhaps the most common source of technical scale-up difficulties. For instance, thorough mixing is not instantaneous in batch reactors; applying washes evenly to a wet product cake on a filter may require care. Secondly, extended processing can create impurities; for instance, hydrolysis or transesterification can occur, along with the co-crystallization of a side product or a polymorph transition. Third processing outside the PAR can produce impurities. Mixing and operational difficulties are best assessed on-site; problems due to extended processing or processing outside of the PAR can be diagnosed long-distance by reviewing batch records and analyzing in-process samples.

Often the best approach to remedying a problem is the simplest one, applying the easiest possible fix first. For instance, if the quality of a product is low, decreasing the temperature of the dryer would be easier than developing another crystallization, so the first in-process samples to be analyzed in troubleshooting a batch might be those of the product going into the dryer.

II. PERSPECTIVE

The barriers to identifying the true cause of problem are conflicting data, confounding data, incomplete data, opinion, and time. Sometimes applying a patch to rework the product is all that can be done, given the amount of time available. Engineers have also reviewed troubleshooting [6] and offered 10 valuable troubleshooting tips [7]. Additional examples may be found in a recent review on process validation and implementation [8].

Laboratory investigations afford us the opportunities to anticipate and avoid scale-up disappointments. Use-tests, abuse tests, extended processing, preparing potential impurities and checking for their presence, all these help. Unfortunately operations on scale often don't conform to those anticipated or desired based on laboratory experience. Predicting the behavior of processes on scale in order to mimic them in the laboratory can be difficult. For instance, a suspension of crystals may be readily de-liquored by suction filtration when the height of the filter cake is only 1 cm, but filtrations may be unacceptably slow if the filter cake is 10 cm high. Learning of others' experience can help us anticipate and avoid problems.

Troubleshooting is carried out in layers. For instance, can this batch be saved? How can problems be prevented in the next batch, and throughout the campaign? How can similar problems be avoided in other processes? Are the problems due to lack of training, poor wording in the log sheet, an unforeseen problem, or due to inattentiveness, over-work, personal problems, or substance abuse?

Troubleshooting affords opportunities to deepen process understanding. In recounting the development of finasteride Williams described troubleshooting a process in the pilot plant, a story that became legend within Merck. A DDQ-mediated oxidation was quenched with 1,3-cyclohexanedione, but during workup the level of the starting alkane in the alkene product was found to be about 1%, up significantly from the 0.1% expected. Purging this level of starting material was known to be extremely difficult. The engineers were confident that the equipment was not the cause of the problem, and suggested that the problem might be due to the chemistry. The root cause was the presence of residual Pd (10 ppm) in that batch of starting material, carried over from the previous step. In the presence of 1,3-cyclohexanedione as a reductant the residual palladium catalyzed the reduction of some of the olefin back to the starting material. Ultimately the problem was solved by quenching with another reagent, methyl acetoacetate [9]. In another example of troubleshooting, researchers from Toray Fine Chemicals and Iwaki Meisei University found difficulties in resolving (R)-2-phenethylamine. In manufacturing the (R)-2-phenethylamine salt with (R)-mandelic acid the chiral purity was variable at 97–99% ee, whereas >99% ee was expected based on laboratory studies. Closer examination showed that during centri-fugation some lower-quality crystals retained a significant amount of mother liquor. Those crystals were long, thin hexagonal plates, which probably arranged themselves with the long dimension perpendicular to the passage of liquids flowing through the pores in the filter cloth of the centrifuge, impeding the deliquoring of the mother liquor and washes. Such long, thin crystals were formed when newly produced racemic phenethylamine was used in manufacturing. However, when the mandelate salt was prepared from racemized phenethylamine a different crystal habit was formed, which allowed for more effective washing of the wet cake and removal of the salt of the (S)-enantiomer of phenethylamine. The modified crystal habit was thicker hexagonal plates with nearly equal hexagonal edges; some impurities formed during the race-mization of the (S)-enantiomer of phenethylamine inhibited growth in the "long" dimension. This unintended "crystal engineering" improved product quality on scale [10].

Seeing the "big picture" is important when troubleshooting, and later in assessing the remedies developed through troubleshooting. The mistakes and knowledge gaps uncovered by troubleshooting provide lessons to be learned with gratitude.

REFERENCES

1. http://www.analects-ink.com/weekend/020308.html. A savvy politician, Donald Rumsfeld assembled a list of guidances for newcomers to the Washington D.C. political scene. This list, dubbed Rumsfeld's Rules, was compiled prior to discussions of Weapons of Mass Destruction that led to the invasion of Iraq in 2003.
2. Viadyanathan, R.; Kalthod, V. G.; Ngo, D. P.; Manley, J. M.; Lapekas, S. P. *J. Org. Chem.* **2004,** *69*, 2565.

3. Clark, J. D.; Weisenburger, G. A.; Anderson, D. K.; Colson, P.-J.; Edney, A. D.; Gallagher, D. J.; Kleine, H. P.; Knable, C. M.; Lantz, M. K.; Moore, C. M. V.; Murphy, J. B.; Rogers, T. E.; Ruminski, P. G.; Shah, A. S.; Storer, N.; Wise, B. M. *Org. Process Res. Dev.* **2004,** *8*, 51.

4. Fray, M. J.; Gillmore, A. T.; Glossop, M. S.; McManus, D. J.; Moses, I. B.; Praquin, C. F. B.; Reeves, K. A.; Thompson, L. R. *Org. Process Res. Dev.* **2010,** *14*, 263.

5. Michael Pollan has discussed the limitations of reductivist thinking regarding food: Pollan, M., *In Defense of Food: An Eater's Manifesto*; Penguin, 2009.

6. Bonem, J. M. *Process Engineering Problem Solving*; Wiley: New York, 2008.

7. Marshall, R. *Chem. Eng.* **2011,** *118*(4), 5.

8. Anderson, N. G.; Burdick, D. C.; Reeve, M. M. *Org. Process Res. Dev.* **2011,** *15*, 162.

9. Williams, J. M. In *The Art of Process Chemistry*; Yasuda, N., Ed.; Wiley-VCH, 2010; p 77.

10. Sakurai, R.; Sakai, K. In *Pharmaceutical Process Chemistry*; Shiori, T., Izawa, K.; Konoike, T., Eds.; Wiley-VCH, 2010; pp 381–399.

acac acetyl acetonate, anion of 2,4-pentanedione

ALARP as low as reasonably possible, an approach to limiting impurities in APIs

API active pharmaceutical ingredient

AUC area under the curve

AZ AstraZeneca

BINAP bis-1,1'-binaphthalene-2,2'-diylbis(diphenylphosphine)

BINOL binaphthol, or 2,2'-dihydroxy-1,1'-binaphthyl

BMS Bristol-Myers Squibb

Bn benzyl

Boc *t*-butoxycarbonyl

BSE bovine spongiform encephalopathy, aka "mad cow disease"

Bz benzoyl

Cbz carbobenzyloxy or benzyloxycarbonyl, aka Z

CDI *N,N'*-carbonyl diimidazole

cGMP current good manufacturing practice

CIAT crystallization-induced asymmetric transformation

CIDR crystallization-induced dynamic resolution

CMC chemistry, manufacturing and controls, part of an NDA

CMO Contract manufacturing organization

COD 1,5-cyclooctadiene

CoG cost of goods (estimate)

CPP critical processing parameter

CQA critical quality attribute

CRO contract research organization

CSA camphorsulfonic acid

Cy cyclohexyl

DABCO 1,4-diazabicyclooctane

dba dibenzylidene acetone

DBDMH 1,3-dibromo-5,5-dimethylhydantoin

DBU 1,8-diazabicyclo[5.4.0]undec-7-ene

DCM dichloromethane

de diastereomeric excess, i.e., % of desired diasteromer A – % of undesired diastereomer B

DEA US Drug Enforcement Agency

DEAD diethyl azodicarboxylate

DIAD di-isopropyl azodicarboxylate

dioxane 1,4-dioxane

DIPHOS 1,2-bis(diphenylphosphino)ethane

DMAc dimethyl acetamide

DMAP 4-dimethylaminopyridine

DME 1,2-dimethoxyethane, aka glyme

DMF dimethyl formamide

DMSO dimethyl sulfoxide

DoE statistical design of experiments

dppe 1,2-bis(diphenylphosphino)ethane

dppbenz 1,2-bis(diphenylphosphino)benzene

dppf 1,1'-bis-diphenylphosphinoferrocene

DQ design qualification, a part of process validation

dr diastereomeric ratio

DSC differential scanning calorimetry

EDCI ethyl dimethylaminopropyl carbodiimide, a peptide coupling reagent. Aka EDC.

ee enantiomeric excess, i.e., % of desired enantiomer A – % of undesired enantiomer B

ELSD evaporative light-scattering detector

EMA European Medical Agency, formerly known as EMEA

EPA Environmental Protection Agency

er enantiomeric ratio

ESI electrospray ionization, a technique used in mass spectrometry

ether diethyl ether

FBRM™ focused beam reflectance measurement

FDA US Food & Drug Administration

FMEA failure modes and effects analyses

fp flash point, temperature at which vapors will ignite in presence of a source of ignition

FTE full-time equivalent, a CRO researcher working under contract for a client

FTIR Fourier transform infrared spectroscopy

GC gas chromatography

GLP good laboratory practice

GMP good manufacturing practice

GRAS generally recognized as safe

GSK GlaxoSmithKline

GTI genotoxic impurity

HILIC hydrophilic interaction liquid chromatography

HMDS hexamethyldisilazane, or hexamethyldisilazide anion

HPLC high pressure liquid chromatography, or high performance liquid chromatography

HPLC-MS HPLC – mass spectrometry

HTS high-throughput screening

HTST high-temperature short-time processing, generally applied to continuous operations

ICH International Conference on Harmonization of Technical Requirements for Registration of Pharmaceuticals for Human Use

ICP-MS inductively coupled plasma-mass spectroscopy

ID internal diameter

IND investigational new drug application

IP intellectual property

IPC in-process control

IQ installation qualification, a part of process validation

IR infrared

IRMS isotopic ratio mass spectrometry

KF Karl Fischer titration, an analysis for moisture content. Aka KF titration

Lasentec™ device to monitor crystallization by focused beam reflectance measurement

LC liquid chromatography

LD$_{50}$ lethal dosage for 50% of animals tested

LEL lower explosive limit, aka LFL

LFL lower flammable limit, aka LEL
LOD limit of detection; also loss on drying
LOQ limit of quantitation
M% mol%, a yield corrected for the output quality and perhaps input quality
MCR multi-component reaction
2-Me-THF 2-methyl tetrahydrofuran
MIBK methyl isobutyl ketone; 5-methyl-2-pentanone
MOM methoxymethyl
mp melting point
MS mass spectrometry
Ms methanesulfonyl
neat solvent-free
NBS *N*-bromosuccinimide
NCS *N*-chlorosuccinimide
NCE new chemical entity, a compound in development as a potential API
NDA new drug application
NIR near infrared analysis
NLT not less than
NMM *N*-methylmorpholine
NMO *N*-methylmorpholine *N*-oxide
NMR nuclear magnetic resonance spectrometry
NMT not more than
NOR normal operating range
OD outside diameter
OFAT one factor at a time, aka OVAT
OQ operation qualification, a part of process validation
OVAT one variable at a time, also known as OFAT
PAR proven acceptable range
PAT process analytical technology
PDE permitted daily exposure
PEG polyethylene glycol
PFD process flow diagram
PGI potentially genotoxic impurity
p*I* isoelectric point, the pH at which charges on a molecule are internally neutralized
PMB *p*-methoxybenzyl
PPE protective personnel equipment
PPQ process performance qualification
PQ performance qualification; efforts to validate equipment and software for process validation
PSD particle size distribution
PTC phase-transfer catalysis, or phase-transfer catalyst
PTS surfactant derived from α-tocopherol
QA quality analysis department
QbD quality by design
QC quality control department
RCM ring-closing metathesis
Re Reynolds number
RFQ request for quotation, asking a CRO or CMO to submit a bid on a project
RI refractive index
ROI residue on ignition
RT room temperature

sc CO$_2$ supercritical carbon dioxide
SFC supercritical fluid chromatography
SiO$_2$ silica gel
SOP standard operating procedure
SPC statistical process control
TBAB tetra-*n*-butylammonium bromide
TBDMS *tert*-butyldimethylsilyl
TCCA 1,3,5-Trichloroisocyanuric acid
TD$_{50}$ dosage that produces tumors in 50% of animals tested
TEMPO (2,2,6,6-Tetramethylpiperidin-1-yl)oxyl, a radical catalyst used for oxidations
TES triethylsilyl
Tf triflate, trifluoromethanesulfonyl
TFA trifluoroacetic acid
THF tetrahydrofuran
THP tetrahydropyran-2-yl
TLC thin-layer chromatography
TMS trimethylsilyl
TMT 2,4,6-trimercapto-1,3,5-triazine, reagent used to remove multi-valent cations
tox batch batch of material prepared for toxicological screening
Ts tosyl, or *p*-toluenesulfonyl
TTC threshold of toxicological concern
TWA time-weighted average, permitted exposure over an 8-hour day
UEL upper explosive limit, aka upper flammability limit
UFL upper flammability limit, aka upper explosive limit
UHPLC ultra high-performance liquid chromatography
USP U.S. Pharmacopeia
UV ultraviolet
WFE wiped film evaporator
XRPD X-ray powder diffraction
Z benzyloxycarbonyl, or carbobenzyloxy, aka Cbz

General Index

Note: Page numbers followed by *f* indicate figures and *t* indicate tables.

Reactions Index

Note: Page numbers followed by *f* indicate figures and *t* indicate tables.

Note: These entries indicate pages where a compound is mentioned as a reagent, intermediate, or formed as a byproduct or side product. Solvents are listed where certain properties or impurities may be significant. General properties of solvents may be found in Chapter 4. Page numbers followed by *f* indicate figures and *t* indicate tables.

Drug Substance Index

Note: Page numbers followed by *f* indicate figures and *t* indicate tables.